Fundamentals of
PLANT PATHOLOGY

About the Authors

A PhD from the University of Saugor, (now Dr. Hari Singh Gour University), Sagar, **R S Mehrotra** has a teaching and research experience of more than 35 years. He retired from Kurukshetra University as a Professor in 1997. He was a Post-Doctoral Fellow at the University of Western Ontario, London; Ontario, Canada (1968–1970) and a Fulbright Fellow at the University of California, Riverside Campus (1979). He is a Fellow of National Academy of Sciences, Allahabad; and was elected President of the Indian Phytopathological Society (1990), President of the Indian Society of Plant Pathologists (1992), President of the Mycological Society of India (1996), and President of the Botany Section of Indian Science Congress (1997). He was the recipient of the Birbal Sahni Gold Medal of the Indian Botanical Society in 1998. Professor Mehrotra delivered the Dastur Memorial Lecture of the Indian Phytopathological Society in 2010.

Dr Mehrotra, along with his research associates, has published around 200 research papers in national and international journals and has authored 5 books—*Plant Pathology*, co-authored with Prof. Ashok Aggarwal, published by McGraw Hill Education (India) Private Ltd., New Delhi; *An Introduction to Mycology* with Prof. K R Aneja; *A Monograph on Phytophthora Diseases in India* (co-authored with Prof. Ashok Aggarwal); *Principles of Microbiology*, co-authored with Prof. Geeta Sumbali and published by McGraw Hill Education (India) Private Ltd., New Delhi; and *Fungal Diversity and Biotechnology* with Prof. K R Aneja.

Professor Mehrotra and his research associates have mainly worked on diseases of *Piper betle* caused by *Phytophthora parasitica* var. *piperina* and other Phytophthora diseases along with several root rot and wilt diseases of various crop plants and their control, especially biocontrol.

Ashok Aggarwal is currently working as Professor of Botany and teaches Plant Pathology, Mycology, and Microbiology at the Department of Botany, Kurukshetra University, Kurukshetra, where he has been a faculty member for the last 25 years. He is a member of several scientific organizations and has published more than 140 research papers and review articles in national and international journals. He has written a monograph in collaboration with Dr R S Mehrotra on the topic entitled "Phytophthora Diseases in India" in 2001. He has also edited another book entitled *Glimpses of Plant Sciences* in 2002. In addition, he has written a textbook entitled *Plant Pathology*, 2nd edition (published by McGraw Hill Education (India) Private Ltd., New Delhi) with Professor R S Mehrotra in 2003. Dr. Aggarwal has guided many MPhil and PhD students on diverse aspects of mycorrhizal fungi.

Fundamentals of
PLANT PATHOLOGY

R S Mehrotra
Retired Professor of Botany
Kurukshetra University
Kurukshetra, Haryana

Ashok Aggarwal
Professor and Former Chairman
Department of Botany
Kurukshetra University
Kurukshetra, Haryana

McGraw Hill Education (India) Private Limited
NEW DELHI

McGraw Hill Education Offices
New Delhi New York St Louis San Francisco Auckland Bogotá Caracas
Kuala Lumpur Lisbon London Madrid Mexico City Milan Montreal
San Juan Santiago Singapore Sydney Tokyo Toronto

McGraw Hill Education (India) Private Limited

Published by McGraw Hill Education (India) Private Limited
P-24, Green Park Extension, New Delhi 110 016

Fundamentals of Plant Pathology

ISBN-13: 978-9-38-328649-2
ISBN-10: 9-38-328649-0

Vice President and Managing Director: *Ajay Shukla*

Head—Higher Education (Publishing and Marketing): *Vibha Mahajan*

Publishing Manager (SEM & Tech. Ed.): *Shalini Jha*
Assistant Sponsoring Editor: *Smruti Snigdha*
Editorial Researcher: *Amiya Mahapatra*
Manager—Production Systems: *Satinder S Baveja*
Senior Production Manager: *P L Pandita*
Assistant Manager—Editorial Services: *Sohini Mukherjee*

Assistant General Manager (Marketing)—Higher Education: *Vijay Sarathi*
Product Specialist: *Sachin Tripathi*
Senior Graphic Designer—Cover: *Meenu Raghav*

General Manager—Production: *Rajender P Ghansela*
Manager—Production: *Reji Kumar*

Typeset at The Composers, 260, C.A. Apt., Paschim Vihar, New Delhi 110 063 and printed at
Sheel Printers Pvt. Ltd., D-132, Hoisery Complex, Phase-II Noida

Cover Printer: Sheel Printers Pvt. Ltd.

RXXYYRCODCBDY

Dedicated to

Our parents

CONTENTS

PREFACE

OVERVIEW

With increase in the human population and its requirements, the responsibility of protecting economically important plants and crops from various pathogenic microorganisms is increasing. The quest to protect crops from various diseases has led to the development of disease-resistant varieties and provided opportunities to identify new virulent races of plant pathogenic microorganisms. With development of innovative molecular techniques in the field of plant pathology for pathogen detection, identification and cloning of genes for resistance seems very promising to achieve this goal. At the same time, with intensification of agriculture, environmental conditions become more favourable for the pathogen to infect and cause more losses on various crops.

However, the use of fungicides/pesticides in agriculture cannot be ruled out. But this has led to the emergence of fungicide/pesticide resistant pathogens and also causing environmental pollution. These environmental changes have further increased problems for plant pathologists by the introduction of new diseases and making some minor diseases into major diseases. Attention is required to know the status of these problems and find ways and means to develop management strategies to protect our crops and plants in an eco-friendly manner.

AIM OF THIS BOOK

Keeping in view these facts, we have written ***Fundamentals of Plant Pathology*** mainly for the undergraduate students of Plant Pathology, Agriculture, and Plant sciences. The present book is a joint effort made after years of teaching and research, and months of focused thinking so that it may help students to develop a deep understanding of the subject.

We have presented the latest information, leavened the lessons learned from our own experiences as well as colleagues, students, and others on precise and well-defined topics in a concise form. We have put in much hard work to make the matter neither too elaborate—so that we do not lose the interest of students—nor too concise so as to leave vital topics uncovered.

SALIENT FEATURES

- Exactly tailored as per UG curricula requisites
- Covers both general and specialized topics like physical, physiological, biochemical, and molecular aspects of plant pathogens as well as phytopathological techniques and applications of biotechnology
- Integrates latest information on local plant diseases (pertinent to Indian scenario)
- Text in confirmation with latest classification of Fungi, Bacteria, and Viruses
- Profusely illustrated with more than 180 figures in the form of line diagrams and photographs
- Over 200 Review Questions to test the subjective understanding of students
- Close to 300 Multiple-Choice Questions to help students prepare for competitive exams.

CHAPTER ORGANIZATION

The book is unique as it contains parts on general concepts as well as parts on specific diseases, that is in the beginning from **Chapters 1 to 10**, we discuss the host-pathogen interaction in detail; and in the later chapters, we give an account of diseases caused by fungi, bacteria, phytoplasma, viruses, and nematodes. Latest information regarding disease control has been added from various journals, review articles, and books to make the subject up to date.

In each chapter (**Chapters 13 to 24**), major diseases are discussed in detail, and minor diseases are given in a tabular form. The changes in pathogen taxonomy and nomenclature are incorporated in each chapter to help both students and teachers attain most recent information. Special attention has been given to highlighting the historical development in plant pathology (**Chapter 2**), and Indian scientists and others who contributed significantly to these developments.

A new feature that has been added to the book for the benefit of students is the *phytopathological techniques* for practical work in the laboratory and field. Recent classification of fungi is given in a precise manner. Some new topics and new information on molecular biology and biotechnology are discussed in the respective chapters to provide better understanding of the latest developments in the subject.

The text is profusely illustrated with line diagrams and photographs. Disease cycles of important plant pathogens are adequately covered. The book is intentionally written with a minimum of references without sacrificing essential information and accuracy of the subject.

ONLINE LEARNING CENTER

The text is accompanied by an Online Learning Center available at *https://www.mhhe.com/mehrotra/fpp1* which contains

- 2 Sample Question Papers
- Extra Reads—Information about local diseases
- Web References (updated with latest links)

ACKNOWLEDGEMENTS

We would sincerely like to acknowledge all those reviewers who went through the initial manuscript and sent their reviews. Some of their feedback helped us immensely in shaping this book. The reviewers are

Deeba Kamil, Indian Agricultural Research Institute, New Delhi; Sekhar Bandyopadhyay, Uttar Banga Krishi Viswavidyalaya, Cooch Behar, West Bengal; R Velazhahan from Centre for Plant Protection Studies, Tamil Nadu Agriculture University, Coimbatore, Tamil Nadu; S G Mantur, University of Agricultural Sciences, Gandhi Krishi Vignyan Kendra (GKVK), Bangalore; and Nikhil Babruvan Gaikwad from Shivaji University, Kolhapur, Maharashtra.

We are thankful to our research scholars (Neetu and Chhavi Mangla) and colleagues for their help in various ways. The authors express their appreciation and thanks to their family members for their unwavering support and encouragement.

FEEDBACK

We hope this book will be appreciated by teachers, undergraduate, and postgraduate students alike. As authors, we seek suggestions for further improvement. You may send us your comments and views on the publisher's website mentioned below.

R S Mehrotra
Ashok Aggarwal

Publisher's Note

Do you have any further request or a suggestion? We are always open to new ideas (the best ones come from you!). You may send your comments to *tmh.sciencemathsfeedback@gmail.com*

Piracy-related issues may also be reported!

1

INTRODUCTION

1.1 NATURE AND CONCEPT OF PLANT DISEASES

It is imperative to define and discuss the meanings of certain terms commonly used in phytopathology before we discuss the basic principles and concepts of plant pathology. It is essential to include these definitions and terms in an introductory chapter, since an understanding of these terms is required before the reader can be made aware of the details of the processes involved in plant diseases.

Phytopathology (*phyton*—plant; *pathos*—suffering; *logos*—knowledge) is a science that studies the nature, development, and control of plant diseases.

According to modern conception, disease is an interaction among the host, parasite, and environment. A simple dictionary meaning of disease is **any departure from health, presenting marked symptoms, malady, illness, and disorder**. Diseased plants are distinguished by changes in their morphological structures or physiological processes, which are brought about by unfavourable environment or by parasitic agencies. Several definitions of diseases in plants have been proposed. Some examples are "a series of harmful physiological processes caused by continuous irritation of the plant by a primary agent"; "a harmful deviation from the normal functioning of physiological processes"; "a continuous impairment of metabolism"; "a physiological disorder or structural abnormality that is harmful to the plant or to any of its part or products that reduces the economic value".

1.1.1 Simple Interest Disease

This denotes a disease, the increase of which is mathematically analogous to simple interest in money. There is only one generation of disease in the course of one epidemic.

1.1.2 Compound-Interest Disease

This denotes a disease, the increase of which is mathematically analogous to compound interest in money. There are several or many generations of the pathogen in the life of the crop, that is, the capital is increased by the amount of interest. Late blight of potato, stem rust of wheat, or powdery mildew of wheat are examples of compound interest diseases.

If we know that a particular disease is of the simple-interest type, suitable control measures can be applied. For example, sanitation affects the simple-interest disease more than compound-interest type. If the disease is of former type, sanitation, including the eradication of the inoculum of the pathogen from planting material and sites, should be insisted upon. If it is of the compound-interest type, and the epidemic spreads rapidly and lasts long, the infection rate should be reduced by a timely application of fungicides. The stress should shift from sanitation to reducing the infection rate in proportion to the increase in the rate of spread and duration of the epidemic.

1.2 CAUSE OF DISEASE

When a parasitic microorganism enters into the causal complex of a disease, it is commonly considered the cause of the disease. However, we know that sometimes a variation in the environment also makes a plant liable to infection. The microorganism is thus not the sole causal factor. After a microorganism has successfully infected a susceptible plant, the subsequent interaction between the plant and the microorganism (which results in a disease) is also subject to environmental influences. Thus, strictly speaking, it is not correct to refer to a microorganism as the cause of a given disease since this implies that it is the sole cause. The term 'causal organism' is proper since it implies that the organism is a part of the causal complex. Most causal organisms are parasites. A parasite is not synonymous with the pathogen. In some cases, organisms are parasitic without becoming causal factors in the disease. Many parasitic establishments do not become pathogenic if the parasite compensates for the damage done by its presence. The mycorrhizal fungus is certainly parasitic on the roots of plants but it is not pathogenic. The root-nodule bacterium (*Rhizobium leguminosarum*) is another example. In other cases, the by-products of a strict saprophyte are an important part of the causal complex. Leaf-tip injury to oats may be due to the absorption of Griseofulvin from soil. Milo root rot is caused by a powerful toxin produced by the fungus (*Periconia circinata*) growing in the root zone.

1. Pathotype

This is a subdivision of a species distinguished by the common characters of pathogenicity, particularly in relation to the range of hosts.

2. Pathovar

In bacteria, a subspecies or group of strains that can infect only plants within a certain genus or species is called a pathovar.

3. Serotype

A serotype is a population of a pathogen (usually a bacterium or virus) in which all individuals possess a given character of serology in common. On the basis of serological tests, differences between apparently similar organisms or viruses may be found and the subdivisions thus formed are called serotypes or strains. For example, only one serotype was designated amongst 45 isolates of *Xanthomonas oryzae* (*X. oryzae* pv. *oryzae*) on the basis of agglutination reactions and gel-diffusion tests.

4. Physiotype

It is a population of pathogens in which all individuals have a particular character of physiology (but not pathogenicity) in common (Robinson, 1969).

5. Pathogenicity

It is the quality or characteristic of a pathogen capable of causing disease.

6. Pathogenesis

It is the sequence of progress in disease development from the initial contact between a pathogen and its host to the completion of the syndrome. Horsfall and Dimond (1960) likened the relationship between the microorganisms and higher plants to a situation in which your mother-in-law lives with you. If she contributes to the harmonious running of the household, your relation is symbiotic; if you support her and she does not do her share of work, she is a parasite and you are the host; if she stirs up trouble between you and your wife/husband, she is a pathogen. Her actions, your reactions, and her reactions constitute pathogenesis.

7. Host

An organism that harbours or supports the activities of a parasite is known as the host.

8. Inoculum

Inoculum is the infectious material that can cause disease, and it is that portion of individual pathogens that is brought into contact with the host.

9. Inoculum Potential

The term 'inoculum potential' describes the degree of infectivity. It is defined as "the energy of a fungal parasite available for infection of a host at the surface of the host organ to be affected".

10. Primary Inoculum

The overwintering or oversummering stage may be in or on the seed of the host, in the perennial wild host, in debris of an infected plant of the previous season, in dormant spores, sclerotia or other organs in the soil, or in the form of an inhabitant of the soil flora. Whatever the source, it causes primary infection. It may serve as an infective agent and thus becomes the primary inoculum, or it may multiply as a saprophyte (or as a parasite on the perennial host) to produce spores which become the primary inoculum. The rust fungus oversummers in India in the uredinial (uredial) stage on summer wheat in the hills, from where the primary inoculum is carried by air currents to the next crop grown at the foot of the hills and then to the plains. When the host becomes diseased, the organism commonly produces another crop of spores or infective bodies which are known as the secondary inoculum and cause secondary infection. This last cycle may be repeated several times.

11. Penetration

Penetration refers to the initial invasion of the host by an organism.

12. Infection

It implies the establishment of the pathogen inside the host following penetration in which a parasitic relationship between the two organisms is established.

13. Incubation Period

There may be a long period between penetration and infection and the first appearance of symptoms on the host and appearance of the disease, but ordinarily one follows the other. This interval of time between the penetration of the host by the pathogen and the first appearance of the symptoms on the host is known as the incubation period. In the case of black stem rust of wheat, the pathogen *Puccinia graminis tritici* enters the host (wheat) a day or so after the urediniospores (urediospores) have germinated, but the actual symptoms of the disease in the form of rust pustules develop only after a week or so at 20°C.

1.2.1 Disease Development

It refers to a series of events which occur between the time of infection and complete development of the disease.

1. Epidemiology

Epidemiology of disease is a study of the factors affecting the outbreak and spread of an infectious disease. The founders of epidemiology maintain that epidemiology deals with population of the host and pathogens under varying environment.

2. Symptoms

The external or internal reactions or alterations as a result of a disease are referred to as symptoms.

3. Syndrome

Diseased plants show several valuable symptoms by which a disease is recognized. These symptoms are collectively known as a syndrome.

4. Disease Cycle

A series of events involved in disease development, including the stages of development of the pathogen and the effect of the disease on the host, is called the disease cycle. When a pathogen is involved, the disease cycle is intimately associated with the organism. It is distinct from the life cycle of the organism.

(a) Monocyclic (Simple Cycle, Simple Interest) Pathogens that complete one or even part of one disease cycle per year are called monocyclic. In monocyclic pathogens, the primary pathogen inoculum is the only inoculum available for the entire season and there is no secondary inoculum. Examples are black leg of potato caused by *Erwinia carotovora*, Verticillium wilt and cereal cyst nematode.

(b) Polycyclic (Multiple Cycles per Year, Compound Interest) Most of pathogens go through more than one (2–30) disease cycles in a growing season are referred to as polycyclic. Only a small amount of sexual spores survive and act as primary inoculum that cause initial infection. Once infection takes place, a large number of asexual spores are produced and act as the secondary inoculum. These spores can produce new (secondary) infections that produce more asexual spores, and so on. With each cycle the amount of inoculum is multiplied manyfold. Examples are downy mildews, powdery mildews, late blight of potato, leaf spots, blights, and rusts.

(c) Polyetic (Multi-year) Cycle Some pathogens take several years before the inoculum they produce can be disseminated and initiate new infection. It may not cause many new infections over a given area in a year and amount of inoculum does not increase greatly within a year. The inoculum may increase from year to year and can cause severe outbreaks when considered over several years. Examples are Dutch elm disease, Citrus tristeza, fungal vascular wilts, and mycoplasmal yellows.

5. Infectious Organism

An infectious organism is one that may be transmitted from a diseased host plant to a healthy plant and which is capable of inciting the disease in a favourable environment.

6. Infectious Disease

An infectious disease is one in which a transmissible causal organism is part of the causal complex. In other words, on infection, the disease can spread from a diseased to a healthy plant.

7. Iatrogenic Disease

A disease which results from or is increased in severity by the use of a specific crop protection chemical may be referred to as an iatrogenic disease.

1.3 VARIOUS LEVELS OF PARASITISM

1. Symbiosis

Two living organisms may live in association with one another and in a sense, each may be parasitic upon the other. However, they are not only nonpathogenic to one another but are also essential to each other's development. This phenomenon is known as symbiosis. Lichens are classical examples of symbiosis in which a fungus and an alga live in close association. The alga, through its photosynthetic activity, converts the kinetic energy of the sun's rays into the potential energy of carbohydrates, which is useful for its own development and essential to the life of the fungus. The latter, through its property of producing hydrolytic enzymes, breaks down the material in the substrate, and thus makes energy available and provides minerals and organic nutrients essential to itself and the alga. Another good example of symbiosis is that of the *Rhizobium* species forming root nodules in leguminous plants.

2. Obligate Parasites

Parasites which are restricted to living tissues (also known as biotrophs) are called obligate parasites. Obligate is also used to denote a parasite which has not been grown in axenic cultures, that is, free from any other organism.

3. Facultative Saprophytes

Facultative saprophytes ordinarily have a life cycle similar to that of obligate parasites in that they pass most of the cycle in association with the host and do not subsist as saprophytes. They may, however, subsist as saprophytes for part of their life cycle but do not complete their life cycle away from the host. Many smut fungi are good examples of this.

4. Facultative Parasites

This term implies that they are ordinarily saprophytic but under certain conditions, they may become parasitic. Many damping-off fungi, such as *Pythium* and *Rhizoctonia*, are good examples of facultative parasites.

5. Saprophytes

Saprophytes are those organisms that have no relation with the living cell and secure their nutrients from dead organic tissues or from available inorganic materials.

In fact, a number of fungi, once considered to be obligate parasites, have been cultured away from their hosts on artificial media. This makes the term *obligate* inappropriate.

There are three broad categories of parasitism which are as follows:

(a) Biotrophs These are organisms which, regardless of the ease with which they can be cultured in the laboratory or not, in nature (*in vivo*) obtain their food from the living tissues on which they complete their life cycles. Some typical examples are rusts, smuts, and mildews.

(b) Hemibiotrophs These are organisms which attack living tissues in the same way as biotrophs but continue to develop and sporulate after the tissue is dead. Typical examples of these are leaf-spotting fungi.

(c) Perthotrophs These are organisms which kill host tissues in advance of penetration and then live saprophytically. *Sclerotium rolfsii* is a typical fungus included in this category.

1. Virulence

Aggressiveness and virulence are ambiguous terms. Many pathologists equate virulence with the pathogenicity; nevertheless, some shade of difference in usage is usually apparent. Thus, the statement that an organism is pathogenic merely implies that it can cause a disease. But a virulent organism is considered to have a high capacity to do so. Aggressiveness is equally ambiguous and has been used to describe the capacity of a parasite to invade and grow in its host plant and to reproduce on or in it. High aggressiveness can be combined with low pathogenicity as in the case of some obligate parasites which invade plants efficiently but cause minimal damage to it, at least in the early stages of attack. Aggressiveness is also used as a measure of pathogenicity.

2. Predisposition

It may be defined as the effect of one or more environmental factors which makes a plant vulnerable to attack by a pathogen. It is a process which antedates penetration and infection.

3. Susceptibility

It is the inability of a plant to resist the effect of a pathogen or any other damaging factor. Susceptibility of a given individual may be increased or decreased by environmental factors. It may also vary between individuals, varieties, or species because of differences in the inherited characteristics which affect susceptibility.

4. Resistance

Resistance can be described as the inherent ability of a plant to prevent or restrict the establishment and subsequent activities of a potential pathogen. It may be regarded as the opposite of susceptibility and may also be influenced by environmental factors. It is the ability of the plant to withstand, oppose, or overcome the attack of a pathogen completely or to some degree. There are two types of resistance, i.e. Vertical Resistance (*VR*) and Horizontal Resistance (*HR*). Resistance may be defined as **horizontal** (uniform) when it is evenly spread against all races of a pathogen or as **vertical** (differential) when it is completely effective against some races of a pathogen but not against others. It may be **polygenic** (determined by many genes), **oligogenic** (several genes) or **monogenic** (one gene). The genes involved may be major or minor depending on their effectiveness in conferring resistance.

5. Hypersensitivity

This results in violent local reaction to attack by a pathogen and the prompt death of tissue around the points of entry prevents further spread of infection. Hypersensitivity, thus, confers high resistance and is sometimes used in the sense of immunity.

6. Symptomless Carriers

Symptomless carriers are those plants in which the agent-inciting disease is present but shows no symptoms. This is an extreme case of tolerance in some plants.

7. Active Resistance

Active (or dynamic) resistance is "due to reactions incited by the attack", while passive (or static) resistance is "due to qualities innate in the host prior to the attack". These terms describe resistance mechanisms and do not necessarily connote any epidemiological significance.

8. Pathodeme

Pathodeme is that population of a host in which all individuals have a particular character of resistance in common.

9. Biotype

A biotype is a population of individuals which are genetically identical (literally any species subdivision which is based on criteria other than morphological).

10. Immunity

This implies exemption from infection by a pathogen and in this sense is an absolute quality.

11. Aetiology/Etiology

Aetiology is the science of the cause of disease. When used in a broader sense, it includes the study of causal factors, their nature, and their relation with the host.

12. Fungitoxic, Fungicidal, Fungistatic

Fungitoxic substances are harmful to fungi and include fungicides. The word **fungicide** has originated from two Latin words, *fungus* and *caedo*. The word *caedo* means 'to kill'. Thus, literally speaking, a fungicide is any agency which has the ability to kill a fungus. According to this meaning, physical agencies like heat and ultraviolet light should also be called fungicides. However, this is not so. In common usage, the meaning of the word *fungicide* is restricted to chemicals. Hence, fungicide means a chemical capable of killing fungi. Some chemicals do not kill fungi but inhibit fungus growth temporarily. Such a chemical is called a **fungistat** and the phenomenon of temporarily inhibiting growth is called **fungistasis**.

1.4 INTEGRATED DISEASE/PEST MANAGEMENT

In view of environmental and health concerns, a determined effort was made in the last few decades to reduce and rationalize the use of chemicals (pesticides, fungicides, and nematicides, etc.) worldwide in order to manage pests/pathogens more effectively. This approach leads to the emergence of a new discipline called IDM/IPM (Integrated Disease Management/Integrated Pest Management). In the IDM/IPM, a fungicide/pesticide is applied only when the pathogen population reaches a level inducing economic crop losses. In other words, IDM/IPM concentrates on reducing the number of the noxious agents with agrochemicals. So, IDM/IPM can be defined as "a disease-management system that in the context of the associated environment and the population dynamics of the pest/pathogen species utilizes all suitable techniques and methods in a manner as compatible as possible and maintains the pest/pathogen population at levels below this causing economic injury". IPM/IDM in the broad sense has been further defined as "the optimization of pests/pathogens control in an economically and ecologically sound manner, accomplished by coordinated use of multiple tactics to assure stable crop production to maintain pathogen/pest damage below the economic injury level while minimizing hazards to humans, animals, plants, and the environment." Health of plants depends upon their interaction with pathogens/pests and a further complex cohort of other organisms, as well as with the physical and chemical environment in which the plant is growing. Moreover, the basic concern of IPM/IDM is with designing and implementing pest/disease management, practices that meet the goals of farmers, consumers, and government in reducing pest/disease losses while at the same time safeguarding against the longer term risks of environmental pollution, hazards to human health, and reduced agricultural sustainability.

1.5 EFFECT OF PLANT DISEASES ON HUMAN AFFAIRS

An example of the tremendous impact that plant disease can have on the course of human history is given by the **late blight of potatoes**, caused by the fungus *Phytophthora infestans*. Between 1840 and 1847, it resulted in a famine among the working-class population of Ireland who had come to depend on potato as their major source of food. In 1845 the potato blight devastated millions of acres in Europe, the USA, and Canada. So sudden and so complete was the catastrophe that in only a few days, fields which had promised abundant harvests, were transformed into blackened wastes of vegetation overlying foul and putrefying masses of rotten tubers. As a result of this total failure of potato crop in Ireland, there was a great famine and about a million people died and almost the same number of people migrated to other continents. In the early 1870s, another fungal disease, the **coffee rust** (*Hemileia vastatrix*), wiped out the coffee plantations of Ceylon (Sri Lanka). Coffee growing was therefore abandoned, at least for many decades, and much of the world's coffee production was shifted to the western hemisphere where the disease did not occur. The tea and rubber industry replaced coffee in Sri Lanka and other eastern areas. In the 1930s, the entire banana industry in Central and South America was threatened with extinction by **Sigatoka** (*Mycosphaerella musicola*) and the industry could be saved only by 1940 when Bordeaux mixture was used on a massive scale.

The eating habits of people are very difficult to change and plant diseases appear to be responsible for many deeply implanted eating habits. While in northern and southern Europe, wheat bread is more common; in central Europe people eat rye bread. In the southern parts of the USA, people eat corn bread while in the north, wheat bread is more common. This situation resulted due to the ravages of **wheat rust**. In France, between 1878 and 1882, the wine industry was threatened due to the introduction of **downy mildew** (*Plasmopara viticola*) from the USA. In India, the 1942 Bengal famine was perhaps largely due to the **Helminthosporium disease of rice** caused by *Helminthosporium oryzea* Breda de Haan. About two million people died of starvation. According to Padmanabhan (1973), nothing as devastating as the Bengal epiphytotic of 1942 has been recorded in plant pathological literature. In 1946-47, the wheat-rust epidemic was responsible for food shortage as well as seed shortage in Madhya Pradesh. The **bacterial blight of rice**, thought to be confined to Japan, was reported in 1951 in Mumbai and in a short time became very serious and widely distributed, attacking the commonly cultivated varieties. The disease broke out in an epidemic form in the Shahbad district of Bihar in 1963. With the introduction of Taichung Native-1, the disease appeared in a severe form in 1966 throughout the country wherever the crop was grown. Thus, with the introduction of one susceptible variety, a disease which was unknown in all but two states earlier became pandemic within two years. At present, **bacterial blight** is a major hurdle in stepping up rice yields. Epidemics of **ergot disease of bajra** caused by *Claviceps fusiformis* (Syn. *C. microcephala*) have occurred throughout the whole country. In the USA (1969-70), there occurred one of the most devastating epidemics in the recent history of plant pathology, the **southern corn-leaf blight** caused by *Helminthosporium maydis*. In 1970, over 70 percent of the North American maize crop carried Texas male sterile cytoplasm because a single type of cytoplasmic male sterility (Tms) was used in the production of hybrid seed. A virulent race of *Helminthosporium maydis*, known as race T, was present and the weather conditions favoured the development of the epidemic. The epidemic led to a great crisis in North America at that time. **Coffee rust** (*Hemileia vastatrix*) was discovered in 1970s in Brazil where the pathogen has spread rapidly over the entire area. The coffee-rust epiphytotic caused colossal losses in Brazil during 1974-1975. Thus, the disease has crossed the Atlantic Ocean to Brazil and Columbia where it is spreading.

Chestnut blight, caused by the fungus *Cryphonectria* (*Endothia*) *parasitica*, nearly wiped out the American chestnut. It is believed the causal organism was introduced from eastern Asia at the beginning of the 20th century. Although the causative agent of the disease was soon discovered, no suitable control measure could be found to control the disease. The result that about 30 billion board feet of wood, used for furniture, flooring, poles, railroad tiers (sleepers), and coffins, was lost. The tannin industry, which was dependent on this tree, collapsed and this forced the leather makers of the USA to obtain this item from other sources. The loss of one species created far-reaching effects that have not been completely overcome till now.

Another instance of serious loss by a fungus disease is due to **red rot of sugarcane** (*Glomerella tucumanesis*) which reached its peak in 1938-39 in the white sugar belt of northern India—Bihar, Punjab, and Uttar Pradesh. In the badly affected areas, most of the mills could crush only 33 percent of their normal quantity. The farmers suffered tremendously.

Catastrophic plant diseases are ably summarized by Klinkowski (1970) in which he mentions diseases like peach yellow, phony peach, sugar cane mosaic, pear decline, swollen shoot of cacao, plum pox, sugarbeet yellow, tobacco veinal necrosis, late blight of potatoes, blue mould of tobacco, hop downy mildew, Dutch elm disease, St. Anthony's fire, cereal rusts, white pine blister rust, brown spot of rice, red rot of sugar cane, and certain catastrophic nematode diseases which had disastrous consequences for humans and made a dramatic impact on their affairs.

1.6 CLASSIFICATION OF PLANT DISEASES

Diseases may be classified in various ways on the basis of

1. Host plant affected, such as cereals, millets, fruits, trees, vegetables, etc.
2. Part of the plant affected, such as root, stem, leaf, flower, fruit, etc.
3. Symptoms produced in the host plants, such as wilt, blight, soft rot, anthracnose, rust, smut, mildew, damping off, etc.
4. The mode of spread and severity of infection, such as epiphytotic, endemic, sporadic, and pandemic
5. Causal organism/factor

1.6.1 Classification of Plant Diseases on the Basis of Spread and Severity of Infection

1. Endemic Diseases

A disease is classified as endemic when it is constantly present in a moderate or severe form and is confined to a particular country or district, for example, the wart disease of potato caused by *Synchytrium endobioticum* is endemic in Darjeeling. In plant pathology, this term is generally applied to simple-interest diseases which are either indigenous or of ancient introduction.

2. Epidemic, or Epiphytotic, Diseases

An epidemic, or epiphytotic, disease usually occurs widely, but periodically, in a destructive form. The pathogen may be present, as in endemic diseases, but the environmental factors responsible for the development of the disease occur only periodically, for example, cereal rusts, powdery mildews, etc.

3. Sporadic Diseases

Sporadic diseases, in reality, belong to the epidemic group. The term is applied to those diseases which occur at very irregular intervals and locations and in relatively fewer instances. Angular leaf spot and blotch disease of cucumber are sporadic diseases (Pathak, 1972).

4. Pandemic Diseases

Pandemic outbreak is an epidemic which occurs over vast and extensive areas causing severe losses. These occur all over the world and result in mass mortality, for example, late blight of potato.

1.6.2 Diseases Classified According to Major Causal Factors

It is mainly of three types:

1. Nonparasitic diseases
2. Parasitic diseases
3. Mycoplasmal and viral diseases based on the major causal factors

REVIEW QUESTIONS

1. Define disease.
2. Distinguish the following pairs of terms:
   (i) Simple-interest disease and compound-interest disease
   (ii) Parasite and pathogen
   (iii) Pathotype and serotype
   (iv) Pathogenicity and pathogenesis
   (v) Host and suscept
   (vi) Primary inoculum and secondary inoculum
   (vii) Penetration and infection
   (viii) Disease cycle and life cycle
   (ix) Symptoms and syndrome
   (x) Obligate and facultative parasites and saprophytes
   (xi) Biotrophs and perthotrophs
   (xii) Vertical resistance and horizontal resistance
   (xiii) Pathodeme and pathotype
   (xiv) Epidemiology and aetiology
   (xv) Fungicidal and fungistat
   (xvi) Monocyclic and polycyclic
3. Name some plant diseases that resulted in grave consequences for many nations.
4. Give the definition and examples of the following types of disease:
   (i) Epidemic or epiphytotic disease
   (ii) Sporadic disease
   (iii) Endemic disease
   (iv) Pandemic disease
   (v) Integrated disease/pest management
   (vi) Polyetic cycle

MULTIPLE-CHOICE QUESTIONS

1. A parasite which attacks living tissues but continues to grow and reproduce after the tissue is dead is called
   (i) facultative parasite (ii) facultative saprophyte
   (iii) necrotroph (iv) biotroph
2. The study of the cause of diseases is
   (i) epidemiology (ii) etiology
   (iii) pathogenesis (iv) symptoms
3. The incubation period and sporulation period of a pathogen is short in
   (i) simple-interest disease (ii) compound-interest disease
   (iii) slow epiphytotics (iv) none of these
4. A disease developing in many individuals of a population, widely but periodically in a destructive form is called
   (i) endemic (ii) sporadic
   (iii) epidemic (iv) pandemic
5. Highest rate of epidemics are known to occur in
   (i) self-pollinated crops (ii) cross-pollinated crops
   (iii) vegetatively propagated (iv) both (i) and (ii)
6. Horizontal resistance is best controlled by
   (i) one or few genes (ii) many genes
   (iii) both (i) and (ii) (iv) none of these
7. A population of a host in which all individuals have a particular character of resistance in common is called
   (i) pathodeme (ii) biotype
   (iii) pathovar (iv) biotroph
8. Name the fungus responsible for Bengal Famine in 1942.
   (i) *Helminthosporium orygae* (ii) *Alternaria alternata*
   (iii) *Rhizoctonia solani* (iv) *Phytophthora infestans*
9. A disease cycle in which a pathogen that takes several years to initiate new infection is called
   (i) polyetic cycle (ii) monocyclic
   (iii) polycyclic (iv) all of these

Answers

1. (ii) 2. (ii) 3. (i) 4. (iii) 5. (iii) 6. (ii) 7. (i) 8. (i) 9. (i)

2

HISTORY OF PLANT PATHOLOGY

It may be asked as to why it is necessary to have a historical account of plant pathology. A study of the history of any science helps make the subject clearer. By studying the history of a science, we get a better perspective of the subject; we come to know the contributions made in that field; the problems that were encountered and the manner in which they were tackled. We can follow the guidelines set by previous scientists while studying a particular branch of science.

It is interesting to note that the largest number of botanists specialize in mycology and plant pathology. This is primarily due to the fact the fungi, bacteria, and actinomycetes yield antibiotics; they have a role in the decomposition of organic matter, fermentation, and nitrogen fixation; they are sources of food (mushroom culture); and lastly, some of these organisms are responsible for plant diseases which cause great annual losses throughout the world. It is estimated that in India there is an annual loss of 5000 crores of rupees as a result of plant diseases. In the USA, the annual loss is estimated to be 9.1 billion dollars.

The earliest known writings make references to the ravages of plant diseases. There can be little doubt that plant diseases preceded humans on earth and that ever since they have practiced agriculture, diseases have taken their toll of crops. We must agree with the late Professor Whetzel (1918) of Cornell University, that phytopathology, like all natural sciences, had its beginnings at the dawn of human civilization. There are references in the Vedas (1500 BC), Buddhist literature, Kautilaya's *Arthasastra*, Krishi-Parashara, Sangam literature of Tamils, Agnipurana, *Brhat Samhita* of Varahmihira, Surapala's *Vrikshayurveda*, *Viswavallabha* of Chakrapani Misra and some documents of the medieval and pre-modern period to plant diseases and the methods of control (Nene, 2003). There are references in the *Old Testament* of blasting, mildew, and insect pests. The Romans used to celebrate a festival, Robigalia, to ward off rust. The annual festival of Robigalia is reported to have been initiated by king Numa Pompilius, probably in about 700 BC, and was continued with modifications in the Christian era. Theophrastus, who lived from 370 BC to about 286 BC, was the first botanist to study and write about the diseases of trees, cereals, and legumes, although his approach was observational and speculative rather than experimental. He noticed that some kinds of plants are more liable to certain diseases than are others. He was not aware that diseases are caused by microorganisms, and it was generally accepted that fungi and various other organisms observed in association with diseased or decaying plants, arose spontaneously from plants or from the environment. Comparatively little was added to the knowledge of botany or plant pathology for nearly two thousand years after the time of

Theophrastus. From the 5th century AD until the Renaissance, there was lack of intellectual activity and little was written on botany and plant pathology. In his writings on the plant pathology of that period, Whetzel (1918) stated that one bright spot in that dark period was the work, in the 10th century AD, of an Arabian landed proprietor, Ibnal-Awam, who was said to have described accurately the symptoms of many diseases of trees and vines and to have given extensive consideration to their control. The period between 300 AD and 1300 AD is said to be the middle or 'dark ages' in history. There were petty wars, religious dogmas, and superstitions, and there was practically no increase in the knowledge of plant pathology during that period. Then came the Renaissance from the 14th to the 16th century AD. It is during this period that printing (1440) and the microscope (1685) were invented. The compound microscope was invented by Anton van Leeuwenhoek, a linen draper, surveyor, and wine gauger of Delft, Holland. Robert Hooke, an English scientist, observed the cell for the first time under the microscope in 1660.

2.1 THE HERBALISTS

The beginning of the revival of interest in botany is reflected in the writings of the herbalists and these are more important from the mycological standpoint.

2.2 THE SYSTEMATISTS

C H Persoon's *Synopsis Methodica Fungorum* (1793) is the chief starting point for the nomenclature of the Uredinales, Ustilaginales, and Gasteromycetes. Persoon, a native of South Africa, and educated in Germany and Holland, did most of his work in France. Although a doctor of medicine, he devoted most of his time to the study of fungi.

Elias Magnus Fries (1794–1878) was Swedish and is regarded as the 'Linnaeus' of mycology. His *Systema Mycologicum* (1821–1832) is the chief starting point for the nomenclature of fungi. *Systema Mycologicum* had put the taxonomy of most fungi on a more scientific footing than any other earlier work. He was one of the most respected mycologists of his time.

Matheiu Tillet (1714–1791), director of the Mint at Troyes, France, and an amateur botanist and agriculturist, published an important paper on wheat bunt. In 1755, he tested the effect of dusting seed with the black mass contained in the diseased kernels. Thus, he demonstrated experimentally that the bunt of wheat was contagious and thought that the spores contained a poisonous entity. Felice Fontana (1730–1805), an Italian physicist and naturalist, published a paper in 1767 in which he expressed the view that grain rust was a distinct parasitic entity. He distinguished the red and black stages of wheat rust and made microscopic drawings.

Benedict Prevost (1755–1819), a Swiss professor of philosophy at the Academy of France in 1807, observed the germination of the spores of the wheat-bunt organism and conceived the idea that this organism penetrated the young wheat plant and was the actual cause of the disease. Prevost's experiments provided the first proof and interpretation of the role of a microorganism in the causation of disease.

2.3 THE BEGINNING OF THE MODERN PERIOD

In 1845, the devastating epidemic of the potato blight in North Europe, and particularly in Ireland, focussed attention on plant diseases, giving a great fillip and stimulating interest in the subject of plant pathology. Much of the early literature on this disease is found in the *Gardener's Chronicle*.

Scientists at that time were unable to agree on the cause of the disease or to suggest a successful remedy. The autogenetic theory of diseases was still dominant, although opposition to this was rapidly increasing. Dr John Lindley, Editor of the *Gardener's Chronicle* and Professor of Botany at University College, London, was the leader of the autogenetists. M J Berkeley (1803–1889) who was then the most prominent British mycologist, was at first somewhat reserved in supporting the parasitic theory, but advocated it strongly in 1846. Montagne in France, in 1945, described the fungus as *Botrytis infestans*. Morreu in Belgium experimented with the disease in 1845. He described the disease and fungus and regarded the fungus as the major cause of the disease, although he lacked convincing evidence. That the fungus was the real cause of the disease and not the result, was proven experimentally by Speerschneider in 1857 and de Bary in 1861 and 1863 (Jones *et al*, 1912).

Anton de Bary (1831–1888) became an outstanding figure in the botanical field. Much of his research was on fungi and the relation of parasitic fungi to their hosts. de Bary was born in 1831 in Frankfurt-am-Main, Germany. His mycological interests were more biological and physiological than systematic. From his investigations on life histories, parasites and saprophytes, the nature of lichens, etc. he made new and important discoveries. Heteroecism in the Uredinales was made clear by his experiments and he gave accounts of the development and sex in a number of Phycomycetes and Ascomycetes. He was the first to indicate the nature of obligate and facultative forms. His well-known textbook, *Morphologie und Physiologie der Pilze, Flechten, und Myxomyceten* in 1866 and its second edition in 1884, records the broad classification of fungi. He is truly regarded as the 'founder of modern mycology'. A large number of students from different parts of the world came to his laboratory. More than sixty of them became prominent in their fields and carried his techniques and influences to many parts of the world. Famous among them were M S Woronin of Russia, O Brefeld of Germany, P A Millardet of France, H M Ward of England, W G Farlow of the USA and A Fischer of Switzerland.

In his studies on the *Sclerotinia* disease of carrots and other vegetables, de Bary noted that the host cells are killed in advance of the pathogen and that the juice from the rotted tissue can break down healthy tissues of the host. On the other hand, if the juice is boiled, it has no effect on the healthy tissues. de Bary concluded that the pathogen produces enzymes that bring about the degradation and killing of plant cells. de Bary's noteworthy contribution to plant pathology were his studies on the Peronosporaceae and the diseases they incite, especially the late blight of potato, the discovery of heteroecism in the rusts and his work on *Sclerotinia sclerotiorum* (1886) which opened up an entirely new field of investigation, that is the physiology of parasitism.

2.4 CONFIRMATION OF PREVOST'S WORK

In the decade of the 1840's, the Tulasne brothers (Louis Rene Tulasne and Charles Tulasne) who were botanists at the Museum of Natural History in Paris began extensive morphological studies on fungi. They made illustrated drawings of rusts, smuts, and Ascomycetes which are regarded as classics in mycology. L R Tulasne (1815–1885) is often called the 'Reconstructor of Mycology'. In 1847, the two brothers published studies on wheat fungus and confirmed the observations of Benedict Prevost with regard to the causal organism of the bunt disease of wheat. Julius Gotthelf Kuhn (1825–1910) was a contemporary of de Bary. He was initially a farm manager. He publised his famous textbook for farmers in 1858 entitled *The Diseases of Cultivated Crops: Their Causes and Their Control*. This was the first textbook to be published in which fungi were regarded as causal factors in plant disease. The diseases were grouped according to causes: unfavourable climatic and soil conditions; influence of animals; and parasitic plants. Nematodes were also included as they were known as plant parasites since the time of Needham (1774). Their economic importance at the end of the 19th century promoted nematological studies as a part of plant pathology (Raski, 1959).

2.5 DOCTRINE OF SPONTANEOUS GENERATION

This theory held that living organisms could arise spontaneously from inanimate or nonliving matter. In 1748, Needham, an English clergyman and a naturalist, supported the theory of spontaneous generation. To prove his point, he boiled meat in a flask which was corked. He demonstrated that the contents had spoiled due to microorganisms. We now know that while he killed the vegetative bacteria, the bacterial spores were not killed at the temperatures used. In 1775, Spallanzani, an Italian investigator, challenged Needham and proved that the decay of meat by bacteria and other lower forms of life could be prevented by heating the material in a flask which was sealed in such a way as to exclude contamination from the air. His experiments supported his contention. Others failed to get similar results consistently.

Schwann (1837) and Cagniard Latour (1838) discovered the yeast cell and maintained that it caused fermentation. It was in 1860 that Pasteur furnished irrefutable evidence that microorganisms arise from pre-existing living entities and that fermentation is a biological phenomenon rather than a purely chemical one. It may be mentioned that the autogenetic theory of disease continued to be dominant throughout the first half of the 19th century. The most influential proponents were Unger, Meyen and Liebig. Unger was an Austrian physician and Professor of Botany. He thought that fungi, associated with plant diseases, arose from the diseased plant because of abnormalities in plant juices and were thought to be products rather than the causes of diseases. His best known work on plant pathology was *Die Exantheme de Pflanzen* (1833). Pasteur was born in 1822 of humble parents in the village of Dole in France. He began his life as a chemist. He worked as a teacher at Strassburg and then at the Ecole des Beaux Arts in Paris. He never studied medicine, though his discoveries about the protozoan nature of the silkworm and the confirmation of the bacterial origin of anthrax buried for ever the controversy over the theory of spontaneous generation. The germ theory in relation to diseases of humans and animals was established in 1876 by Robert Koch. Koch was a Germen physician and a co-worker of Pasteur. The first final proof of the causal relation of the anthrax bacillus to anthrax disease was published by Robert Koch in Germany in 1876. Koch (1843–1910) was trained as a physician; however he turned his attention to the new science of bacteriology. He began his work by establishing laboratory procedures and techniques that would yield reproducible results. Koch established the principles of pure culture technique although it was Lister, the English surgeon, who pioneered work in the field of aseptic surgery. He was impressed by Pasteur's work and kept it as a basis of his own research. He was the first person to culture a bacterium capable of souring milk. In the medical field, Robert Koch discovered the bacterium that causes tuberculosis. He enunciated certain rules or criteria that should be satisfied before the identity of the disease-producing organism in a particular disease can be established.

Brefeld (1875, 1881, and 1883) in Germany worked on fungi and was mainly responsible for the early development of modern techniques of growing microorganisms in a pure culture. Along with the improvement made by Koch, Petri *et al.*, his techniques are the basis for the pure culture methods currently employed. He was the pioneer in tracing the life cycles of the cereal smut fungi and their role in the causation of diseases. He made important contribution both to pure culture of smut fungi and to the methods of inoculation.

2.6 FOREST PATHOLOGY

Robert Hartig was born in Braunschweig in Germany in 1839. He was an enthusiastic and successful teacher. He is well known for his contributions to many aspects of forestry, botany, and entomology. He is aptly called the 'Father of Forest Pathology'. His foremost contributions are *Wichtige Krankheiten der Waldaume* in 1874 and *Lehrbuch der Baumkrankheiten* in 1882. The latter publication was an authoritative work for many years.

2.7 DISCOVERY OF BORDEAUX MIXTURE

The discovery of this fungicide has an interesting story. It was discovered by Pierre Marie Alexis Millardet in 1882. Millardet was born in Montmery-la Ville in France in 1838. He studied medicine but was more interested in botany. He learnt botany from the great botanists Hoftmeister at Heidelberg and de Bary at Freiburg. He returned to France and obtained his doctorates both in medicine and science. In 1876, he joined the University of Bordeaux, France, and served there till his retirement in 1899.

The downy mildew of the vine was first reported in Europe in 1878. Millardet and Planchon discovered it at about the same time in France, where it had evidently been introduced from the USA somehow or the other during the importation of resistant root stocks of grape vines for root aphid *Phylloxera*. The disease spread rapidly and threatened the vineyards of Europe. Millardet began a thorough study of the disease and the methods of control. In 1882, he noticed that the vines that had been treated with a mixture of copper sulphate and slaked lime to prevent the grapes from being stolen retained their leaves, whereas the untreated vines were defoliated. This fact suggested to Millardet a means of curing the disease. In 1883 and 1884, he performed extensive spraying experiments with many preparations of copper, calcium, and iron salts, used singly and in various mixtures and also arranged tests by viticulturists. Both the seasons were dry and little mildew developed. Being a conservative scientist, Millardet preferred to delay the publication until he could recommend a thoroughly tested spraying programme. However, the news of his work spread and others began to publish the effects of copper preparation on mildew. In May, 1885, Millardet published his works and gave details of spraying with a mixture of copper sulphate and slaked lime, which was later known as the Bordeaux mixture.

2.8 PLANT PATHOLOGY IN THE TWENTIETH CENTURY

The major trends initiated in the latter half of the 19th century progressed in several directions. The rediscovery of Mendel's laws in the beginning of the century had its impact and the genetics of the host as well as the pathogen received the attention of plant pathologists. Some of the major trends were in the following areas:

1. Physiological plant pathology
2. Genetics of the host and the pathogen
3. Fungicidal research
4. Environment and its relation to plant disease
5. Nature of disease resistance in plants
6. Biochemistry and physiology of the diseased state
7. Molecular biology of pathogenesis and induced systemic resistance
8. Tissue culture in plant pathology

9. Biotechnological approach towards disease control
10. Ecological studies of plant pathogens
11. Bacteriology
12. Virology
13. Mycoplasma group of pathogens, etc.
14. Nematology

These major trends are now discussed in brief.

2.9 PHYSIOLOGICAL PLANT PATHOLOGY

Sorauer, born in Germany in 1839, initiated modern studies in physiogenic diseases. His best known work is *Handbuch der Pflanzenkrankheiten*, written in 1909. Sorauer laid tremendous emphasis on physiogenic diseases.

Prevost (1807) demonstrated experimentally and interpreted clearly the indirect or secondary causal role of the environment in the etiology of parasitic disease. H M Ward (1880) emphasized the role of the environment in the epidemiology of coffee rust. He may be reckoned as the father of tropical plant pathology. L R Jones (1926) at Wisconsin initiated a new field of research on the effects of soil temperature and moisture and other environmental factors upon the epidemiology of the disease.

de Bary (1886), in his last work, pioneered research on the physiology of parasitism. de Bary worked on *Sclerotinia sclerotiorum* and the disease it causes to carrots and other plants. He demonstrated that the pathogen killed the cells of the host in advance of the invading hyphae of the pathogen. He found that juice from rotted tissue could rapidly break down healthy host tissues. The capacity to break down tissues was completely lost after boiling. L R Jones did pioneering work on pectic enzymes secreted by soft rot bacteria. William Brown's papers on the physiology of parasitism form the foundation of further researches in the role of enzymes in plant pathogenesis. After the Second World War mainly on the initiative of the American, Japanese, Soviet, English and Indian researchers, the so-called physiological era commenced. The outstanding names of this era are J C Walker, K O Müller, J G Horsfall, A E Dimond, L P Miller, R G Ovens, T Akazawa, I Uritani, W H Fuchs, E Gaumann, R K S Wood, T S Sadasivasn, H E Wheeler, J M Duniway, J A Kuć, F Fric, G F Pegg, R Heitefuss, R P Scheffer, K Rudolph, D F Bateman, R N Goodman and I A M Cruickshank.

2.10 GENETICS OF THE HOST AND THE PATHOGEN

In England, as early as 1905, R H Biffen began breeding cereals resistant to yellow rust. The genetics of the host plant and pathogen were studied extensively with regard to the inheritance of resistance and virulence, when Biffen's rust-resistant varieties succumbed to infection in Australia, and the fact of genetically bound disease resistance was doubted. E C Stakman, as early as 1914, contributed valuable information on delineating and cataloguing physiological races of the pathogen. It has been demonstrated by crossing formae specialis (Stakman *et al.*, 1930) and pathotypes (Newton *et al.*, 1930) of *Puccinia graminis* that the virulence of the pathogen is also inherited, according to Mendel's laws. The same is true for various other pathogens (Person and Sidhu, 1971). The demonstration by Biffen (1905) of the Mendelian nature of the inheritance became the starting point for the systematic genetic analysis of resistance. An increase in the knowledge of the sexuality of fungi by Blakeslee (1904), Craigie, and Dodge (1927) enabled a genetic study of pathogenic fungi. Gaümann (1946) proposed that in many host-parasite combinations, plants remain resistant due to hypersensitivity. Flor (1955) explained host-parasite interaction in flax rust caused by *Melampsora lini*. He gave a gene for gene hypothesis to explain this reaction. Van der Plank (1963, 1968) gave the concept of horizontal and vertical resistance in plants. Other pioneers in the field of breeding disease-resistant varieties are W A Orton and L R Jones of the USA and B P Pal and K C Mehta of India.

2.11 FUNGICIDAL RESEARCH

The earliest known fungicide is sulphur. The history of the development of fungicides is fascinating. In the 20th century, several biochemists and organic chemists along with plant pathologists started investigations on this subject. Fungicides in the 20th century, especially in the 1930 period, were mercury or Bordeaux mixture. From the Bordeaux mixture developed the inorganic insoluble copper fungicides, such as cuprous oxides or oxychlorides. The emphasis shifted from the oxides of copper to the organic fungicides, the organo-mercurial, the dithiocarbamates, and bisdithiocarbamates. W H Tisdale and I Williams (1934) revolutionized the study of organic fungicides by discovering alkyldithiocarbamates. Alkylene bisdithiocarbamates were discovered by Dimond, Heuberger, and Horsfall in 1943. Interest in the chemotherapy of plants has been generated and works on an interesting group of compounds, such as chloranil and systemic fungicides. Strobilurins are of current interest to workers in this field.

2.12 ENVIRONMENT IN RELATION TO PLANT DISEASE

L R Jones pioneered the study of the effect of soil temperature on cabbage yellows at Wisconsin. He developed soil-temperature tanks to see the effect of soil temperature on the development of disease. For the type of experimental approach, a thermostatically controlled water bath was devised into which water-tight vessels containing plants growing in soil or other substrates could be inserted. This apparatus is known as the **Wisconsin soil-temperature tank**.

Before the acceptance of the germ theory in relation to plant disease, the environment was considered to be the major factor. It is now well known that in cases where a parasitic microorganism is part of the casual complex of a disease, the environment plays an important part. Experimental studies in disease development in relation to various environmental factors have now become quite extensive and the predisposing factors are considered to play an important role in the development of any disease.

2.13 NATURE OF DISEASE RESISTANCE

The fact that many plants show a natural resistance to disease has long been used by plant geneticists in their breeding programmes. Rapid advances in modern biochemistry have now made it possible for plant pathologists to investigate the biochemical basis of this resistance. Valuable contributions in this field had been made by G L Farkas and Z Kiralay (1962), J Kuc' (1963), I Uritani (1963), and others who investigated the role of phenolic compounds in natural chemical resistance.

The biochemical nature of induced resistance is still shrouded in obscurity. It is an interesting and important phenomenon in both virus and bacterial diseases. A significant discovery by Loebenstein showed that as a consequence of viral infection, interferon-like substances are produced in the host plant. In infections by animal viruses, similar antiviral substances appear in the host plant. Müller (1956) and later Cruickshank (1963) proposed that in the reaction to infections caused by microbes, a mechanism, leading to the formation of antibody-like materials, operates in plants. This recognition led to a demonstration of the substances called *phytoalexins* which are currently known to be synthesized post-infectionally in the host plant and have a role in the tissue-resistance phenomenon. An ever-increasing body of research on this class of compounds that has been accumulating during the last fifty years but more convincing evidence as to how phytoalexins play a deciding role in host pathogen interaction, leading to resistance, is still lacking.

2.14 BIOCHEMISTRY AND PHYSIOLOGY OF THE DISEASED HOST PLANT

In recent years, much of the earlier symptomatological approach to problems of plant disease has given way to an integrated study of cause and effect. It is now becoming increasingly clear that a sufficient knowledge of plant physiology and biochemistry is essential for interpreting the metabolic changes in plants during pathogenesis. As a matter of fact, the biochemical and pathological trends of research started about 55 years ago and were founded on the physiological research of host-parasite relations. These modern trends of research intend to unravel the physio-biochemical processes of the diseased plant (such as photosynthesis, respiration, nitrogen metabolism, and growth). They wish to determine the true causes of cell proliferation, the possibilities for the inhibition of virus multiplication, the mechanism of fungal toxicity; the mode of action of toxins; the role of hormones or hormone-like substances in pathogenesis, as well as to elucidate the virus biosynthesis. The importance of extracellular enzymes in the etiology of the disease is emphasized by workers currently interested in this area of investigation.

2.15 MOLECULAR BIOLOGY OF PATHOGENESIS AND INDUCED SYSTEMIC RESISTANCE

The present era of plant pathology has seen much progress in molecular biology of pathogenesis and induced systemic resistance (Vidhyasekaran, 1998). Recent advances in molecular plant pathology have paved the way for development of innovative techniques to manage crop diseases. Crop plants are endowed with numerous defence genes which encode various proteins responsible for synthesis and accumulation of defence arsenals in plants. The defence genes are involved in the production of antimicrobial compounds like phenolics, phytoalexins, Pathogenesis-Related proteins (PR proteins), and active oxygen species or involved in reinforcement of cell wall by accumulating hydroxyproline-rich glycoproteins, callose, lignin, and wall phenolics. Many of the proteins produced by defence genes are enzymes but some of them are not enzymes. Some of them show antifungal activity, some show elicitor-releasing activity, some increase mechanical strength of the host wall and some others are involved in release of second messengers and systemic signals (molecules). It is now believed that signals are required to activate defence genes which are commonly present in both susceptible and resistant plants. Signal molecules have been detected in fungi (Schaffrath *et al.*, 1995, Vidhyasekaran, 1988a,b, 1993, 1997 and 1998), bacteria (He *et al.*, 1993) and viruses (Culver and Dawson, 1991). These are produced by pathogens as well as nonpathogens and are released by the host enzymes when these pathogens come into contact with plants. The signal transduction system plays a key role in disease resistance. Thus, the use of molecular genetic techniques has provided new insights into how plants defend themselves against attack. In turn this new understanding is suggesting novel ways in which plant diseases may be controlled.

2.16 TISSUE CULTURE IN PLANT PATHOLOGY

Tissue culture has had the widest impact on the study of 'crown-gall'. It has enabled plant pathologists to grow organs, tissues, and free cells alive *in vitro* for almost an indefinite period. In the words of White (1968), "Tissue culture is a method of simplifying organisms in order that they may be handled more effectively". In the early 1920s, an offshoot of conventional tissue culture—**organ culture**—made its modest appearance and has since established itself as a method of choice in may areas of plant physiology and plant pathology, biological and medical research, biochemistry of drugs, antibiotics, hormones, virus research, etc.

2.17 BIOTECHNOLOGICAL APPROACH TOWARDS DISEASE CONTROL

One of the frontier areas of plant pathology is the biotechnological approach towards plant-disease control. The genetic modification of organisms and application of modern techniques to study cellular and molecular processes are now expanding our knowledge of interactions between organisms and allowing a greater degree of control over biological processes. Specific genes may now be transferred between plants of different cultivars, lines, and species, from microbes to plants and from plants to microbes using suitable vectors such as plasmids, to integrate new characters into the plant genome (Issac, 1992). Transgenic plants are current interests of both plant pathologists and agriculturists throughout the world.

2.18 ECOLOGICAL STUDIES OF SOIL-BORNE FUNGAL PLANT PATHOGENS

Initially it was thought that knowledge of the life cycle of a plant parasite was sufficient to ensure its speedy control. Pathologists have now realized that a wider outlook is required which embraces entire dynamics of microbial ecology, and that knowledge of life cycles is only a prerequisite to the proper understanding of a soil-borne fungus pathogen in relation to its environment. The surroundings of an organism in the soil are biologically complex and the interactions of the organisms themselves produce balance. The most important worker in the field of root disease investigations was Dr S D Garrett. To paraphrase Garrett (1970), his subject has opened for him, and for others too, a window upon a world wider than that of the diseased plant. This world is that of the living soil, which is the source of all terrestrial plant life, and therefore, of direct concern to every human being and of interest to every biologist. Inside the soil, many kinds of living organisms are interacting continuously with one another and with the root systems of green plants. Thus, diseases of the root system cannot be studied in isolation as a simple interaction between the root system and the pathogen; hence, the root-disease investigator should be concerned with everything in the soil, such as changes in the soil texture, temperature, moisture, pH and other biological effects.

2.19 BACTERIOLOGY

Thomas J Burrill (1839–1916) of the University of Illinois, USA, was the first to describe a bacterial disease of plants in 1878, and he showed that the fireblight of apples and pears is caused by the bacterium known as *Erwinia amylovora*. Several other plant diseases were described shortly afterwards. Wakker (1883) showed that the yellow disease of hyacinth is caused by a bacterium. Savastano (1887) worked on the olive knot disease and E F Smith worked on the bacterial wilt of cucurbits. Smith is regarded as the most outstanding man who used the best available methods and standards in the study of bacterial diseases of plants and in establishing the modern bacteriological trend in plant pathology. Smith's name is still remembered as he resolved the controversy with the German bacteriologist, Alfred Fisher (1897, 1899), who did not think that bacteria were the primary cause of disease in plants.

2.20 VIROLOGY (VIRUSES, VIROIDS, AND PRIONS)

The German-born Adolph Mayer (1886) worked in Holland on a serious disease of tobacco (tobacco mosaic) and this work can be said to be the alpha of virology. He showed that the juice from infected plants, if applied to healthy plants, could reproduce the disease. In the same way, E F Smith (1891) showed that peach yellows was a contagious disease and could be bud-transmitted. In 1862, Dimitri Ivanowski, a Russian, showed that the causal agent of tobacco mosaic could even go through a Chamberland filter that withholds bacteria. It was Beijerinck, a distinguished Dutch microbiologist (1898) and Professor of Microbiology at the Technical University of Delft, Holland, who demonstrated that it could pass through a porcelain filter and diffuse through an agar-agar membrane, and concluded that the tobacco mosaic was caused by a noncorpuscular *contagium vivum fluidum* which he called **virus**. In 1935, the first major contribution regarding the nature of viruses was made when Stanley obtained a crystalline protein which he thought to be gamma globuline. It was in 1938 that Bawden and Pirie discovered the real nature of the tobacco mosaic virus. In 1956, Gierrer and Schramm (1956) proved that the nucleic acid fraction of the virus is actually the infectious agent. Within the attacked cell, the virus somehow mobilizes nitrogenous and other compounds of the plant and replicates itself. The relationship between an insect and a virus was discovered by a Japanese farmer, Hashimoto, who worked in 1894 with the rice-dwarf disease and the leaf hopper *Nephotettix apicalis* var. *cinticeps*. This information was, however, not available to the western world for quite some time (Smith, 1974). In about 1907, three American workers—Ball, Adams and Shaw—proved the connection between the curly top of sugar beet and the leaf hopper *Eutettix tenella*.

During the last few decades, much more information of the relationship of plant viruses with their vectors has been accumulated. It is now clear that in addition to the insects which transmit plant viruses, other organisms such as mites, nematodes, worms, and fungal spores have all been associated as vectors. Other advances in the study of plant viruses include an increased knowledge of incomplete viruses, and the discovery of satellite viruses (Kassanis, 1966) and 'viroids' (Diener and Raymer, 1967). Potato-spindle tuber virus (Diener and Raymer, 1967) and the rice-dwarf virus (Hiura *et al.*, 1966) are shown to contain double-stranded RNA. These are termed **viroids**. Viroids differ from viruses by the absence of a dormant phase (**virions**) and by genomes that are much smaller than those of viruses and lack a protein coat. The algal viruses which form a very distinct group of DNA viruses are another important and interesting field of research.

The rod-shaped tobacco mosaic virus was first reconstituted from its protein and RNA components (Fraenkel-Conrat and Williams, 1955). Since then several spherical viruses have been reconstituted. "The tide of molecular biology has now reached the shores of plant virology", says Dr Kenneth M Smith; and more precise knowledge of the ultra-structure of viruses has been obtained by electron microscopy and X-ray diffraction techniques. A photographic rotation technique has been developed to increase the details of high-resolution electron micrographs. The mechanism of replication of the viral nucleic acid and other constituents within the cells is being looked into critically, and in the case of the tobacco mosaic virus, the work of K K Reddi (1964) stands out prominently. By modifying the purine bases of RNA, Mundry (1959) succeeded in inducing TMV mutations; thus, he was the first to elucidate the biochemical mechanism of mutation.

2.21 PRIONS

Prions were first proposed in 1972 by Prusiner who was later awarded the Nobel Prize in Physiology and Medicine in 1997. Prions are at first normal, small protein molecules which are produced in nerve cells and other cells of the brain. Prions are infectious proteins which cause diseases in animals and humans. Prions have not been observed in plants or other organisms.

2.22 MOLLICUTES (PHYTOPLASMAS)

Before 1967, only fungi, bacteria, nematodes, and viruses were considered as the main incitants of plant diseases. However, as early as 1898, Nocard and Roux had discovered another agent of diseases caused in animals which was similar to viruses in size but which could be cultured on artificial media. This group was known as mycoplasma. *Mycoplasma mycoides*, the causal agent of infectious pneumonia, had been cultured by that time. In 1967, Japanese scientists found that mycoplasma-like organisms could be responsible for most of the disease of the yellows type which were earlier thought to be caused by viruses in plants. MLOs have been reclassified as Phytoplasms/Spiroplasms.

Mycoplasma-like organisms (MLOs) found in certain leaf hopper-transmitted–diseases are morphologically similar to animal mycoplasmata, but Ploaie and Maramorosch (1969) were unable to grow the leaf-hopper organisms in culture. The cells of mycoplasma lack a cell wall and hence are pleomorphic. They do have a cell membrane (10 nm unit membrane). They are very small in size (from 100 nm to 1 mm in diameter) and because of the lack of cell walls, can easily pass through bacteria-proof filters. They contain ribosomes and thin stranded DNA and are sensitive to antibiotics of the tetracycline group.

2.23 NEMATOLOGY

The first report of nematode association with a plant disease was made in England by T Needham in the year 1743 AD with some microscopical observations on the farina of red lilly, and of worms discovered in smutty corn. He described the wheat-gall nematode now known as *Anguina tritici*. Almost 100 years after Needham, C Devaine described the life cycle of *Anguina tritici*. There are some important landmarks in the development of phytonematology. For example the root nematode was the second phytonematode to be discovered and reported. In England, M J Berkeley (1855) noticed galls on the roots of cucumber plants growing in a greenhouse. However, the credit goes to Cornu in 1879 who, for the first time, made a specific mention of a root-knot nematode which he described under the name *Anguillula mationi*. *Ditylenchus dispaci*, a stem nematode, was first reported by Kuhn in 1857 while the beet-cyst nematode was first noticed in Germany by H Schacht in 1859. N A Cobb studied the structure of many nematodes, classified them, and coined the term **nematology**.

The association of nematodes with diseases caused by other organisms had been noted as early as 1892 when Atkinson reported that Fusarium wilt of cotton was more severe in the presence of root nematodes. Similarly, the bacterial wilt of tomato was facilitated by root-knot nematodes. According to Singh (2008), an outstanding discovery was made by Hewitt in 1858 that the grapevine fan-leaf virus was transmitted by a nematode. Several plant viruses are transmitted through soil by means of nematodes.

2.24 HISTORY OF THE DEVELOPMENT OF PLANT PATHOLOGY IN INDIA

Some of the important events in the development of plant pathology in India are as follows:

- Barclay's (1886) *Uredinales of Simla.*
- Publications of two journals—*Memoirs of the Department of Agriculture* and *Agriculture Journal of India.*
- Appointment of E J Butler in 1901 as Cryptogamic Botanist to the Government of India and then later transferred to the then Imperial Agricultural Research Institute established in Pusa in 1905 as the first Imperial Mycologist. Butler's work on Phytophthora blight of potato from Assam, Bengal, and Eastern Himalyas and Nilgiris in 1903 and fusarial wilt of pigeonpea caused by a new species of *Fusarium* in 1906 and 1910 are pioneering contributions in phytopathology in India.
- Among Butler's contemporaries, F J F Shaw's contribution in 1912 along with S L Ajrekar on the genus *Rhizoctonia* with special reference to its morphology and parasitism are one of the landmarks in plant pathological research in India (Raychaudhri, 1991).

The following are important plant pathologists of the past and present eras in India.

1. E J Butler (1874–1943)

He came to India in 1901 and was appointed Cryptogamic Botanist to the Government of India. Butler's monograph of the genus *Pythium* and his book *Fungi and Diseases in Plants* (1918) are still regarded as classics. In 1901, he discovered an important genus, *Allomyces,* in India. Among his numerous contributions to science was *Fungi and Diseases in Plants* published in 1918 from Calcutta. Butler is regarded as the father of Indian mycology and plant pathology. He was appointed as the first director of Imperial Bureaus (now Institute) in 1921.

2. J F Dastur (1886–1971)

He made valuable contributions on cotton wilt, foot rot of betel vine, ripe rot of banana, pink disease of citrus, anthracnose, and *Nematospora* on cotton.

3. S R Bose (1888–1970)

He initiated studies on the effect of radiations on polypores and systematics of Bengal Polyporaceae, and nature of colouring substances in coloured polypores. At the Department of Medical Mycology, Calcutta School of Tropical Medicine, he initiated work on the physiology of skin diseases. He built up a herbarium of about 4000 specimens of polyporaceae in this department.

4. K C Mehta (1892–1950)

He was born in 1892 at Amritsar, Punjab. He studied botany under Professor S R Kashyap of Government College, Lahore. He was Principal of St. John's College, Agra. He carried out comprehensive studies on cereal rusts and was awarded a PhD in 1922. Mehta's main contribution was that he declared that spores of wheat and barley rusts die in the plains of India during the hot summer, and the origin of infection is from hills where they oversummer on self-grown plants of wheat and barley. These cause early infection in local crops which are then blown down to the plains to cause epidemics there.

5. Manoranjan Mitra (1895–1942)

He got his earlier training in mycology under the guidance of Sir E J Butler. He was the first to discover the Karnal bunt of wheat caused by *Tilletia indica* in 1930 at Karnal in the then Punjab (now Haryana), India.

6. B B Mundkur (1896–1952)

He made studies on smuts of Indian cereals. He took interest in smuts of India which was later published in the form of a book by the CMI, Kew, Surrey in 1952. He also published a supplement on *Fungi of India* by Butler and Bisby. He published a textbook in 1949 entitled *Fungi and Plant Diseases*. Mundkur laid the foundation of the Indian Phytopathological Society in 1946. His last work was *Genera of Rusts* with M J Thirumalachar.

7. M N Kamat (1897–1980)

He has contributed mainly in the area of Ascomycetology. He discovered several new Ascomycetes genera, worked on mechanism of the sex and ascocarp development, and the cytology of ascus. He has written important books entitled *Introductory Plant Pathology* (1956), *Practical Plant Pathology* (1953), and *Handbook of Mycology*, volumes 1 (1952) and 2 (1961).

8. K D Bagchee (1898–1973)

He did pioneering work on mycology and forest pathology with special reference to biology and pathology of rust fungi of conifers, soil-borne diseases and their control at Forest Research Institute, Dehradun.

9. M K Patel (1899–1967)

He was born at Bardoli, Gujarat in 1899, and did his PhD work on crown-gall bacterium at the University of Iowa, USA. He was known for his work on Phytopathogenic bacteria and described more than 50 bacterial diseases from India.

10. S N Das Gupta (1902–1990)

He worked mainly on aquatic fungi. He also concentrated on diseases of complex etiology, especially tip necrosis of mango at Lucknow University, Lucknow.

11. R N Tandon (1903–1999)

He established a strong school of fundamental pathology at the University of Allahabad by his active research in nutritional studies on fungi, particularly leaf-spot pathogen and storage fungi. His notable publications are *Mucorales of India*, *Physiological Studies on Some Phytopathogenic Fungi* and *A Supplement to the List of Indian Fungi*.

12. R S Vasudeva (1905–1987)

He obtained a PhD in plant pathology under the able guidance of Professor William Brown in 1930. On his return to India, he was appointed as a mycologist for the investigation on root rot of cotton, a soil-borne disease. He recommended the growing of moth (*Phaseolus aconitifolius*) as an intercrop with cotton, which reduced soil temperature because of its spreading nature and inhibited the growth of soil-borne fungi.

13. B L Chona (1906–1994)

He was a world-renowned sugar cane pathologist who worked on the physiology of fungi. Dr Chona confirmed the earlier observations of the sugar cane research workers that the main and the primary role of red rot and smut of sugar cane was diseased vegetative seeds, contrary to the observations in other countries.

14. R Prasada (1907–1992)

He did commendable work on the epidemiology of cereal rusts in association with Dr K C Mehta. His work was published in the form of scientific monographs of ICAR in 1940 and 1952. He contributed significantly on rust of grasses, maize, millets, legumes, oilseeds, and fiber crops. His memorable work on linseed rust bestowed him with a DSc degree from Agra University in 1943.

15. P R Mehta (1907–1982)

He worked with Prof. E C Stakman of the University of Minnesota, USA. He served as a lecturer in Allahabad, Banaras Hindu University. He was plant pathologist to the UP Government. In 1954, he was appointed Deputy Director, Plant Protection Quarantine and Storage, New Delhi. While working in the directorate, he was mainly responsible for certain important projects such as Establishment of Plant Protection, Central Institute of Plant Protection and Training, Hyderabad.

16. T S Sadasivan (1913–2001)

He developed a strong school of mycology and plant pathology at Madras University, Madras, and it was here that the foundation of research on soil-borne disease in crops such as cotton, pigeonpea, and rice—aimed at controlling the survival of their pathogens—in soil was laid. His main contribution is on the taxonomy of Hyphomycetes. He was the first to demonstrate fusaric acid, a metabolite produced by *Fusarium* in wilting cotton.

17. M J Thirumalachar (1914–1999)

He created over 20 new genera and 300 new species of fungi. His joint work with B B Mundkur on *Genera of Rusts* is a classic and reference work. He published a monograph on the Ustilaginales of India. He is credited with the discovery of more than 100 antibiotics and some antifungal chemicals. He was responsible for the development of the systemic fungicide KT-1982 for the control of Dutch elm disease and oak wilt. While in USA, he isolated a new genus, *Geotrichum marinum*, which can grow on the oil in the ocean and uses oil as a carbon and nitrogen source.

18. S Y Padmanabhan (1916–1985)

He worked mainly on rice diseases and their control. He wrote a famous review article on the Bengal famine of rice in 1942 in the *Annual Review of Phytopathology* and brought out the role of the disease in the Bengal famine.

19. S P Raychaudhary (1916–2005)

He made extensive studies on viral diseases and the virus-vector relationship of the viruses affecting corn, large cardamom, mulberry, tea, etc. at the virus research laboratory at Kalimpong (WB). He established a strong school of virology at IARI, New Delhi, from where he retired as the Head of the Division of Mycology and Plant Pathology.

20. K S Thind (1917–1991)

He began his distinguished carrier in the field of mycology and plant pathology. He worked mainly on fleshy fungi, especially the Aphyllophorales. His monographs *Indian Clavariaceae* and *Myxomycetes of India* are published by ICAR, New Delhi.

21. S B Saksena (1917–1988)

He laid the concept of ecology of higher plants to soil fungi and used the terms 'frequency', 'abundance', and 'density' while studying the microflora of Sagar soil for the first time. He discovered a large number of new taxa such as *Saksenaea*, *Monicillium*, and *Gliocladopsis*.

22. K S Bhargava (1918–2010)

He initiated cytological work on the members of Saprolegniaceae. His contributions were mainly in the area of plant virology.

23. Abrar M Khan (1919–2004)

He was Professor and Head, Department of Botany at Aligarh Muslim University, Aligarh. He obtained a PhD with Professor Stakman. He worked out conditions conducive for the development of *Sphareotheca* and *Erysiphe* infecting cucurbits in tropical climates. In India, he pioneered research on applied aspects of phytonematology with a view to evolving the most promising controlled techniques which are stable and safe from the standpoint of pesticide use and are also economical. He was also considered as the father of nematology in India and initiated the Nematological Society of India.

24. M S Pavgi (1919–2005)

He was Professor of Plant Pathology at Banaras Hindu University, Varanasi. He investigated on the ontogeny and cytology of *Protomyces*, *Taphrina*, and rusts. He is credited with the discovery of several new genera and species.

25. R L Munjal (1920–2008)

He made significant contribution on the Karnal bunt of wheat and mushroom cultivation in India.

26. K Ramakrishnan (1920–1978)

He mainly worked on the taxonomy of fungi of Madras where he has described many rust, smuts, and imperfect fungi.

27. A P Mishra (1920–1999)

He worked mainly at Rajendra Agricultural University, Pusa, Bihar. He and his students worked on the Helminthosporium diseases of crop plants.

28. H K Saksena (1922–2007)

He did his associateship with distinguished mycologist B B Mundkur. He also worked with Professor R Prasada and showed that the weed *Trigonella polyceratia* is a collateral host of gram rust. He worked on Rhizoctonia diseases and established a strong school of Rhizoctonia diseases of crop plants at Kanpur (now C S Azad University of Agriculture and Technology).

29. J S Chouhan (1922–2009)

He served at PAU, Ludhiana, and did pioneer work on the Aflatoxin problem in groundnut and maize, and the etiology of viral diseases in oilseed crops and vegetables.

30. C V Subramanian (b1924)

He retired as Director, CAS Botany, University of Madras. He is a noted authority on Hyphomycetes and his two important monographs—*Hyphomycetes* (1971) and *Hyphomycetes: Taxonomy and Biology* (1983)—are outstanding. He is the discoverer of a large number of new genera and species of fungi.

31. L M Joshi (1925–2010)

He initiated a national wheat-disease survey and surveillance programme in 1967. The monitoring system rejuvenated the interest in the epidemiological studies of wheat research in India. He was instrumental in explaining the "Puccinia Path" in the Indian subcontinent.

32. G Rangaswami (1925–2005)

He was one of the leading pioneers of biochemistry and soil microbiology in India. He has written two important books, namely *Pythiaceous Fungi* and *Diseases of Crop Plants in India.*

33. R S Singh (b1927)

He has done considerable work on soil-borne diseases. His widely read books are *Plant Diseases*, *Principles of Plant Pathology*, *Diseases of Vegetable Crops*, *Diseases of Fruit Crops*, *Plant Disease Management* and *A Handbook of Economic Nematology*.

34. M M Payak (1928–2010)

His work encompassed both basic and applied methods in various areas like cereal rusts, diseases of maize, development of multiple disease resistance, and management of major diseases of crops in India.

35. G P Agarwal (b1929)

He served as Professor and Head of Department of Botany, R D University, Jabalpur. He has made valuable contributions on the nutritional aspects of several fungi, on fungal infection of humans, entomogenous fungi, aquatic fungi and biological control.

36. H S Sohi (1930–2007)

He worked at the Indian Institute of Horticulture Research at Bangalore, Punjab University, Chandigarh, and was Director of Mushroom Research Centre at Solan, HP. He worked mainly on fleshy fungi, control of fruit-tree diseases, and mushroom cultivation.

37. J N Chand (1930–1992)

He worked for his PhD degree with Professor C J Walker at the University of Wisconsin, USA. He worked on diseases of piper betel at Jabalpur and mainly on bacterial diseases of various crop plants at the Haryana Agriculture University, Hisar.

38. R K Grover (1931–2007)

He worked mainly at Haryana Agriculture University, Hisar, on fungicidal control of plant diseases specially caused by Rhizoctonia diseases of crop plants. He introduced the concept of environmental factors of soil and aerial parts on the effectiveness of fungicides, both in the soil as well as on the aerial parts.

39. K S Bilgrami (1933–1996)

Initially worked on leaf-spotting fungi and later did extensive work on different aspects of mycotoxins contamination of food and feed and its biological significance and management at Bhagalpur University, Bihar.

40. K G Mukerji (1934–2010)

He was Professor of Botany at Delhi University, Delhi. He contributed significantly on the taxonomy of Xylariaceae and Chaetomiaceae. He also worked on biocontrol of plant pathogens, mycorrhizal fungi, and microbial ecology with special reference to fertility of soil.

41. R P Purkayastha (1935–2011)

He contributed mainly in the areas of phytoalexin, and PR proteins in defence mechanism in plants. He has written the *Handbook of Phytoalexin Metabolism and Action* published by CRC Press, USA.

42. Y L Nene (b1936)

He did his PhD from the University of Illinous in 1960. He was the first professor of plant pathology in the Department of Plant Pathology, UP Agricultural University (now G B Pant University of Agriculture and Technology), Pant Nagar. He retired as Deputy Director General of ICRISAT, Patancheru, Hyderabad, in 1996 and devoted all his time to the Asian Agri-History Foundation. Nene's main contribution was that he solved the mystery of Khaira disease of rice and established that it was caused by zinc deficiency. He also solved the mystery of wilt complex of chick pea and developed multiple disease-resistance screening techniques and disease-resistant varieties of pulses. His two books *Fungicides and Plant Disease Control* and *The Pigeon Pea*, and an excellent pictorial bulletin *Tomato Disorders—Identification Handbook* are useful contributions.

43. Anupam Varma (b1940–)

He made notable contribution to fundamental and applied aspects of virology by developing technologies and probes for quick and reliable diagnosis of viral diseases of crop plants. He demonstrated the etiology of sandal spike and mango malformation which are of national importance. He has also developed virus-resistant transgenic lines of potato, tomato, papaya, and various cucurbits.

44. A N Mukhopadhyay (b1940)

He did his PhD from Banaras Hindu University. He served as Professor and Head of Plant Pathology at Gujarat Agriculture University in 1982 and G B Pant University of Agriculture and Technology in 1984. He is known for his work on biocontrol of plant diseases. He is credited with the publication of four edited volumes entitled *Plant Diseases of International Importance*.

45. M P Srivastava (b1942)

He worked mainly on post-harvest diseases of fruits and vegetables, on mycoherbicides, biocontrol, integrated pest management, and mushroom technology at CCS, Haryana Agriculture University, Hisar.

46. Y R Sharma (b1942)

He retired as Director of Indian Institute of Spices Research, Marikunnu, Calicut, Kerala, and has done extensive work on *Phytophthora* diseases of *Piper nigrum* and was responsible for National Network on Phytophthora Diseases of Horticultural Crops (PHYTONET).

47. C Manoharachary (b1944)

He was Professor of Botany, Osmania University, Hyderabad, and now Emeritus Professor there. He contributed extensively on the various aspects of soil fungi, aquatic fungi, forest fungi, fungi from diverse habitats and mycorrhizal fungi. He has also discovered several new genera and new species of fungi.

48. T N Lakhanpal (b1944)

He was Professor of Botany at Himachal Pradesh University, Simla, and now Emeritus Professor there. He initiated research on myxomycetes and later concentrated mainly on edible mushrooms and ectomycorrhizal fungi of Himachal Pradesh.

49. S M Pal Khurana (b1944)

He did his PhD in 1968 from Gorakhpur University and has made tremendous contribution to fundamental and applied aspects of virology, especially on potato viruses. He was Director of CPRI, Simla, Vice-Chancellor of Rani Durgawati University, Jabalpur, and at present Professor of Biotechnology at Amity University, Haryana.

50. S Nagarajan (b1945)

He has contributed significantly on the epidemiological studies, and disease management and wheat improvement programme. He has also developed a weather-based disease prediction system, remote sensing, and modeling pathogen monitoring. He wrote an important book entitled *Plant Disease Epidemiology* in 1963.

51. C D Mayee (b1946)

He served as Director, Central Institute of Cotton Research, Nagpur; Vice-Chancellor, MAU, Parbhani; Chairman, Agricultural Scientist Recruitment Board (ASRB); Director, Central Institute of Cotton Research, Nagpur; and Agriculture Commissioner, Government of India, New Delhi. His important books are *Phytopathometry*, *Stem and Pod Rot of Groundnut* and *Studies on Wheat Mosaic Streak Virus Diseases: Effect of Nutrition and Cytopathology*.

52. S S Chahal (b1947)

He contributed mainly in the area of cereal smuts and bunts. He was professor and head of Plant Pathology at PAU, Ludhiana, and later became Vice-Chancellor at Maharana Pratap University of Agriculture and Technology, Udaipur.

53. R P Thakur (b1947)

He joined ICRISAT as plant pathologist in 1976 soon after obtaining his PhD degree from the IARI. He has strong and wide research background on various aspects of diseases of sorghum, pearl millet, maize, groundnut, and pulses.

54. R D Khulbe (1949–2002)

He worked mainly on the taxonomy of aquatic fungi. He wrote two monographs—*Parasitic Water Molds* and *A Manual of Aquatic Fungi*—which are considered monumental contributions in the field of Indian aquatic mycology.

55. T S Thind (b1952)

He has done considerable work on the control of late blight of potato, and downy and powdery mildew of grapes with fungicides. He also worked on the development of resistance in pathogens to fungicides. Presently, he is Professor Emeritus at Punjab Agriculture University, Ludhiana.

56. A Mahadevan (1932–2002)

He was Director, Centre for Advanced Studies in Botany, University of Madras. He and his research associates have done extensive work in the area of physiological plant pathology and was the author of the book entitled *Growth Regulators, Microorganisms and Diseased Plants*.

57. R S Dwivedi (b 1932)

He was Professor of Botany at Banaras Hindu University and contributed significantly in area of soil fungi, microbial interactions, nodulation and mycophagous soil amoebae in relation to biocontrol.

2.25 SOME OUTSTANDING PLANT PATHOLOGISTS OF THE WORLD

1. Elvin C Stakman (1885–1979)

He is considered the world authority on rust (*Puccinia graminis*). His researches in plant pathology and mycology were principally on genetic variation and physiologic specialization in parasitic fungi, and in stem rust epidemiology. His discoveries in physiologic specialization changed the concept of a species and revolutionized the concept of developing resistant varieties of crop plants. He was the co-author of the book *Principles of Plant Pathology*.

2. J C Walker (1893–1994)

He reported the presence of protocatechuic acid and catechol in the dry pigmented skins of onion bulbs resistant to *Colletotrichum circinans*. His widely read texts are *Plant Pathology* and *Vegetable Diseases*.

3. E A Gaumann (1893–1963)

He carried out outstanding and extensive research in plant pathology (including antibiotics, defense reaction, and resistance), rust biology, and taxonomy of downy mildews. He published some important books like *Principles of Plant Infection*, *Comparative Morphology of Fungi* and a monograph on the genus *Peronospora.*

4. H H Flor (1900–1991)

His seminal scientific work was with flax and its rust disease fungus *Melampsora lini.* He suggested that for every gene for rust resistance in flax, there was a corresponding gene for pathogenicity in the flax rust fungus. This led to his famous "gene-for-gene" interaction hypothesis.

5. George A Zentmyer (1913–2003)

After joining the university of California, Riverside, Zentmyer confirmed that the cause of the wide-spread decline disease of avocado trees was due to root rot caused by *Phytophthora cinnamomi.* He made many fundamental discoveries on the biology of *P. cinnamomi* and noted that zoospores of *P. cinnamomi* were attracted to the zone of elongation of roots and that the germ tube produced by encysted zoospores were oriented to grow toward this area of roots. He also worked on black-pod disease of cacao and reported that it is not only due to *P. palmivora* but also due to *P. megakarya*, *P. capsici*, and *P. citrophthora.* In 1980, he published a monograph entitled *Phytophthora cinnamomi and the Diseases it Causes.*

6. Norman E Borlaug (1914–2009)

Dr Borlaug's team in Mexico developed high yielding wheat varieties (dwarf varieties) and improved agronomic practices which when introduced in South Asia in 1960's brought revolution in wheat production and saved hundreds of millions of people from starvation. Known as the Green Revolution and Dr Borlaug as its father, it became a scientific base for agriculture in developing countries. In 1970, he received Nobel Peace Prize for this unparalleled achievement.

2.26 PLANT-DISEASE CLINICS

According to Barnes (1994), the evolution of the plant-disease clinic arose from a perceived need for an organized systematic, and professional effort to assist in the identification of plant-disease problems. In the United States of America for many years, most states operated a plant-disease clinic through their departments of plant pathology (Agrios, 1997). Growers, country extension agents, and homeowners would send diseased plants, and soil from areas infested with diseased plants, and sometimes insects, to the plant-disease clinic and the pathogen or insect would be identified and some suitable control measures suggested by the extension plant pathologists.

The plant-disease clinic frequently represents "the tie that binds", bridging various aspects of basic and applied features of plant pathology into a meaningful, functional, responsive, and accountable component of agriculture (Barnes, 1994).

In our country too, several plant-disease clinics have now come up in almost all agricultural universities after a modest beginning was made by Punjab Agricultural University, Ludhiana, some years ago.

REVIEW QUESTIONS

1. Mention some important milestones in the development of the subject of plant pathology.
2. List five major breakthroughs in phytopathology during the 20th century and the scientists responsible for them.
3. Name five contemporary scientists of plant pathology.
4. Name the person who introduced plant pathology in India. Mention his important contributions.
5. Mention the most important contribution of the following:
   (i) L Spallanzani (ii) A Leeuwenhoek
   (iii) P A Micheli (iv) M Tillet
   (v) B Prevost (vi) A de Bary
   (vii) T J Burrill (viii) M S Woronin
   (ix) O Brefeld (x) L Pasteur
   (xi) P A Millardet (xii) H M Ward
   (xiii) J G Kühn (xiv) W G Farlow
   (xv) R Koch (xvi) R Hartig
   (xvii) A Mayer (xviii) D Ivanowski
   (xix) V M Biejerinck (xx) E Gäumann
   (xxi) W Brown (xxii) K O Müller
   (xxiii) S A Waksman (xiv) E J Butler
   (xxv) R H Biffen (xxvi) E C Stakman
   (xxvii) J E Vander Plnak (xxviii) J C Walker
   (xxix) J G Horsfall (xxx) S R Bose
   (xxxi) T S Sadasivan (xxxii) K C Metha
   (xxxiii) J F Dastur (xxxiv) M J Thirumalachar
   (xxxv) R N Tandon (xxxvi) T O Diener
   (xxxvii) T O Diener (xxxviii) S P Raychaudhuri
6. What are the recent trends in phytopathological research and training worldwide?
7. What are prions? Who discovered them for the first time? Do they cause any plant disease?
8. Name at least four important virologists of the country.
9. Name at least four bacteriologists of the country.
10. Name at least four nematologists of the country.

MULTIPLE-CHOICE QUESTIONS

1. What was the actual profession of Anton van Leeuwenhoek, who invented compound microscope?
   (i) A linen draper (ii) Surveyor
   (iii) Wine gauger (iv) All of these
2. Who wrote Species Plantarum in 1753?
   (i) C Linnaeus (ii) P A Micheli
   (iii) G Bauhin (iv) Robert Hooke
3. Who is called the father of botany?
   (i) C Linnaeus (ii) Anton van Leeuwenhoek
   (iii) Theophrastus (iv) C H Persoon
4. Anton de Bary was born in 1831 in
   (i) Germany (ii) England
   (iii) Switzerland (iv) Russia
5. Name an English clergyman and a naturalist who supported the theory of spontaneous generation.
   (i) Needham (ii) Schwann
   (iii) Tulasne (iv) Prevost
6. Who established the germ theory in relation to diseases of humans and animals?
   (i) Robert Koch (ii) Charles Tulasne
   (iii) L R Tulasne (iv) H M Ward
7. Who is called the father of forest pathology?
   (i) E C Stakman (ii) R K S Wood
   (iii) T S Sadasivan (iv) Robert Hartig
8. Who discovered Bordeaux mixture?
   (i) P M A Millardet (ii) D M Millardet
   (iii) A Millardet (iv) None of these
9. Which of the following contributed valuable information on delineating and cataloguing physiological races of pathogens?
   (i) E C Stakman (ii) R N Goodman
   (iii) J G Horsfall (iv) D F Bateman
10. Who gave gene-for-gene hypothesis to explain host-pathogen interaction in flax?
   (i) W A Orton (ii) H H Flor
   (iii) R W Marsh (iv) L R Jones
11. Who gave the concept of horizontal and vertical resistance in plants?
   (i) G L Farkas (ii) Z Kiralay
   (iii) J Kuć (iv) Van der Plank
12. Name the scientist who proposed prions and was later awarded the Nobel Prize in Physiology and Medicine in 1997?
   (i) Kassanis (ii) Prusiner
   (iii) Mc Manus (iv) Markham

Answers

1. (iv) 2. (i) 3. (i) 4. (i) 5. (i) 6. (i) 7. (iv) 8. (i) 9. (i) 10. (ii)
11. (iv) 12. (ii)

3

PATHOGENESIS

3.1 PENETRATION AND ENTRY BY PLANT PATHOGENS

3.1.1 Contact

Pathogenesis caused by an infectious agent begins as soon as the pathogen comes into contact with any portion of a susceptible plant. Most of the bacterial and fungal pathogens come into contact with their hosts accidently in the form of wind-borne spores or propagules. Some fungi, many bacteria, and most of the viruses are carried to their host plants by insects, nematodes, and other vectors. Motile propagules of fungal pathogens are attracted to root exudates such as the zoospores of many plant pathogens. Attraction and accumulation of zoospores to root in the soil just behind the root tip has been demonstrated by Mehrotra (1970) and the evidence of the attraction of nematodes to roots has been ably summarized by Webster (1969).

It has been demonstrated that the hyphae of the fungus *Rhizoctonia solani* aggregate on the outside of cellophane bags in contact with the roots indicating that root exudates induce the formation of infection cushions by the fungus. Electrotaxis has been the subject of conflicting reports for both zoospores and nematodes. Electric current and intensities within the range (0.3–0.6 μA) around plant roots are capable of initiating electrotaxis resulting in accumulation of zoospores, etc.

3.1.2 Recognition Between Host and Pathogen

It is still not clear as to how pathogens recognize their hosts and vice versa.

Recognition plays a central role in interactions between plants and their pathogens.

Pathogens must be able to recognize the presence of their host plants in their environment and often must recognize specific surface features of their host in order to effect successful penetration and infection. Successful pathogens must also be able to recognize and overcome plant-defense responses. Since plants are sessile, they must, in return, have evolved sophisticated mechanisms to detect the multitude of potential pathogens in their environment and to activate various diverse defense mechanisms. The mutual signaling between plant hosts and potential pathogens has been the focus of extensive researches during the last few decades. The emphasis has been to identify and characterize the molecules involved in recognition between the two organisms.

Host components acting as signals for activation of pathogens are numerous. These may include fatty acids of the plant host cuticle that activate production of cutinase enzymes. These enzymes break down cutin of the host galacturonan molecules of pectin which stimulate the production of pectin lyase enzymes by the fungal or bacterial pathogen, phenolic compounds which stimulate the activation and germination of propagules of some pathogens, and phenolics and sugars released from plant wounds that activate a series of genes in certain pathogens leading to infection.

The term *elicitor* was originally used to refer to the molecules and other stimuli that induce the synthesis and accumulation of antimicrobial compounds (phytoalexins) in plant cells, but is now commonly used for molecules that stimulate any plant-defense mechanism. Examples of plant-defense mechanisms induced by elicitors include the synthesis and accumulation of antimicrobial phytoalexins, the production of glycosyl hydrolases capable of attacking surface polymers of pathogens, the synthesis of proteins that inhibit degradative enzymes produced by pathogens, the production of activated oxygen species (oxidative burst), and the modification of plant-cell walls by the deposition of callose, hydroxy proline-rich glycoproteins and/or lignin.

Pathogen components that act as elicitors of recognition by the host plant and subsequent mobilization of plant-defenses are still poorly known. The induction of these plant defense reactions or otherwise is mediated by an initial recognition process between plants and pathogens, which involves detection of certain unique structural features of incompatible or compatible pathogens by recognition molecules in plants, thereby setting of a cascade of biochemical events leading to resistance or susceptible reaction. When the initial signal by the pathogen favours growth and development of the pathogens, disease may be induced. If the signal suppresses the growth and development of the pathogens, disease does not occur. On the other hand, if the initial recognition elicitor received by the host triggers a defense reaction, the growth of the pathogens and its activity is slowed down or completely and will not develop at all. If the elicitor either suppresses or bypasses the defense reactions of the host, disease may develop.

3.1.3 Host Penetration and Infection

When a pathogen comes in contact with a host plant, a complex combination of interacting factors, including temperature, moisture, susceptibility of plant tissues, effects of other microorganisms present, aggressiveness of the pathogen, etc. determine whether infection will occur or not. **Pathogenicity** is the property of a microorganism whereby it becomes a part of the causal complex. **Pathogenesis** is a process or the chain of events by which disease development takes place. The events of pathogenesis will be discussed under the following heads:

1. Pre-penetration
2. Penetration
3. Post-penetration or further colonization of the plant by the pathogen

3.2 PRE-PENETRATION

The propagules of bacteria and fungi have mucilaginous substances on their surface or at their tip. These substances consist of mixtures of water-insoluble polysaccharides, lipids, etc. which when moistened become sticky and help the pathogen adhere to the plant. These propagules have certain mechanisms that prevent their germination until they get some stimulus. Once the stimulation for germination has been received by the propagules, the latter mobilizes its stored food, and directs the propagules towards germ-tube formation. The germ tube also perceives stimulus from the host and if it does not receive the right stimulus, the germ tube remains undifferentiated and stops growing. When appropriate signals are present, **germ-tube differentiation**, i.e. **formation of appressorium**, takes place. The perception of signal by the pathogenic fungi seems to be controlled by the cyclic Adenosine Monophosphate (cAMP) and Mitogen-Activated Protein Kinase (MAPK). In response to a signal from the host plant, the fungus perceives the extracellular signal and its transmission via the plasma membrane. It then accumulates intracellular signaling molecules and induces a phosphorylation cascade. In some fungi, the receptor of the signal is a protein in the plasma membrane of the fungal spores. After attachment of the propagule to the host surface, germ tubes also produce mucilaginous materials that allow them to adhere to the cuticular surface of the host. Once appressoria are formed, they stick tightly to the host surface. After that, the appressoria secrete some extracellular enzymes and generate physical pressure to bring about penetration of the cuticle. The turgor pressure inside an appressorium is quite high and it is due to the accumulation of glycerol in the appressorium, which due to its high osmotic pressure draws water into the cell and generates hydrostatic pressure, that the appressorial penetration peg is pushed further.

In fungal pathogens, this includes spore germination and growth of the resulting tube on the surface of the host plant. Germination is essentially the change from a low metabolic rate to a high metabolic rate and involves a change from near-dormancy to intense activity. For this, an energy source is needed such as a carbohydrate or fat reserve in the propagule. Fungal invasion is chiefly by germ tubes or structures derived from them. There are two well-known situations involving hyphae acting in a concerted way to achieve host penetration. Hyphae of *Rhizoctonia solani* often aggregate to form an infection cushion from which multiple penetrations occur by means of appressoria and penetration pegs (Fig. 3.1a, b). The entry of bacteria and viruses is achieved in various ways.

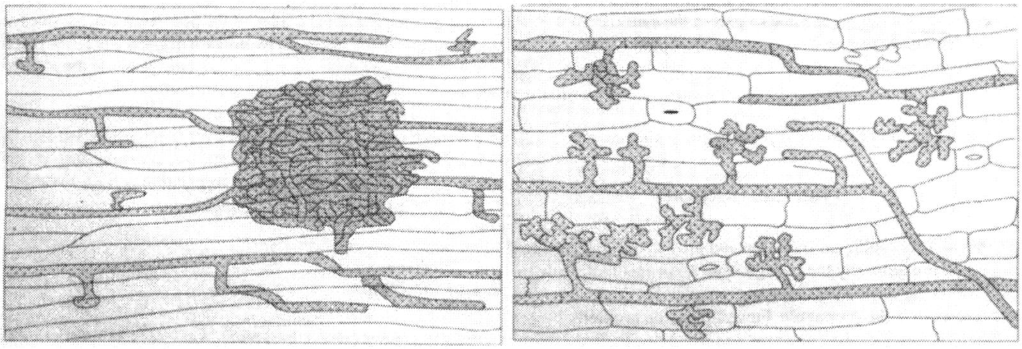

(a) (b)

Fig. 3.1 (a) Formation of infection cushion by crucifer isolate of *R. solani* on radish hypocotyl
(b) Production of lobate appressoria by solanaceous isolate of *R. solani* on tomato hypocotyl

Some spores germinate easily on release and sometimes even before release, while resting spores have a dormancy period during which they do not normally germinate. In order to procure food, plant pathogens have to establish an intimate connection with their hosts; this means that they have to enter into tissues or send haustoria into surface cells. Entry is mostly affected by the germ tube from spores, but can also often be accomplished by mycelial hyphae. The pathogen invades plants through natural openings (stomata, lenticels, nectarthodes, hydathodes), through wounds (natural or due to insects or nematodes), or by direct penetration. While some enter in one way only, others enter in more than one way. In Uredinales, the basidiospores penetrate directly while urediniospores generally, but not invariably, invade through stomata. *Botrytis cinerea* may invade directly or through wounds and some fungi which normally grow through stomata may penetrate directly under certain conditions.

3.3 ENTRY THROUGH NATURAL OPENINGS

The process by which fungi enter through stoma will be illustrated by the stem-rust fungus *Puccinia graminis tritici* (Fig. 3.2a). Urediniospores of pathogen will germinate in water on a leaf surface in an hour under optimum condition. The germ tube from the urediniospore grows over the leaf and approaches the stoma. The protoplasm of the germ tube accumulates at the tip to leave the older parts empty. On reaching stoma, the end of the germ tube swells to form an appressorium in the stomatal aperture. The empty germ tube is cut off by the formation of a cross wall. From the appressorium, a bladelike wedge grows through the stomatal slits and swells inside to form a substomatal vesicle into which the contents of the appresorium pass, leaving the latter empty. One or more hyphae grow from the vesicle towards the cell lining the substomatal cavity and after making contact proceed to penetrate the cell wall. After penetration, a haustorium is produced in the cell. In diseases caused by obligate parasites and facultative saprophytes, and in a few caused by facultative parasites, there is no massive penetration of the cell cavities of the host by ordinary hyphae of the parasite. Instead, the cell wall is penetrated by short branches which usually grow to a limited extent in the lumen of the cell before their ends are transformed into structures called haustoria, the forms of which may be simple or relatively complex depending on different parasites. Haustoria are presumed to have an absorptive function and are the main source of metabolic exchange between the host and parasite. A detailed account of haustoria will be given later.

There are variations in the behaviour of the germ tube at the time of penetration through the stomata, as in *Peronospora destructor* infecting onion leaves. Here, the germ tube continues to grow after the formation of the first appressorium (Fig. 3.2b).

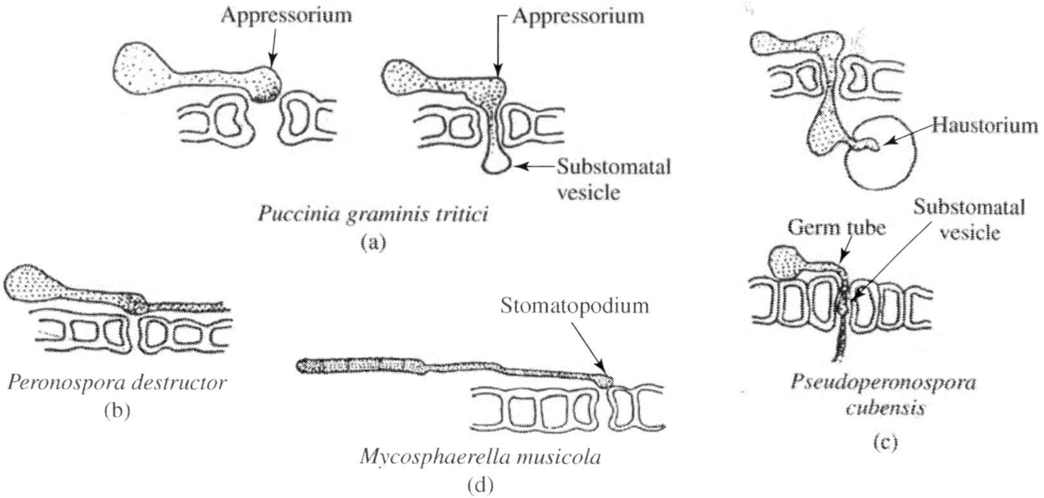

Fig. 3.2 Penetration through natural openings

On the other hand, zoopores of a species of *Phytophthora* infecting fennel (*Foeniculum vulgare*), almost always enter through stomata while *Phytophthora infestans* almost always enters plants directly, but at times through stomata after the formation of an appressorium. In *Pseudoperonospora cubensis* (Fig. 3.2c), the zoospores swim along the boundary lines of the epidermal cells towards the stomata and encyst above the lines separating the guard and epidermal cells. Hyphae penetrate the stomatal aperture and swell to form a substomatal vesicle from which, in turn, other hyphae may grow to form haustoria in the adjacent cells of the leaves. *Mycosphaerella musicola* (Fig. 3.2d), responsible for Sigatoka disease of bananas, enters leaves in a similar manner as do the rusts. The germ tube from a conidium grows slowly over the leaves for a few days before its penetration. This indicates that it can withstand a period of desiccation. A small structure called a stomatopodium is formed over the pore of a stoma. A hypha then arises from it which grows into the substomatal chamber and swells to form a vesicle, which in turn gives rise to hyphae which invade palisade tissues.

The mode of entry into plants by *Sclerotinia cinerea,* which causes stone fruits to rot depends on the host species infected. It enters plums through the stomata, apricots through the stomata or cuticle, peaches through hair sockets and cherries and nectarines through the cuticle.

Stomata are the most important or the only pathways of entry for many parasitic bacteria such as *Ralstonia* (*Pseudomonas*) *tabaci* on tobacco, *Erwinia amylovora* infecting apples and pears, *Xanthomonas campestris* pv. *phaseoli* infecting *Phaseolus* spp., *Xanthomonas campestris* pv. *malvacearum* infecting cotton, and *Pseudomonas phaseolicola* infecting *Phaseolus* spp. Bacteria enter stomata only through films of water that extend from outside through the stomatal aperture into the inside of the shoots. Because of this, the water content of shoots is of great importance in determining whether or not they will get infected. The stomatal morphology may determine whether successful infection will take place or not. In case of citrus fruits, the conformation of the cuticle around the stoma either prevents or allows the passage of water droplets containing the bacterial pathogen *Xanthomonas campestris* pv. *citri.*

3.3.1 Lenticels

Lenticels are openings on stems, tubers, and on fruits that are filled with loosely connected cells that allow free passage of air into the internal tissues.

Streptomyces scabies, the cause of the common scab of potato, enters young developing tubers through stomata or lenticels.

Oospora pustulans, which causes the skin scab of potato tubers, also enters through lenticels. *Penicillium expansum,* which is responsible for storage rot of apples and *Gloeosporium perennans*, the cause of another storage rot, also enters through stomata or lenticels.

3.3.2 Nectarthodes, Hydathodes, and Trichomes

Erwinia amylovora enters through nectaries of apple and pear flowers. In pears, nectaries occur in saucer-shaped tissues between the point of emergence of the style and stamens.

Droplets of nectar are exuded through stoma-like nectarthodes at the bottom of depressions on the surface of the nectaries. Within 24–28 hours of adding suspensions to open pear blossoms, the bacteria are found locally in great number beneath the nectarthodes.

Hydathodes are structures through which water is discharged from the interior of leaves, usually more or less directly from bundles of terminal tracheids which end in thin-walled parenchyma known as epithem. Hydathodes are the only means of entry for certain bacteria. Bacteria, which enter through hydathodes, are *Erwinia amylovora* and *Xanthomonas campestris.*

During periods of high humidity, bacteria accumulate in water from hydathodes and are drawn into epithem and tracheids, as the water recedes when conditions reduce water congestion, for example, in rice leaves and *Xanthomonas oryzae* pv. *oryzae*. *Clavibacter* (*Corynebacterium*) *michiganense*, which causes the bacterial canker of tomato plants, often enters through trichomes.

3.4 DIRECT PENETRATION

The physicochemical characteristics of cuticles and cell walls are important factors in direct penetration.

3.4.1 Structure of Cell Walls and Cuticles

The external layers lining the intercellular spaces and between the adjacent cells (middle lamella) consist largely of pectic substances, particularly calcium and magnesium pectates, and can be disintegrated by the fungal and bacterial pathogens which produce pectolytic enzymes. Proteins have also been reported to occur in the middle lamella. Within this is a primary cell wall consisting largely of cellulose embedded in a matrix of pectic substance and hemicelluloses. The cellulose may be present as micelle. Other substances, such as proteins, lignin, suberin, and cutin have been reported in the primary cell wall. In some tissues, a secondary cell wall consisting chiefly of cellulose is laid down within the primary wall. The cuticle is a noncellular membrane which covers most of the aerial parts of the higher plants. The outer waxy layer is hydrophobic and largely impermeable to water but the inner layer can absorb moisture, and swell and force the wax aggregates apart. The innermost layer of the cuticle passes gradually into a predominantly pectic layer which in turn merges into the largely cellulose outer walls of the epidermal cells.

The intact or uninjured plant surface or epidermis which is covered by a well-defined layer of cuticle is penetrated directly by a mechanical process. In most cases, penetration is preceded by the formation of a special structure, namely, penetration hypha or appressorium, which exerts a mechanical pressure on the intact wall of the host. An **appresorium** is the swollen tip of hypha or germ tube that facilitates attachment and penetration of the host by a fungus. An **infection peg**, arising from the appressorium, pierces the host wall to cause infection. The penetration of the cuticle is regarded by many as purely mechanical although there is evidence of the chemical degradation of the cuticle. Many root pathogens accumulate on the surface of roots and hypocotyl and other sub-aerial parts and penetrate directly. The penetration may or may not be preceded by the formation of an appressorium. It has been shown that barley leaves induce the conidial germ tubes of *Blumeria* (*Erysiphe*) *graminis* to grow longitudinally on leaves and stimulate the production of appressorium. The papilla, a result of host reaction to the fungus, begins to form in the epidermal cell below the appressorium arm as the appressorial infection pore develops. The tip of the blunt infection peg lacks a cell wall and the host epidermal wall dissolves only in the region of the tip. Mechanical pressure is thought to force the peg through the host cell wall and papilla, and to invaginate the cytoplasm.

The primary infection of wheat and barley by *P. graminis* consists of a number of morphologically indentifiable stages of development, that is, spore germination, formation of appressoria, penetration into host cells, formation of haustoria in host cells, and formation of elongating secondary and tertiary infections. The wax layer of the cuticle surface is a major factor for the formation of appressoria.

In direct penetration of a leaf, the infection hypha (peg or thread) which arises from the underside of an apressorium has to pass through the cuticle and underlying tissue and sometimes through the thickened outer wall of the epidermal cell, although at times it passes between the epidermal cells. The process has been studied in many fungi, including *Cladosporium carpophilum* [Fig. 3.3(a, b)] on almond, *Venturia inaequalis* and *Venturia pirina* on pear, *Phytophthora infestans* on potatoes, *Puccinia graminis* on barberry, *Puccinia psidii* on pimento, and *Ravenelia humphreyna* on *Caesalpinia*. The smut fungi usually penetrate directly into young ovaries, shoot buds, young seedlings or young tissues, such as the loose smut of wheat and barley, sugar cane smut (*Ustilago scitaminea*), bunts of wheat, flag smuts of wheat and rye, covered smuts of barley and oats, and common smut of maize (Fig. 3.3c).

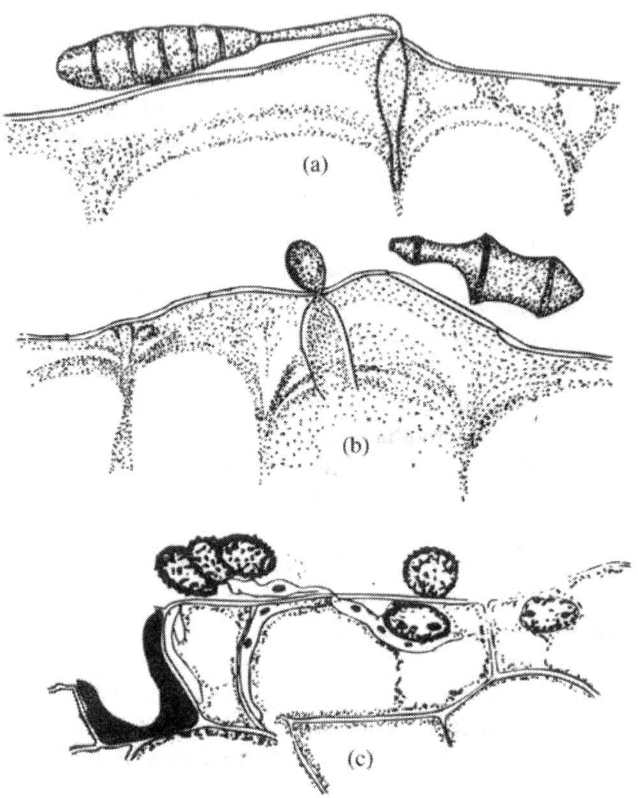

Fig. 3.3 Direct penetration of the cuticle. Penetration of almond leaves by germ tubes of *Cladosporium carpophilum*, the penetration hyphae passing (a) between epidermal cells, (b) passing into the epidermal cell, and (c) direct penetration of corn cells by the smut fungus, *Ustilago maydis*, the germ tube penetrating directly through an epidermal cell and making contact with host nucleus

In the case of infection due to *Blumeria* (*Erysiphe*) *graminis* on wheat leaves, a fine stylar process from the germ tube (or from a small subapical appressorium) pierces the cuticle, and in its passage through the cellulose layer, it is preceded by a local thickening of the layer into a papilla which it eventually pierces at the apex. Inside the host cell, it becomes dilated into a branched haustorium (Fig. 3.4).

The main role of appressoria is the direct penetration of a host. To achieve this, appressoria must attach them firmly to the host surface and produce infection pegs that pierce the cuticle and usually the epidermal cell wall. Appressoria produced by germinating urediniospores of rusts and spores of *Mycosphaerella citri* penetrate through stomata. Few authors disagree that this is the primary role of the appressorium, but there are suggestions that appressoria may play a secondary role, and not an auxiliary role. It may be that slight localized softening of the cuticle occurs around the infection hypha, thus facilitating penetration. In the presumed absence of cuticle-softening enzymes, it is probable that penetration is largely by a physical force.

Resistance of the leaves of *Berberis* and *Odostemon* spp. to basidiospores of *Puccinia graminis* was partly associated with their resistance to the combined thickness of the epidermal walls and cuticle. Cuticle thickness has been correlated with resistance to foot rot of peas caused by *Ascochyta pinodella*. It has been reported that older barley leaves with cuticles 2.5–5 μm thick were resistant to mildew

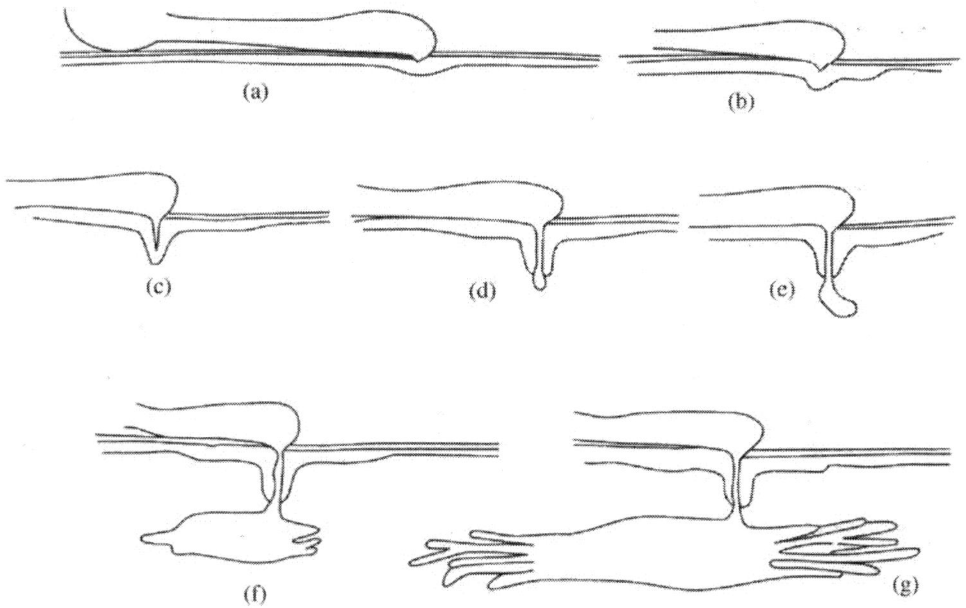

Fig. 3.4 Germination of conidia of *Blumeria* (*Erysiphae*) *graminis* on the epidermis of wheat: (a) Germination of conidium (b) A fine stylar process from the germ tube (or from a small sub-apical appresorium) pierces the cuticle (c) Local thickening of the layer to form a papilla (d) Papilla eventually pierces at the apex (e) Dilatation inside the host cell (f) Development of a lobed haustorium (g) Branched haustorium and invagination of the inner layer of cytoplasm (Redrawn from S A J Tarr, *Principles of Plant Pathology*, The Macmillan Press, London and Basingstoke, 1972)

as compared to younger leaves with 0.4–1.5 μm cuticles in thickness. Resistance of the strawberry varieties to *Sphaerotheca macularis* has been shown to vary with cuticle thickness.

3.4.2 Penetration of the Cell Wall

After passing through the cuticle, the infection hypha has to grow through the outer wall of the epidermal cells which consists of a network of cellulose microfibrils in a matrix consisting principally of pectic materials. The infection hyphae of some fungi pass between the radial walls of the epidermal cells without entering them. In *Diplocarpon rosae*, the cause of the black spot of rose leaves—the mycelium—develops between the cuticle and epidermis and enters the cells of the latter. The penetration of the walls is largely enzymic. The infection hyphae grow quickly through the softened materials. Pectolytic enzymes are important in this but there is doubt as to the part played by cellulases. Penetration of the cell walls may occur within a minute and is probably a combination of physical pressure and enzymic softening. Swelling of the invaded cell wall and of walls some distance away may occur due to the diffusion of enzymes. Such swellings vary considerably in extent, ranging from sharply localized (*Ustilago segetum* on wheat) to pronounced thickening which almost obliterates the lumen of the cell, as in the case of *Venturia inaequalis* attacking the epidermal cells of the stem of apple.

Infection can be as readily caused by mycelial hyphae of many fungi as in germ tubes from spores. In *Armillaria mellea*, a root parasite of many trees, infection is restricted to the rhizomorphs.

The germ tubes of aeciospores of *Puccinia menthae* (mint rust) produce infection structures resembling appressoria when exposed to thymol at its own saturated vapour pressure on glass but not on glass in the absence of thymol. Thymol also reduces spore germination and elongation of germ tubes. It is thought that it acts as a morphogenetic switch in bringing about the formation of appressoria. Studies on different strains of *Corticium solani* on roots, stems, and leaves of several plants indicated that appressoria developed on susceptible but not on resistant plants. The formation of appressoria is probably affected by the nature of the surface on which the hypha grows and also by diffusable substances which cause the arrested growth and multiple branching of side branches leading to the development of appressoria.

In the case of infection by *Rhizoctonia solani*, infection cushions are seen on the roots or hypocotyls of the infected plants. Multicellular infection cushions are produced from a branch hypha in which normal elongation has ceased but prolific side branching now takes place. Strains of *Rhizoctonia solani*, nonpathogenic to seedlings of a particular species, failed to become attached to the hypocotyl and organize the infection cushions.

3.5 ENTRY THROUGH WOUNDS

All bacteria, most fungi, some viruses and all viroids can enter plants through various types of wounds. Penetration of viruses, mollicutes, fastidious bacteria, and protozoa through wounds depends on the deposition of these pathogens by their vectors in fresh wounds created at the time of inoculation.

Many so-called low-grade pathogens are not able to penetrate the host surface in the absence of a wound. However, once they are in the plant, they are often able to cause severe disease symptoms. Wounds may be natural or they may be caused by the activities of humans or other agencies, such as insect bites or nematodes. Many horticultural and agricultural practices involve accidental or even deliberate wounding. Entry through natural wounds (lenticels and abscission areas) has been dealt with earlier. Other natural wounds are those that occur due to the emergence of laterals from main roots. This is a means of entry for the tobacco root-rot fungus *Thielaviopsis basicola* and for *Fusarium solani f. phaseoli* into bean roots. Many important forest pathogens also enter through wounds. *Heterobasidion (Fomes) annosus*, a destructive pathogen of conifers, naturally colonizes wounds caused by high winds, snow, or other natural agencies. It becomes a very damaging pathogen in plantations where it enters through the stumps left after felling. Human activities are responsible for many of the wounds through which fungal penetration occurs. Many fruit and vegetable products are quite succulent when they are ripe for marketing. In this condition, they are most susceptible to wounds by rough and careless handling. The Rhizopus rot of sweet cherries and peaches develops in wounds in this manner. The size of the wounds necessary for the infection of oranges by *Penicillium digitatum* varies with the degree of maturity and variety of the fruit.

An unusual type of entry of a variety of microorganisms occurs through wounds in flax and maize seeds. The damage is caused mechanically during or after harvesting. Somewhat similar is the infection of wheat seedlings by *Cephalosporium gramineum* which enters through wounds caused by the breaking of the pericarp and coleorhiza as the root emerges on germination. Wounds caused by insects are important in many diseases. In some diseases, the relationship between the insect and parasite is non-specific while in others the relationship may be more specific, the distribution of the disease coinciding with the appearance of a particular insect species.

Among the diseases caused by bacteria, vascular wilt of cucurbits is one where the relationship is highly specific. This wilt is caused by *Erwinia tracheiphila* which overwinters in hibernating adults of the cucumber beetles *Diabrotica vittata* and *D. duodecimpunctata*. When adults feed on plants, they introduce the bacterium into the vascular system, where it spreads to cause a typical vascular disease.

In another disease, the relationship is more casual, for example, in fire blight of apple and pear leaves caused by *Erwinia amylovora* where the bacteria overwinter in cankers. With the return of suitable growing conditions in spring, bacteria multiply in the cankers and are carried to the blossoms by insects.

A large number of fungal diseases are disseminated by insects and are instrumental in their entry into the host. In the ergot disease of rye caused by *Claviceps purpurea*, conidia are disseminated by species of *Diptera* and *Coleptera*. In oak wilt caused by *Ceratocystis fagacearum*, the relationship between the fungus and insects is also relatively nonspecific.

In the well-known Dutch elm disease, there is a highly specific relationship between the fungus and insect vector. The pathogen *Ophiostoma* (*Ceratocystis*) *ulmi* is carried and transmitted to elm trees by feeding beetles (*Scolytus scolytus, S. multistriatus*). Conidia are introduced directly into the sapwood as the beetles feed on the bark. In this example, the feeding tunnels not only breach the external protective layers of bark but also provide direct access to the vascular elements in which the pathogen can flourish.

3.6 WOUNDS CAUSED BY OTHER FUNGI

Resistance to infection can be modified by other fungi. Tubers highly resistant to *Phytophthora infestans* become susceptible when the parasite enters through wounds caused by *Spongospora subterranea*. The apple canker fungus, *Nectria galligena*, may enter through lesions on a stem caused by *Venturia inaequalis*. Similarly, *Fusarium* species which rot potato tubers may enter through wounds caused by *Spongospora subterranea* and *Phytophthora infestans*.

3.7 WOUNDS CAUSED BY NEMATODES

It is now well known that infection of plants by some bacteria and fungi depends on or is increased by damage caused by nematodes. Wilt of solanaceous plants caused by *Ralstonia* (*Pseudomonas*) *solanacearum* is introduced by *Meloidogyne incognita* var. *acrita*. In the case of yellow ear rot of wheat caused by the bacterium *Clavibacter* (*Corynebacterium*) *tritici*, recently renamed as *Rathayibacter tritici*, another nematode *Anguina tritici* plays an important part in the transmission and infection process. Several vascular wilt pathogens enter into the vascular plants through the injuries caused by the attack of nematodes; for example, *Fusarium oxysporum* f. sp. *vasinfectum* gains entry into the roots of susceptible cotton plants through the injuries caused by the root-knot-nematode, especially *Meloidogyne* species.

3.8 ENTRY THROUGH ROOT HAIRS

Root hairs are particularly vulnerable to invasion by pathogens. Among the pathogens which enter through root hairs, are *Plasmodiophora brassicae* (finger and toe disease of crucifers), *Phymatotrichum omnivorum* (Texas root rot of cotton), *Fusarium oxysporum* f. sp. *lini* (flax wilt), *F. oxysporum* f. sp. *conglutinans* (cabbage yellows). The mode of entry by *Plasmodiophora brassicae* appears to be quite unique. This process involves several unique structural features. Zoospore of the pathogen encysts on the wall of a root hair. The contents of the zoospore are injected into the host cell by a fine puncture or hole made in the cell wall of the host. The puncture is the result of the production of a bullet-shaped structure which suddenly forces its way from within the cyst through the wall into the interior of the root-hair cell. This is an example of mechanical penetration and the actual infection process takes only a few seconds. Another interesting example of entry through root hair of leguminous plants is by the root-nodule bacterium *Rhizobium* which enters through root hairs in the form of an infection thread.

3.9 ENTRY THROUGH BUDS

Some rust fungi, including *Uromyces pisi*, enter through unfolding buds. *Taphrina cerasi* attacks cherry buds and causes them to develop into typical witches' brooms. The peach-leaf-curl fungus *T. deformans* enters its host through the unfolding leaf buds.

3.10 DEVELOPMENT INSIDE HOST TISSUES

Following penetration, there are very wide variations in the subsequent pattern of development and colonization of host tissues. In certain pathogens, such as powdery mildews, the growth of the pathogen is superficial (i.e. ectotrophic). The mycelial development of the apple-scab fungus (*Venturia inaequalis*) is restricted to the subcuticular layer of the leaf and fruit, at least until the onset of general tissue senescence. Some parasitic fungi colonize the interior of the plant and also produce an external aerial mycelium which grows superficially and enters the host repeatedly, for example, *Corticium solani* responsible for stem canker of potato and *Gaeumannomyces* (*Ophiobolus*) *graminis* responsible for take-all diseases of wheat. Many pathogens colonize the parenchyma of mesophyll and cortex; their further spread into the vascular bundles is prevented by the endodermis, although there may be an entry into the pith. In *Taphrina*, the hyphae are entirely intercellular, obtaining nutrients through the membranes of the adjacent cells and this indirect method perhaps resuls in some degree of systemic development. The hyphae of less specialized pathogens are often intracellular, progressing with branches from cell to cell until the tissue is more or less completely colonized. Toxins and enzymes may be secreted and bring about rapid softening and necrosis of tissues ahead of the pathogen. Some parasites enter living cells, which however, may die soon after penetration. In diseases caused by obligate parasites and facultative saprophytes and in a few caused by facultative parasites, the cell wall is penetrated by short branches which usually grow to a limited extent in the lumen of the cell before their ends are transformed into structures which are simple to relatively complex in form in different parasites. These structures are haustoria, they are presumed to have an absorptive function and are the main means of metabolic interchange between the host and parasite. Haustoria almost always occur in parenchyma but in most powdery mildews, they are confined to epidermal cells.

The haustorium originates as a slight bulge on the inside wall of the haustorium mother cell and it becomes closely appressed to the host cell wall at the point it is to be penetrated. Here, there is formed a conical opening, wide at the base and tapering at the apex, through which the haustorium passes. As it emerges, it is bounded only by a membrane continuous with that of the haustorium mother cell. The fact that the haustorium is not surrounded by a cell wall at this stage suggests that penetration is primarily a chemical and not a mechanical process and that the haustorium mother cells produce substances that dissolve both their own walls and those of the host cells. After penetration, the haustorium elongates to form a neck around which a wall is formed. The apex of the neck expands into the characteristic final structure. There is some (but not conclusive) evidence of the invagination of host plasma membrane by the developing haustorium. The mature haustorium containing nuclei, mitochondria, and endoplasmic reticulum is more complex than in the haustorium mother cell. The plasma membrane is invaginated at certain places. Outside is the cell wall which is thick and probably consists of at least two zones. Between the cytoplasm of the host and the wall of the haustorium is an encapsulation or "zone of apposition", which is not found around the parent hyphae or around the developing haustorium. The encapsulation is enclosed by a membrane which may be the host plasma membrane or a new structure. Ehrlich and Ehrlich (1963) interpret this to mean that haustorial protoplasm probably oozes out through ultramicroscopic pores in the walls and comes in intimate contact with the protoplasm of the host cell. At some points, the surface of the encapsulation is disrupted by small vesicles which project into the host cytoplasm and which contain particulates similar to those of encapsulation. These vesicles or secretary bodies are formed in the host cytoplasm, which then move to the encapsulation and discharge their contents.

Encapsulation is also well developed in downy mildews. Berlin and Bowen (1964) suggest that this term should be used instead of 'sheath' which was used by Fraymouth (1956) and earlier workers and that 'sheath' should be applied to the inward projection of the host wall which surrounds the base of the haustorium as a collar.

REVIEW QUESTIONS

1. What is the recent concept of recognition by host of pathogens and nonpathogens?
2. What is chemotaxis and electrotaxis?
3. Distinguish between penetration and infection.
4. Name at least two pathogens which enter their hosts through the following:
   (i) Stomata
   (ii) Lenticels
   (iii) Wounds caused by natural agencies
   (iv) Directly through cell walls
   (v) Wounds caused by nematodes
   (vi) Wounds caused by fungi
   (vii) Root hairs
   (viii) Through unfolding buds
5. Name some pathogens which grow superficially (i.e. ectotrophic), subcuticularly, and endophytically in their hosts.
6. Distinguish between appressoria and hyphopodia. What is the function of appressoria? What are the different types of appressoria produced by plant pathogens?
7. Give examples of plant diseases in which insects play an important role in the entry of pathogen.
8. How do the following pathogens enter their hosts?
   (i) *Heterobasidion (Fomes) annosus*
   (ii) *Gaeumannomyces (Ophiobolus) graminis*
   (iii) *Penicillium expansum*

MULTIPLE-CHOICE QUESTIONS

1. Penetration of post-harvest pathogen occurs commonly through
   (i) stomata (ii) lenticels
   (iii) hydathodes (iv) injuries
2. Appressoria are usually produced by those fungi that penetrate their host plants
   (i) through stomata (ii) through wounds
   (iii) directly (iv) through hydathodes
3. A thread-like structure constituting the mycelium of the fungus is known as
   (i) Hypha (ii) Thallus
   (iii) Paraphysis (iv) Rhizome
4. The fungi *Armillaria mellea* penetrates root periderm by
   (i) appressorium (ii) runner hyphae
   (iii) rhizomorph (iv) stomatopodium
5. Name the structure which exerts a mechanical pressure on the host wall to cause infection
   (i) Appressorium (ii) Stomatopodium
   (iii) Rhizomorph (iv) Hypopodia
6. The perception of signal by the pathogenic fungi for prepenetration is controlled by
   (i) cyclic Adenosine MonoPhosphate (cAMP)
   (ii) Mitogen-activated protein kinase (MAPK)
   (iii) both (i) and (ii)
   (iv) none of these

Answers

1. (iv) 2. (iii) 3. (i) 4. (iii) 5. (i) 6. (iii)

4

ENZYMES AND TOXINS IN PLANT DISEASES

4.1 ENZYMES IN PLANT DISEASES

For successful infection by a pathogen, it should first enter the body of the host on which it feeds. Although some pathogens may use mechanical force or enter their hosts through natural openings or mechanical injury, the activities of pathogens in plants are largely chemical in nature which includes enzymes, toxins, growth regulators, and polysaccharides. Here, in this chapter, the role of enzymes and toxins will be discussed.

4.2 COMPOSITION OF CELL-WALL MATERIALS AND MIDDLE LAMELLA

4.2.1 Cuticular Wax

Plant waxes are formed as granular or rodlike projections or as a continuous layer outside or within the cuticle of many aerial plant parts. Cuticular waxes are made up of long-chain molecules of paraffin hydrocarbon, alcohols, ketones, esters, and acids. They differ from fats in that glycerol is replaced by monohydric alcohols.

4.2.2 Cutin

Cutin is the main component of the cuticular layer. Cutin is believed to be an insoluble polymer of hydroxylated octa- and hexadecanoic acid. The upper part of the layer is admixed with waxes, while in its lower part, in the region where it merges with the outer walls of the epidermal cells, cutin is admixed with cellulose and pectin.

4.2.3 The Cell Wall

The cell wall of higher plants comprises a middle lamella and primary cell wall, and in some cases, a secondary cell wall. Pectic substances are the basic constituents of wall layers and form almost the entire middle lamella, whereas in the other layers, cellulose is also found in good amounts. Besides these two major groups of compounds, other components are hemicelluloses, lignins, proteins (rich in hydroxy proline), and limited amounts of other materials. Functionally, the cell wall may be divided into three regions: middle lamella, primary wall, and secondary wall. The middle lamella acts as an **intercellular cement** that binds cells together in the tissue system.

4.2.4 Pectic Substances

Pectic substances are a group of complex, colloidal carbohydrate derivatives which occur in or are prepared from plants. The basic structure of these is a linear polymer of D-galacturonic acid units in α 1-4 glycosidic linkages, esterified with a methyl group or linked with other carboxyl groups in calcium or magnesium salt bridges. The esterification varies with the age, organ, or species of the plant. The term **protopectin** is applied to substances which are insoluble in water and which, upon restricted hydrolysis, yield pectinic acid. Pectinic acid contains a variable quantity of methoxyl groups, though less than pectin, which may have 75% or more of the uronic carboxyls methylated. Pectic substances, essentially free from methyl ester groups, are classified as pectic acids.

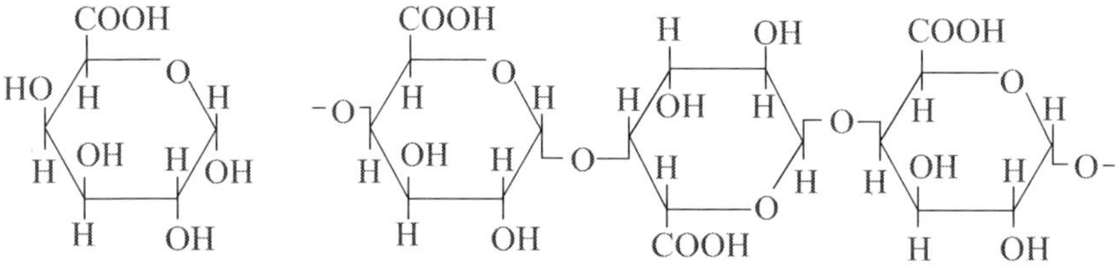

α d-galacturonic acid   Pectic acid

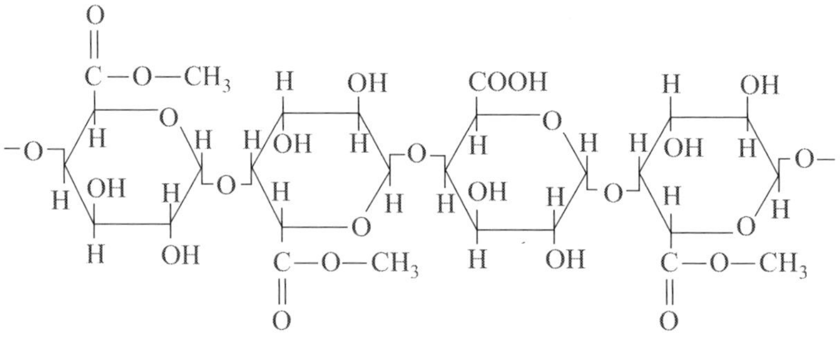

Pectinic acid

The proportions of carboxyl and methyl ester groups have profound effects on the properties of polymers. Calcium salts of pectin occur in the middle lamella of plant cells and represent a large portion of the pectic substances in plants.

4.2.5 Cellulose

Cellulose is a polysaccharide which consists of an aggregation of β-D-glucopyranose in which C_1 is linked to C_4. Cellulose is the major constituent of the cell walls of higher plants. Cellulose is an insoluble, crystalline substance in its native form.

Callose differs from cellulose in that it consists of (1-3) β-D glucane chains that can form duplexes and triplexes.

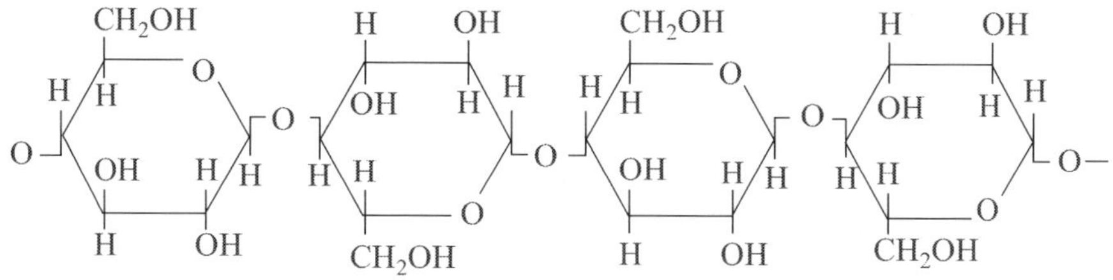

Cellulose

Structure of cellulose showing β-D 1, 4 linkages of glucopyranose form

4.2.6 Hemicelluloses

In addition to celluloses, these also occur in cell walls. Hemicelluloses include several polymers. **Xyloglucan** is an important hemicellulose in the primary-wall region. It is made up of a β-1, 4 linked glucopyranose chain with terminal branches of 1, 6 linked D-xylopyranose. Other hemicelluloses include **glucomannan**, a heteropolymer of D-glucopyranose and D-mannopyranose linked β-1, 4; and **glactoglucomannan**, a glucomannan with 1, 6 linked D-gluctopyranose side branches.

4.2.7 Lignin

This is a high molecular-weight, structurally complex, amorphous polymer that is different from carbohydrates and proteins. Its structure is still not known but the basic unit which forms a polymer is **phenylpropanoid**.

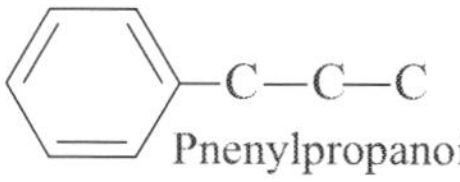

Pnenylpropanoid unit

Lignin is found in the middle lamella, the cell walls of xylem vessels and the sclerenchyma fibres as secondary thickening that strengthen plants. It may also be formed in other cells such as bundle sheath cells and occasionally epidermal cells. The lignin content of mature woody tissues may vary from 15% to 38% and is second only to cellulose.

4.2.8 Cell-Wall Protein

Proteins are also present in all plant cell walls in the region of the cuticle and possibly in the middle lamella. The structural-wall proteins seem to be similar to other cell-wall proteins in all aspects, except that an amino acid, **hydroxyproline** is also present, which is not present in other plant proteins. **Mucopolysaccharides** are formed as a result of the covalent bonds between the protein molecules and cell-wall polysaccharides. These give a complex called **extension** and are envisaged as playing a part in cell-wall extension. Five classes of structural proteins have been formed in cell walls: (i) Extensions, which are glycoproteins. There are up to 0.5 percent of the cell-wall mass in healthy tissues but may increase to 5 to 15 percent in infected tissues. (ii) Proline-rich proteins, (iii) glycine rich proteins, (iv) solanaceous lectins, and (v) other structural proteins (Agrios, 1997, 2005).

4.2.9 Lipids

Phospholipids and glycolipids, along with proteins, are the main constituents of all plant-cell membranes. The common characteristic of all lipids is that they contain fatty acids that may be saturated or unsaturated. Oils and fats are formed in many cells especially as storage compounds in seeds. Wax lipids are formed on aerial epidermal cells.

4.3 ENZYMES FOR WAXES AND CUTINS

It was earlier thought that no pathogens are known to produce enzymes that can degrade waxes. Wax layers are mainly penetrated by mechanical force alone. However, electron-microscopic studies suggest that some pathogens such as *Puccinia hordei* and *Pestalotia malicola* produce enzymes that can degrade waxes (Agrios, 1997, 2005).

The significance of cuticle-softening enzymes in plant diseases is uncertain and needs more investigation. The cuticle consists of a cutin framework with waxes embedded within it and extruded from its surface to give a waterproof 'bloom'. Until 1961, no enzymes were known which could degrade cutin. Heinen and Linskens (1961) demonstrated that the fungus *Penicillium spinulosum* from rotting leaves produced at least two types of cutinolytic enzymes.

4.3.1 Cutin Esterase

This catalyzes the hydrolysis of ester bonds occurring between free hydroxyl and carboxyl groups of cutin bonds.

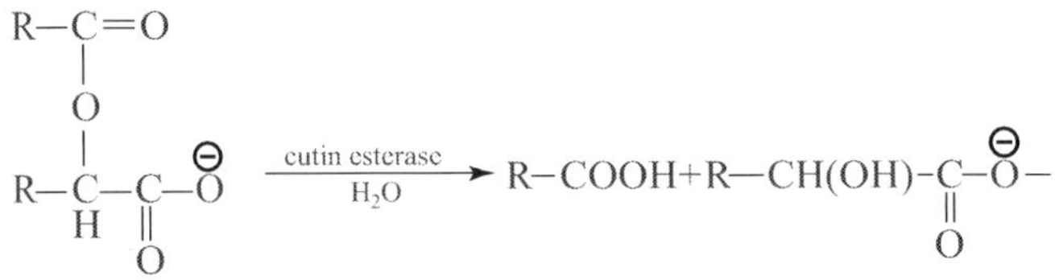

4.3.2 Carboxycutin Peroxidase

This catalyzes the hydrolysis of the peroxide group of cutin. The germinating spores of *Colletotrichum gloeosporioides* degrade the cuticle of orange leaves. Cuticular breakdown in olive leaves by *Spilocaea oleagini* has been reported. It is possible that in organisms like *Spilocaea oleagini*, which live sub- or intracuticularly, the entire cuticle is also used as a food source. There is a considerable reduction in the content of cutin in rose leaves infected by the fungus *Venturia inaequalis*. However, some fungal pathogens may not produce sufficient quantity of cutin-dissolving enzymes so as to play an important role in the penetration of cuticle. A localized secretion of cutin-dissolving enzymes is significant as it may aid in the penetration process. The work of Edwards and Allen (1970) is significant in this connection. They visualized a localized secretion of cuticle-degrading enzymes from the tip of the infection peg by the fungus *Blumeria* (*Erysiphe*) *graminis* f.sp. *hordei* attacking barley leaves. It is believed that fungal pathogens constantly produce low levels of cutinase which on contact with cutin release small amounts of **cutin monomers** (single molecules). These small manomers gain entry into the fungal pathogen which then trigger further expression of the cutinase gene and stimulate the fungus to produce enormously large amounts of the cutinase enzyme. Fatty acids present in the wax can also stimulate the production of cutinase enzyme. Glucose, on the other hand, suppresses the production of cutinase enzyme. Involvement of cutinase in the penetration of host cuticle by plant pathogenic fungi comes from several indirect evidences. It has been shown that the production of the enzyme is highest at the point of penetration of the germ tube of the pathogen.

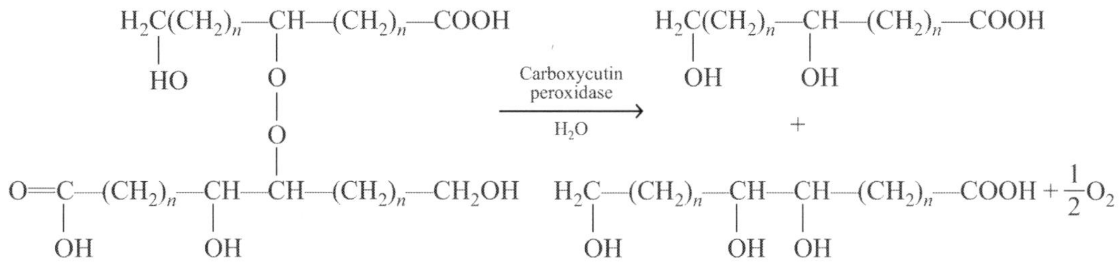

To prove the involvement of cutinase in penetration of pathogen into its host, secretion of cutinase by the penetrating fungus must be conclusively shown, and conclusive evidence that cutinase is required for infection must be provided. Both of these have been accomplished for infection of pea by *Fusarium solani* f. *pisi* under controlled conditions.

4.4 PECTIC ENZYMES

The capacity to produce pectic enzymes is widespread in fungi, bacteria, and nematodes. The pectic enzymes which are currently known are classified into two major groups.

4.4.1 Pectinesterases (PE or PME)

Pectin esterases or pectin methyl esterases catalyze the hydrolysis of methyl ester groups of pectinic acids to methyl alcohol and pectinic acids of reduced methoxy content and eventually to pectic acid. PME is also found in most of the higher plants and in microorganisms. The microbial PME and plant PME may, however, differ in their chemical and physical properties. It is of interest that the rate of cleavage of the α-1, 4 bond of the pectic substance is enhanced by the removal of the methoxyl groups. This is illustrated by the fact that endopolygalacturonase prefers demethylated pectin.

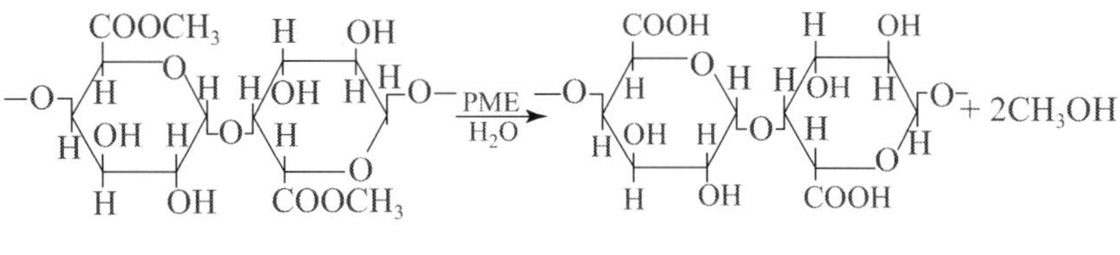

4.4.2 Polygalacturonases (PG)

Polygalacturonases, or pectic glycosidases and layses, are chain-splitting enzymes which break the links between adjacent galacturonic acid units in pectic substances. This can occur by a hydrolytic mechanism (hydrolases) or an eliminative mechanism (lyases). There has been some confusion in the names of various chain-splitting enzymes of pectinous materials. However, Wood (1960) followed the classification proposed by Demain and Phaff (1957) who distinguished between the polygalacturonases

(PG) which prefer pectic acid of the de-esterified parts of pectinic acid as substrates, and Pectin Methyl Galacturonases (PMG) which prefer esterified chain parts. They further distinguished these polygalacturonases according to whether they cause cleavage in glycosidic linkages at random by an endo-type of polygalacuronase or polymethyl-galacturonase, or by an exo-type of action in which the end linkages are preferred.

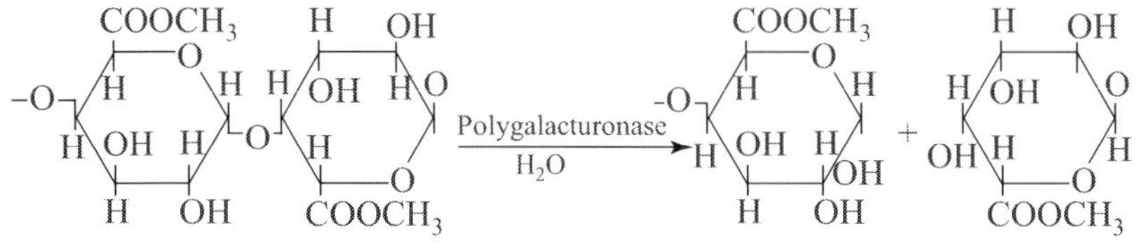

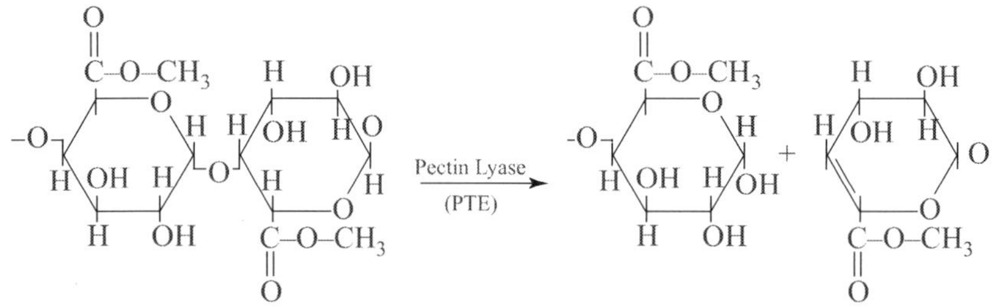

A significant advance was the discovery of the Pectin Transeliminases (PTE) which split the pectin chain by a transeliminative mechanism. This enzyme breaks the glycosidic linkage of pectin at C_4, accompanied by a simultaneous elimination of H^+ from C_5. Wood (1960) further classified these enzymes which preferably attack esterified or non-esterified parts of the chain and determined whether enzymes attack chains terminally (exo-) or at random (endo-).

4.5 CELLULOLYTIC ENZYMES

These enzymes have been investigated in those organisms which bring about the deterioration of textiles and wood but not much attention has been given to their significance in plant diseases. The enzymic degradation of cellulose is imperfectly understood. Cellulose-degrading enzymes are produced by fungi, bacteria, and nematodes and also by parasitic higher plants.

1. They cause the break-up of the wall barrier for entry.
2. They provide food in the form of glucose from cellulose on hydrolysis.

The enzymes cellulases degrade cellulose and its derivatives. Cellulases which degrade native cellulose by destroying its crystalline structure are designated as C_1 enzymes. A second cellulase also attacks native cellulose and breaks it into shorter chains and exposes the glucan chains to a third group of cellulases, termed C_x enzymes, which degrade the glucan chains to cellobiose. The conversion of cellulose to glucose also requires a cellobiase or β-glucosidase. The presence of cellulolytic enzymes and the degradation of cellulose by bacteria and fungi have been reviewed by several workers. Wood (1967) proposed a scheme of cellulolytic enzymes when native cellulose is degraded. However, this has been slightly modified which is as follows

Native cellulose

↓ Cellulase C_1 attacks native cellulose cleaving cross-linkages between chains

Native cellulose
(modified without cross-linkages)

↓ Cellulase C_2 breaks native cellulose or modifies cellulose into short chain of cellulose

Shorter chain of cellulose

↓ C_x (third group of cellulases which degrade shorter chains into disaccharide cellobiose)

Cellobiose

↓ β-glucosidase

Glucose

Cellulases are generally adaptive in their formation because the latter is stimulated by the presence of cellulose. The capacity to degrade native cellulose is confined to a few organisms, while soluble cellulose is degraded by a large number of microorganisms.

It is generally believed that cellulases are not of much importance in the early stages of pathogenesis; perhaps cellulose is first protected by the enveloping matrix. Many soft-rot organisms produce cellulases as well as pectolytic enzymes. Members belonging to the family Pythiaceae do not form cellulases *in vitro*, whereas others, such as *Rhizoctonia solani, Fusarium moniliforme* are reported to produce both cellulolytic and pectic enzymes. *Sclerotium rolfsii* produces cellulolytic enzymes and has the capacity to cause soft rot as well as degradation of hardened tissues. In the root rot and foot rot of many plants and in leaf-spot diseases, cellulases seem to play an important part in pathogenesis.

Cellulolytic enzymes have been implicated in wilt diseases caused by *Pseudomonas solanacearum* (now *Ralstonia solanacearum), Verticillium albo-atrum* and *Fusarium oxysporum* f. sp. *lycopersici*. Wood (1960) suggested that the conditions in the trachea favour the production of cellulases by vascular-wilt pathogens and that local action of these enzymes might result in the formation of the compounds of high molecular weight which could impede the flow of sap in the transpiration stream.

The production of cellulolytic enzymes *in vitro* is often reduced by sugars and by various inhibitors, such as polyphenols. Germinating urediniospores of *Puccinia graminis tritici* are known to produce a cellulase but its role is uncertain.

There is increasing evidence that a number of fungi, particularly the wood-rotting fungi, produce several cellulolytic enzymes. *Polyporus versicolor* produces a multiple-component cellulase system.

4.6 HEMICELLULASES

Many microorganisms—saprophytic and parasitic—produce hemicellulases and can convert hemicelluloses into pentoses and uronides. Depending on the manomer released from the polymer on which they act, the particular enzymes are called xylanase, galactanase, arabinase, mannase, and so on. Activated oxygen, hydroxyl and other radicals produced by attacking pathogens also bring about non-enzymatic breakdown of hemicelluloses (Agrios, 1997 and 2005). Although it is well known that pathogens or nonpathogens produce hemicellulases and oxidative agents, yet the exact manner in which these enzymes contribute to the breakdown of the cell walls or the ability to cause disease is not well understood. In the case of sunflower hypocotyls infected with *Sclerotinia sclerotiorum*, the production of xylanase and arabinase by the pathogen has been reported. An arabino-furanosidase has been reported in culture filtrates of *S. fructigina*, while Strobel (1963) reported the production of at least two extracellular xylan-degrading enzymes by *Diplodia viticola* when grown on autoclaved grapes. It has been found that the pathogen *Sclerotium (Corticium) rolfsii* produced exogalacturonase, endomannase, galactosidase, and endoxylanase in culture and these enzymes have also been detected in bean tissues infected by the pathogen.

4.7 LIGNOLYTIC ENZYMES

Lignin is a resistant material and only about 500 species of basidiomycetous fungi are capable of degrading it. Bacteria are seemingly not important in the degrading of lignin. The nature and action of enzymes which decompose lignin have not been properly understood. It has been recognized for many years that the white-rot fungi, together with carbohydrates, decompose lignin in plant tissues. *Polyporus versicolor* can destroy over 97% of the lignin in sweetgum wood, and many reports of the removal of most of the lignin from wood by white-rot fungi is found in the literature. There are several hundred species belonging to a number of families of Hymenomycetes, namely, Agaricaceae, Corticiaceae, Hydnacece, Polyporaceae, and Thelephoraceae that can bring about white rot. In addition, a few wood-inhabiting Ascomycetes in the order Sphaeriales cause a white-rot type of decay (e.g. *Hypoxylon deustum* and *Xylaria polymorpha*). There is very little experimental data available on the role of lignolytic enzymes in disease syndrome.

It appears that white-rot fungi secrete an extracellular polyphenol oxidase, possible laccase, which enables them to use lignin. On the other hand, all types of wood-rotting fungi seem to produce **transmethylase** that removes methoxyl groups form lignin, thus "modifying" lignin and increasing the exposure of the embedded cellulose and other polysaccharides to the action of fungal enzymes. Then there are brown-rot fungi, like most white-rot fungi, they are the members of Holobasidiomycetes and unlike white-rot fungi, which decompose both carbohydrates and lignin in wood, brown-rot fungi decompose the carbohydrates with only a slight decrease in lignin content. It has been pointed out that lignolytic enzymes might be playing an important role in many foot rot diseases of monocotyledonous plants which contain a high degree of lignin content in plant tissues.

4.8 PROTEOLYTIC ENZYMES

Cell walls contain a protein rich in hydroxyproline. Pathogens are known to degrade proteins enzymatically, and since no particular enzyme responsible for the degradation of cell-wall protein is known, it is presumed that its breakdown follows the same steps as the degradation of cytoplasmic proteins. Some of the pathogens which are known to produce the cell-wall-dissolving enzymes have also been shown to produce proteolytic enzymes.

One of the well-documented studies on the secretion of proteolytic enzymes was on *Pyricularia oryzae* (now *P. grisea*) by Otani (1959). Protease was detected in the mycelium of the pathogen and in culture fluids. The enzyme was also located inside the diseased tissues. Degradation of RNA by *Phytophthora infestans*, the late-blight pathogen, has been demonstrated. Mahadevan (1970) reported the production of proteolytic enzymes and the existence of an RNAase in *Pyricularia oryzae* (now *P. grisea*). He also reported the production of proteolytic enzymes by the cotton-wilt fungus (*F. oxysporum* f. *vasinfectum*) and *Helminthosporium oryzae* responsible for the brown leaf spot of paddy.

4.9 LIPOLYTIC ENZYMES

Various types of lipids occur in plant cells, the most important of them being the neutral lipids functioning as energy sources, while phospholipids and glycolipids are the main components of cell membranes. Several fungi, bacteria, and nematodes are capable of degrading lipids. Plant pathogens produce enzymes that attack major membrane constituents. Lipolytic enzymes, called lipases, phosholipidases, etc. hydrolyze the liberation of fatty acids from the lipid molecule by hydrolyzing the acyl ester and phosphate ester bonds of phospholipids. Phospholipase A is not known to be produced by plant pathogens. Phospholipase B is produced by *Thielaviopsis basicola, S. rolfsii,* and *Botrytis cinerea.* These enzymes and proteases might affect cell membranes since phospholipids and proteins are the major constituents of plant-membrane systems.

4.10 INACTIVATION OF ENZYMES

In some host parasite combinations, resistance may be related to inactivation or inhibited synthesis of enzymes by the parasite.

In several instances of fungal enzyme inactivation, there is evidence that the inhibitors are of phenolic origin. Phenols and their oxidation products inhibited a wide range of enzymes. In *Sclerotinia* attack on apple fruits, the hyphae grow through the tissue, macerating and killing the cells as they advance. From the time any cell is killed, the intermingling of phenolic oxidizable substances contained in the cell with the oxidizing enzymes supplied by both the host and fungus, leads to the formation of oxidized phenolic substances that inactivate the macerating enzyme.

Varietal resistance of certain varieties of cider apples to brown rot has been attributed to their high content of polyphenols in parenchyma of the fruit. Oxidized phenols, quinones, and related substances have been employed as chemotherapeutic agents in wilt diseases.

Turner (1961) made an interesting contribution in the field of molecular phytopathology while studying the take-all disease of oats caused by *Ophiobolus graminis*. Oat leaves and roots contain a fluorescent glucoside (avenacin) which inhibits the growth of several fungi. *Ophibolus graminis* var. *avenae* is pathogenic because it produces a specific enzyme (glucosidase) which hydrolyzes avenacin. Thus, the pathogenic capability of *O. graminis* var. *avenae* is dependent on the production of an inhibitor.

Bateman (1964) had shown that in the case of bean seedlings infected by *Rhizoctonia*, infection causes accelerated respiration and this results in accumulation of Ca^{++} and other cations in the infected region. The increased level Ca^{++} activates pectin methylesterase activity which in turn de-esterifies the carboxyl groups of polygalacturonide, converting them to pectic acid. Pectic acid then reacts with available Ca^{++} to form calcium pectate. The invaded tissue then becomes resistant to attack by polygalacturonases. This limits the size of the lesion.

4.11 TOXINS AND PLANT DISEASES

There is an enormous amount of literature on toxins and their role in plant diseases.

A toxin can be defined as a substance of microbial origin involved in host pathogenesis. Toxins are normally proteins which are antigenic in nature and have antitoxins. In addition, by introducing small doses of a toxin, it is possible to develop immunity in the host against the toxin, and the organism producing it. However, this was not known when Roux first visualized the presence of a toxin in the disease caused by the bacterium *Corynebacterium diphtheriae*.

Plant pathologists have raised doubts of there being toxins in phytopathology. Actually, we do not have toxins in the real sense, playing their part in the plant-disease syndrome, as in human or animal pathology, because none of these satisfy the conditions of a toxin in mammalian pathology.

Toxins are different from enzymes in that they do not attack the structural integrity of the tissues but affect the metabolism in a subtle manner. It is the subtlety of action that differentiates toxins from enzymes. Toxins act directly on the protoplasts of cells, and substances with the same end effect but an indirect method of action are not toxins. The toxin of *Clostridium welchii*, lecithinase, even though an enzyme, is a toxin because it disrupts cell membranes. Such enzymes are toxins.

4.12 CLASSIFICATION OF TOXINS

There are three broad classes of toxins, namely, phytotoxins, vivotoxins, and pathotoxins according to their source of origin.

4.13 PHYTOTOXINS

Any compound produced by a microorganism which is toxic to plants is a phytotoxin. Phytotoxins are nonspecific, incite few or none of the symptoms that are incited by the pathogen, and as happens in most cases, show no relation between toxin production and pathogenicity. Lycomarasmine and alternaric acid are examples of phytotoxins.

4.14 VIVOTOXINS

It is defined as a substance produced in the infected host by the pathogen and/or its hosts which functions in the production of the disease but is not itself the initial inciting agent of the disease. The requisites of a vivotoxin are the following:

1. It must be isolated from the diseased plant but should not be present in the healthy plant.
2. It must be characterized chemically.
3. When introduced in a pure form into a healthy host, it must produce the symptoms of the disease or a portion of the syndrome. Fusaric acid and piricularin are typical examples of vivotoxins.

4.15 PATHOTOXINS

Pathotoxins play a causal role in disease production, and produce symptoms characteristic of the disease in susceptible plants. These toxins may be produced by the pathogen, host, or interaction between them. Piricularin produced by *Pyricularia grisea* (*oryzae*) is a typical pathotoxin. The following few toxins are discussed briefly.

4.16 FUSARIC ACID (A VIVOTOXIN)

Fusaric acid (5-*n*-butyl pyridine-2 carboxylic acid) is produced by several species of *Fusarium*, mainly belonging to the group Elegans Woolenweber [*F. oxysporum* (Schl.) Sny. et Hans.] and causing wilt diseases in tomato, cotton, pea, banana, and other plants. Fusaric acid is a vivotoxin and it has been detected in the host plants and is involved in wilt syndrome.

Fusaric acid was first isolated as the metabolic product of *Fusarium heterosporum* Nees, a nonspecific parasite, by Yabuta *et al.* in 1934. In 1952, it was recognized as an important toxin of *Fusarium oxysporum* f.sp. *lycopersici*, *Fusarium vasinfectum*, and *Gibberella fujikuroi*. It is now known to play a pathogenic role in the wilt of tomato and cotton. Subsequently, there has been evidence to show that fusaric acid is involved in the Panama disease of banana caused by *Fusarium oxysporum* f. sp. *cubense*. A possible correlation has been demonstrated between pathogenicity and the amount of fusaric acid production *in vitro* in respect of several strains of *F. oxysoporum* f. sp. *niveum* and *F. oxysporum* f.sp. *pisi*.

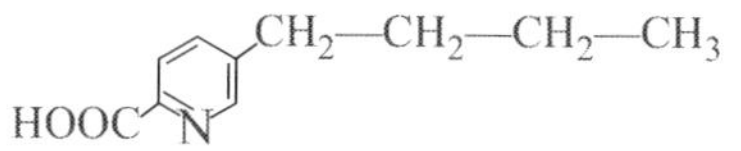

Fusaric acid

Fusaric acid is called a wilt toxin as it is now known to play a more distinct role in the wilt of tomato and cotton plants than in any other disease. Even when the pathogen is restricted to the roots, symptoms like vein clearing and epinasty are manifested on the leaves. This appears to be due to toxic metabolites translocated to distant parts of the plant from the locus of infection. Some of the toxic symptoms on diseased plants could be produced by pure toxin fusaric acid. Moreover, it has been firmly established that a concentration of 10^{-4}M of fusaric acid is reached in the infected plants, a concentration lethal to the host tissues. The symptoms of fusaric acid injury appear first in the stems and soon after in the leaves in the form of intercostals necrosis of leaves.

Fusaric acid forms stable chelate rings, catching a metal ion between the N atom of the pyridine ring and carboxyl radical in the α-position, forming a 5-membered coordinate link. Of the metal ions, the ferric ion is most strongly chelated with it. On the basis of the antagonism reported between ferrioxamine B and fusaric acid, it could be conjectured that fusaric acid specifically interferes with the incorporation of Fe^{+++} in the porphyrin skeleton. The mode of action of fusaric acid has been discussed by Kalyansundram (1970). Sanwal (1961) described the physiological effects of this toxin, and suggested that it could cause a partial uncoupling of oxidative phosphorylation. According to Deverall (1964), it plays no part in the development of wilt symptoms, although it might act in conjunction with other metabolites.

4.17 LYCOMARASMIN (HOST SPECIFIC BUT NOT A VIVOTOXIN)

Lycomarasmin is a wilt toxin of the tomato wilt pathogen *F. oxysporum* f.sp. *lycopersici* Sacc. This toxin, unlike fusaric acid, appears to be specific to *F. oxysporum* f.sp. *lycopersici.* It has not been detected in the infected host tissues.

Lycomarasmin, when administered to tomato plants, produces injury in the intercostals fields of the leaves, and rolling of the leaf tips. The symptoms of tomato wilt simulate the toxic injury manifested by the synergistic action of lycomarasmin and fusaric acid. While fusaric acid has been demonstrated in infected tomato plants, no such direct evidence exists for lycomarasmin.

Lycomarasmin is a polypeptide with molecular weight 277.3 and empirical formula $C_9H_{15}O_7N_3$.

The toxicity of this substance seems to arise from the fact that it can form metal chelates.

Piricularin (A Pathotoxin)

Pyricularia grisea (formerly *P. oryzae*) belonging to the group of Hyphomycetes, is the causal organism of the blast disease of rice and has a wide distribution in all rice-growing areas of the world. The symptoms of rice-blast disease indicate that the causal fungus produces some very effective toxins.

It appears that two of the metabolic products of the fungus are responsible for the characteristic symptoms of the disease. One of them is identified as α-picolinic acid and the other is piricularin.

The toxic action of piricularin appears to be due to its inhibitory effect on the activity of several enzymes such as peroxidases, catalase, cytochrome oxidase, and ascorbic acid oxidase, even at low dilutions (1/100,000/200,000).

A third substance, pyriculol ($C_{14}H_{16}O_4$), was isolated from cultures of *P. oryzae* which causes natural blast lesion and also inhibits the growth of rice seedlings.

4.18 ALTERNARIC ACID (A PATHOTOXIN)

Alternaric acid occurs in the early blight of tomato and potato and is caused by *Alternaria solani*. Some of the lesions appear to be devoid of the fungus under conditions, suggesting that a mobile toxin is involved.

4.19 TABTOXIN OR WILDFIRE TOXIN (A NON-HOST-SPECIFIC TOXIN)

Pseudomonas syringae pv. *tabaci* produces a pathotoxin called tabtoxin which is nonselective (nonspecific). The pathogen causes a disease known as wildfire disease of tobacco.

Tabtoxin is a dipeptide composed of a common amino acid threonine and the amino acid tabtoxinine (not known previously). Tabtoxin itself is not toxic as such but in the host cells, the toxin is hydrolyzed with the release of the toxic moiety tabtoxinine (which is the active toxin) and threonine. Tabtoxinine is toxic to cells because it inactivates the enzyme glutamine synthetase which leads to depleted levels of glutamine with subsequent accumulation of ammonia which reaches toxic levels. Ammonia accumulation leads to the uncoupling of photosynthesis and photorespiration. The thylakoid membranes of the chloroplasts are destroyed which result in the chlorosis and necrosis of the tissues in the leaves.

4.20 CERCOSPORIN TOXIN (A NON-HOST-SPECIFIC TOXIN)

It is a nonspecific toxin and is produced by most of the species of *Cercospora* such as *C. zea maydis* causing gray leaf spot of *Zea maydis*, and *C. zinniae* causing leaf spot of *Zinnia elegans*, along with leaf spot and blight diseases of many vegetable and ornamental plants. This toxin is activated in the presence of light and becomes toxic to plants by generating single oxygen, i.e. reactive atomic oxygen in the cell. This generated single oxygen destroys the membrane of infected host plants which in turn provides nutrients to the intercellular pathogen. **Cercosporin** is a photosensitizing perylenequinone that absorbs light energy and forms activated oxygen. This activated oxygen interferes with the host lipid, protein, and nucleic acid synthesis, thereby increasing the virulence of the pathogen that ultimately kills the plant cells. The fungal spores and mycelium are resistant to Cercosporin toxin due to production of pyridoxine (Vitamin B_6) by the fungus. This pyridoxine reacts with activated single oxygen and is neutralized during the course of infection. Thus, this toxin incites the production of reactive oxygen in the cells which causes disruption of the cell membrane and loss of electrolyte from cells.

4.21 PHASEOLOTOXIN (A NON-HOST-SPECIFIC TOXIN)

Phaseolotoxin is a modified ornithine-alanine-arginine tripeptide (host nonspecific toxin), produced by the bacterium *Pseudomonas syringae* pv. *phaseolicola*, which causes halo blight of beans and some other legumes. The localized and systemic cholorotic symptoms produced in infected plants are identical to those produced on plants treated with the toxin alone. Plants treated with the toxin alone or those infected with the bacterial pathogen show reduced growth of newly formed and expanding leaves, disruption of apical dominance and the accumulation of the amino acid ornithine.

4.22 TENTOXIN (A NON-HOST-SPECIFIC TOXIN)

Tentoxin (host nonspecific toxin) is produced by the fungus *Alternaria alternata* (previously called *Alternaria tenuis*). The pathogen causes chlorosis in seedlings of many plant species. Tentoxin is a cyclic tetrapeptide that binds to and inactivates a protein (chloroplast-coupling factor) involved in energy transfer into chloroplasts, and also interferes or inhibits the light-dependent phosphorylation of ADP to ATP.

In addition to this, tentoxin also inhibits the activity of polyphenol oxidases, enzymes involved in resistance mechanisms of plants. There are several areas such as the molecular site of action of tentoxin, and the exact mechanism by which it brings about these effects are still not known.

There are several other non-host-specific toxins such as fumaric acid produced by *Rhizopus* spp. in the almond hull disease, oxalic acid produced by *Sclerotium* and *Sclerotinia* spp. in various host plants in which these infect and by *Cryptonectria parasitica,* the cause of chestnut blight disease, Cerato-ulmin produced by *Ceratocystis ulmi* in Dutch elm disease, fusicoccin produced by *Fusicoccum amygdali* in twig blight disease of almond and peaches, cercosporin produced by *Cercospora* spp. and several others, which are involved in host pathogenesis, but are not host specific.

4.23 HOST-SPECIFIC OR HOST-SELECTIVE TOXINS

A host-specific or host-selective toxin is one which only affects plants susceptible to the pathogen producing it. There are several host-specific toxins such as victorin (HV toxin) on victoria variety of oats, T-toxin produced by *H. maydis*, now called *Cochliobolus heterostrophus* on maize lives having Texas male-sterile (Tms) cytoplasm, HC-toxin produced by *Helminthosporium (Cochliobolus) carbonum* on specific maize leaves, AM toxin produced by apple pathotype of *Alternaria alternata* (previously called *A. mali*), HS-toxin produced by *H.* (*Cochliobolus) sacchari* affecting sugar cane, ACL-toxin produced by *Alternaria citri* (lemon race) affecting rough lemons, ACT-toxin produced by *A. citri* (tangerine race) affecting Dancy tangerine, AK-toxin on Japanese pear, AT-toxin on tobacco, PC-toxin produced up *Periconia circinata* in sorghum, PM-toxin in corn that has Texas male cytoplasm, and CC-toxin produced by *Corynespora cassicola* in tomato.

Some of the evidences that a particular host-specific toxin is involved in disease can be deduced from the following:

(i) Purified toxin reproduces all disease symptoms in the host.

(ii) Toxin affects susceptible but not resistant genotypes.

(iii) Tox^- strains are nonpathogenic

(iv) In crops between Tox^+ and Tox^- strains, there is co-segregation among progeny for toxin production and pathogenicity

Some of the host-specific toxins will be discussed here in some detail.

4.24 VICTORIN OR HV-TOXIN (A HOST-SPECIFIC TOXIN)

In the late 1940s, oat crops in the USA were affected by a seedling blight caused by *Cochliobolus (Helminthosporium) victoriae* in the Victoria variety or its derivatives. Although the pathogen was typically localized in the basal portion of infected plants, symptoms appeared on the leaves which often collapsed. It has been found that the culture filtrates of the fungus contained a toxin that can reproduce accurately the visible symptoms of the disease. What was particularly interesting about this toxin was that resistant oat cultivars were not affected by it. The name "victorin" was given to this toxin by Wheeler and Luke (1954). The toxin is highly specific in its activity and is claimed to produce the symptoms only in the pathogen-susceptible variety of oats. Thus, victorin was the first host-specific toxin to be described. Since then a number of other toxins with host-specificity and an ability to incite all of the symptoms associated with a particular disease have been identified. These are known as host-specific toxins to distinguish them from those which are nonspecific.

Its chemical structure has been determined. It is a complex chlorinated, partially cyclic pentapeptide (tricyclic secondary amine), with molecular weight of around 2000.

The earliest effect detected was in permeability. This led to the suggestion that the disruption of cell permeability could be an initial event which triggers subsequent pathological changes in many diseased plants. Victorin causes a decrease in C_6/C_1 ratio of susceptible oat leaves. This provides additional evidence that victorin induces metabolic alterations similar to those found in other diseased plants.

Victorin is able to incite toxic symptoms only in the blight susceptible variety of oats. Perhaps this could be attributed to the presence of a specific enzyme on the surface of the cells of the susceptible and not the resistant variety. Cell-wall modifications and increased secretory activity were the earliest ultrastructural changes found in victorin-treated tissues. The primary target of the toxin seems to be cell membranes (plasma membranes) where the toxin is believed to be binding to several proteins. But the exact mechanism of its action is still not clear.

The widespread use of new crop varieties often results in new disease problems. Thus, the black-spot disease of Japanese pears (*Pyrus serotina*) appeared after the new natural mutant variety *Nijisseiki* was introduced. The causal organism was identified in 1917 as a species of *Alternaria*, later named as *A. kikuchiana*. Tanaka, in 1933, reported a host labile—a host-selective factor produced by the causal fungus in culture. The host-specific toxin was extracted from naturally infected leaves.

The effect of AK-toxin on pear appears to be similar to the effect of HV-toxin on oats. AK-toxin caused an increase in electrolyte loss from susceptible but not resistant leaves within 20 minutes of initial exposure.

4.25 T-TOXIN (A HOST-SPECIFIC TOXIN)

T-toxin (host-specific) is produced by the race T of *Helminthosporium (Cochliobolus) heterostrophus*, the cause of southern corn leaf blight in the USA. It attacks corn that have Texas male-sterile (Tms) cytoplasm. Corn with normal cytoplasm is resistant to the pathogen.

REVIEW QUESTIONS

1. What do you understand by cell-wall-degrading enzymes? How are these involved in the production of disease syndrome? Explain with reference to soft rots in vegetables or fruits.
2. How will you classify cell-wall-degrading enzymes on the basis of their substrate utilization and mode of action?
3. What do you understand by the mechanism of disease development in soft rots?
4. What are "macerating principles"? Explain their role in disease development.
5. What are toxins? How are these involved in the development of vascular wilt syndrome?
6. Explain the following:
   (i) Host-specific toxins
   (ii) Vivotoxins, phytotoxins, and pathotoxins
   (iii) Role of proteolytic enzymes in pathogenesis
   (iv) Role of cutinolytic enzymes in pathogenesis
   (v) Role of lignolytic enzymes in pathogenesis
7. What is the role of toxins in host specificity?
8. What do you understand by mycotoxin, phytotoxin, pathotoxin, and vivotoxin? Explain by giving suitable examples.
9. Write in brief about the chemical 'weapons' utilized by the pathogens in disease development.
10. What do you know of the detoxification of enzymes and toxins?
11. What is the mechanism of action of the following host-specific (selective) toxins?
    (i) Victorin (ii) T-toxin (iii) Ak-toxin
12. The discovery of host-specific toxin raised hopes that biochemical basis of pathogenicity, and conversely host resistance, might turn out to be simple, determined by a right compound of major effect. Overall, these hopes have not been fully realized. Why?
13. What can be the evidences for involvement of a host-specific toxin in plant disease?

MULTIPLE-CHOICE QUESTIONS

1. If a pathogen initiates the dissolution of certain constituents of the cell wall, and subsequently the loss of tissue coherence, it is due to
   (i) toxins (ii) enzymes
   (iii) antigens (iv) none of these
2. If a pathogen kills the protoplast, it is due to
   (i) proteolytic enzymes (ii) toxins
   (iii) cellulases (iv) cutin esterases
3. Name the pathogen that can degrade waxes:
   (i) *Erysiphe graminis* (ii) *Penicillium spinulosum*
   (iii) *Puccinia hordei* (iv) *Spilocaea oleagini*
4. Which of the following are pectolytic enzymes?
   (i) Pectin methyl esterases (ii) Polygalacturonases
   (iii) Pectin trans eliminases (iv) All of these

5. Pectic enzymes are of great importance because they cause
   (i) soft-rot diseases (ii) brown-rot diseases
   (iii) yellow-rot diseases (iv) black-rot diseases
6. Which of the following is not a cellulolytic enzyme?
   (i) β-1,4 endo-gluconases (ii) Cellobiose
   (iii) Protopectinases (iv) C_1 enzyme
7. Which of the following is not a hemicellulase enzyme?
   (i) Xylanase (ii) Arabinase
   (iii) Galactanase (iv) β-1,4 endo-gluconases
8. Which of the following fungus has the ability to produce lignolytic enzymes?
   (i) *Phytophthora* (ii) *Sclerotium*
   (iii) *Polyporus* (iv) *Penicillium*
9. Name the substance that may be related to inactivation or inhibition of enzyme synthesis in host-parasite combination:
   (i) Proteins (ii) Hydroxyprolines
   (iii) Mannase (iv) Quinones
10. Which of the following is not a toxin?
   (i) Fusaric acid (ii) Lycomarasmin
   (iii) Piricularin (iv) Arabinase
11. Which of the following is a host-specific toxin?
   (i) Fusaric acid (ii) Alternaric acid
   (iii) Victorin (iv) Piricularin

Answers

1. (ii) 2. (ii) 3. (iii) 4. (iv) 5. (i) 6. (iii) 7. (iv) 8. (iii) 9. (iv) 10. (iv)
11. (iii)

5 ALTERATION IN PLANT PHYSIOLOGICAL FUNCTION DUE TO PLANT PATHOGEN INTERACTION

5.1 PERMEABILITY CHANGES IN DISEASED PLANTS

Permeability phenomena in plant diseases have been reviewed by Wheeler (1976), and Wheeler and Hanchey (1968). In addition, a number of reviews have dealt completely or in part with the effects of phytotoxins on permeability.

Membranes are composed primarily of lipids and proteins, with small amounts of carbohydrates. These are arranged as a bilayer of phospholipid molecules with the ionic groups facing outwards and the nonpolar ends of the phospholipids facing inwards with embedded protein molecules protruding on either side. These proteins may move within the fluid bilayer. It is generally accepted that all water relations are controlled by the plasma membrane although in plant cells, the rigid cell wall is also important in maintaining turgor. Due to its semipermeable nature, the plasma membrane is of central importance in regulating the passage of ions and organic molecules into and out of the cell. Thus, the plasma membrane is a very complex and dynamic structure which maintains a suitable intracellular environment for various activities of the cell. Similar membranes also occur in cell cytoplasm as endoplasmic reticulum and delimiting various organelles. Membranes are responsible for ion transport. There are two methods of transport of ions. One is by passive diffusion in response to concentration gradients. The other method of transport of ions is against concentration gradient and against electrochemical gradients by active transport in which ions are pumped across membranes using metabolic energy, in the form of ATP, and mediated by ATPases. Membrane permeability plays a vital role in this inward and outward movement of the substances.

Among the several disturbances that a sick plant shows some are primary while others are the secondary effects of the former. Regardless of the disease type or the nature of the pathogenic agents, changes in permeability are present in all diseased tissues which have been examined. Changes in permeability have been detected before any other symptom of the disease appeared or in advance of the pathogen. This, plus the fact that changes in permeability can be caused by a variety of biological, chemical, or physical agents, led to the suggestion that a change in cell permeability might be the initial event in pathogenesis.

The first case of altered permeability was reported by Hutchinson (1913) for Rangpur tobacco wilt caused by *Pseudomonas solanacearum* (now *Ralstonia solanacearum*). Thatcher (1939, 1942, 1943), however, made the first comprehensive study of the extent of permeability changes in plant diseases, that is rusts, powdery mildews, soft rots, dry rots, and wilts and concluded the following:

1. Cells of the fungus pathogens have higher osmotic pressure than those of their hosts.
2. Altered permeability of the host cells is consistently associated with disease.
3. For rusts, which were most extensively studied, an increase in host-cell permeability characterizes susceptibility, while decreased permeability results in resistance.

Biotrophic fungus, which forms haustoria, establishes a very specific relationship with the host-cell membranes. These fungi do not cause massive membrane damage since this would kill their host cells. The plasma membrane of the host and the pathogen remain in very close contact near the point at which the cell wall is penetrated. As the fungal haustorium grows, the host membrane becomes invaginated (extrahaustorial membrane) to accommodate the fungus and seals it at this point. The extra haustorial membrane lacks ATPase, membrane-bound enzymes which control pumping of H^+ ions out and K^+ ions in, in exchange across the membrane. The haustorium of the pathogen can deplete solutes from the matrix which surrounds it and maintains a high concentration gradient. The fungus thus maintains a tight control over membrane permeability. Thus biotrophic pathogens also increase the permeability of host tissues. Cells penetrated by haustoria can still be plasmolysed, so the integrity of the membrane is maintained but there is a definite change in the semipermeable properties of the plasma membrane which is related to the uptake of nutrients by the pathogen. The extra haustorial membrane loses control of nutrient transport so that sugars pass freely to the pathogen.

Ionic imbalance has been reported in Verticillium wilt of cotton and blast disease of rice. Several workers have related calcium to resistance in plants infected by *Rhizoctonia solani*. Membrane damage has been observed in the electron-photomicrographs of infected tissues. It has been reported that **endo-PGTE**, a pectic enzyme, damages plasma membranes and causes the leakage of ions. The rupturing of plasma membrane occurs at plasmodesmata as a result of excessive influx of water from the hypotonic solution (formed by the enzymic dissolution of middle lamella) into the host cells. The maceration and cell death were closely correlated when potato discs were treated with a purified pectin, methyl trans-eliminase, produced by *Penicillium expansum*. Tissues exposed to pectic enzymes undergo changes in permeability before cells are separated or killed; the relationship among these phenomena is not clear at all. If tissues are plasmolysed in hypertonic solutions, cell death and permeability changes can be greatly delayed during exposure to pectic enzymes, but such conditions have little effect on the maceration.

Attempts to correlate tissue permeability with disease resistance have yielded variable results. It had been shown that when the leaves of Minden wheat were exposed to chloroform vapour, there was an increase in the permeability of the tissue and there was definitely an increase in the susceptibility to rusts. The susceptibility of several varieties of wheat to rust and cotton of *Fusarium* wilt has been correlated to the quantities of organic substances secreted.

Changes in cell permeability are often the first noticeable response of cells to infection by pathogens, to most host-specific toxins, several nonspecific toxins to certain pathogen enzymes and to toxic chemicals such as air pollutants. In an interesting work done by Stall and Cook (1979), they recorded the levels of conductivity, measuring leakage of electrolytes released from leaves of pepper plants inoculated with three races of the bacterium *Xanthomonas campestris* pv. *vesicatoria*. Disruption of membranes and electrolyte leakage occurred much earlier and at a much greater rate when the leaves were inoculated with bacterial races carrying virulent genes while release of electrolytes occurred later and at a slower ratio with a virulent race of the pathogen.

Some phytoalexins have been shown to cause membrane damage as well as other effects on the host plant. The phytoalexin **pisatin** influences stomatal opening by affecting membrane function. Leakage of electrolytes has also been reported in virus-infected tissues, especially roots but this does not seem to be a common symptom in other diseases.

5.2 EFFECT OF PLANT PATHOGENS ON TRANSLOCATION OF WATER AND NUTRIENTS IN THE HOST PLANTS

Many plant pathogens interfere in one or more ways with the absorption of water and uptake of nutrients and their translocation in the plants. In many diseases such as the root rotting fungi, bacteria, nematodes, and viruses where the pathogens bring about extensive damage or destruction of the roots, absorption of water is considerably reduced. Root injury directly affects the amount of functioning roots and proportionately decreases the amount of water absorbed by the roots. In vascular wilt diseases, there is resistance to water flow which may be due to physical blockages which obstruct the vascular system by the fungal biomass (as in *Verticillium spp. Fusarium oxysporum* and *F. solani*), or action of fungal toxins and enzymes. There may be host-plant responses which may lead to xylem-vessel blockage. The formation of tyloses is often reported, particularly as host response to vascular wilt infections. Xylem parenchyma cells balloon through pit membranes and very effectively clog or close xylem vessel lumen. Gels and polysaccharide gums are produced by the host plant, probably in response to the release of cell-wall degrading enzymes by the invading pathogens. These may also obstruct xylem vessels, often accumulating at perforation pits.

Ion transport is also interfered with as a result of infection. Ions diffuse differently through solution than through pure water and their diffusion coefficient is greatly reduced in soil. As far as organic nutrients are concerned, these are produced in the leaf cells through photosynthesis and move through plasmodesmata into adjoining phloem elements. From there they move down the phloem sieve tubes and eventually again through plasmodesmata into the protoplasm of living photosynthetic cells, where these are either utilized or stored into storage tissues. Plant pathogens may interfere with the movement of organic nutrients from the leaf cells to the phloem, with their translocation through the phloem elements, or possibly, with their movement from the phloem into cells that will utilize them.

Biotrophs such as powdery mildew fungi and rusts cause an accumulation of photosynthetic products, as well as inorganic nutrients, in the areas invaded by the pathogens. As indicated earlier in these diseases, the infected areas are characterized by reduced photosynthesis and increased respiration. However, it has been seen that the synthesis of starch and other compounds as well as dry weight are temporarily increased in the infected areas. In some virus diseases such as leaf curling type and some yellows diseases, starch accumulation in the leaves is mainly the result of degeneration of the phloem of infected areas. There is another possibility in the interference with translocation of starch in that the virus inhibits the enzymes involved in the breaking down of starch into smaller translocable molecules.

5.2.1 Effect of Pathogens on Translocation of Nutrients and Water

In order to survive and carry out various physiological and metabolic processes in the plant, all living cells require water and organic and inorganic nutrients. The roots of the plant help in the absorption of water and minerals from the soil. These substances are generally translocated upward through the xylem vessels of the stem and into the vascular bundles of leaves. As a result of photosynthesis, these organic nutrients are then translocated downward and distributed to all the living cells through the phloem tissues. As a result of pathogen infection to the plant, the upward and downward translocation of nutrients is inhibited and thereby not enabling the plant to carry out their normal functions and thus causing disease of the entire plant.

5.2.2 Interference with Upward Translocation of Water and Inorganic Nutrients

Some of the pathogens affect the roots and cause them to absorb less water or some of pathogens growing in the xylem vessels interfere with the translocation of water through the stem or some pathogens cause excessive transpiration. Many pathogens, such as root rot of black pepper caused by *Phytophthora capsici*, wilt of leguminous plant with the fungus *Fusarium*, galls caused by the bacterium *Agrobacterium tumefaciens* and root knot galls caused by the nematode *Meloidogyne* sp. cause destruction of the root before any symptoms appear on aerial parts of the host. Such abnormality in the roots will interfere with the normal absorption of water and nutrients by the roots. Apart from this, some pathogens also inhibit root-hair production and also alter permeability of root cells which will ultimately interfere with the normal absorption of water by roots.

In case of damping off, wilts and stem rots, bacterial and fungal pathogens may reach the xylem vessels in the area of infection and may cause death of the young plants. In vascular wilts caused by *Fusarium* spp., the reduction in water translocation may vary from little to complete depending upon the susceptibility or resistance of the host. Apart from this, the affected vessels may be filled with the microconidia, macroconidia, or chlamydospores of *Fusarium* spp. and with the substances like Fusaric acid or polysaccharides secreted by the pathogen and may become clogged. Such clogged vessels cease to function normally and allow little or no water to pass through them. Some bacteria and nematodes cause root galls or root knots and these enlarged and proliferating cells around the xylem exert pressure on the xylem vessels which may be crushed, thereby becoming less efficient in water uptake.

In all cases, in infected hosts the flow of water is reduced through reduction in the size or death of xylem vessels. This happens due to infection or development of tyloses in the xylem vessels or release of some molecules in the vessels as a result of breakdown of cell walls by the enzymes of the pathogens.

5.3 PHOTOSYNTHESIS

Photosynthesis is a basic function of green plants that enables them to transform light energy into energy of chemical bonds—literally, photosynthesis means "synthesis with the help of light".

In the basic reaction of photosynthesis, carbon dioxide from the atmosphere and water from the soil are brought together in the chloroplasts in the presence of light which react to form glucose with the concurrent release of oxygen.

$$6CO_2 + 12H_2O \xrightarrow[\text{Chlorophyll}]{\text{Light}} C_6H_{12}O_6 + 6O_2 + 6H_2O \qquad \Delta G^{0\prime} = 686 \text{ kcal}$$

Carbon dioxide is reduced to the oxidation reduction level of carbohydrate, as symbolized by CH_2O. Each molecule of CO_2 fixed into carbohydrates requires a net input of 1,18,000 calories from light. As an inherent feature of this process, light energy is also converted to chemical energy by a process known as **photosynthetic phosphorylation**, where light energy is stored in the useful high energy organic phosphate compound – **Adenosine TriPhosphate (ATP)**.

In view of the fundamental positions of photosynthesis in the life of plants, it is very clear that any interference by pathogens with photosynthesis results in a diseased condition. In leaf spot, blight, and other kinds of diseases, there is destruction of tissues, e.g. in cereal rusts and fungal leaf spots, bacterial leaf spots, viral mosaics, and yellowing and stunting of tissues in many diseases. In certain cases there is defoliation of leaves, thus resulting in reduced photosynthesis in diseased plants in the diseased areas. It has been observed that not only is the photosynthetic area of the plant reduced, plant pathogens reduce photosynthesis, especially in the late stages of infection by affecting chloroplast directly and causing their degeneration. It is interesting to note that although overall chlorophyll content of leaves in many fungal pathogenic combination is reduced but the photosynthetic activity of the remaining chlorophyll seems to remain unaffected. In the fungus *Alternaria alternata*, which causes spots and chlorosis in many plants of many species, the pathogen produces a toxin—**Tentoxin**—which is a cyclic tetrapeptide that binds to and inactivates a protein (chloroplast-coupling factor) involved in energy transfer into chloroplasts. The toxin also inhibits the light-dependent phosphorylation of ADP to ATP.

Similarly, **Tabtoxin** produced by the bacterium *Pseudomonas syringae* pv. *tabaci* which causes wildfire disease of tobacco, is toxic to the cells. It uncouples photosynthesis and photorespiration and destroys the thylakoid membranes of chloroplast, thereby causing chlorosis and eventually necrosis. In most wilt diseases, stomata remain closed, thus indirectly reducing photosynthesis. Various degrees of chlorosis and stunting is caused by many viral, bacterial, phytoplasmal, and nematode infections resulting in reduced photosynthesis.

Symptoms of viral infection in leaves of plants often include characteristic symptoms of mosaics of green and yellow areas in which chloroplasts are reduced in number, or show ultrastructural abnormalities such as swelling and fewer lamellae which are the actual sites of photochemical reactions.

5.4 RESPIRATION

Respiration is the *sine qua non* for life. In normal healthy plant cells, the function of respiration is twofold—to provide (1) energy, and (2) carbon skeletons. Respiration is the oxidation (aerobic or anaerobic breakdown) of organic material, primarily carbohydrates, to simple compounds, H_2O and CO_2 with the production of utilizable energy for the cell in the form of ATP. In aerobic respiration, this takes place with the uptake of oxygen from the air. In fermentation, the degradation of organic material is anaerobic. Energy is liberated through the process of respiration. Actually, the oxidation of carbohydrates and related compounds is the main source of energy but energy derived from photosynthetic phosphorylation is also a characteristic of plant metabolism. In the process of respiration, energy is not "liberated" as such, it is first redistributed in molecules formed by oxidative reactions, creating bond energy. The energy-rich group generated by oxidation is capable of initiating reactions requiring energy. Hence, the coupling of the chemically bound energy to other systems is the key reaction in the biosynthetic processes of plants.

One of the simplest substrates for oxidation is glucose. The overall reaction involved in the respiration of glucose is as follows:

$$C_6H_{12}O_6 + 6O_2 = 6CO_2 + 6H_2O + \Delta G° = 686 \text{ kcal}$$

5.5 GLYCOLYSIS

Glycolysis is the conversion of glucose to pyruvic acid by any biochemical pathway. The most common glycolytic pathway is the **Embden-Meyerhof-Paranas** (EMP) **Pathway**, although the **Hexose Monophosphate Pathway** (HMP), and the **Entner Doudoroff (ED)** Pathway may provide alternate mechanisms.

The EMP and HMP pathways occur in animals, plants, bacteria and fungi, while the ED pathway is restricted to bacteria and fungi. The EMP pathway in fungi is the principal glycolytic pathway as in most other organisms and usually accounts for at least 50% of the glucose dissimilated; in some fungi it accounts for all the glucose dissimilated. However, the EMP pathway may be secondary in importance to either the HMP or ED pathway in a few fungi or totally lacking, as in the case of *Cladariomyces fumago*. The EMP pathway becomes increasingly important under anaerobic conditions as the HMP pathway accounts for all the glucose dissimilated.

The HMP pathway is common in plants and animals. The enzymes required to carry out the HMP pathway are widely distributed in fungi and may be universal. The presence of the enzymes, however, is not a proof that the HMP pathway occurs. It is found in a number of fungi, usually accounting for the less than 40% of the glucose dissimilated. The relatively low level of the HMP reactions is perhaps due to limiting amounts of NADP.

5.6 RESPIRATION IN PLANTS UNDER PATHOGENESIS

Plant respiration as an index of disease has been a subject of investigation for a number of years. Increased respiration in tissues generally accompanies a variety of pathological conditions except in plants systemically infected with viruses. A study of the altered respiration of infected plants is also of interest to plant physiologists and biochemists investigating the nature of the respiratory rate. There is some evidence that genotypically controlled resistance to disease depends on differences in metabolism and respiration and respiratory rates as indices of such differences seem to open up new lines of interest both to the pathologist and the plant breeder.

Prior to 1953, studies on respiratory stimulation in infected plants were confined mainly to the quantitative aspect of the increase. It was Allen (1953, 1954) who first pointed out that disease was the result of disturbances in host metabolism induced by the diffusible metabolites (toxins) of the parasite. He postulated that increased respiration in mildew disease was the result of a toxin produced by the fungus which uncouples phosphorylation from electron transfer in host cells. Attempts have been made by several workers to isolate toxic agents inducing stimulation of respiration.

5.7 MECHANISM OF RESPIRATORY INCREASE

In a diseased plant there are a number of changes which accompany increased respiration, including changes to pathways used, increases in the concentration of enzymes involved and changes to the uncoupling of respiration to other related cellular processes. There are a number of theories which seek to explain the enhanced respiration rate in diseased plants. These are

1. Uncoupling of oxidative phosphorylation as in *Helminthosporium maydis* on maize
2. Augmented synthetic processes as in safflower infected with *Puccinia carthami*
3. Changes in respiratory pathways as in TMV infected in *Datura stramonium*
4. Increased activity of terminal oxidases as in diseases caused by *Phytophthora*, *Ceratocystis* and *Fusarium* spp.

The direct contribution of the parasite in the increased respiration rate is negligible.

5.8 RELATIONSHIP OF RESISTANCE AND SUSCEPTIBILTY

In infected resistant cultivars (incompatible reaction), oxygen uptake has been shown to rise sharply quite early in infection, although at a later stage subsequently the level is reduced to normal. In infected susceptible cultivars (compatible reaction), there is much less change initially and a rise in oxygen uptake occurs a few days after infection.

It has been observed that in wheat rust, the resistant variety recorded a steep rise followed by an equally steep fall in respiration while the susceptible variety had a continuous rise in respiration.

It was found that the respiratory rate of three varieties of rice differing in their degree of susceptibility to *Fusarium moniliforme* and the respiratory rates of both root and shoot tissues were directly proportional to the degree of susceptibility.

Studies on potato-*Phytophthora*, cabbage-*Botrytis*, citrus-*Penicillium* and other diseases showed that resistant varieties respire more intensively than susceptible ones.

Effect on Transpiration

In infected plants, transpiration is usually increased due to an increase in the permeability of leaf cells and the dysfunction of stomata. The wheat-leaf-rust pathogen *Puccinia recondita* produces numerous lesions (Uredinia) and causes many breaks in leaf epidermis through which transpiration goes on uncontrolled. In grape berries infected with the powdery mildew fungus *Uncinula necator*, the mycelium of which penetrates and forms haustoria in almost every epidermal cell, the destruction of cuticle and epidermis results in uncontrolled loss of water from the affected areas.

Similarly, the apple scab fungus *Venturia inaequalis* grows between the cuticle and the epidermis, causing the cuticle to break in numerous places, allowing transpiration to occur. If the transpiration is more than absorption of water, wilting of leaves occur and lead to death of the host plant.

5.9 CHANGES IN NITROGEN METABOLISM OF INFECTED PLANTS

In many diseased tissues, especially in those characterized by overgrowths, galls or other growth abnormalities, increased metabolic activity is accompanied by marked increases in nucleic acid and protein synthesis. Cytochemical evidence for major changes in nuclei, nucleoli, and changes in ribosomes is available.

Nucleic Acid Metabolism

No field of biological science has developed and progressed to such an extent during the last few decades as have nucleic acids because they are involved not only in the transfer of genetic information but also in the regulation of gene activity. Changes in the nucleic level have been reported in diseases caused by all kinds of pathogens. The host RNA content in TMV infected tobacco leaves increases shortly (8 hours after inoculation) but afterwards (70 hours), the viral RNA increases together with the total increase of RNA. The total RNA increases in infected tissues, while it decreases in the uninoculated control tissues. In various virus-host plant combinations, transitory increases, moderate declines, or no significant changes in DNA have been reported. Such variations probably reflect differences among the plants and viruses studied. Gall cells induced by various fungi, bacteria, and nematodes have been shown to contain a very large quantity of nucleic acids. In the case of *Plasmodiophora brassicae*, causing club-root disease of crucifers, the infected cells each contain an enlarged nucleolus which may be as much as 30 or more folds in volume over nucleoli in nonparasitized cells and the DNA content may be sixteen-fold greater than in the non-infected cells. Hypertophy of host cells together with increases in the sizes of nuclei have been reported for various other host-parasite combinations. In the same disease, the nucleolus of root-hair cells begins to enlarge within two hours after penetration and after twenty hours this volume has increased sixfold. The increase in nucleolar volume is accompanied by an increased nuclear RNA and nonhistone protein. The lysine-rich portion of histone decreases. Since histones have a regulatory role, any change in their quantity is related to a change in the regulation of gene action. It seems that the normal transcriptional processes of invaded cells are altered by parasites.

5.10 CHANGES IN PROTEIN METABOLISM

The total content of the protein (that of the plant and pathogen) usually increases during the early stages of infection. When the contribution of the pathogen is seen, little change, or in some cases, a decrease in plant protein has been found in most diseased tissues. In the case of crown-gall tumors, there is much more protein in them that could be expected from synthetic activities of the pathogen. In late degenerative stages of disease, large decreases in protein, sometimes accompanied by increases in free amino acids, are often found.

Levels of protein synthesis in infected plants in susceptible plants are usually increased over levels in healthy or resistant plants. Much of this increase in protein level can be attributed to the synthesizing activities of the invading pathogen.

It is generally agreed that there are marked differences between the nitrogenous constituents of normal resting cells and actively dividing tumour cells incited to division by the crown-gall bacterium *Agrobacterium tumefaciens*. Both normal and tumour tissues contain about the same amounts of non-protein nitrogen but tumours have three times the protein of normal tissues. This was interpreted as a greater tendency and capacity on the part of tumour tissues to synthesize protein.

5.11 PHENOLS

Phenols are compounds, which contain one or more aromatic (benzene) rings with one or more phenolic hydroxyl groups. The simplest compound is phenol. Compounds having more than one hydroxyl group on the benzene ring are called **polyphenols**. Phenolic compounds are widely distributed in higher plants and fungi. Anthocyanins, leucoanthocyanins and anthoxanthins, hydroxybenzoic acids, glycosides, sugar esters of quinones and shikimic acid, esters or hydroxycinnamic acids, and coumarine derivatives make up a vast class of compounds known as phenolics. Formerly, the term *tannin*, and more recently, the term *polyphenol*, has been used to refer to these compounds. Phenolic compounds function as hydrogen donors or acceptors in oxidation-reduction reactions and they play an essential role in lignification. Some phenolics show anti-auxin activities and are potent growth inhibitors. Others act synergistically with auxins in growth stimulation. Phenolic compounds have been shown to play key roles in the resistance of plants to disease-causing microorganisms. In addition, phenolics may also serve as protectants against leaf-eating insects, for example, they may influence the availability or digestibility of proteins which insects obtain from leaves. Phenolics may also interfere with growth and other energy-dependent activities by uncoupling oxidative phosphorylation. The oxidation of phenols yields highly reactive quinones which inhibit enzymes by complexing with metal ions, reacting with sulfahydryl groups or binding nonspecifically to proteins. It is obvious that even small changes in phenol metabolism may severely disrupt many processes which are essential for normal plant growth and development.

Evidences from ^{14}C tracer elements indicate that the biosynthesis of phenolic compounds in plants may proceed via one or more of three different pathways:

1. Shikimic acid pathway
2. Acetate malonate pathway
3. Acetate-mevalonate pathway

The most important of these appears to be the shikimic acid pathway in which the initial reaction involves the condensation of phosphoenol pyruvate and erythrose-4-phosphate to yield a seven-carbon intermediate which is converted by a series of reactions into shikimic acid.

5.12 PHENOL METABOLISM

Phenol oxidizing enzymes—laccase and tyrosinase—are the two most important phenol oxidases. These occur both in higher plants as well as in fungi. The enzymes laccase catalyzes the aerobic oxidation of *p*-hydroquinones or *p*-phenylenediamines to *p*-quinones of *p*-quinonediamines. The other enzyme, **tyrosinase**, is actually composed of two enzymes. One of the enzymes possesses only catecholase activity, whereas the other shows cresolase activity. This enzyme (tyrosinase) complex is also known as phenolase, phenoloxidase, polyphenoloxidase or DoP A-oxidase. Tyrosinase is also an aerobic oxidase which is able to oxidize many phenol compounds by oxygen. However, neither para nor meta-dihydroxyphenols are suitable substrates for this enzyme since it is able to oxidize only ortho-dihydroxyphenols.

Changes in the activity of phenol-oxidizing enzymes, peroxidase and phenolase may also play a role in the regulation of metabolic pathways in diseased or injured tissues. Phenolase is a copper-containing enzyme which catalyzes the oxidation of mono- and O-dihydroxyphenols to phenols. The resistance to several fungal pathogens has been ascribed to a higher concentration of fungitoxic phenolic substances and their oxidation products, to increased polyphenol oxidase activity which generally, but not invariably results from infection. More than 75 years ago, Dufrenoy (1936) observed that when the progress of some pathogens is checked, phenolic compounds, mostly tannins of the gallic group, develop in abundance within the vaculoes of adjacent cells, whereas in specialized parasites, tending to become systemic, tannin is formed very slowly in small amounts and only in vacuoles close to the parasite. Subsequently, it was suggested that the establishment of *Puccinia coronata* in oat leaves depended on the supply of phosphorus from the host tissue; on the depletion of this element, relatively innocuous host phenolic compounds were dehydrogenated by tyrosinase into toxic quinones, leading to the localized necrosis of the tissue.

5.13 IMPORTANCE OF PHENOL OXIDATION IN PLANT DISEASES

The phenol oxidase system plays a role in respiration by transferring electrons from respiratory substrates to other hydrogen or electron acceptors. Oxidation products of phenols (quinones) may be reduced to their original phenol form by respiratory carriers. $NADH_2$ and $NADPH_2$, generated by the oxidation of respiratory substrates, may reduce quinones in the presence of quinone reductase enzyme. Polyphenolases oxidize monophenol to polyphenols by means of cresolase activity in which the hydroxylation of monophenol leads to the formation of complicated polyphenols, such as flavonoids, tannins, lignins, and others, whereas the catechol activity has a role in the browning reaction of plant tissues.

Phenolase may influence the amino acid metabolism of plants. It has been seen in many cases that a close correlation exists between the enhanced activity of phenolase and peroxidase and the concentration of phenolic substances. Phenolic substances in turn have been correlated with resistance to pathogens. The increased phenolase activity in diseased or wounded areas of plant tissues is generally accompanied by an increased concentration of phenolic substances.

Polyphenols, like chlorogenic and caffeic acids, are known to play a role in the inhibition of the activity of indole acetic acid oxidase. Phenolics and their oxidation products seem to play a role in the auxin metabolism of plants. Polyphenols, such as chlorogenic and caffeic acid, known to be powerful inhibitors of IAA-oxidase, may in turn be able to reverse the enzyme activity and confer resistance via an auxin-spring action on plants that produce them in response to infection.

The phenols synthesized by plant tissues after infection appear to have a broad antifungal spectrum. Oxidized polyphenols inhibit the pectinolytic enzyme produced by various fungi and they have early been associated with the resistance of apples to brown rot.

It has been shown in many experiments that a correlation may exist between the degree of resistance and the phenol level in healthy plants. It has been shown that varieties of *Amaranthus tricolor* resistant to *Alternaria* sp. accumulate O-D phenols and the amount of the latter was many a times that of the susceptible varieties. Similarly, onion varieties resistant to *Colletotrichum circinans* accumulate flavones, anthocyanine and simple phenolics, such as protocatechuic acid and catechol in the dried bulb scales which are red pigmented.

It has been shown that oxidized phenols as well as oxidized phenolic glucosides are even more effective. In the case of apple leaves, resistance to *Venturia inaequalis* has been attributed to phloretin and that of pea shoots to *Erwinia amylovora* is attributed to arbutin. This is due to the action of β-glucosidases, polyphenol oxidases, and peroxidases.

Accumulation of higher amounts of chlorogenic acid in the potato roots of varieties resistant to Verticillium wilt disease has been reported. Mahadevan (1965) showed that phenols and their oxidation products showed inhibitory activity on the pectolytic and cellulolytic enzymes of the pathogen *Cladosporium cucumerinum*. He suggested that the defense mechanism in resistant plants should involve the importance of phenolic constituents and their oxidation products on the cell-wall degrading enzymes of the parasite. The inactivation of extracellular microbial enzymes by the phenols or quinones present in plants after penetration is certainly a defense mechanism.

The periderm of tubers from a variety resistant to scab (*Streptomyces scabies*) showed much more chlorogenic and caffeic acid and were largely confined to the outermost tissues where the scab organism proliferate. In the Verticillium wilt of potato, the resistance is ascribed to a greater quantity of chlorogenic acid in root tips, sprout tips, and the vascular system of resistant varieties.

In the case of *Phytophthora infestans* infection, the leaves of resistant potato varieties show a more intense accumulation of toxic oxidative derivatives of polyphenols than do the leaves of susceptible varieties. The resistance to *Phytophthora* is reversed by copper enzyme inhibitors that interfere with the formation of toxic quinones via PolyPhenol Oxidases (PPO).

Following inoculation with *Phytophthora infestans*, both polyphenol and PPO increased in resistant potato varieties but not in susceptible varieties which suggests a resistance mechanism to the enzyme. It has been reported that when streptomycin is absorbed through the root, it gives resistance to potato against *Phytophthora*. This is explained by the fact that antibiotic activity generally enhances PPO activity, and hence, the resistance in plants.

The activity of PPO would seem to be important in that it can oxidize phenolics to quinones which may be more fungitoxic. Polymerization of quinones to dark-coloured tannins may occur. In tomato wilt (*F. oxysporum* f. sp. *lycopersici*), conjugated phenols such as glucosides, tannins and lignins may be hydrolyzed by enzymes to toxic quinones which then polymerize to cause characteristic vascular discoloration. PPO produced by the pathogen might oxidize host polyphenols to more highly fungitoxic substances which prevent further development of the fungal pathogen for example, a PPO inhibitor reduces the toxicity of catechol to *Cochliobolus miyabeanus* (leaf blight of rice) and prevents the development of melanin presumably by preventing enzymic conversion of catechol into fungitoxic quinones. The melanin is apparently fungitoxic although in other cases these insoluble pigments are relatively nontoxic to fungi. Such inhibition can thus result from the production of PPO by the pathogens.

The germinating urediniospores of *Puccinia graminis* release phenols and PPO and the resistance of barely plants to *Blumeria* (*Erysiphe*) *graminis* appears to be associated partly with an accompanying release of a phenolic substance which accumulates around the haustoria and inhibits further development of the fungus. Aromatic substances, such as polyphenols, phenolic glucosides, flavonoids, anthocyanins, aromatic amino acids, and coumarin derivatives tend to accumulate in and around infected plant tissue and also in tissue adjacent to wounds, where presumably they might exert a fungitoxic effect.

5.14 GROWTH REGULATORS IN PLANT DISEASES

5.14.1 Role of Growth Regulators in Plant Disease

The growth regulators play vital roles in the growth and development of the plant. The concentration of various growth regulators in the plants is not uniform as it rises and declines very quickly as a result of the action of growth-hormone inhibitory systems present in the plant. These growth regulators ultimately regulate the synthesis of messenger RNA molecules which lead to the formation of specific enzymes that finally control the biochemistry and physiology of the plant.

Plant pathogens, as a result of infection in the host plants, may produce more of growth regulators or more of growth inhibitors or may produce new or different growth regulators or inhibitors or may produce substances that stimulate or retard the production of growth regulators or growth inhibitors by the plants. Whatever the mechanism of action involved, pathogens often cause an imbalance of hormonal system of the plant which is incompatible for the healthy development of the plant. As a result of this abnormal secretion of growth regulators or inhibitors, the infected plant shows a variety of symptoms such as stunting, overgrowth, rosetting, stem malformation, leaf epinasty, and suppression of bud development in healthy plants. Here important growth regulators, their functions and role in disease development are discussed briefly.

Symptoms, such as galls, overgrowth or malformation of host tissues, epinasty, elongation, witches' broom effects, etc. indicate abnormal growth activities in many plant diseases. The extracts of tissues which exhibit such abnormal growth patterns are often ten times more active in the tests for growth-regulating substances than are the extracts of healthy plants. Even in the absence of obvious growth abnormalities, diseased tissues often contain elevated levels of growth regulators. Four classes of compounds are known to be involved in plant diseases. All the major types of plant-growth regulators (auxins, gibberellins, cytokinins, and volatile hormone ethylene) are produced by viral, bacterial, and fungal pathogens.

5.14.2 Auxins

The naturally occurring auxin in plants is Indole-3-Acetic Acid (IAA). It is continually produced in growing plant tissues in which it occurs in concentrations of about 5 μg per kg of green material. Perhaps, no other single compound in plants has received as much attention from plant biologists as has β-3-Indole Acetic Acid (IAA or auxin). Of all the endogenous plant-growth regulators, IAA is perhaps the nearest to being ubiquitous. Auxin is an organic substance that promotes growth along the longitudinal axis when applied in low concentrations to the shoots of plants which are freed, as far as possible, from their own inherent growth-promoting substance.

Went (1926) first showed the effect of IAA on *Avena*, following which it was demonstrated that IAA has a widespread distribution in higher plants and especially in diseased tissues. Auxin is also produced freely by many species of bacteria and fungi. The synthesis of IAA in extremely small quantities takes place presumably in young leaves and root tips. IAA is required for cell elongation and differentiation. The absorption of IAA to the cell membrane also affects the permeability of the membrane.

Many primary functions have been attributed to IAA. It has been linked with cell metabolism, phenol metabolism, increase in respiration, enhancement of the pentose phosphate pathway activity, and membrane permeability. Evidence has also accumulated of the primary site of action of IAA being on the nuclear control of RNA synthesis or on RNA itself, thus affecting the level of certain key enzymes in cell growth.

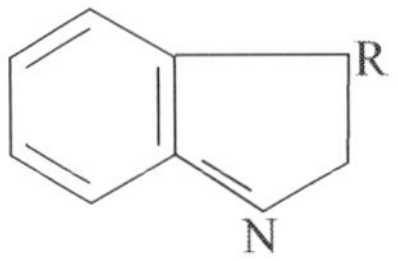

The indole nucleus

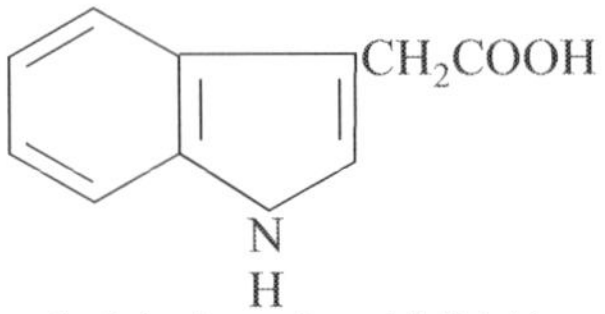

Indole-3 acetic acid (IAA)

Changed Level of Auxin in Diseased Tissues

The study of plant-growth substances in diseased plants is comparatively recent. This is somewhat surprising, since in many diseases, symptoms typically illustrate the effect of enhanced growth-regulatory substances.

(a) Viral Diseases Grieve (1943) attributed the reduction in the growth of tomato plants infected with tomato spotted wilt virus to a decrease in auxin content and to reduced rate of translocation through the stem. Surprisingly, Jones (1956) did not find a decreased auxin content in the tomato tissues infected with spotted wilt virus. This can be explained by the fact that such infected plants contain a high level of scopoletin which is a known inhibitor of IAA oxidase. Symptoms of dwarfing in many viral infections are believed to be due to the lower auxin content of the tissues. Lower auxin content has been associated with TMV and potato leaf-roll virus. Lower endogenous levels of auxins have also been found in *Lycopersicon esculentum, Phaseolus vulgaris,* and *Beta vulgaris* plants infected with curly top virus rather than in healthy plants.

(b) Bacterial Diseases The production and role of auxins in plant disease has been studied more extensively in some bacterial diseases of plants such as *Agrobacterium tumefaciens, Pseudomonas savastoni,* and *Pseudomonas solanacearum.* One of the best known plant diseases, displaying overgrowth symptoms, is crown gall caused by *Agrobacterium tumefaciens.* There is ample evidence that the abnormal amounts of auxins found in crown-gall tumor are produced by the host. Tumours develop when the crown-gall bacteria enter fresh wounds on a susceptible host. The cells around the wound produce various phenolic compounds specially (acetosyringone) and are activated to divide. The pathogenic bacteria do not invade or enter cells of the host but simply attach themselves to the cell walls. In response to the phenolic compounds and other signals, they begin processing the DNA in their Ti-plasmid (Tumor-inducing plasmid). Intense cell divisions begin to take place after wounding and second or third days after wounding, the plant cells are somehow conditioned and made receptive to a piece of bacterial plasmid DNA (T-DNA, for transfer DNA). A part of the bacterial plasmid DNA is cut out from the Ti-plasmid and is transferred into the host plant cells as T-DNA-protein complex. This T-DNA then becomes incorporated or integrated into the nuclear plant DNA and transforms normal plant cells into tumor cells (Agrios, 1997). Tumor cells subsequently grow and divide independently of the bacteria, and their organization, rate of growth, and rate of division can no longer be controlled by the host plants. The transferred (or T) region of plasmid DNA carries several genes which, when expressed in the host cell, cause the characteristic tumour. One set (as in oncogenic—cancer causing) of genes specify the synthesis of auxins and cytokinins leading to uncontrolled cell division, while other genes code for the production of unusual amino acids known as **opines**. These unusual amino acids cannot be utilized by the plant, but are instead catabolized by the pathogen.

(c) Fungal Diseases Many fungal diseases of plants cause an increase in the levels of auxin in tissues (hyperauxing) in often very high concentrations. The best known example of increased levels in plant trees infected by fungal pathogens is club-root disease of crucifers caused by *Plasmodiophora brassicae*. The pathogen on entry to the roots through mobile zoospores upsets the plant's normal regulatory control of IAA synthesis which leads to root distortion in infected individuals. Increased level of IAA has also been found in rust-infected tissues of safflower with *Puccinia carthami*. A tenfold increase has been measured in susceptible potato tubes tissues infected by *Phytophthora infestans* and in corn infected by *Ustilago zeae*.

5.14.3 Gibberellins

Gibberellins are a group of chemically similar growth-regulating substances. They can be considered normal constituents of green plants. The existence of gibberellin in plants might still have been unknown if the "bakanae" or "foolish seedling" disease (which had a devastating effect on the rice economy of Japan) had not broken out. Japanese farmers noted that plants affected with this disease outgrew normal plants but were thinner and paler than their normal counterparts and were sometimes devoid of fruits. In the early part of the 20th century, Japanese pathologists first demonstrated the connection between the bakanae disease and the fungus *Gibberella fuzikuroi* (the teleomorph of *Fusarium moniliforme*). Yabuta and Hayshi (1939) were able to isolate the crystalline gibberellin from the infected rice plants. Since that time, gibberellins and gibberellin-like substances have been found in higher plants and implicated in disease symptoms. They have a gibbane skeleton.

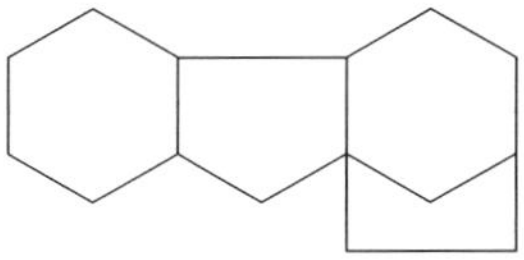

Gibbane skeleton

Following the classical discovery of the bakanae disease, the first truly experimental investigations on endogenous gibberellins and infected plants were made by Radley (1961). It was shown by her that pea nodules infected with *Rhizobium* contain an increased level of GLS (Gibberelin-Like Structure) as compared to healthy roots. Gibberellic acid has effects in infection of cassava (*Manihot esculenta*) caused by *Sphaceloma manihotica*, which initially causes the formation of necrotic leaf spots and stem canker, but subsequently gives rise to massive elongation of internodes as a characteristic disease symptom.

Gibberellins have been studied more extensively in relation to viral diseases. The spraying of diseased plants with gibberellin has been shown to overcome some of the diseases caused by viruses. Thus, the stunting of corn plants infected with corn-stunt virus and tobacco plants infected with severe etch virus was reversed after treatment with gibberellin. Axillary bud suppression caused by Sour Cherry Yellows Virus (SCYV) on cherry and leaf-curl virus on tobacco was also overcome by gibberellin sprays. The fact that stunting caused by some viruses can be reversed by applying exogenous gibberellins suggests that the reduced growth of these virus-infected plants might be due to decreased gibberellin content.

5.14.4 Cytokinins

Cytokinins are potent growth factors necessary for cell growth and differentiation. The first cytokinin discovered was **kinetin** by Skoog and Miller in the early 1950s. The main action of cytokinin, an exogenously applied kinetin, is on cell division. It also seems to be capable of orienting plant growth induced by auxin.

Cells in cultures show no response when they are supplied with kinetin alone, and they increase in size but do not divide when supplied with only IAA. When both substances are supplied together, the cells enlarge and divide. The relative concentrations of kinetin seem to determine the organs that will differentiate, low concentrations inducing root formation, higher concentrations inducing bud formation. The mechanism of action of cytokinins is not yet known but the available evidence indicates that it is similar to that of gibberellins. A sixfold increase in cytokinin activity was detected in extracts from the galled roots of root knot nematode-infected tobacco plants compared with extracts from healthy plants; the cytokinin, however, appears late in the infection since they can be detected in galled roots 75 days, and not 35 days, after inoculation. Root cells near a nematode's head display hypertrophy with repeated nuclear divisions together with the incorporation of neighbouring cells, and this results in large, thick-walled multinucleate syncytia or giant cells. The endogenous levels of several hormones are reported to increase in several hosts infected by nematodes. Cytokinin activity also increases in rust-infected beans and broad bean leaves and seems to be related to both the juvenile features of the green islands around the infection centers and the senescence outside the green island. In the Helminthosporium blight disease of oats, cytokinins do not increase the toxicity of the toxin victorin but they increase the quantity of toxin absorbed. It appears that cytokinins may enhance the effectiveness of the pathogens that produce pathotoxins. On the other hand, cucumber leaf discs treated with kinetin become resistant to the powdery mildew fungus, and tobacco leaves infected with the wildfire toxin and treated with kinetin fail to develop the typical toxin-induced chlorosis. It will be recalled that kinetin tends to arrest the breakdown of proteins and nucleic acids which occurs when a leaf is removed from a plant and it tends to mobilize the accumulation of amino acids at the point of its application to a leaf. Osborne has suggested that the new RNA and proteins are even synthesized under these conditions which, without kinetin, greatly favour catabolism. A decrease has been observed in TMV production in the leaves and leaf discs of a local lesion host *Nicotiana glutinosa* treated with kinetin (40-50 mg/litre). The kinetin treatment to be effective has to be applied before or shortly after the infection. Kinetin reduces virus multiplication in a systemic host.

5.14.5 Ethylene

Ethylene (C_2H_4) is a plant-growth regulator which is biologically highly active. It has an important role in regulation of plant growth and is implicated in a number of disease symptoms, e.g. tissue swelling, stimulation of root formation, leaf abscission, epinasty, and fruit ripening. It is produced by plants and some species of pathogenic organisms, is effective in very low concentrations and is synthesized from methionine in the presence of light.

Ethylene stimulates the production of a number of enzymes. In healthy plants, ethylene is reported to influence the levels of hydroxyproline-rich proteins in cell walls which are important to wall structure. It has been seen that ethylene treatment increases the levels of cellulase and β-1, 3-glucanase which are implicated in the degradation of cell-wall polymers. The role of ethylene in disease resistance may be due to nonspecific induction of phytoalexin synthesis and the stimulation of enyzmes which are active in the synthesis of phytoalexin, e.g. PAL, peroxidase, polyphenol oxidase, and pectin esterase. These enzymes oxidize phenolic compounds and are involved in resistant reactions of the host plants. Ethylene is ubiquitously present in plant tissues and is produced in increased amounts in damaged or diseased plants. Applied exogenously, it is effective in very low concentrations (below 0.1 ppm) and produces a variety of plant responses. Some of these responses—epinasty, leaf abscission, premature senescence, fruit ripening, and shedding of leaves—are similar to the symptoms of certain plant diseases. A number of bacterial or fungal pathogens produce ethylene and in the case of fusarial wilt of tomato caused by *F. oxysporum* f. sp. *lycopersici*, ethylene production by the pathogen is sufficient to account for the epinastic symptoms of a disease. The defoliating symptoms caused by Verticillium wilt of hop plants is attributed to increased levels of ethylene in infected hop plants. In Verticillium wilt disease, ethylene has been attributed a dual role; acting directly as a toxin on plant tissues covering leaf drop, epinasty, and premature senescence, and also inducing disease resistance in infected plants.

In a number of viral diseases, leaves with necrotic local lesions produce more ethylene than those of systemically infected plants without local lesions. Necrosis induced by toxic chemicals also results in increased ethylene evolution. These results suggest that ethylene is a product rather than the cause of necrosis.

Erwinia amylovora does not produce ethylene in culture but when cauliflower florets are inoculated with this bacterium, it stimulates ethylene production. This indicates that the evolution in diseased tissue most probably is of host origin arising as a result of cellular damage in a still-living tissue. A pectate lyase in the culture filtrate seems to be responsible for the increased production by the plant.

Ethylene undoubtedly interacts with other growth-regulating substances. It had been pointed out that changes in the balance of growth regulators as a result of such constantly changing interactions may be more important in symptom development than the quantity of any single component. It must be emphasized that it is still not clear whether ethylene is a primary factor in triggering metabolic changes or it is merely produced as a result of these changes.

5.15 TRANSCRIPTION AND TRANSLATION IN DISEASED PLANTS

All living cells hold genetic information in DNA molecules which can be transcribed into RNA and translated into polypeptides according to the code of particular gene. Mechanisms involved in transcription and translation are now fairly well understood, at least in prokaryotes. Transcription is the process by which genetic information encoded in DNA is copied during the process of mRNA synthesis. Translation is the process by which the information is used to specify certain amino acid sequences during peptide synthesis. Several pathogens, particularly viruses and fungal biotrophs such as rusts and powdery mildews, affect the transcription process in infected cells. In diseases caused by viruses, the pathogen through its own enzyme or by modifying the host enzyme (RNA polymerase) that makes RNA, utilizes the host-cell nucleotides and machinery to make its own RNA.

5.15.1 Effect on Transcription

In some cases, the pathogens affect transcription by changing the composition, structure of, or function of the chromatin associated with the cell DNA. Ribonucleases have been investigated during obligate biotrophic infections such as rusts and powdery mildews. Ribonucleases in infected plants have been shown to exhibit different substrate preferences form those in infected plant tissues (host RNase) and from those of invading fungal mycelium (fungal Rnase). Additionally, this type of novel activity was not detected in resistant hosts which, therefore, suggests that the changes in RNases may be related to the expression of resistance in the host or to the modification of host RNase by the pathogen. It is generally believed that the greater RNA levels and, therefore, increased transcription in cells, indicate increased synthesis of substances involved in defense mechanism of plants.

5.15.2 Effect on Translation

Infected plant tissues often show increased activity in several enzymes, particularly those associated with respiration or with the production of various phenolic compounds, some of which may be involved in defense mechanism of the plants. Although a certain amount of these enzymes (proteins) may be present in the cell at the time of infection of plants by pathogens, several new enzymes or proteins are synthesized *de novo*. Thus, there is an increased level of transcription and translation activity in the infected tissues. It has been commonly seen that the increased level of proteins in hosts are those which are resistant to the pathogens. If the resistant tissues are treated with Actinomycin-D, an inhibitor of protein synthesis, their resistance to the pathogens is reduced. Thus, it is very clear that the increased activity of the protein synthesis in plants attacked by pathogens is related to the defense mechanism of the plants.

5.15.3 Polysaccharides

Production of polysaccharide appears to be necessary for many pathogens to cause normal symptoms in the plant or by promoting colonization of the pathogen in the host or by increasing survival rate of the pathogen.

In vascular wilts, slimy polysaccharides appear to play an important role as these polysaccharides released by the pathogen in the infected tissue (especially xylem) may be responsible for mechanical blockage of vascular bundles and thus cause wilting. Such a type of possibility may occur rarely in nature, but when macromolecules (substances) are released in the xylem vessels as a result of breakdown of host substances by the infecting pathogen through its enzyme activity, the possibility of polysaccharide involvement in the blockage of vessels during wilt becomes obvious.

5.15.4 Detoxification of Antimicrobial Molecules

Several kinds of low-molecular-weight antimicrobial molecules like saponins which include the avenacins and tomatines are present in the plants or are produced as a result of host-pathogen interaction that provide plants with some degree of protection against fungal pathogens.

Roots and leaves of oat contain **avenacins** that are resistant to *Gaeumannomyces graminis* but this fungus infects other cereals that contain no avenacins. But a strain of fungus that infects oat, *G.graminis* f.sp. *avenae,* produces the enzyme **avenacinase** that detoxifies the avenacins by secreting some enzymes. Another saponin, **tomatine**, is present in tomatoes, that protect the plant against infection but the fungus *Septoria lycopersici* produces the tomatinase enzyme that detoxifies tomatine and infects tomato plants. Similarly, some strains of *Botrytis cinerea* can detoxify tomatine and can infect, but not all the strains can detoxify tomatine.

REVIEW QUESTIONS

1. What do you know about the changes in the permeability of membranes as a result of infection by a pathogen? Suggest various methods by which you can measure the altered permeability.
2. Define photosynthesis. How is this process affected as a result of infection by pathogens?
3. Define respiration. How is this process affected due to infection by pathogens?
4. What are the various mechanisms responsible for augmented respiration in a diseased plant?
5. Explain the differences in respiration between resistant and susceptible varieties.
6. What are laccases and tyrosinases? How are these involved in the resistance mechanism?
7. What is transcription and translation? How does the disturbance in any of these lead to drastic changes in structure and function of the affected cell/cells?

MULTIPLE-CHOICE QUESTIONS

1. In vascular wilt diseases, which of the following factors do not play an important role in xylem-vessel blockage?
   (i) Tyloses (ii) Toxins
   (iii) Polysaccharide gums (iv) Antibiotics
2. Which of the following is not the primary function of IAA (Indole Acetic Acid)?
   (i) Enhancement of PPP (Pentose Phosphate Pathway)
   (ii) Phenol metabolism
   (iii) Photosynthesis suppression
   (iv) Increase in respiration
3. Tabtoxin, which causes wildfire disease of tobacco is produced by
   (i) *Pseudomonas syringae* p. v. *tabaci* (ii) *Alternaria solani*
   (iii) *Trichoderma viride* (iv) *Phytophthora infestans*
4. Under anaerobic conditions, pyruvate undergoes fermentation and finally yields
   (i) lactic acid and alcohol (ii) sulfuric acid
   (iii) tabtoxin (iv) hydrochloric acid
5. Which one of the following is a powerful inhibitor of IAA-oxidase?
   (i) Chlorogenic acid (ii) Phosphoric acid
   (iii) Mevalonic acid (iv) NADP
6. The disease "axillary bud suppression" caused by Sour Cherry Yellows Virus (SCYV) on cherry can be overcome by:
   (i) gibberellin spray (ii) cytokinin spray
   (iii) auxin spray (iv) ethylene spray
7. Which one of the following is not the pathway of biosynthesis of phenolic compounds?
   (i) Shikimic acid pathway (ii) Acetate-malonate pathway
   (iii) Acetate-mevalonate pathway (iv) Glycolysis
8. In crown-gall disease caused by *Agrobacterium tumefaciens*, which type of DNA causes infection from one host plant to the other?
   (i) T-DNA (ii) m-DNA
   (iii) c-DNA (iv) None of these
9. In Shikimic acid pathway of phenol biosynthesis, reduction of 5-dehydroshikimic to shikimic requires
   (i) NADPH (ii) K^+
   (iii) K^+ and NADPH (iv) K^+ and ATP
10. Melanin, a dark-coloured pigment produced in enzymatic and non-enzymatic reactions is produced by
    (i) o-quinone (ii) lactose
    (iii) glucose (iv) mevalonic acid
11. The disease "clubrot of crucifers" is caused by
    (i) *Plasmodiophora brassicae* (ii) *Pseudomonas syringae*
    (iii) *Claviceps purpurea* (iv) *Fusarium oxysporum*

Answers

1. (iv) 2. (iii) 3. (i) 4. (i) 5. (i) 6. (i) 7. (iv) 8. (i) 9. (i) 10. (i)
11. (i)

6 HOW PLANTS DEFEND THEMSELVES AGAINST INFECTION

A great variety of plant and animal life exists in a common environment but the parasitism of one form by another is rare. All plants have resistance to some pathogens under certain conditions. Host-parasite relationships may be inhibited or completely hindered if the host is able to ward off the pathogen by passive or active processes. Disease resistance appears to be the rule rather than the exception in nature. Its genetic constitution and the environmental conditions under which the genes operate govern the capacity or ability of a plant to defend itself against a pathogen. The attributes of the hosts that reduce the chances of infection or the development of the pathogen are considered defense mechanisms. Different plants defend themselves against pathogens in different ways. The defense mechanisms of plants against pathogens fall into two categories.

1. Morphological or structural defense mechanism
2. Biochemical defense mechanism

6.1 MORPHOLOGICAL OR STRUCTURAL DEFENSE MECHANISM

The first line of defense against pathogens is the surface barrier, which a pathogen must penetrate before it can cause infection. The entry of the pathogens may either be through the epidermal cell walls directly or through the natural openings in the epidermis, such as stomata, lenticels, and hydathodes, or through injured areas caused by living or nonliving agencies. When once a pathogen has gained entry into the host, its movement and spread is further checked between or through succulent parenchymatous cells, and in the vascular diseases through the lumen of the xylem vessels. Certain structural features of the epidermis or its interior may greatly affect the ability of the pathogen to penetrate or to invade a host plant. Such morphological defense structures may be present before penetration or infection or may be produced afterwards as a result of the interaction of the host and the pathogen.

6.1.1 Defense Structures Existing Before Infection

Waxes and Cuticles

The cuticle, which consists of cutin and waxes, forms the outermost covering of the epidermal cells and appears as a noncellular, membranous layer. Waxes play a defensive role by forming a hydrophobic surface which acts as a water repellent and thereby prevents the retention of water on sloping plant surfaces on which pathogens might be deposited and subsequently further invade the host. The cuticle acts as a physical or chemical barrier to infection in *Berberis* spp. infected with *Puccinia graminis*. There are several examples where cutin acts as a physical barrier and enables a greater degree of resistance, for example, *Linum usitatissimum* (linseed) attacked by *Melampsora lini* and coffee attacked by *Colletotrichum coffeanum*.

6.1.2 Structure of Epidermal Cell Wall

The thick and tough outer wall of epidermal cells forms an important barrier for certain pathogens. **Lignification**, or the presence of silicic acid in the epidermis of some plants, acts as an important structural defense mechanism. In the case of the rice-blast fungus *Pyricularia grisea* (*oryzae*), the outer walls of most of the epidermal cells of the leaves are lignifed and the entry of the pathogen inside the host is generally through the thin-walled motor cells having pectin rather than lignin. Suberin deposition in potato tuber cells enables better resistance to penetration by some pathogens.

6.1.3 Structure of Natural Openings

Most pathogens enter plants through natural openings. In the case of the stem rust of wheat caused by *Puccinia graminis tritici*, the varieties in which stomata open late in the morning are resistant because the germ tubes of the spores germinating in the night dew desiccate. This is due to the evaporation of the dew before the stomata begin to open. This is sometimes called **functional resistance**. The kind of structure of the stomata of the host gives resistance to penetration in certain pathogenic bacteria. *Xanthomonas campestris* pv. *citri* is responsible for the citrus canker disease and affects grapefruits and mandarin oranges when artificially introduced into them. The mandrin variety, Szinkum, is resistant to the citrus canker bacterium because it possesses a broad cuticular ridge projecting over the stomata and a narrow slit leading to the stomatal cavity so that the raindrops containing fungal spores and bacteria rarely reach the mesophyll of the leaf (Fig. 6.1).

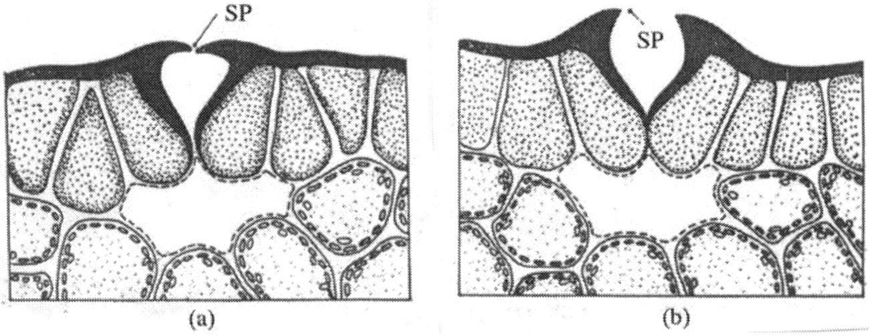

Fig. 6.1 Stomata of Szinkum Mandarin: (a) Resistant to the citrus canker bacterium (*Xanthomonas campestris* pv. *citri*) of a variety (b) Susceptible to the pathogen. SP = stomatal pore

6.1.4 Internal Structural Barriers to Pathogen Invasion

In certain varieties of wheat, the presence of bundles or increased areas of sclerenchyma cells prevent infection in cereal rusts.

6.1.5 Defense Structures Formed After Infection

Defense structures are produced inside the host plant to block or prevent the spread of the pathogen such as the formation of cork layers, tyloses, and abscission layers; deposition of gums, resins, and tanninlike substances; swelling of the cell walls, sheathing of hyphae, etc.

6.1.6 Formation of Cork Layers (Suberization of Tissues)

In the Rhizoctonia diseases of potato tubers, following infection, cork layers are produced just below the areas of infection. This layer prevents further invasion by the pathogen (Fig. 6.2). It blocks the spread of toxic substances produced by the pathogen and checks the flow of nutrients and water from the healthy to the infected area and this results in the starvation of the pathogen. The speed of the formation of the defense barrier is also important. Cork layers are also formed between healthy and diseased areas on *Prunus domestica* leaves attacked by *Coccomyces prunophorae* (Fig. 6.3), common scab of potato (*Streptomyces scabies*) and necrotic lesion of tobacco mosaic virus.

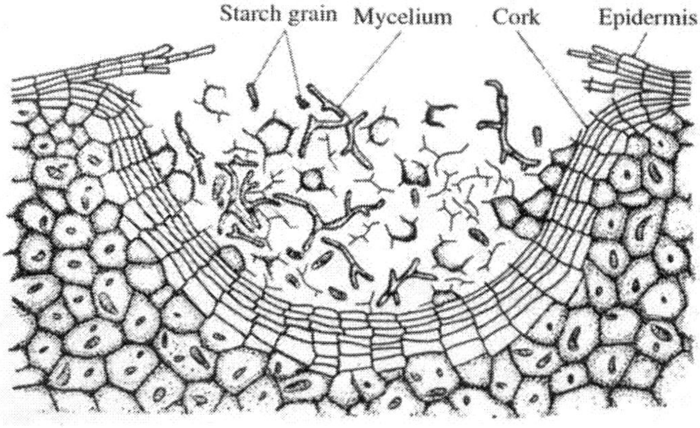

Fig. 6.2 Formation of cork layer on potato tubers following infection with *Rhizoctonia*

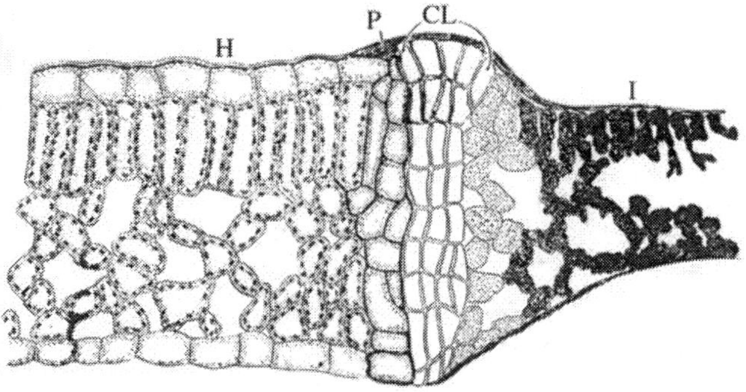

Fig. 6.3 Formation of cork layer between infected and healthy areas of Prunus leaf. CL = Cork layer, H = Healthy tissue, I = Infected, P = Phellogen

6.1.7 Formation of Tyloses

Tyloses are the overgrowths of the protoplast of adjacent living parenchymatous cells which protrude into xylem vessels through half-bordered pits and effectively clog xylem vessels. Tyloses are usually considered to be one of the factors which cause wilt. The time and rapidity of the formation of tyloses determine whether their role will be defensive or whether they themselves will be one of the factors for the causation of the disease (Fig. 6.4). In sweet potato wilt caused by the fungus *Fusarium oxysporum* f. sp. *batatas*, tyloses in some varieties are formed abundantly and quickly before the pathogen, and thus bring about resistance as this prevents the further spread of the pathogen. Evidence for the importance of vessel tyloses as a defense mechanism is available for vascular wilt of tomatoes.

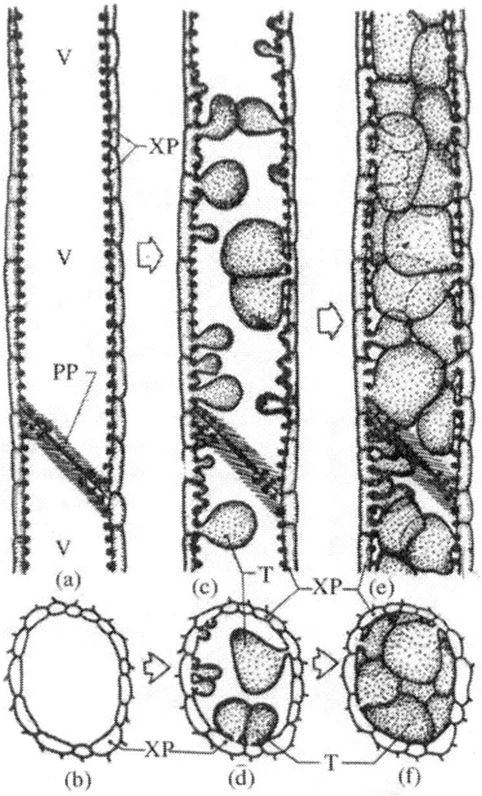

Fig. 6.4 Development of tyloses in xylem vessels: (a) Longitudinal and (b) Cross section view of healthy vessels (c) Longitudinal and (d) Cross section of vessels with tyloses (e) Longitudinal and (f) Cross section of vessels completely clogged with tyloses. PP = perforation plate, V = xylem vessels, XP = xylem parenchyma cell, T = tyloses

6.1.8 Formation of Abscission Layers

As a result of infection by several fungi, bacteria, or viruses, abscission layers are formed in the young active leaves of stone fruits. Due to the formation of abscission layer (upon infections) in peach leaves by the bacterium *Xanthomonas arboricola* (formerly *X. campestris*) pv. *pruni* or the fungus *Clasterosporium carpophilum* and in sour cherry trees due to infection with necrotic ring spot and other viruses, there is the swelling of the two layers of cells surrounding the infected spot. They become thin-walled while the pectic materials of the middle lamellae get dissolved, which results in the formation of abscission layers (Fig. 6.5). A shot-hole effect results from the development of an abscission layer which extends from the upper to the lower epidermis around the infected tissue and causes the latter to drop out, as in the shot holes of peach caused by *Clasterosporium carpophilum*.

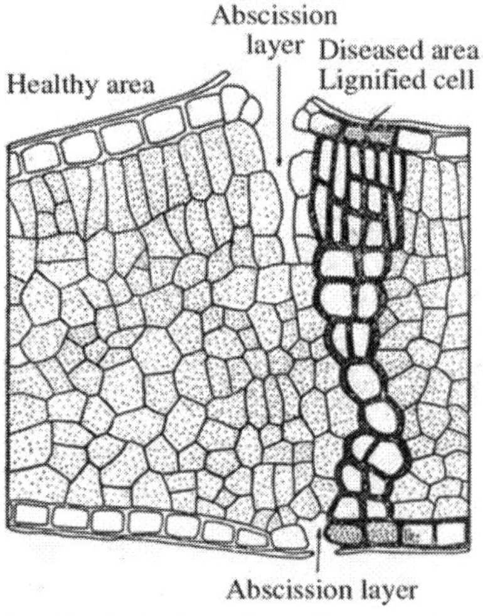

Fig. 6.5 Abscission-layer formation around a diseased spot of a *Prunus* leaf

6.1.9 Gum Deposition

Gum deposition along the borders of diseased lesions often serves as a protective demarcation and constitutes a type of mechanical resistance. In the silver-leaf disease in plums caused by *Stereum purpureum*, the wood of the host resistant variety,

Pershore, produces so much gum in advance of the fungus in a relatively short time that the fungus becomes completely enclosed by an impenetrable gum barrier in comparison to the susceptible Victoria variety. Hesler (1916) has found a similar gum barrier in the diseased part of the apple twig infected with *Physalospora cydoniae* (Fig. 6.6). In varieties of rice resistant to blast disease or *Helminthosporium* leaf spot, gum deposits are formed in the intercellular spaces that aid in restricting the fungus to the area of primary invasion. These deposits are found to be highly developed in Shoemed rice (resistant to helminthosporiose).

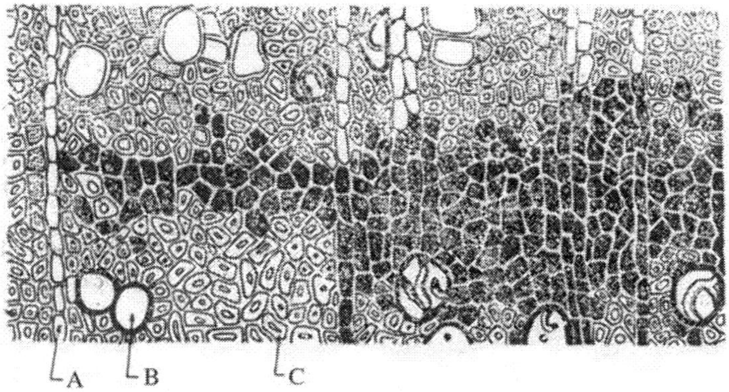

Fig. 6.6 Gum barrier in the apple twig infected by Physalospora cydoniae.
A = wood parenchyma, B = xylem vessels, C = wood fibre

6.1.10 Sheathing of Hyphae

It has been seen that the hyphae penetrating the cell wall are often enveloped in a sheath formed by the extension of the cell walls and prevent the pathogen from spreading (Fig. 6.7). The nature of the sheath is not clear. It consists of cellulose, callose substances, or other materials, and in some cases, a sheath

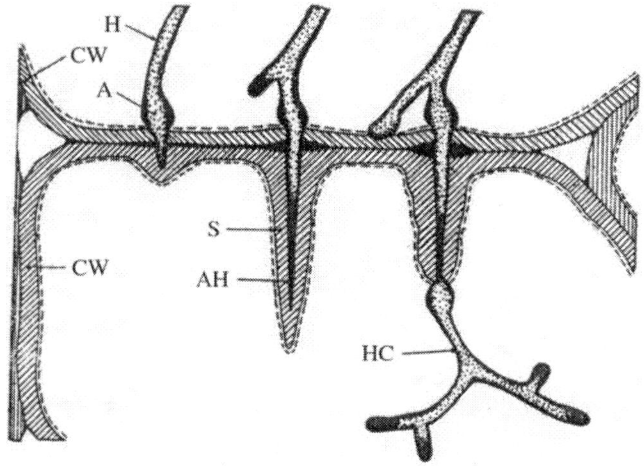

Fig. 6.7 Formation of sheath around hypha penetrating cell wall. CW = cell wall, H = hyphae, A = appresorium, AH = advancing hyphae still enclosed in sheath, HC = hypha in cytoplasm, S = sheath

may be the deposition product of the cytoplasm rather than the cell wall. Lignitubers may develop as protuberances on the inner wall opposite the site of penetration. Sometimes, the cell wall being penetrated by the infection peg swells and becomes suberized or lignified.

6.2 BIOCHEMICAL DEFENSE

6.2.1 Pre-existing Biochemical Defense Mechanism

1. Inhibitors Released by the Plant in its Environment

In onion smudge caused by *Colletotrichum circinans*, varieties with red scales were resistant in contrast to varieties with white scales. This resistance is due to the fact that the onion varieties with pigmented outer scales contain catechol and protocatechuic acids which are phenolic compounds and which diffuse out from the dead red scales and prevent spore germination. The blight-resistant varieties of *Cicer arietinum* to *Mycosphaerella* have more glandular hairs, which secrete malic acid on the leaves instead of the susceptible varieties. A higher concentration of malic acid inhibits spore germination and retards the hyphal growth of the fungus.

Fungitoxic substances have been detected in plant waxes, and cutin acids of citrus lime are toxic to *Gloeosporium limetticola*. The root exudates also protect plants against diseases. It was reported that the root exudates from the resistant varieties of flax excrete a glucoside, which on hydrolysis produces hydrocyanic acid—inhibitory to the soil inhabitants but less toxic to *Trichoderma viride*. Resistant varieties have *T. viride*, which helps flax to resist wilt caused by *Fusarium oxysporum* f. sp. *lini*. *T. viride* was earlier reported to produce two antibiotic substances, gliotoxin and viridian, but now it is known that the antagonistic effect of *T. viride* is not due to these two antibiotic substances which, in fact, it does not produce (Webster and Lomas, 1964). The antibiotic action of *T. viride* is due to the production of a volatile antibiotic substance known as **trichodermin**.

2. Inhibitory Substances Present in the Plant Cells before Infection

In potato scab caused by *Streptomyces scabies*, higher concentrations of chlorogenic acid have been reported in potato tubers and in citrus lime leaves. It was suggested that isopimpinellin (a furocoumarin) gives resistance to citrus lime leaves and protects them against *Gloeosporium limetticola*.

Turner (1961) reported that oat leaves and roots contain a fluorescent glucoside (avenacin) which is inhibitory to the growth of several fungi. The growth of *O. graminis*, the cause of the take-all disease of wheat, is also inhibited by the sap exuded from the oat roots, but the growth of *O. graminis* var. *avenae* is not. Yet the purified inhibitor is equally toxic to both the strains. On the other hand, Turner (1961) has shown that a specific glucoside (avenacinase), produced by the strain var. *avenae*, is pathogenic to oats while the wheat race completely inhibited by 5 μg/ml avenacin, lacks this enzyme. This enzyme destroys the biological activity of the inhibitor and hence the fungus is able to infect the host. Thus, the pathogenic ability and specificity of *O. graminis* var. *avenae* to oat is dependent on the inactivation of the inhibitor. According to Bergman (1966), the infection of tulips by *F. oxysporum* f. sp. *tulipae* mainly takes place during the last week before harvesting. The phase of susceptibility of tulip bulbs to *F. oxysporum* coincides with the drop in the tuliposide content.

In the case of resistance of pears to fire blight caused by *Erwinia amylovora*, it was seen that resistance in certain tissues is related to the presence of the phenolic glucoside **arbutin**. This compound is hydrolyzed by β-glucosidase of host origin to yield glucose and an antimicrobially active semiquinone. Arbutin is uniformly distributed in the tissues of the pear tree, yet these tissues vary in their susceptibility to *E. amylovora*. These investigators positively correlated those portions of the tree that were resistant (floral bracket) with the high levels of β-glucosidase activity. Likewise, those parts of the tree that were the most susceptible (nectaries of flowers) had decreased levels of β-glucosidase activity.

According to Kuć (1963), **phloridzin**, the glucoside of phloretin, is present in the leaves of both resistant and susceptible apple varieties. There is no correlation between the contents of phloridzin and resistance to pathogens.

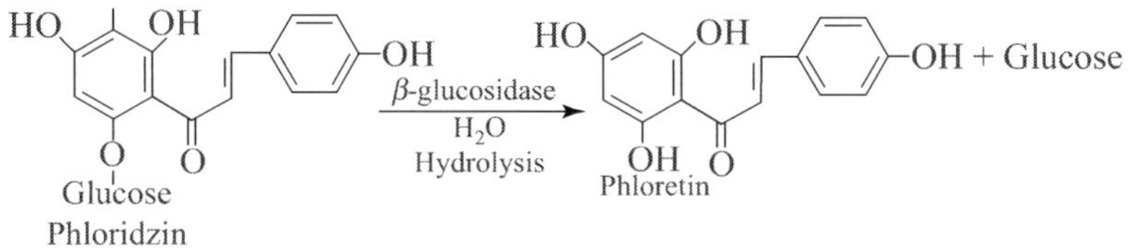

Hydrolysis of phloridzin to phloretin by β-glucosidase

When the pathogen penetrates the leaves of a resistant variety, the plant cells around the point of penetration collapse and die. Apparently, the nontoxic glucoside phloridzin is hydrolyzed by a β-glucosidase to yield phloretin and glucose. The phloretin in turn is oxidized by phenoloxidases to yield highly fungitoxic compounds. In susceptible varieties, the fungus penetrates the leaves and grows extensively below the cuticle for 10 to 14 days without initiating the collapse of the host cells. Thus, phloridzin is not hydrolyzed and the pathogen is not contained.

6.2.2 Phenolic Substances

Resistance to several fungal plant pathogens has been ascribed to higher concentrations of fungitoxic phenolic substances and their oxidation products and to increased Polyphenol-Oxidase (PPO) activity which generally, but not invariably, results from infection. In the case of *Puccinia coronata*, Dufrenoy (1936) observed that when the progress of some pathogens is checked, phenolic compounds (mostly tannins of the gallic group), develop in abundance within the vacuoles of adjacent cells, whereas in specialized parasites tending to become systemic, tannin is formed very slowly, in small amounts and only in vacuoles close to the parasite.

The activity of PolyPhenolOxidase (PPO) seems to be important because it can oxidize phenolics to quinones which may be more fungitoxic. Polymerization into dark-coloured melanins may then occur. PPO produced by the pathogen may oxidize the host polyphenols to more highly fungitoxic substances which, may prevent further development of the pathogens. For example, a PPO inhibitor reduces the toxicity of catechol to *Cochliobolus miyabeanus* (leaf blight of rice) and prevents melanin development, presumably by preventing enzymic conversions of catechol into fungitoxic quinones. The melanin is apparently fungitoxic although in other cases insoluble pigments are relatively nontoxic to fungi. Germinated urediniospores of *Puccinia graminis* release phenols and PPO and resistance of barley plants to *Blumeria* (*Erysiphe*) *graminis* appears to be associated partly with the collapse of the mesophyll cells with an accompanying release of a phenolic substance which accumulates around the haustoria and inhibits further development of the fungus. Chlorogenic acid is partly responsible for the resistance of potato tubers to scab (*Streptomyces scabies*), perhaps by stimulating protective cambium activity.

6.2.3 Absence of Nutrients Required by the Pathogen

Unless the host tissue supplies all the nutrients necessary for the growth of a pathogen, the latter will not grow and cause infection. Walker and Stahmann (1955) suggested that in such pathogens, the availability of highly elaborate growth factors, such as vitamins, polypeptides, amino acids and proteins, and the presence of complex enzyme systems may be essential for successful parasitism, and that resistant plants may lack one or more of these factors. Some mutants of *Venturia inaequalis,* which were deficient in choline and ribloflavin, were unable to cause scab unless these substances were supplied. A total of 54 biochemical mutants including those in which synthesis of various vitamins, organic bases, and amino acids has been impaired have been studied in *Venturia inaequalis*, and the effect of each mutation on the pathogenicity of the fungus has been determined. Of the vitamins requiring mutants, only those in which metabolic reactions involving choline or riboflavin have been affected were found to have lost in pathogenicity. All mutants in which the metabolism of purines, pyrimidines, and amino acids had been affected, incited only the fleck reaction and were nonpathogenic.

Lukezic and De Vay (1964) have correlated the resistance of plum varieties to *Rhodosticta quercina* (canker) with the amount of myoinositol in the bark of trees. This fungus is naturally deficient in myoinositol, and more of this was present in the bark of the susceptible cultivar (President) than in the bark of the two resistant cultivars (Beauty and Duarte); further, the resistance of Duarte was reduced by adding myoinositol to the bark tissue.

High sugar content of the tissues favors some pathogens, such as certain rusts and powdery mildews, whereas the reverse is true for other, including *Alternaria solani* (early blight of potato) and *Ophiostoma* (*Ceratostomella*) *ulmi* (Dutch elm disease). Plants presumably become resistant to high-sugar pathogens when the sugar content of their tissues falls to a certain level and this may be brought about, for example, by treatment with certain growth-regulating substances. Boron facilitates the translocation of sugars so that these tend to accumulate in the leaves of boron-deficient plants and the latter are apparently resistant to *Cochliobolus sativus* (barley leaf spot), a low-sugar pathogen, but susceptible to *Blumeria* (*Erysiphe*) *graminis*, a high-sugar pathogen. *Erwinia carotovora* var. *atroseptica*, which causes bacterial soft rot of potatoes, is less severe on potatoes with low reducing sugar content than on potatoes high in reducing sugars.

Beever (1970) found that the xylem sap of trees (peach, plum) susceptible to *Stereum purpureum* contained more nitrogen and carbohydrate and was more favourable for the growth of the fungus than was the sap of resistant trees (apple, pear). There were also qualitative differences in that the main sugar and sugar alcohol in the sap from resistant trees were sucrose and sorbitol, neither of which supported good growth of *S. purpureum*, whereas the sap of susceptible trees also contained glucose and fructose which were more favourable for growth of the fungus.

6.2.4 pH

There are few convincing cases of disease resistance directly attributable to pH, although the resistance of ripe tomatoes to *Xanthomonas campestris* pv. *vesicatoria* (*Bacterium vesicatorium*) may be due to their high acidity (pH 4-4.5). The pH of the plant tissues influences many physiological processes in both plants and parasites, notably enzyme activities, and resistance to certain pH sensitive pathogens might be related to unsuitable pH of the plant tissue.

6.2.5 Osmotic Pressure and Permeability Effects

The cells of parasitic fungi have unusually higher osmotic pressure than that of the surrounding host cells, this being necessary for absorption of water from the latter. In the case of *Erysiphe cichoracearum* (lettuce mildew), cells of resistant tissues or varieties have a higher osmotic pressure than those of the susceptible ones. Further, resistance can be modified by altering the mineral nutrition of the plant and shaded leaves or those cultured on distilled water become susceptible, whereas young lettuce seedlings with high osmotic pressure are resistant.

6.2.6 Defense through Absence of Common Antigens

The gene-for-gene concept of the relationship between the host and pathogen evolved from the work of Flor (1955) with flax rust. The idea that the antigens shared by the host and pathogen may determine the nature of disease reactions has been explored by Doubly *et al.* (1960). Tests of globulin antigens from four races of rust and four from near isogenic lines of flax suggested that susceptibility depended on the presence, in the host, of a protein serologically similar to the one possessed by the pathogen. Some support for the common antigen hypothesis has been obtained from work on non-obligate parasites. Pathogenic races of *Xanthomonas campestris* pv. *malvacearum* which cause the angular leaf spot of cotton, had more antigens in common with those of cotton leaves than did the nonpathogenic races of bacteria and the black-rot fungus *Ceratocystis fimbriata* which attacks sweet potato.

6.2.7 Phytoalexins

The term was introduced by Müller and Börger (1940) to describe the substances that inhibit fungus development formed when living plant tissues are invaded by a fungal parasite. It has since been redefined and broadened to include substances formed by host tissue "....... *in response to injury, physiological stimuli, infectious agents or their products ..... (that) accumulate .... to levels which inhibit the growth of microorganisms*" (Kuc, 1972). Day (1973) proposed to include substances that are formed in response to infection that are toxic to, or which repel insect or nematode parasites. Phytoalexins can be described as low-molecular-weight antimicrobial compounds that accumulate in plants as a result of infection or stress. They may be associated with host-cell necrosis (hypersensitivity) at the infection site but not necessarily so when induced by chemicals. It was envisaged that the factor which induces the formation is released by the parasite into the host cell.

The initial work of Müller and Börger (1940) laid the foundation of the concept of phytoalexins. The conclusions are given below:

1. A principle designated as phytoalexin which inhibits the development of the fungus in hypersensitive tissues is formed or activated only when the host cells come into contact with the parasite.
2. The defense reaction occurs only in the living cells.
3. The inhibitory material is a chemical substance and may be regarded as a product of necrobiosis of the host cell.
4. The phytoalexin is nonspecific in its toxicity towards fungi; however, fungal species may be differentially sensitive to it.
5. The basic response that occurs in resistant and susceptible hosts is similar. The basis of differentiation between resistant and susceptible hosts is the speed of formation of phytoalexin.
6. The defense reaction is confined to the tissue colonized by the fungus and its immediate neighbourhood.
7. The resistant state is not inherited.

Phytoalexins are probably fungistatic rather than fungicidal but this no doubt depends partly on the quantity present.

6.2.8 Ipomeamarone

This is an abnormal sesquiterpinoid (tetrahydrofuran, $C_{15}H_{22}O_3$, mol. wt. 230) which seems to be induced in sweet-potato (*Ipomea batatas*) tissue infected with the black-rot fungus *Ceratocystis fimbriata* by a possible disturbance of carbohydrate and fatty acid metabolism of the host. The biosynthesis of ipomeamarone may involve mevalonic acid as a precursor and could perhaps also be induced by poisonous chemicals and pectinase. Ipomeamarone increases substantially in diseased tissues adjacent to infected tissues. Such tissues show terpenes which are also produced by slices of sweet potato in response to a chemical stimuli. The interesting part is that ipomeamarone is toxic to both nonpathogenic and pathogenic fungi and to some bacteria, and there seems to be no correlation between its inhibitory effect and pathogenicity.

6.2.9 Pisatin

Pisatin has the chromocoumarin ring system and is a phenolic ether ($C_{17}H_{14}O_6$, mol. wt. 314). It was isolated from the pods of the garden pea inoculated with fungi, using the drop-diffusate technique.

When different fungi, pathogenic and nonpathogenic to pea, are used for inoculation, they induce the production of different concentrations of pisatin. The nonpathogenic fungi induce the formation of pisatin at concentrations high enough to inhibit their growth. If pea pathogens induce the formation of pisatin, it is at concentrations much below these that are toxic to the pathogen. Production of pisatin by pea pods inoculated with *Monilinia fructicola*, a nonpathogen, is reduced or delayed at high temperature on anaerobic storage, and the pea pods become susceptible to this fungus, while storage at normal temperature results in the accumulation of high levels of pisatin and resistance.

Pisatin can also be formed when dilute solutions of certain metallic salts (especially those of mercury, silver, and copper) are placed on pea-pod endocarp tissues and it is possible that phytoalexins are produced when metallic fungicides are applied to crops. Certain metabolic inhibitors, such as sodium fluoride and sodium monoiodoacetate, are reported to have similar effects. Other substances reported to induce pisatin formation by pea pods include ethylene which at a concentration of 1 ppm induces a slight production. At certain concentrations, a number of antibiotics (actinomycin D, cycloheximide, chloramphenicol, and others) and ribonuclease stimulate pisatin formation.

Some other important phytoalexins are Phaseollin, Glyceollin, Trifolirhizin, Orchinol, Hircinol, Isocoumarin, Rishitin, and Gossypol.

6.3 DEFENSE THROUGH INDUCED SYNTHESIS OF PROTEINS AND ENZYMES

In addition to the phenolic substances produced in plant tissues in response to infection, induced synthesis of proteins and enzymes seems to play a role in disease resistance. In the case of the black rot of sweet-potato caused by the fungus *Ceratocystis fimbriata*, sweet-potato tissues inoculated with the pathogen and with a nonpathogen that induces immunity, showed marked changes in 9 of the 13 enzymes. Nucleic acid and protein changes in wheat-leaf nuclei during rust infection were studied and there were decrease in histones and marked increase in RNA and acidic proteins as early as two days after inoculation in the nuclei of the mesophyll cells of a susceptible wheat infected with *Puccinia graminis tritici*. Similarly, pectic enzyme formation by *Fusarium* or *Verticillium* was greatly reduced when the pathogens were grown on tissues from resistant varieties as compared to that from susceptible varieties.

In numerous incompatible host-parasite combinations, peroxidase activity is often several times higher than in compatible ones while in some cases peroxidase activity is greater in compatible reactions. It is, therefore, clear that the activation of peroxidase in the host and parasite complex need not always be accompanied by the incompatible reaction.

Plants defend themselves from pathogen infection through a variety of mechanisms that can be either local or systemic, constitutive or inducible. In most cases, the extent to which these various systems affect disease progression is poorly understood and is an area of extensive biochemical and genetic research (Kessman *et al.*, 1994). One particular inducible systemic response, known as **Systemic Acquired Resistance (SAR)** has become a subject of great importance and enquiry. SAR is a broad, physiological immunity that develops from infection with a necrogenic pathogen. In addition to this, certain natural and synthetic chemical compounds can trigger similar plant responses. A carefully controlled laboratory study of SAR was published in 1961 when Ross (1961a, 1961b) used Tobacco Mosaic Virus (RMV) on a local lesion host (*Xanthi-nc* tobacco) and found that subsequent infections were reduced in severity. Ross coined the term SAR for the resistance that developed in the distal, unheated portions of TMV-inoculated plants. Resistance that developed in the same leaf that had been TMV-inoculated, was named **Local Acquired Resistance (LAR)** by him. Cruickshank and Mandryk (1960) showed high level of resistance in field-grown tobacco against blue mold (*Peronospora tabacina*) when plants were injected in the stem with the spore suspension of the same fungus. Cohen and Kuc (1981) subsequently showed that resistance takes approximately 3 weeks to develop. Heat-killed conidia and other pathogens of tobacco failed to induce SAR. Other workers have shown that inoculation of tobacco leaves with *Thielaviopsis basicola* leads to necrosis and subsequent systemic resistance to other pathogens, TMV, and Tobacco Necrosis Virus (TNV). The SAR response in tobacco gives a broad spectrum disease resistance to fungal, bacterial, and viral pathogens but is ineffective against aphids and tobacco horn worm (*Manduca sexta*).

In monocolyledonous plants, there are well-developed SAR systems established both in rice and wheat. In rice, Smith and Metraux (1991) reported that the inoculation of the first, fully developed leaf with *Pseudomonas syringae* pv. *syringae*, which causes a rapid hypersensitive necrosis results in systemic resistance against *Pyricularia grisea* (*P. oryzae*), the causal agent of blast of rice. There are several other examples of SAR and LAR in various host-pathogen combinations.

6.3.1 Mechanism of SAR (Self-Acquired Resistance)

In 1970, the groups of Van Loon (1970) and Gianinazzi (1970) showed that viral infection of tobacco induced the accumulation of a distinct set of proteins called Pathogenesis-Related proteins (PR proteins)

6.3.2 PR-Proteins

Pathogenesis related proteins, often called PR-proteins, are generally a group of plant proteins that are toxic to invading fungal pathogens or other pathogens. Several groups of PR proteins have been classified according to their function, serological relationship, amino acid sequence, molecular weight, and certain other properties. PR proteins are either extremely acidophilic or extremely basic and, therefore, are highly soluble and reactive. Some better known PR proteins are PR1 proteins, β,1,3,glucanases, chitinases, lysozymes, PR4 proteins, thaumatin-like proteins, osmotin-like proteins, cystein-rich proteins, glycine-rich proteins, proteinase inhibitors, proteinases, chitosanases, and peroxidases (Agrios, 1997).

Although healthy plants may contain trace amounts of several PR proteins, when attacked by plant pathogens. Treatment with defense elicitors, wounding or stress induces transcription of a battery of genes that code for PR proteins. It has been shown that β-1,3 glucanase and chitinase diffuse towards and affect (the breakdown) the chitin-supported structure cell walls of several but not all pathogenic fungi, whereas lysozymes degrade the glucosamine and muramic acid component of the cell wall of bacterial plant pathogens. In those plants which have been genetically engineered and in which the chitinase gene has been introduced, show good resistance against *Rhizoctonia solani.*

Some characteristics of systemic acquired resistance (SAR) are the following:

1. It is induced by agents/pathogens causing necrosis, e.g. local lesions.
2. There is a delay of several days between induction and full expression.
3. Production is conferred on tissues not exposed to induced inoculation
4. It is expressed as reduction in lesion number, size, spore production, etc.
5. Production is long lasting.
6. Production is nonspecific.
7. Development of SAR is associated with expression of several gene families, e.g. PR proteins.
8. The signal for SAR is translocated and graft transmissible.
9. The protein is not passed on to the next progeny, and the transmission to clonal tissues is unresolved so far.

Most current interest in SAR concerns the possible identity of the translocable signal. It has been known for some time that the signal is graft transmissible and presumably moves in the vascular system.

In addition to developing as a result of a primary infection, resistance can be induced by treatments of plants with such compounds such as viral coat, proteins, bacterial or fungal proteins, lipoproteins, polysaccharides, and yeast RNA; and by synthetic molecules mostly polyanions such as polyacrylic acid, salicylic acid (Aspirin) and 2-chloroethyl phosphonic acid, 2,6-dichloro-isonicotanic acid (CGA-41396) and its methylated ester (CGA-41397), both referred to as INA. In general, the greater the concentration of the inducer, the faster and more effective the induced resistance in the host.

Localized induced resistance in inoculated (treated) leaves develops 2-3 days after primary infection (treatment) while systemic induced resistance, if present, usually develops 7 or more days after the primary infections and may persist for 3 to 5 weeks. It is interesting to note that induced resistance may be prevented if the plants are treated with Actinomycin D. Actinomycin D prevents transcriptions of the cell DNA to messenger RNA and, therefore, prevents productions of new proteins (enzymes).

Further evidence showing that salicylic acid plays an important part in the signal pathway has come from the work on *Arabidopsis* mutants altered in SAR expression and transgenic plants engineered to break down salicylic acid.

The signal molecules implicated in the induction of plant defense are salicylic acid, acetyl salicylic acid, dichloro-iso-nicotinic acid, ethylene, xylanase, the polypeptide systemin, jasmonic acid, and probably others.

6.4 DEFENSE THROUGH FORMATION OF SUBSTRATES RESISTING THE ENZYMES OF THE PATHOGEN

The resistance of green immature apple fruit to several fungi which can infect mature fruit has been attributed to the presence in the green fruit of a pectin-protein-polyvalent cation complex which is resistant to hydrolysis by fungal hydrolytic enzymes and suppresses the production of pectin methyl esterase and polygalacturonase by the fungi.

An induced mechanism of resistance through increased resistance of the substrate to enzymes by the accumulation of polyvalent cations has been reported for *Rhizoctonia* infected bean tissues. In the resistant variety, calcium ions accumulate in and immediately around developing *Rhizoctonia* lesions, inhibit further tissue maceration by polygalacturonase, and therefore, play a role in confining the pathogen to lesions of a limited size. Fungal PME demethylates pectic substances near and in advance of the pathogen and results in the formation of pectate salts with polyvalent cations, particularly calcium, which in turn are resistant to hydrolysis by fungal polygalacturonase.

Disease resistance introduced through artificial application of auxins to infected plants has been reported for Fusarium wilt of tomato and late blight of potato. The resistance to disease in auxin-treated plants is accompanied by higher levels of insoluble pectates and lower levels of water-soluble pectins that are found in untreated, susceptible plants.

The resistance of cider apples to wound infection by the brown-rot fungus *Sclerotinia fructigena* seems to be due to the ability of phenolic oxidation products in these apples to inactivate extracellular pectinolytic enzymes produced by the fungus. The phenolics produced in the juice of the resistant variety do not inhibit the growth of the fungus, but do inhibit the activity of its polygalacturonase.

6.5 DEFENSE THROUGH DETOXIFICATION OF PATHOGEN TOXIN

Detoxification may operate in resistance of vascular wilt diseases caused by *Fusarium oxysporum*. Its significance depends on the importance properly ascribed to toxins in these diseases. Lakshminarayanan (1955) reported that varieties of cotton resistant to *F. oxysporum* f. *vasinfectum* contain relatively high concentrations of cystine and that cystine reduces the toxicity of the complex formed when ferric iron is added to filtrates from the cultures of the parasite. When shoots are allowed to absorb cystine from solutions before the complex is transferred to the former, they are less damaged than are plants not pre-treated with cystine.

In the case of the toxin piricularin, Tamari (1959) found that chlorogenic acid could detoxify the piricularin when applied to rice seedlings in equimolar concentrations. Tamari *et al*. (1962) also showed that piricularin was detoxified with ferulic acid, which was oxidized by peroxidase, and its oxidized products were also shown to detoxify the piricularin. Those varieties of rice which have more chlorogenic or ferulic acid are more resistant to the blast disease.

6.6 DEFENSE THROUGH ALTERED RESPIRATION

The relation of increased respiration to disease resistance and susceptibility is unclear. In some incompatible host pathogen combinations (resistance), the increase is initially greater than in compatible combinations (susceptibility), although the increase may be of greater total magnitude and longer in duration in susceptible reactions.

From observations, however, indicating that substances (narcotics) inhibiting respiration also decrease resistance to disease in the host, it is suggested that increased respiration is often a necessary condition for the development of defense reaction in the host. A marked increase in oxygen uptake has been reported for the hypersensitive defense reactions. This is also accompanied by the accumulation and oxidation of phenolic compounds and by the activation of phenoloxidases.

6.7 DEFENSE THROUGH HYPERSENSITIVE REACTION

Hypersensitivity reaction, or hypersensitivity, functions as a defense reaction in plants against pathogens. It occurs only in incompatible host-parasite combinations. This is a significant and common form of reaction in plants. Usually, a small part of the plant tissue quickly dies after infection, resulting in a minor or local lesion or one as extensive as to cause the death of the plant. Hypersensitive reaction is believed to be caused by the disturbance of the balance between the oxidative and reductive processes in the cells resulting in excessive oxidation of phenolic substances and leading to the breakdown of cellular structures (Fig. 6.8).

The hypersensitive response, the culmination of the plant-defense response, indicates the recognition by the plant of specific pathogen-produced signal molecules (elicitors). Recognition of the elicitors by the invaded host plant elicited a cascade of biochemical reactions in the attacked plant and surrounding cells leads to altered cell function and new or greatly activated defense-related compounds.

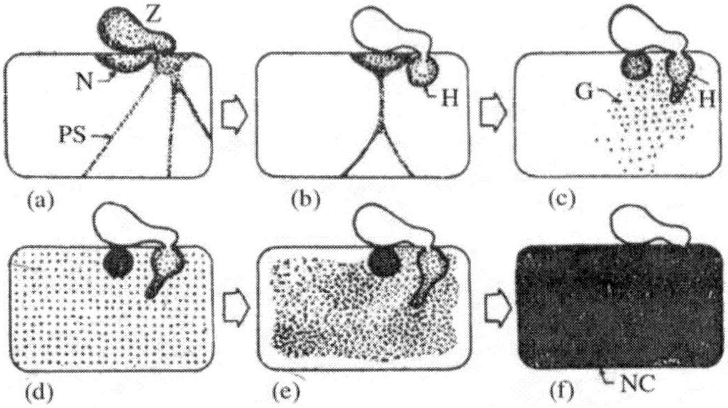

Fig. 6.8 Different stages in the development of necrotic defense reaction (hypersensitivity) in a cell of a very resistant potato variety infected by *Phytophthora infestans*. N = nucleus, Z = zoospores, PS = protoplasmic strands, H = hypha, G = granular material, NC = necrotic cell (Redrawn from K. Tomiyama (1956), *Ann. Phytopath. Soc.*, Japan. 21:54-62).

6.8 THE CONCEPT OF PHYTONCIDES

Phytoncides, by definition, include phytoalexins, and pre-formed and post-infectionally formed substances which give resistance to plant pathogens, in addition to those which deter or destroy insect pests.

It is being felt by plant pathologists that the term is too all-embracing to be of much use and the individual categories of chemical substances known so far are adequately covered by more specific terms.

6.9 DEFENSE THROUGH GENES

It must be clear that whatever the kind of defense or resistance a host plant shows against infection, or against abiotic factors, it is finally controlled by genes or genetic material of the host plant and of pathogen.

It is easy for the plant to defend itself when it comes in contact with a pathogen to which the plant is not a host and it is called **nonhost resistance**. For example, apple trees are not infected by the pathogens of wheat, tomato, or others because genetic constitution of apple is different for such infection to the pathogens of wheat and tomato. All such unsuccessful infection is due to nonhost resistance. It has been seen in some related pairing, e.g. when powdery mildew of wheat caused by *Blumeria graminis* f.sp. *tritici* inoculated on barley, the fungus produces haustoria as a result of infection, the host plant in response to infection produces hydrogen peroxide and as a result of hypersensitive response, the cells die and further infection is controlled. This is due to again nonhost resistance.

When a plant is attacked by its own pathogen, it may develop no disease, only mild disease or severe disease, depending on the genetic makeup (genes) of the plant and of the pathogen that attacks it. This difference of disease severity is mainly due to the large number of genes being involved in keeping a plant protected from pathogenic attack. Some of the genes provide protection to the plant for general upkeep and normal functioning of the plant. Some of the latter plant genes code for those chemicals that are toxic to pathogen or neutralize the toxin. The plant, in addition to this, has genes that slow down or inhibit the growth of the pathogen that cause disease. These structures, present in a plant naturally or may be produced as a result of host-pathogen interaction or by abiotic agents, are important in the defense mechanism. When a pathogen attacks a host plant, the genes of the pathogen are activated and release all their weapons at their command, i.e. toxins and enzymes against the plant, and infect it. As plants have some pre-existing chemical substances, plants defend themselves either partially or completely. Some of the plants that survive and produce yield, and have this type of defense are known as **polygenic**. This type of resistance which is present at different levels against pathogens in plants is called partial, horizontal, multigenic, or minor gene resistance. In some polygenic plant-pathogen combination like early blight of tomato caused by *Alternaria solani*, the resistant varieties accumulate pathogen-induced pathogenesis related (PR) proteins more rapidly than in susceptible varieties. These PR-proteins include some antifungal isozymes of chitinase and β-1,3-glucanase. In resistant varieties, total enzymes release elicitors of the Hypersensitive Response (HR) whereas enzymes of susceptible varieties do not. This suggests that production of hydrolytic enzymes may act as a mechanism of elicitor release in tomato resistance to the early blight.

In many plant-pathogenic combinations like downy mildews, powdery mildews, many bacteria, nematodes and viruses, defense of a host plant against pathogens is governed by a pair of genes for the disease in the host plant and the pathogen. The host plant carries one or few resistance genes (R) per pathogen while each pathogen carries matching genes of avirulence (A) for each of the R genes of the host plant. These avirulence genes of the pathogen trigger the host R genes, this in turn starting the defense mechanism of host plants that finally neutralize and eliminate the specific pathogen. During this process, only a few cells die. This type of resistance or defense is called **race-specific**, **major gene**, **R gene** or **vertical resistance**, or **monogenic**.

REVIEW QUESTIONS

1. The cuticle acts as a physical or chemical barrier to infection. Comment.
2. What are the different morphological defense mechanisms in plants? Give suitable examples.
3. What is biochemical defense mechanism?
4. Explain why in onion smudge caused by *Colletotrichum circinans*, varieties with red scales are resistant in contrast to varieties with white scales.
5. Give an example of a soil-borne disease in which root exudates are implicated in the protective mechanism.
6. What is the role of phenolic substances in the resistance mechanism in plants?
7. Give an example of a host-pathogen combination where
   (i) Absence of nutrients brings about resistance to disease
   (ii) Resistance is attributable to pH
   (iii) Resistance is attributable to osmotic pressure
8. Is the defense mechanism in plants akin to that in animals?
9. Who suggested the theory of gene-for-gene responsible for host specificity?
10. What are phytoalexins? Are phytoalexins responsible for host specificity?

11. What is hypersensitivity? How does it bring about resistance to pathogens?
12. How does altered respiration bring about resistance?
13. Give a summary of the potential ways in which metabolic pathways altered by infection play a role in disease and pathogen development.
14. What is the concept of phytoncides?
15. What is induced resistance? What is its possible mechanism and what are some of the induced chemical signals?

MULTIPLE-CHOICE QUESTIONS

1. Which of the following do not act as morphological defense mechanisms?
   (i) Waxes (ii) Cuticles
   (iii) Epidermal cells (iv) Phenolics
2. Which of the following do not act as internal structural barriers to pathogen invasion?
   (i) Cork layer (ii) Tyloses
   (iii) Gum deposition (iv) Pisatin
3. In onion smudge caused by *Colletotrichum circinas*, why were varieties with red scales resistant in contrast to varieties with white scales due to the presence of
   (i) Catechol (ii) Protocatechuic acid
   (iii) Both (i) and (ii) (iv) Gliotoxin
4. Which of the following inhibitory substance is present in the potato cells before infection in resistant varieties that account for resistance against *Streptomyces scabies*?
   (i) Avenacinase (ii) Chlorogenic acid
   (iii) Arbutin (iv) Phloridzin
5. Why does the activity of polyphenoloxidase (PPO) seem to be important? Is it because it can
   (i) oxidize phenolics to quinones
   (ii) oxidize phenolics to coumarin derivatives
   (iii) oxidize phenolics to aromatic amino acids
   (iv) oxidize phenolics to anthocyanins
6. Which of the following is not a phenolic compound involved in the defense mechanism of plants?
   (i) Caffeic acid (ii) Ferulic acid
   (iii) Phloretin (iv) Gliotoxin
7. Which of the following fungitoxic substances are not a steroid-glucoalkaloids present in the potato peel?
   (i) α-chaconine (ii) α-solanin
   (iii) Tomatine (iv) Both (i) and (ii)
8. Who coined the term *phytoalexins*?
   (i) Muller and Borger (ii) Cruickshank
   (iii) Kuć (iv) Hammerschmidt
9. Which of the following statement is not correct for phytoalexins?
   (i) It inhibits the development of the fungus.
   (ii) The defense reaction occurs only in the living cells.
   (iii) It is a chemical substance.
   (iv) The resistance is inherited.

10. Which of the following is not a phytoalexin?
    (i) Ipomeanmarone
    (ii) Pisatin
    (iii) Phaseollin
    (iv) Actinomycin-D

Answers

1. (iv) 2. (iv) 3. (iii) 4. (ii) 5. (i) 6. (iv) 7. (iii) 8. (i) 9. (iv) 10. (iv)

7

GENETICS OF PLANT PATHOGEN INTERACTION

7.1 GENETICS OF HOST-PARASITE INTERACTION

Each host-parasite interaction is a struggle for survival between two organisms. In nature, there exists a state of balance and surviving plants and their parasites/pathogens are capable of coexistence. The attack and defense mechanisms discussed earlier are determined genetically in the pathogen, some of which have different pathogenic abilities and may be able to attack hitherto resistant plants. In both organisms, the more effective races and varieties tend to survive, the less effective ones tend to disappear. Effective disease develops when a virulent race meets a susceptible variety but not when the pathogen is avirulent or the plant is resistant. The genetics of the host-parasite interaction, including molecular aspects, have been discussed by several workers in the field.

7.2 RESISTANCE AND SUSCEPTIBILITY

A plant is either immune to a pathogen, that is, it is not attacked at all by the pathogen even under the most favourable conditions, or it may show varying degrees of resistance ranging from immunity to complete susceptibility. Resistance may be conditioned by a number of internal and external factors which reduce the chance and degree of infection. There are several kinds of resistance: it may be physical or chemical, it may depend upon the lack of a nutrient or a substance that the parasite needs for development, or it may be due to a toxic or repellent substance that is either preformed or formed only in response to infection. Irrespective of whether the morphological or biochemical defense mechanism is involved in the resistance of a particular host-parasite combination, it is ultimately the genetic makeup of the two that determines whether a particular host will be susceptible or resistant under given environmental conditions. The mechanisms by which genes control the physiological processes that lead to disease resistance or susceptibility are not yet clear, but they are presumably no different from the mechanisms controlling any other physiological process in a living organism. Under the influence of the genetic material (DNA), a specific protein, which may be a structural protein or an enzyme, may be produced. The produced enzyme, of course, under certain conditions will participate or initiate biochemical reaction related to one or other cellular process, and may result in the production of certain morphological characteristics or accumulation of a certain chemical substance. Most of the approximately 100,000 known species of fungi are strictly saprophytic and degrade dead organic material for their nutrients. Only 8000 can cause disease to one or more species and only 100 are pathogenic to humans and animals. It is the same for all other groups of plant pathogenic organisms such as bacteria, viruses, and mycoplasmas. What then distinguishes this small minority of pathogenic fungi or other pathogens from the majority of saprophytic fungi or other microorganisms? The fungus *Fusarium oxysporum* f.sp *lycopersici* that causes tomato wilt attacks tomato and has absolutely no effect on wheat or mango. Similarly, the fungus *Phytophthora infestans*, causing late blight of potato and tomato, does not infect wheat or apple; and the fungus *Puccinia graminis tritici* infects wheat only and not apple or potato.

Interactions between a host and pathogen are mediated by environmental conditions but are ultimately determined by genotype. Flor (1956) studied the genetics of flax cultivars and isolates of the linseed rust pathogen *Melampsora lini* in field observations. He demonstrated a close genetic relationship between the host and pathogen and formulated the gene-for-gene concept, which is still regarded an important hypothesis in plant pathology. The concept states that for each gene conditioning avirulence or virulence in the pathogen, there is a corresponding gene conditioning resistance or susceptibility in the plant. Usually, although not always, genes for resistance in the host are dominant (R) and those for susceptibility (lack of resistance) are recessive (r). In the pathogen, genes for avirulence are dominant (A) and those for virulence (a) are recessive. Where the gene-for-gene hypothesis operates, therefore, four gene combinations are possible in host-pathogen interaction. These combinations are R-a, r-A, r-a, and R-A. Three combinations R-a, r-A, and r-a give rise to compatible reaction (susceptible) and infections are successful. The combination R-A results in an incompatible reaction and no infection occurs (resistant).

Table 7.1 *Quadratic Check of Gene Combinations and Disease Reaction Types in a Host–Pathogen System in Which the Gene-for-Gene Concept for One Gene Operates.*

	Resistance or susceptibility genes in the plant	
Virulence or avirulence genes in the pathogen	R (resistant) dominant	r (susceptible) recessive
A (avirulent) dominant	AR (–)	Ar (+)
a (virulent) recessive	aR (+)	ar (+)

Minus sign indicate incompatible (resistant) reaction and therefore no infection. Plus sign indicate compatible (susceptible) reaction, and therefore infection develops.

Genes for virulence appear in pathogens in response to new genes for resistance in the host. New resistance in the host leads to a situation which selects new virulence in the pathogen. Hosts and pathogens are in a continuing state of co-evolution. Exceptions to this hypothesis have been seen, particularly where virulence or resistance is controlled or modified by more than just a few genes.

It is also possible that not all host-pathogen interactions are represented in this way. According to Issac (1992), the gene-for-gene concept remains largely theoretical and provides information, which is important in epidemiological terms but it does not consider the biochemical or molecular basis of resistance.

Although a number of theories have been put forward to explain the specificity of host-pathogen interactions in biochemical terms, very little is known about the genes involved and their regulation. Few gene products have been analyzed but modern molecular biological techniques now provide means by which these factors may be investigated.

In an incompatible reaction, molecular recognition leads to the activation of a range of defense mechanisms. In a compatible reaction, however, recognition does not occur, or is suppressed, and successful host penetration and colonization of the host may be achieved by this invading pathogen. In an incompatible reaction, structural and biochemical host-defense mechanisms are initiated by the activation of the defense-response genes. Crop plants are endowed with numerous defense genes which encode various proteins responsible for synthesis and accumulation of defense arsenals in plants. The defense genes are involved in the production of antimicrobial compounds such as phenolics, phytoalexins, pathogen-related (PR) proteins, and active oxygen-species, or involved in the reinforcement of the cell wall by accumulating hydroxylproline-rich glycoproteins, callose, lignin, and wall-bounded phenolics. Some of them show antifungal action, some show elicitor-releasing activity, some increase the mechanical strength of host cell walls, and some others are involved in the release of secondary messengers and systemic signal molecules. While defense genes are functional genes, resistant genes are regulatory genes which regulate the functions of the defense genes. Defense genes are quiescent in healthy plants, but activated when the pathogens come into contact with the plants, releasing signals. These signal molecules are transferred to the host-plant nucleus through a signal-transduction pathway activating defense genes.

It is now established that signals are required to activate defense genes which are commonly present in both susceptible and resistant plants. There is no plant in the world without defense genes. These signal molecules have been detected in fungi, bacteria, and viruses, irrespective of whether they are pathogens or nonpathogens and are released by host enzymes when the pathogens come into contact with plants. Endogenous elicitors of host origin, **pectate fragments**, are released by pathogen enzymes. These signals have to be transferred to the plant nucleus to generate mRNA for the synthesis of various defense-related proteins.

Evidence for Specific Inducers of Resistance

Despite extensive research, there is little evidence for materials produced under the control of specific avirulence genes which, on the basis of Flor's concept, should have selective effects on resistant plants. Work in Holland which indicated that filtrates of avirulent cultures of a certain strain of *Venturia inaequalis* were selectively toxic to resistant plants was not confirmed in later work (Day, 1974).

7.3 THE COMMON ANTIGEN HYPOTHESIS

The idea that plants and pathogens which interact to give susceptible (compatible) reactions have more antigens in common than those that give resistant (incompatible) reactions was explored first with the flax-rust system. Further attempts to explore the hypothesis that antigens shared by the plant and pathogen may account for specificity have been carried out with several types of pathogens. Pathogenic races of *Xanthomonas campestris* pv. *malvacearum,* which cause angular leaf spot of cotton, had more antigens in common with those of cotton leaves than did nonpathogenic races of bacteria. However, in another system, Fusarium wilt of cotton, such a relationship was not observed. In this case, loss of virulence in the pathogen did not result in a marked change in antigenic relationships between the plant and pathogen. Taken together, serological data suggest that common antigens may play a role in pathogen specificity.

7.4 THE PHYTOALEXIN-INDUCED HYPOTHESIS

Phytoalexins are low-molecular-weight antimicrobial compounds produced *de novo* in plants as a result of infection or abiotic stress. The degree of resistance of hosts has been correlated with the rate of production/accumulation of phytoalexin in the host or detoxification/suppression of phytoalexin by the pathogen. Although in a few cases (bean—*Colletotrichum lindemuthianum* and potato—*Phytophthora infestans*) there was no correlation between concentration of phytoalexin and resistance of the host (Agrios, 1997) but in most of the cases, a positive correlation did exist between the two.

7.5 VERTICAL AND HORIZONTAL RESISTANCE

True host resistance is genetically controlled. It has been shown that resistance may be controlled by a single gene or very few (2-3) genes (oligogenic resistance) or many genes (polygenic). Van der Plank (1963) introduced the concept of vertical resistance to describe the condition in which plants showed a very high level of resistance to a particular physiological race of the pathogen but showed very little resistance to other races of the pathogen. Such hosts very often show a hypersensitive response usually very early in the infection, and the pathogen is not able to establish or multiply within host tissues. This is also known as **race-specific resistance**. Thus, vertical resistance is complete but not permanent in nature as any change in the race flora will make the host plants susceptible if the host does not possess genes for resistance against the new races which may be more virulent than the other race or races of the pathogen present in that particular area.

Some plants have a general level of resistance to pathogens, which is race nonspecific and controlled by a large number of genes (polygenic resistance). Equal resistance is shown to all pathogen races. Although this resistance, called **horizontal resistance** may not totally protect plants from infection, it does reduce susceptibility of the host plants and slows down the establishment of the pathogen. It is also known as **field resistance**. Thus, horizontal resistance is incomplete but is of a permanent nature as some resistance in the host will always be present to all the races of the pathogen. Breeders try to bring in this resistance into the crop plants.

The mechanisms which bring about variations in plants, principally hybridization and mutation, will not be considered here. Those which bring about variations in fungi are perhaps more diverse and include mutation, hybridization, heterokaryosis, parasexual recombination, and possibly adaptation.

7.6 MUTATION

Mutations are abrupt changes in the genetic material of a cell, which is then transmitted in a hereditary fashion to the progeny. Mutations occur spontaneously in nature in all living organisms which reproduce only sexually or only asexually and also in those which reproduce both sexually and asexually. Mutations can also be induced at a high rate artificially by exposing the organisms to either physical agents, such as ultraviolet light, X-rays, γ (gamma)-rays and extreme temperatures, or to highly reactive chemicals such as the nitrogen and sulfur mustards, epoxides, peroxides, phenols, alkaloids, and alkylating agents (e.g. ethyl methane-sulfonate, methyl methane-sulfonate, diethyl sulfate, etc.). One highly potent mutagenic chemical, often styled as a magic bullet, is N-methyl-N nitro-*N*-nitrosoguanidine.

The pathogenic abilities of parasites, as well as their morphological and other characteristics, can be changed in this way and these are important in the development of new strains. In a haploid fungus, such mutations should have immediate expression but this may be modified by epistatic or other effects. If the cell in which gene mutation occurs contains more than one nucleus, the expression will be modified by the other nuclei present and in such heterokaryotic fungi, pathogenicity and other characteristics will have a composite effect.

New races attributed to mutations have been reported in *Cladosporium fulvum* (tomato-leaf mould) where step mutation (each mutation overcoming the protective effect of a single gene) is thought to have occurred and similar mutations have been achieved experimentally by ultraviolet irradiation. Pathogenic variation due to mutation has also been reported in *Phytophthora infestans*, *Puccinia graminis*, *Melampsora lini*, and various smuts, to mention but a few.

This could be of considerable significance in Fungi Imperfecti and in those Chytridiomycota, Oomycota, Ascomycota, and Basidiomycota in which the perfect stage is rarely formed as in some Erysiphaceae in which cleistothecia fail to develop under certain climatic conditions. Although the inadequate information available suggests that mutations affecting pathogenicity are fairly rare, they are doubtless an effective mechanism for variations when the enormous numbers of spores produced by many pathogens are taken into account. For example, Watson (1957) reported increased pathogenicity due to mutations in *P. graminis* when large numbers of urediniospores were involved. Mutation can also bring about reduced pathogenicity or other changes disadvantageous to the pathogen. The practical implications of mutation in the aggressiveness of pathogens are difficult to assess. The sudden appearance of new races of increased pathogenicity in a pathogen which has hitherto been of only moderate pathogenicity has been ascribed to mutation and this may well be true. The appearance of a new, highly destructive race of *Helminthosporium maydis* (southern leaf blight of maize) in the USA in 1970 may be cited as an example of this.

7.7 HETEROKARYOSIS

It has been defined as a term which "...precisely describes the condition of a cell containing two or more genetically different nuclei....". In a later clarification of the use of the term *heterokaryosis*, Hansen (1942) pointed out that the term may be used whenever such a condition exists, not only in a single cell, but also when it exists in any or all cells of an individual thallus. This clearly implies that heterokaryosis involves the occurrence of dissimilar nuclei in a single cell, even if it is only one cell in a thallus.

Nelson *et al.* (1955) used the term *heterokaryosis* to designate the association of more than two genetically different kinds of nuclei in a vegetative cell or spore. It is agreed by most workers that heterokaryosis provides a haploid organism with many of the advantages of heterozygosity enjoyed by diploid organisms. Nuclear changes could provide additional somatic flexibility with changing environments. However, the main advantage of heterokaryosis appears to be the fact that it is a requisite step for genetic recombination brought about by interchanges of whole chromosomes or through mitotic crossing over. Heterokaryosis may thus play an important role in homothallic and imperfect fungi.

Heterokaryosis can arise in several ways like gene mutation, by fusion of vegetative cells, and heterokaryosis which may take place at the time of spore formation in many fungi, such as inclusion of + and – nuclei in the same ascospore of *Neurospora tetrasperma* or *Podospora anserina*.

Apart from dikaryotics in smuts, rusts, and Hymenomycetes, heterokaryosis has been reported in pathogens belonging to the Fungi Imperfecti, Ascomycota, and to a lesser extent, in lower fungi.

Heterokaryosis appears to be necessary for pathogenicity in heterothallic rusts, but not in infection by basidiospores as the latter are normally uninucleate and hence necessarily homokaryotic. There is considerable evidence that heterokaryosis not only conditions pathogenicity but also enables the effect of mutation on pathogenicity (often recessive) to be expressed. Buxton (1956) produced nearly an avirulent race of *Fusarium oxysporum* f. sp. *pisi* by ultraviolet irradiation and showed that heterokaryons were recovered by culturing from the diseased plants and concluded that some of the increases in the virulence of pathogens in nature may be due to heterokaryosis between weakly virulent strains.

Nutritional deficiencies can affect pathogenicity as shown by Hrushowvitz (1957), who demonstrated that repeated culturing of *H. sativum* on media supplemented with certain amino acids led to a progressive loss in pathogenicity in some cases. The mechanism involved is uncertain, but such repeated culturing may have led to the selection of deficient nuclei, the thalli containing such deficient nuclei being less pathogenic. The increased pathogenicity due to heterokaryosis in the mycelium would not be inherited by the spores unless the latter contained a full complement of the parental nuclei and would not be obtained with uninucleate conidia. Heterokaryosis is thus unlikely to bring about any permanent inheritance change in pathogenicity which would persist after sporulation, although parasexual recombination in which heterokaryosis is the first step, could do so.

Heterokaryosis is certainly a way in which avirulent strains may acquire virulence. One of the best documented cases is that of *Thanatephorus cucumeris*. Heterokaryosis between avirulent strains, both spontaneous and induced, or between avirulent and virulent strains were all highly pathogenic, in the latter case more so than either component. Heterokaryosis may give rise to races or strains with novel characteristics. One of the best documented examples is *P. striiformis*.

7.8 PARASEXUAL RECOMBINATION

This process in which genetic recombination occurs in the vegetative thallus in the absence of a sexual stage has been investigated chiefly in species of *Aspergillus* and *Penicillium* in the laboratory of Pontecorvo and his co-workers (Pontecorvo *et al.*, 1963).

The sequence of events in a complete parasexual cycle is as follows:

1. Formation of heterokaryotic mycelium
2. Fusion between two nuclei:
   (a) Fusion between like nuclei
   (b) Fusion between unlike nuclei
3. Multiplication of diploid nuclei alongside haploid nuclei
4. Occasional mitotic crossing over during the multiplication of diploid nuclei
5. Sorting out of diploid nuclei
6. Occasional haploidization of the diploid nuclei
7. Sorting out of the new haploid strains

The results are similar to those achieved by meiosis, a much more precise and regular process. The frequency of these events appears to be low and there is little information about the significance of this somatic recombination in nature, especially about fungi which attack plants. Parasexuality, however, produced new races of *F. oxysporum* f. sp. *pisi* and has also been reported in *Ascochyta imperfecti* responsible for black stem of alfalfa, *Verticillium albo atrum*, *F. oxysporum* f. sp. *cubense* responsible for banana wilt, *Cochliobolus sativus*, and others. There is evidence that somatic recombination may occur in smuts and rusts as in *Ustilago maydis*, *Puccinia graminis*, and *Puccinia recondita*.

7.9 ADAPTATION

The possibility that pathogens adapt themselves to their host plants has been debated for many years. There seems to be no incontrovertible evidence that this occurs although the evidence for adaptation to chemicals is stronger. Apparent adaptation to host plants might be due to mutation or to the selection of pre-existing strains which are better able to attack the host. Changes in the enzymatic capabilities of a pathogen may be mediated by the presence of appropriate substrates (adaptive enzymes) and this might modify its pathogenicity. Many years ago, Ward (1903) suggested the concept of a bridging host by means of which a pathogen is able to pass from a susceptible to a hitherto resistant host: for example, in *Puccinia dispersa*, the pathogen becomes adaptive to resistant species of *Bromus*. After propagation in some species of grass, somewhat similar changes were reported by Salmon (1904) in powdery mildews, and more recent work indicates that passage through some host plants can sometimes affect pathogenicity. Repeated passage through a resistant plant sometimes enhances the aggressiveness of a pathogen and vice versa, as has been demonstrated in several fungi and bacteria: for example, *Phytophthora infestans* and *Xanthomonas stewarti*.

Changes in the pathogenicity of *F. oxysporum* f. sp. *pisi* are brought about by exudates. Although there is increasing evidence in favour of parasitic adaptation, the mechanism involved remains obscure. Host selection of more aggressive biotypes from a mixture of biotypes, together with adaptation, mutation (perhaps by some mutagens in the host), and cytoplasmic variation may be involved in different cases.

7.10 SALTATION

This phenomenon, also described as dissociation or sectoring, refers to the appearance of morphologically different sectors in fungal colonies. It occurs frequently in some fungi including some isolates of *Fusarium* and *Helminthosporium*, but infrequently in others. Some such sectors remain fairly stable when subcultured while others are unstable. Saltation may be influenced by the compaction and thickness of the culture medium and can affect such characteristics as colony color and topography; rate of growth, sporulation, and size and shape of the spore. Pathogenicity might also be affected and this has been studied in *Fusarium* species by several investigators. The results have been somewhat conflicting; saltants tended to be less aggressive than the parent cultures, but occasional ones were more aggressive and these may be of significance where the saltation rate is high. No new pathogenic races with altered host ranges have been reported in saltants. The mechanisms involved in saltation are varied and not well understood.

7.11 CYTOPLASMIC VARIATION

There is a great deal of evidence for the existence in fungi of genetical determinants located in the cytoplasm outside the nucleus. Inheritance of such cytoplasmic determinants differs from that of nuclear genes. Cytoplasmic inheritance has been attributed to plasma genes. Cytoplasmic RNA (possibly some ribosomes) is the genetic material detached from the chromosome (episomes). There is little information as to the possible significance of cytoplasmic factors in bringing about variations in the pathogenicity of plant pathogens, although Johnson (1954) had described the cytoplasmic control of pathogenicity of *Puccinia graminis* f. sp. *tritici*; this is a subject which needs investigation.

7.12 SEXUAL RECOMBINATION

The origin of new physiological races through meiotic recombination is well attested in many pathogenic fungi, rusts, smuts, powder mildews, and the potato-blight fungus. Hybridization is likely to be encouraged by incompatible mechanisms which prevent or reduce inbreeding, and many pathogens which are heterothallic.

Hybridization can be intraspecific, interspecific, or even intergeneric, and the resulting hybrids may have different pathogenic abilities than the parental races. Often the hybrids are intermediate in pathogenicity between the two parental races but some may be more pathogenic than others and similar considerations apply to other inherited characteristics. Extensive recombination of genes occur in autoecious rusts such as *Melampsora lini* (flax rust) in which all the spore stages occur in the same plant.

7.13 TRANSFORMATION

In addition to sexual recombination, there are a number of possibilities for the production of new variants in microorganisms by asexual means. One of these, a sexual type of combination of genetic characters is **bacterial transformation**. The observation of Avery *et al.* (1944) called attention to this process. They were successful in converting a nonpathogenic non-capsule-forming strain of *Diplococcus pneumoniae* into a pathogenic strain which forms a carbohydrate capsule. This was done by culturing the nonpathogenic strain in the presence of the biologically active fraction of the capsule forming a pathogenic strain. It was later demonstrated by these workers that this transformation was brought about by the introduction of purified deoxyribonucleic acid (DNA of the donor). Thus, transformation is a specific form of recombination, involving the naked genes of DNA.

Several workers have demonstrated the production of new strains of plant pathogenic bacteria by means of transformation. Corey and Starr (1957) were able to change the type of colonies and resistance of *Xanthomonas campestris* pv. *phaseoli* (a pathogen of beans) to streptomycin by transformation. Similarly, Klein and Klein (1953) have shown that the tumor-inducing ability of *Agrobacterium tumefaciens* can be transferred from virulent donor strains to avirulent strains. In this case, the material transforming the avirulent cells was also deoxyribonucleic acid (DNA).

7.14 TRANSDUCTION

Another process of recombination different from sexual recombination is transduction. This process has been found in a variety of bacteria where DNA is transferred from cell to cell through the agency of temperate viruses. Bacteria need not necessarily be destroyed by their infective viruses (bacteriophages). In certain cases after infection of the bacterium with a bacteriophage, the phage (temperate) may produce in the host an immunity that prevents multiplication of the phage and lysis of the bacterial cell. Such an association is called the **lysogenic state**. Under the influence of different chemicals and physical treatment (e.g. ultraviolet irradiation), lysogenic bacteria liberate their latent phages by lysis. These are infective viruses with the ability to enter the new bacterial cells. During symbiosis, the phage becomes the genetic material of the host, with page DNA coming into close contact with the host DNA. This makes it possible for the DNA molecule, carrying the genetic factor of the donor bacterium to be transformed to a recipient bacterial cell through the vehicle of a phage particle. Pathogenicity can also be transduced by bacteriophages. Therefore, transduction is an important mechanism for genetic recombination in bacteria. This interesting mechanism of genetic exchange was discovered by Zinder and Lederberg (1952) in *Salmonella typhimurium*. Its importance was also demonstrated in a plant pathogenic bacterium *Ralstonia* (*Pseudomonas*) *solanacearum*.

Viruses also show considerable variability in several characteristics, such as infectivity, longevity, symptoms produced on the host plant, host range, and specificity to vectors. These variations are possibly due to some sort of mutation process and are inherited in that they are perpetuated during the transfer of the virus from one plant to another. The mechanism involved in these changes in not quite well understood. In virus-resistant plants, the relationship is, in effect, between the nucleic acids of the resistant genes of the host plant and RNA of the virus particle.

7.15 PHYSIOLOGICAL SPECIALIZATION

Many factors in host-parasite interactions present the fundamental problems of plant disease. The major factor towards the elucidation of host-parasite interactions is the genetic aspect.

The rediscovery of Mendel's laws of heredity in 1900 and the subsequent rapid development of the basic science of genetics gave a strong impetus to the study of plant pathology. The genetic basis of disease resistance was analyzed in many cases. Although the genetics of the host received most attention initially, advances in the knowledge of the sexuality of fungi led to the development of materials and methods for genetic studies of pathogenic fungi. There is now positive evidence that the general patterns of the laws of inheritance are the same in fungi as in other groups of plants. The fungi undergo heritable variations brought about by segregation and recombination of genes, mutations, heterokaryosis, cytoplasmic inheritance, etc. Not only this, because of their great variety in sexual and asexual reproduction and their unique suitability for experimental manipulation, microorganisms now afford the best available material for many types of fundamental studies in genetics, biochemistry, pathology, and other fields. Despite all these advances, however, knowledge of the actual physico-chemical mechanisms that determine parasitism, pathogenicity, or disease resistance is still very meager. In a nutshell we can say that disease resistance is a genetic character as regularly inherited as any other character. However, the genetic behavior of resistance usually differs in different cases.

Resistance to one disease does not necessarily imply resistance to any other disease. Thus, Thatcher wheat, while resistant to stem rust is highly susceptible to leaf rust (*P. recondita, P.triticina*). Various kinds of genetic behaviour have been observed with respect to disease resistance, for example, simple monohybrid inheritance of resistance, in cabbages to yellows (*Fusarium conglutinans*), lettuce to downy mildews (*Bremia lactusae*), and peas of wilt (*Fusarium oxysporum* sp. *pisi*). In these cases, resistance is a dominant characteristic.

Resistance as a recessive characteristic in simple monohybrid inheritance is seen in powdery mildew of barley (*Blumeria graminis* f. sp. *hordei*), smut of sorghum (*Sphacelotheca sorghi*), and speckled leaf blotch of wheat (*Septoria tritici*). Two gene segregations appear to be involved in resistance to anthracnose of beans (*Colletotrichum lindemuthianum*) and some cereal rusts. Many genes which are more difficult to analyze are known to be involved in a large number of diseases. And finally, there may be a number of alleles (instead of two) for one gene, giving a complex series of resistance relationship when several parents are studied.

7.15.1 Physiological Specialization in Fungi

1. Complexity of Organisms

Like all other organisms, fungi are also complex and variable. This complexity is governed by the genetic material contained in the chromosomes. The crop plants are similarly complex, for example, there are 14,000 varieties of wheat; similarly, there are innumerable varieties of apples, beans, gram, roses, corn, and so on.

2. Gene Action

When fungal species parasitize a number of host varieties, they often develop physiological specialization, which is the development of certain character in the fungus which are not morphological but physiological, and can be detected only when the effect is noted on the host. A special species of fungus such as the black stem rust, *Puccinia graminis*, may include a number of "physiological races" which are similar in morphological structure but may be different in their capacity to produce disease. So the difference between them is physiological and morphological. Such races are commonly encountered in many groups of parasites such as rusts, smuts, powdery and downy mildews, and so on.

Gene action expresses itself in various morphological features and it also always governs the various physiological processes of the organism. The interaction between the host and parasite is actually the interaction between the enzyme systems of the two (now proved by Dodge and others in *Neurospora*) and the result is the effect produced.

The details of this phenomenon can be easily explained by taking the example of the black stem rust, *Puccinia graminis*. The discovery of pathogenic specialization is fundamental to studies on pathogenic variability. Eriksson (1894) showed that *P. graminis* on cereals and grasses is composed of several pathogenically different strains which he described as specialized forms (*formae speciales*). These are now known as varieties. There are at least six varieties of *P. graminis* which can be recognized by the size of the spores and also by their pathogenicity on different hosts.

7.15.2 Production of New Races

All cereal rusts are heteroecious. An aecidial stage is never produced on a cereal host. As long as the fungus produces itself by means of urediniospores, it breeds true except for the occasional appearance of a mutation. On the alternative hosts (barberry, buckthorn, and so forth), the sexual stage of the fungus is developed (Cragie, 1927) and this gives an opportunity for hybridization. This could result in the formation of heterozygous characteristics with a difference in physiological specialization. When such a heterozygous race forms the basidiospores, it segregates again and different recombinations may result again. In this way, the appearance of new races are recorded more in those regions where *Berberis* spp. play an active role, as in the USA and Europe. On the other hand, in countries where *Berberis* spp. play little or no part in the life cycle, the number of races is much less, as in Australia, India, and Kenya. In *P. graminis tritici* alone, there are about 225 races in USA, and 32 races and biotypes in India.

In other species of *Puccinia*, the number of physiological races is lesser. Of *Puccinia recondita*, there are 90 in USA and 28 in India. Of *Puccinia striiformis*, there are 30 in the USA and 14 in India.

Biotypes

A biotype is a population of individuals that are identical genetically. The descendants of a single non-sexually produced stem rust spore constitutes a biotype. The biotype will remain strictly pure unless a mutation occurs. Most species of pathogenic fungi are now known to have many genotypes. The most closely related races are similarly grouped into varieties and finally the varieties into species. Actually, a reverse procedure is followed. The larger groups (varieties) are recognized first and then smaller groups (races and biotypes) are easily distinguished by more refined methods. Examples of the occurrence of biotypes can be given in the case of race 59 in which several biotypes are known to exist. Biotypes show enough similarities to be placed within one race, but when different biotypes are studied very closely, they show minor but consistent differences in their effects on some differentials, proving that they are genetically pure, though with different identities. Such races are, therefore, known to have biotypes designated as 59, 59A, 59B, 59C.

Another instance may be given. Race 15 of wheat-stem rust was discovered in 1918. It was a fairly homogenous rust for a long time and was identified as such. Later, some dissimilarity in isolates from different regions was noted. These differences were consistent enough and 15A and 15B were recognized as biotypes. Race 15B which was known to occur occasionally near barberry bushes became widespread in 1950 and attacked the well-known hitherto resistant varieties. Besides rusts, specialization of parasitism is found in other fungi also. In smuts, as in rusts, races and biotypes are known to occur in different species. As a matter of fact, smuts offer better material for the study of genetics. In smuts (*Ustilago zeae*), the chlamydospores undergo reduction division and the sporidia produced are haploid structures which can be grown into haploid mycelium in culture media and matings done as per plan. The colonies from haploid sporidia are genetically pure and differ from one another in characters such as size, color, and surface pattern. It is now known that in *Ustilago zeae*, there are thousands of haploid biotypes. Biotypes may differ from one another in one or more of all known characters, including pathogenicity. Similar, though less elaborate work, has been done on other smuts, for example, *Ustilago segetum* var. *tritici*, *Ustilago nuda*, *Sphacelotheca sorghi*, *Ustilago avenae*, and *Ustilago hordei*.

Investigations in rusts and smuts are complementary. In rusts only the binucleate uredinial stage can be propagated clonally, while in smuts it is only the haplophase which can be propagated in this way. In both smuts and rusts three distinct phases correspond with the nuclear condition—the **haplophase**, the **dikaryophase**, and the **diplophase**. In rusts, both dikaryophase and haplophase are parasitic in life, but in smuts, the haplophase is generally saprophytic. The dikaryophase, in both smuts as well as rusts, can be hybrid in character. This character is multiplied (in an equational manner) vegetatively through the urediniospore stage in rusts. In many cases, rusts (in India) can go on repeating indefinitely from year to year. In smuts, such multiplication does not occur and sexuality is an indispensable process. Variability in smuts is, therefore, more common, and hence there is a very large number of races, but these races have to be expressed in terms of haploid strains and not in the dikaryotic phase, as in rusts. No sooner is the chlamydospore formed in smuts, than the dikaryotic phase comes to an end and is not propagated clonally (by urediniospore, like conidia) as in rusts.

Blumeria (*Erysiphe*) *graminis* is also strongly specialized into physiological races.

7.15.3 Adaptation of Fungi to Different Hosts

The Gene-for-Gene Theory of Host-Parasite Relations

Flor (1956) in his studies on rust disease of Flax (*Melampsora lini*) came to the conclusion that for each gene that governs resistance in the host, there is a specific and related gene in the pathogen that governs its virulence (pathogenicity). In other words, there is a genetic inter-relationship between the host and pathogen. This is the complementary gene-for–gene theory. The mechanism for the evolution of this relationship in a host- parasite system is an automatic consequence of the distribution of new resistance genes by mutation and selection in the host, and virulence genes in the parasite population. The genetic definition of specificity embodies the so-called "quadratic model" in which genes in the host are clearly defined in terms of resistance or susceptibility and the complementary genes in the pathogen in terms of virulence or avirulence.

The relationship between the host and parasite induces a genetic polymorphism in both partners. This genetic variability, exhibiting a self-regulation automatic mechanism, is the reason for the vicious circle that consists of the alternative prevalence of the resistant host and the virulent pathogen.

The gene-for-gene hypothesis was proposed by Flor (1955 and 1956) as the simplest explanation of the results of studies on the inheritance of pathogenicity in the flax-rust fungus (*Melampsora lini*). In the varieties of flax (*Linum usitatissimum*) that have a gene to resist the avirulent parent race, F_2 cultures of the fungus segregate into monofactorial ratios. In the varieties that possess 2, 3, or 4 genes for resistance to the avirulent parent race, the segregation in the F_2 cultures is in bi, tri, or tetrafactorial ratios.

Flor further summarized the evidence for flax rust in the gene-for–gene hypothesis. This states that during their evolution, the host and parasite develop a complementary gene system so that "for each gene-conditioning rust reaction in the host, there is a specific gene-conditioning pathogenicity in the parasite" (Flor, 1956). This hypothesis is applicable to most host-parasite systems in which resistance is conditioned by major (vertical resistance) genes and virulence increase in a stepwise manner. Hatchett and Gallun (1970) showed that the ability of specific races of Hessian fly to survive on certain varieties of wheat was controlled by genetic systems in the host and insect that was complementary.

There is considerable literature on the inheritance of pathogenicity and resistance, including smuts, *Venturia inaequalis*, and *Phytophthora infestans*. In the case of *Phytophthora infestans*, Black (1952) in Scotland identified four major dominant genes, R1, R2, R3, and R4. At least nine major genes are now known to control resistance. This polygenic resistance is apparently conditioned by different factors in the case of tuber rot compared with foliage blight.

The reactions of host plants in the different pathogenic races are controlled by many genes, and within them, an even large number of alleles exist to enable resistance.

7.15.4 Genetics of Virulence in Pathogens and of Resistance in Host Plants

Infectious plant diseases as a result of interaction between host and the pathogen and properties of these two organisms are governed by their genetic material. The degree of susceptibility or resistance of host to various pathogens is an inherited characteristic. It has been known that every pathogen is having races, each different from others in its ability to attack certain varieties of a plant species but not other or all varieties. So when a variety is inoculated with two races of a pathogen, the variety is resistant to one race and susceptible to the other. This shows that one race possesses a genetic characteristic that enables it to attack the variety of a plant while the other race does not. When one race of the pathogen is inoculated on two different varieties, one is susceptible while the other is resistant. This shows that one variety possesses a genetic characteristic that enables it to defend itself against the pathogen so that it remains resistant while the other variety does not. Studies of the inheritance of resistance versus susceptibility in plants show that single gene-control resistance and their absence allows susceptibility.

Similarly, studies further prove that single gene-control avirulence and their absence allows virulence in pathogens. The number of genes determining resistance or susceptibility varies from plant to plant as the number of genes determining virulence and avirulence varies from pathogen to pathogen. Here we shall be discussing various processes leading towards disease development and how genes are responsible towards resistance or susceptibility of the plants.

1. Nature of Resistance to Disease

A microorganism is pathogenic or a host is resistant because it has the genetic ability to infect or immune a pathogen even under most favourable conditions. The mechanism by which genes control the various physiological processes that lead to disease resistance or susceptibility are not yet very clear but they are, presumably, no different from the mechanisms controlling any other physiological process in living organisms.

It is thought that for the production of enzymes or fungitoxic substances required for defense, a stimulant secreted by the pathogen reacts with a receptor molecule of a host-cell. This then transmits signals to other host-cell molecules, activating plant defenses. However, if a pathogen does not secrete the particular elicitor that activates the defense reaction, infection will occur and further disease development. The success of a fungal pathogen depends to a high degree of its ability to perceive and to respond to signals generated by the plant, especially at the early stages of infection (recognition), but also in later stages involving different cell types/tissues.

2. Genes Involved in Pathogenesis

Pathogenicity genes (pat) and **disease-specific genes (dsp)** are those involved in steps crucial for the establishment of disease. Such genes include those essential for recognition of host by pathogen, further infection, penetration, and colonization of the host tissue. Further pathogens utilize various genes that enable them to break down host-wall macromolecules, produce toxins, neutralize host defenses, move through the host, multiply in the host, disseminate from host to host, and so on.

3. Genes Controlling Production of Infection Structures

Depending how fungi enter either through stomata or any other opening or through direct penetration, they need to degrade the cuticle and the cell wall, for which they may need to form specialized structures called **appressoria**. Appressoria contain glycerol for creating a high turgor pressure that allows the penetration peg of appressoria to break or puncture the plant epidermal cells. A large number of genes have been isolated that affect appressoria differentiation and function. Most of these genes have been isolated from the large appressorium-forming species of *Magnaporthe grisea* and *Colletotrichum* sp. Appressorium-related genes can be divided into three subgroups: (1) Genes that operate prior to and are necessary for appressorium formation, (2) Genes that are uniquely expressed in appressoria or contribute significantly to specific features of appressoria structure and may be regarded as appressorium-specific, (3) Genes that control and affect appressorium germination, i.e. formation of the penetration peg and host penetration. Genes of groups 2 and 3 include melanin biosynthesis genes, which are needed for functional appressoria and appressorium-specific genes that are expressed only in the appressorium.

4. Genes Controlling Degradation of Cell Wall

It is known that enzymes that degrade cell walls and other physical structures essential for pathogenicity are encoded by more than one gene that too is not related. As a result, absence of one such gene does not eliminate pathogenicity of the pathogen because the other genes that encode the same enzyme fill in the need for the enzyme. The enzyme cutinase that degrades cutins and the enzyme pectinases (PG, PMG, and PME) that degrade pectins are encoded by multigene families, and proof of their significance as essential pathogenicity factors is difficult. The disruption of a pectin lyase gene in *Cryphonectria parasitica, Alternaria* sp. and *Glomerella* sp. had no effect on its pathogenicity, whereas disruption of a pectinase gene in *Botrytis* reduced the pathogenicity of the fungus on apple and tomato.

5. Genes Controlling Secondary Metabolites

Apart from the needs that are required for producing infection structures (appressorium) and for degrading cutin and various cell-wall components, pathogens require additional genes that will help them overcome the many secondary metabolites produced by the plant in order to defend themselves. The genes of the pathogen respond to chemical defenses of the host plant that help the pathogen to degrade secondary metabolites produced by the host through their action. Plants produce a vast variety of secondary metabolites, many of which have *in vitro* antimicrobial activity. Antimicrobial compounds include preformed substances (phytoanticipins) and compounds that are synthesized upon a microbial challenge (phytoalexins). Phytoanticipins are secondary metabolites that include saponins, avenacin, and tomatine. Presence or absence of saponins has been found in correlation with disease development in several plant-fungus interactions. Infection of oats and wheat by the take-all fungus *Gaeumannomyces graminis* is correlated with production of the triterpene saponin avenacin A in the host and the avenacin A degrading enzyme avenacinase (Ava1) in the fungus. Ava1 mutant strains are unable to detoxify avenacin; they are nonpathogenic on avenacin A producing oats but are fully pathogenic on wheat, which does not naturally produce saponins and on oat mutants defective in saponin production. Another class of plant protecting molecules is **antifungal proteins**, also known as **PR (Pathogenesis Related) proteins**. As for the secondary metabolites, these are both preformed and induced antifungal proteins that are assumed to play a role in plant protection by direct toxicity to the fungus and by release of fungal elicitors that may activate plant defenses. Recently, a class of Glucanase Inhibition Protein (GIP) has been characterized in the soybean oomycete pathogen *Phytophthora sojae*. The GIPs are homologus with the trypsin class of serine proteases, but lack proteolytic activity. Structural motifs of protein-protein interaction found in GIPs and high affinity binding of the soybean endoglucanases EGase (endo-β-1,3-glucanases) A with GIP1 suggest that they may act as glucanase-specific inhibitors. Thus, the GIPs may represent a novel class of fungal counter-defensive proteins that suppress plant-defense responses.

6. Genes Controlling Fungal Toxins

There are several genes, often clustered, involved in the synthesis of toxin. For example, disruption of toxin genes in *Cochliobolus* shows that a fungus with an altered toxin profile can still be pathogenic. Similarly, disruption of genes involved in the biosynthesis of *Alternaria alternata* host-specific toxin resulted in reduced pathogenicity.

7. Signaling Genes used by Pathogenic Fungi

Fungal signaling genes include the Mitogen-Activated Protein (MAP) kinase gene, cyclic AMP-dependent protein kinase genes and G-protein-coding genes. If these fungal signaling genes are disrupted by mutation, the fungus loses its pathogenicity and as a result of which growth of the fungus, multiplication, and toxin produced is lost or reduced. For example, in case of *Magnaporthe grisea*, three G-protein genes and three MAP kinase genes have been cloned and tested through disruption. Further disruption of the PMK-1 gene of *M. grisea* reduced appressoria formation and lost the ability to infect but had no effect on growth and multiplication of fungus.

REVIEW QUESTIONS

1. What are the various mechanisms which bring about variations in fungi? Give examples of diseases in which mutations have been instrumental in increased pathogenicity.
2. What are heterokaryosis and somatic hybridization?
3. How does the parasexual cycle differ from the normal sexual cycle?
4. Fungi Imperfecti do not have a normal sexual cycle and yet they are quite abundant and successful in nature. Why is this so?
5. What are transformation and transduction?
6. What is physiological specialization?
7. Identify the biotypes in *Puccinia graminis tritici*.
8. What is the gene-for-gene concept?
9. Write in brief what you know of host-pathogen specificity. What are the various hypotheses in this regard?

MULTIPLE-CHOICE QUESTIONS

1. Which of the following factors account for resistance to plants against a number of pathogens?
   (i) Internal factors (ii) External factors
   (iii) Phenolic substances (iv) All of these
2. If the resistance of host is controlled by single or few (2-3) genes, it is called
   (i) horizontal resistance (ii) vertical resistance
   (iii) polygenic resistance (iv) none of these
3. Which of the following phenomena is true for heterokaryosis?
   (i) Mutation (ii) Nutritional deficiency
   (iii) Virulence of pathogen (iv) Increased pathogenicity
4. Which of the following organism has been investigated for parasexual cycle?
   (i) *Polyporus* (ii) *Plasmodiophora*
   (iii) *Pythium* (iv) *Aspergillus*
5. Which of the following statement is true for saltation?
   (i) Rate of growth and sporulation of fungi
   (ii) Thickness of the culture medium
   (iii) Colony colour
   (iv) Appearance of morphological different sectors in fungal colony
6. Which of the following statement is true for biotype?
   (i) They are morphologically identical. (ii) They are genetically identical.
   (iii) Similarity in isolates. (iv) Isolated from one region.
7. Which of the following substance is present in appresorium for creating a high turgor pressure to break the plant epidermal cells?
   (i) Orcinol (ii) Glycerol
   (iii) Saponins (iv) Tomatine

Answers

1. (iv) 2. (ii) 3. (i) 4. (iv) 5. (iv) 6. (ii) 7. (ii).

8 EFFECT OF ENVIRONMENTAL FACTORS AND NUTRITION ON DISEASE DEVELOPMENT

The weather and soil conditions influence the seasonal development and geographical distribution of plant diseases. About 90 years ago, Jones (1924) directed increased attention towards the relation of the environment to the inception and development of disease.

The 'environment' connotes all the external conditions affecting the life and development of an organism, which include temperature, light, and moisture, may also include living factors, such as competing microorganisms or vectors.

The environment can affect the development of plant disease in different ways. It can affect the perpetuation or over-wintering of the pathogen from one growing season to the next, the build-up of both primary and secondary inoculum, the dissemination of inoculum, germination, and host penetration. The environment can affect growth development of the host prior to being infected in such a way as to affect its susceptibility. Finally, it can influence the actual development of disease after the host has become infected. The development of all parasitic diseases also obviously varies with the pathogen, host, interactions between the pathogen and host and environmental factors in the air and soil.

8.1 EFFECT OF TEMPERATURE ON THE DEVELOPMENT OF A DISEASE

Temperature is undoubtedly one of the most important factors, which influences the occurrence and development of many diseases. The late blight of potato is usually considered a disease which occurs in the northern latitudes, but it can become destructively epidemic when potatoes are grown during winter in the subtropics. Some pathogens develop best in low-temperature regions. *Taphrina pruni* and *T. deformans* are confined to the cool, wet parts of the globe. Similarly, *Spongospora subterranea*, the powdery scab, is limited to areas of cold temperature. *Puccinia striiformis* also thrives in cold temperature and in India, it oversummers at heights above 7000 feet above sea level while black rust survives at 5000 feet above sea level. *Cercospora beticola*, the cause of sugarbeet leafspot, needs a temperature 15°C for sporulation, germination, and infection. *Urocystis cepulae,* the cause of onion smut, is spreading throughout the USA on infected onion sets. It is absent in the extensive southern onion-growing regions, but is an established disease in the cooler, northern onion-growing areas of the USA and Europe.

On the other hand, there are diseases which occur in the warm belts of the globe, for example, *Fusarium oxysporum* f.sp *lini, F. oxysporum* f.sp *lycopersici, F. oxysporum* f.sp. *conglutinans* causing wilts of flax, tomato and cabbage yellows respectively, *Erwinia amylovora*, the fire-blight bacterium, *Ralstonia solanacearum* causing bacterial wilt of potato, tomato, tobacco, brinjal, pepper, etc. *Sclerotium rolfsii* on various hosts, *Macrophomina phaseolina* and *Phymatotrichum omnivorum* causing Texas root rot of cotton.

The stages in infection are also affected by change in temperature. Temperature can also affect the rate of penetration of the pathogen into the host. In epidemiology, this is often quantified as incubation period. Both infection and latent periods are influenced by temperature. Latent periods tend to decrease with increasing temperature, provided other factors are favourable. Latent period for *Blumeria* (*Erysiphe*) *graminis* decreases from about 14 days at 5°C to 3 days at 18 to 25°C. In the case of black stem rust of wheat the incubation period is 22 days at 4.5°C, 15 days at 10°C, 9 days at 19°C, and only 5 days at 24°C.

The incidence of cereal rusts depends mainly on humidity and temperature. The incubation period and the germination of urediniospores of *Puccinia graminis tritici* is favoured by a temperature of 20°C. In the Indian plains this usually occurs in the period between February and March when the weather just begins to warm up. If the weather becomes cloudy and moist during this period, the rust incidence is immediately high, causing huge damage. The time required for the production of urediniospores of *Puccinia graminis tritici* on a susceptible wheat cultivar, can range from 85 days at 0°C, to 5 days at 24°C. The incubation period of *Pseudoperonospora humuli* can vary from 3 days at 21-25°C to 23 days at 5°C or 11 days at 29°C.

Thielaviopsis basicola, which causes black root rot of tobacco, is highly damaging when the soil temperature is between 17°C and 23°C but causes little damage when the temperature is between 21°C and 30°C. This has been attributed to the formation of a phellogen layer at the higher temperature, which arrests the spread of the fungus.

Many viral diseases are more virulent when the temperature is between 20°C and 25°C. The most favourable temperature for disease is between 20°C and 25°C; these include the tobacco mosaic virus, potato mosaic virus, cabbage virus, potato yellow dwarf, and others.

Late blight of potatoes generally does not occur in the plains of India, where potatoes are grown as a winter crop. At this time, the conditions necessary for the growth of pathogen do not exist in plains. But in the hills where the crop is grown in the rainy season, late blight is very severe.

Musk melons are relatively resistant to seedling blight and wilt caused by *Fusarium oxysporum* f. sp. *niveum* when soil temperature is at the optimum of 30°C or higher for the growth of melons. However, the disease is abundant and destructive at temperatures below 27°C which is optimum for the growth of the pathogen. The resistance at high temperatures appears to be due to the greater ability of the host to exclude the pathogen from the vascular bundles by the formation of periderm.

8.2 HUMIDITY AND MOISTURE

Atmospheric humidity and precipitation in the form of rain, fog, dew, etc. determine disease incidence to a great extent. Likewise, the amount of soil moisture determines the severity of the disease in soil. The speed of germination of inoculum, its entry into the host, and the period of incubation are very much affected by the amount of moisture available. In case of soil diseases, the amount of moisture will also affect the soil aeration and the concentration of various salts in the soil.

Downy mildew of grapes is very much favoured by moist weather and high temperature. Similarly, late blight of potato is favourably affected by high moisture content. A large number of diseases are favoured by increased humidity, such as the late blight of potatoes, Koleroga of arecanuts, downy mildew of cereals, blast of rice, early blight of potatoes, scab of apples, and circular leaf spot of coffee.

Precipitation and dew play their own part in increasing atmospheric humidity and cannot be considered in isolation. Heavy dewfall favours the germination of the spores of the pathogens and infections in diseases, such as downy mildew and rusts.

Soil moisture appears to have a decisive effect on the amount of the big-vein disease of lettuce caused by a virus that persists in the soil. In experiments performed, 85% of the plants became infected in soil containing 95% of its moisture-holding capacity, and there was a progressive decrease in infection with decreasing moisture, until only 10% of the plants became infected at 35% moisture.

Diseases such as the root rot of pear caused by *Aphanomuces euteiches*, the damping off of seedlings caused by *Pythium ultimun, Pythium debaryanum* and others, the club-root disease of crucifers (*Plasmodiophora brassicae*) are greatly increased by an increase in the moisture content of the soil. But, on the other hand, diseases like common scab of potatoes (*Streptomyces scabies*) and white rot of onion (*Sclerotium cepivorum*) usually spread under dry condition. The severity of damping off caused by *Rhizoctonia solani* increases with increasing moisture up to 65% of saturation and then decreases abruptly. Moderate soil moisture is most favourable for head smut of Sorghum caused by *Sphacelotheca reiliana* and pea-foot rot due to *Fusarium solani* f. *pisi*.

Soil moisture appears to have a decisive effect on the amount of big-vein disease of lettuce, caused by a virus which persists in the soil. The disease is favoured with light moisture content up to 85%.

Moist weather (coupled with suitable temperature) favours the spread of rusts and downy mildew in general. The powdery mildews (Erysiphales), such as powdery mildew of pea caused by *Erysiphe polygoni* and mildew of *Dalbergia sissoo* caused by *Phyllactinia corylea*, are not so dependent on moisture and they are known to flourish even in conditions of drought.

Karnal bunt of wheat caused by *Tilletia indica* (formerly known as *Neovossia indica*), which is airborne, occurs as an epiphytotic only when there is heavy rain and air and soil temperature is around 15-20°C two to three weeks before flowering time. The rainwater soaks the spores thoroughly and the low temperature promotes germination. The disease is rare in the absence of both these conditions.

Colocasia blight due to *Phytophthora colocasiae* occurs only in the rainy season, especially when there is a continuous spell of moist and cloudy weather. Plants recover as soon as dry conditions return. Soil moisture also materially affects the cause of disease in Phytophthora foot rot of *Piper betle* and gummosis of *Citrus*.

Lack of atmospheric humidity is also the reason for the relative absence of certain geographical areas of foliage diseases, particularly those of beans, such as anthracnose (*Colletotrichum lindemuthianum*) and bacterial blight (*Xanthomonas campestris* pv. *phaseoli)*. Beans are grown for seed in the arid portions of the USA to avoid seed infections by those organisms which require free moisture on plant parts. Late blight of potato seldom occurs in drier areas, except sometimes in those under irrigation.

The powdery mildew fungi, a group of pathogens, are affected by moisture in an unusual manner. Germination of conidia of the lettuce powdery mildew fungus (*Erysiphe cichoracearum*) greatly decreases at 100% relative humidity, germinates best at 93% RH and germinates to some extent at 0.1% relative humidity. Thus, it is not hard to see why powdery mildews develop in the driest of climates. It is believed that the reason these conidia germinate under such dry conditions is that they contain about 70% water and they carry their own water supply with them.

Diseases favoured by wet soil or high relative humidity are damping-off of papaya and vegetable seedlings, foot rot of *Piper betle* caused *Phytophthora nicotianae* var. *parasitica* or other *Phytophthora* species, collar rot of citrus caused by *Phytophthora citrophthora,* late blight of potato, cereal rusts, downy mildew of bajra (pearl millet), sorghum or maize bacterial blight, streak of rice and many others. Diseases favoured by dry soil or water-deficient soils are common scab of potato (*Streptomyces scabies*) and seedling blight of cereals caused by *Fusarium* species. Fusarium wilt of tomato, peas and cotton are more severe in wet soils in comparison to dry soils. There is a very close relationship between plant pathogenic nematodes and the amount of soil moisture in the soil.

8.3 SOIL pH

The effect of alkalinity and acidity of the soil is so pronounced in many cases as to level the incidence and distribution of certain diseases and to be the most important factor in their control.

Alkaline soils are conducive to the spread of the scab of potatoes (*Streptomyces scabies*). It usually develops between pH 5.2 and 8.0. Its incidence falls sharply when the soil pH falls below 5.2. Successful attempts have been made to control this disease by acidifying the soil with sulfur or sulfuric acid.

The club-root disease of crucifers (*P. brassicae*) is favoured by acidic soil. It develops most at about pH 5.7 and drops sharply if pH is raised so that it totally fails to develop at 7.8. Liming, therefore, helps to control the disease. The influence of pH in this case is on the swarming myxamoebae.

Other pathogens which are favoured by the alkaline conditions of soil are Texas root-rot pathogen, *Phymatortichum omnivorum*, which attacks more than 1000 spp. of hosts, *G.* (*Ophiobolus) graminis*, the take-all of wheat, and *Verticillium albo-atrum* vascular parasite of many kinds of plants. Other diseases which are associated with acidic soils are *F. oxysporum* f. sp. *lycopersici* and F. *oxysporum* f. *vasinfectum* which causes tomato and cotton wilts respectively, *Synchytrium endobioticum* which causes potato wart, and *Spongospora subterranea,* the cause of powdery scab of potatoes.

High soil acidity is inimical to the development of tobacco root rot caused by *Thielavia basicola*, while such a condition is highly favourable to wilt of tomatoes. The behaviour of *F. oxysporum* f. *vasinfectum* which causes the cotton wilt in the USA and India was studied. It was found that the American strain was favoured under acidic condition, while the Indian strain developed best in alkaline reaction. Root rot of cotton caused by *Rhizoctonia destruens* is favoured by light soils with acid reactions, while in India Fusarium wilt of cotton develops best in heavy alkaline soils. Covered smut of oats develops at a pH value of 7.8 and falls sharply with either a decrease or increase in the soil pH; it is moderate at 6.5 pH.

8.4 SOIL TEXTURE

In the black cotton soils which are clayey and heavy, cotton wilt (*F. oxysporum* f.sp *vasinfectum*) occurs abundantly and root rot due to *Rhizoctonia bataticola* also occurs.

On the other hand, in regions of lighter soils (Gujarat), the case is reversed, that is only root rot occurs and there is no wilt disease. These types of soils are alkaline and the determining influence in this case appears to be texture (clayey or sandy); *Rhizoctonia* cannot survive poor aeration. It is very sensitive to oxygen tension.

8.5 EFFECT OF LIGHT

Although light can influence the development of some diseases, it is rarely a limiting factor in their seasonal or regional occurrence. It probably is not usually a determining factor, neither in the development of epidemics nor in the geographical distribution of diseases.

Light intensity and length of day, however, may affect the survival of inoculum, prepenetration processes, possibly the entrance of some pathogen/s, length of the incubation period, abundance of sporulation, and sometimes the type of pathogenic effect. Light intensity and day length also affect the development of many host plants and can be the determining factors in their viability as media for the growth of pathogens and their relative resistance or tolerance to pathogenic effects.

Although wheat stem rust develops well at light intensities ranging from 500 to 10,000 foot candles, the incubation period is likely to be shorter and sporulation to be more abundant with increasing light intensity. In experiments on *Melampsora lini*, a flax rust, uredinia appeared 9 days after inoculation when plants were kept under normal light and after 14 days under reduced light. It has been found that high light is most favourable for leaf rust of rye, *Puccinia rubigovera* var. *secalis*.

Light is necessary for photosynthesis and thus the general well being of the plant is affected by both the quantity and quality of light. Several virus diseases are known to be influenced by light intensity. Reduced light intensities generally increase the susceptibility of a plant to virus infection. For example, tomato plants will get infected by sugarbeet curly-top virus when they are grown under conditions of low light intensity, that is, shaded areas, whereas in full sunlight they will die. Apparently, the concentration of carbohydrate in the leaf affects virus-symptom expressions. The intensity of light also affects several fungal diseases. Powdery mildews often develop better under shaded conditions. Although it is difficult to separate the response of the host from that of the fungus to light, it has been shown that low light intensities (25 to 50 foot candles) favour maturation of appressoria of *Blumeria* (*Erysiphe*) *graminis* while high intensities (7200 foot candles) and darkness inhibit appressorial maturation.

The quality of light may also affect plant pathogenic fungi. It has been shown that the ultraviolet portion of light spectrum initiates and stimulates the sporulation of several plant pathogens, for example, *Botrytis cinerea* and *Helminthosporium avenae*.

Sclerotia of *Sclerotinia sclerotiorum*, when placed in moist sawdust and in the dark, only produce apothecial initials and not apothecia. They develop apothecia as soon as they are exposed to light.

Ultraviolet light and direct sunlight inhibit the germination of the teliospores of *Puccinia graminis*. If the dishes are covered with green light filters, germination occurs normally. *Colletotrichum lagenarium* (anthracnose of cucurbits) produces perfect stages of the culture if exposed to ultraviolet light. Prolonged exposure to ultraviolet light is harmful to fungus for they have fungicidal properties.

8.6 EFFECT OF OXYGEN AND CARBON DIOXIDE CONCENTRATION

The ratio of oxygen to carbon dioxide in the atmosphere is almost fixed and generally there is no marked affect of these gases on the foliar pathogens. However, in soil-borne diseases, O_2 and CO_2 are the two gases which, are of importance. The oxygen present in the soil also influences the root growth of host plants. Carbon dioxide in the soil also influences the general microbiota and also the growth of roots in the soils of the host plants. The role of available oxygen in the soils also appears very important in determining the population limit reached by the nematode species. Increased oxygen concentration stimulates activity of nematode larvae while the reverse is seen when carbon dioxide concentration is increased.

The amount of oxygen is not that important as the concentration of carbon dioxide. *Rhizoctonia solani*, infecting various crops, is inhibited if the concentration of CO_2 in the soil is more. *Fusarium culmorum* and *Gebberella zeae* cause root rot and seedling blight of wheat. These fungi can tolerate higher CO_2 concentrations. Plants become weak and increased activity of fungal pathogens cause more diseases. Many species of *Phytophthora* require some oxygen for their growth and zoospore production. The availability of oxygen determines the vertical distribution of fungi in soils which has marked relevance to the occurrence of disease.

8.7 EFFECT OF WIND/AIR

Most plant diseases that are spread either directly by wind or indirectly by insect vectors that can be carried over long distances by wind are likely to assume large epidemic proportions. Some conidia, sporangia and zoospores are delicate and do not survive long-distance transport in the wind but urediniospores and teliospores can be transported by the wind. Wind dispersal is important in disease development if it is accompanied by rain. Windblown rain releases spores/bacteria from infected tissue, then carries them through air currents deposits them on wet surfaces of plant, and can cause infection immediately. Wind also influences the number of wounds on host plants by rubbing against one another and this facilitates infection in some mechanically transmitted viruses. Winds, to some extent, prevent infection by increasing the drying of wet surfaces of plants. For example, the inoculum for rusts of wheat builds up in South India and it is disseminated towards North India through air currents.

8.8 EFFECT OF HERBICIDES/FUNGICIDES

Use of herbicides/fungicides is common and widespread in agriculture as they not only result in environmental and safety hazards but also lead to development of new strains of pathogens. In many cases, herbicides have been shown to increase in some diseases on crop plants like Fusarium wilt of cotton and tomatoes or decrease diseases, for example, Phytophthora collar rot of many crops. The direct effects of these chemicals may include stimulation or inhibition of the growth of the pathogen or increase/decrease in the population of beneficial microorganisms, especially arbuscular mycorrhizal fungi. The higher dozes of these chemicals generally suppress the activity of soil microflora or alteration of microclimate of the crop-plant canopy by changing humidity.

8.9 EFFECT OF AIR POLLUTION

As pathogens and infected parts of the plant are both exposed to air pollution, it is not clear whether a particular air pollutant causes an increase or decrease of a disease caused by the pathogen. For example, in the rusts of wheat and oats, ozone reduces hyphal growth, uredial growth, and the number of urediniospores on the infected leaves whereas with the powdery mildew of barley, the rate of infection is increased if exposure is early. With non-obligate parasites, ozone may increase the disease percentage on wheat by *Drechslera* fungus. Similarly, the bacterium *Pseudomonas glycinea* infecting soybean caused a smaller number of lesions on plants exposed to ozone in comparison to unexposed ones.

8.10 EFFECT OF NUTRIENTS

The relation between nutrition and susceptibility to parasitic and nonparasitic diseases in plants has been discussed by many workers. We can distinguish three main types of reactions in the behaviour of the host plants towards the parasite. These are susceptibility, resistance, and reduced receptivity.

These different kinds of reactions to parasite attack are correlated in the following way with the different forms of nutrition. Susceptibility is usually most evident where there is potash deficiency, resistance where there is nitrogen deficiency, and reduced receptivity where there is phosphoric acid deficiency. When one nutrient is given in excess, there is also a relative deficiency of the two other main nutrients. Usually, however, only one deficiency factor predominates and it behaves in line with the excess factor as regards the kind of reaction involved. The effect of different levels of application of lime is considerably less than that of different levels of nitrogen and potash.

Nitrogen abundance results in the production of more vegetative and succulent growth of the plants. It is known that fertilizer application with large amounts of nitrogen increase the susceptibility of the pear to *Erwinia amylovora*, cause of fire blight, rusts of wheat caused by *Puccinia* species, and powdery mildew. While reduced availability of nitrogen increases the susceptibility of tomatoes to Fusarium wilt, early blight of potato and tomato is caused by *Alternaria solani,* wilt of solanaceous plants is caused by *Ralstonia* (*Pseudomanas*) *solanacearum*, and damping-off of seedlings is due to *Pythium* species.

Not only the amount of nitrogen but the form of nitrogen also has its bearing on certain diseases. For example *Plasmodiophora brassicae, Sclerotium rolfsii*, and *Pyrenocheata lycopersici* are more serious when ammonium nitrogen fertilizer is applied; on the other hand, *Gaeumannomyces graminis, Phymatotrichum omnivorum*, and *Streptomyces scabies* are more serious when nitrate nitrogen is supplied in the soil. The effect of each nitrogen form appears to be related to soil pH influences (Agrios, 1997).

Phosphorus has been shown to reduce the severity of take-all disease of barley caused by *G. graminis* and potato scab caused by *S. scabies*. Potassium reduces the severity of several diseases. Stem rust of wheat and stalk rot of corn are reduced if more potassium is available to the plants. However blast disease of rice caused by *Pyricularia grisea* and root knot of nematodes are increased due to higher amount of potassium. Calcium reduces the severity of several diseases such as diseases caused by *Rhizoctonia, Sclerotium*, and *Fusarium* species. But black shank of tobacco and common scab of potato are enhanced by calcium. Micronutrient applications have tremendous impact on the occurrence of several diseases: copper application enhances take-all and ergot diseases caused by *G. graminis* and *Claviceps purpurea*. Addition of silicon in the nutrient solutions has been shown to reduce the blast of rice and brown leaf spot of rice.

8.11 ROLE OF BIOTIC ENVIRONMENT

For the foliar pathogens, the phyllosphere is the region which influences the disease in the aerial parts and for the root diseases, it is the rhizosphere which greatly influences the activity and growth of several pathogens. Phylloplane is the natural habitat on the leaf surface which supports a heterogenous population comprising both pathogens and nonpathogens. Last (1955) termed the niche as phyllosphere (now known as phylloplane). The **phylloplane microbes** are of special interest from various points of view. For example, some of them have antagonistic action against fungal pathogens, degrade plant-surface wax and cuticles, produce plant hormones, decompose plant material, activate plants to produce phytoalexins, and influence growth behavior and root exudates of plants. Thus, phylloplane exerts a great influence on the diseases caused by aerial plant pathogens.

The microbiota in the soil influences the growth of plant roots as well as microbial population in the rhizosphere or rhizoplane. A rich flora and fauna of soil microbiota lessens the danger of epiphytotic outbreaks of diseases caused by soil-borne pathogens. There are antagonistic and associative effects on the plant pathogens in the soil.

8.12 ROLE OF ENVIRONMENTAL FACTORS IN EPIPHYTOTICS

A plant disease is called epidemic or epiphytotic when the amount of disease present increases rapidly from a low level to a high level. In an endemic disease, the disease level remains fairly constant. Environmental factors which predispose the plants to infection, leading to the occurrence of epiphytotics are temperature, light, relative humidity and moisture, wind and the presence of inoculum or introduction of a new pathogen species or race. Thus, a combination of environmental factors both at macro and micro levels has great relevance in the causation of plant-disease epiphytotics. These have been discussed at the appropriate places in Chapter 10.

REVIEW QUESTIONS

1. Give examples of plant diseases influenced by soil temperature.
2. Why does late blight of potato generally not occur in the plains of northern India?
3. How does pH of soil influence the occurrence of a disease? Give examples of a disease which occurs:
   (i) in alkaline soils
   (ii) in acidic soils
4. Name the environmental factors necessary for the development of Karnal bunt of wheat.
5. Explain why wilt of cotton occurs in clayey soil while root rot generally occurs in sandy or loamy soils.
6. Give examples of diseases influenced by light condition.
7. How does nutrition affect the incidence of diseases?
8. Name some plant pathogens requiring high relative humidity throughout their development.
9. Name some plant pathogens causing infection at low relative humidity.
10. Give an example of a disease in which high levels of phosphorus bring about resistance in the plants.
11. How do phylloplane and rhizosphere organisms influence diseases in plants?
12. How do the following affect the incidence of disease?
    (a) Effect of wind/air
    (b) Effect of herbicides/fungicides
    (c) Effect of air pollution

MULTIPLE-CHOICE QUESTIONS

1. Which of the following disease occurs in the warm belts of the globe?
   (i) *Fusarium oxysporum* f. sp. *lini* (ii) *Phytophthora infestans*
   (iii) *Puccinia graminis tritici* (iv) All of them
2. In case of soil-borne diseases, the amount of soil moisture determines
   (i) speed of germination of inoculum
   (ii) entry of pathogen into the host
   (iii) concentration of various salts in the soil
   (iv) all of them
3. Which of the following pathogen is not disseminated from one place to other through the air currents?
   (i) *Rhizoctonia* (ii) *Colletotrichum*
   (iii) *Puccinia* (iv) *Melampsora*

4. Which of the following fertilizer application with large amounts increase the susceptibility of plants to pathogens?
   (i) Nitrogenous fertilizer
   (ii) Dung manure
   (iii) Vermiculite
   (iv) None of these
5. Which of the following disease is favoured by wet soil or high relative humidity?
   (i) Wart disease of potato
   (ii) Common scab of potato
   (iii) Powdery mildew of wheat
   (iv) Root and stem rot of peas
6. Which of the following disease is favoured by alkaline soils?
   (i) Take-all disease of wheat
   (ii) Cotton wilt
   (iii) Bacterial wilt of stone fruits
   (iv) Root rot of tobacco
7. Which of the following herbicides have been shown to reduce the activity of *Rhizoctonia*, *Fusarium*, *Alternaria*, and *Xanthomonas*?
   (i) Propanil
   (ii) Benthiocarb
   (iii) Dinitroaniline
   (iv) All of them
8. The rhizosphere may influence diseases by one or more of the following mechanisms:
   (i) Availability of inorganic minerals and N_2 to the plant and pathogen
   (ii) Production of beneficial microbial metabolites affecting the plant and pathogen
   (iii) Production of harmful microbial metabolites affecting the plant and pathogen
   (iv) All of them

Answers

1. (i) 2. (iv) 3. (i) 4. (i) 5. (i) 6. (i) 7. (iv) 8. (iv).

9

DISPERSAL OF PLANT PATHOGENS

Distribution implies the spread of plant pathogens into new geographical areas and their establishment there. The spread of a plant pathogen within the general area in which it is established is termed **dissemination** or **dispersal**. Contamination—the movement of inoculum (even if only a few inches) to susceptible plant—is a kind of dissemination. Knowledge of the methods by which pathogens are dispersed is usually essential before effective control measures can be devised. There are essentially two types of dissemination of plant pathogens:

1. Direct transmission
2. Indirect transmission

In **direct transmission**, the dispersal takes place along with the seeds and vegetative parts, while indirect transmission may be autonomous, that is, by wind, water, animals, or human beings.

1. Direct Transmission
    (a) Germinative transmission
    (b) Vegetative transmission
    (c) Adherent transmission
2. Indirect Transmission
    (a) Autonomous transmission
    (b) Wind dispersal of pathogerns—anemochory
    (c) Water dispersal of pathogens—hydrochory
    (d) Animal dispersal of pathogens—zoochory
    (e) Human dispersal of pathogens—anthropochory

9.1 DIRECT TRANSMISSION

It occurs when the pathogen is carried externally or internally in the seed or planting material like tubers, bulbs, cuttings, etc.

9.1.1 Germinative Transmission

This takes place along with the seeds or propagules. Bean mosaic is transmitted through pollen grains and carried in the seeds. In the case of loose smut of wheat and barley caused by *Ustilago segetum* (*U. tritici*) and *U. nuda* respectively, the seeds get infected without producing any symptoms in the parent carrier. The seeds carry the intraseminal mycelium. When such apparently healthy-looking seeds are sown, the symptoms appear when the ears emerge. The pathogen develops systemically inside the plants. Alternaria leaf blight of wheat is both internally as well as externally seed-borne. The Helminthosporium blight of wheat and barley is seed-borne. The seed-borne infection is present in the form of mycelium within the seed including the endosperm and embryo. The fungus has been found to remain viable up to seven years on barley seed. Bacterial blight of paddy caused by *Xanthomonas campestris* pv. *oryzae* is believed to be seed-borne.

9.1.2 Vegetative Transmission

Transmission of a large number of plant pathogens takes place through the vegetative parts of plants used as seed, such as tubers, cutting, runners, grafts, etc.

Ring rot of potato caused by *Clavibacter michiganense* subsp. *sepidonicum* (*Corynebacterium sepidonicum*) is carried through infected potato tubers to newer areas as in the case of the pathogen *Pseudomonas solanacearum* responsible for brown rot of potatoes. A large number of viruses are transmitted through vegetatively propagated parts.

Dormant mycelium of *Phytophthora infestans* is carried through the potato tubers. Seed setts of sugar cane carry the pathogens and when such apparently healthy-looking vegetative parts are sown, the pathogen develops systemically. Whip smut of sugar cane caused by *Ustilago scitaminea* and red rot of sugar cane caused by *Colletotrichum falcatum* are seed-borne (i.e. through the diseased cane setts).

9.1.3 Adherent Transmission

In this case, the propagules of the pathogens are carried over the surface of the seed or vegetatively propagated parts. Many pathogens are present on or in seeds, transplants, hard woods, or nursery stock and are disseminated by them as the latter are transported to other fields or areas or are sold and transported to other areas far and wide. Bunt of wheat caused by *Tilletia foetida* and *T. caries* is carried through the seeds externally, as in the case of covered smut of barley, where the smut spores attach themselves to the surface of the seeds at the time of threshing. There are many other fungi whose spores remain adherent, such as *Synchytrium endobioticum* responsible for potato wart and species of *Helminthosporium*, *Rhizoctonia solani* (black scurf) sclerotia on the tuber surface of potato, and *Alternaria* which causes various diseases.

9.2 INDIRECT TRANSMISSION

It occurs when the pathogen is transmitted by natural agencies like wind, water, insects, nematodes, mites, birds, animals, etc.

9.2.1 Autonomous Transmission

Plant pathogens can be spread to short distances by this method. Autonomous transmission takes place by active growth of hyphae or hyphal strands. It is characteristic of wood-rotting fungi, such as *Armillaria*, *Fomes*, *Ganoderma*, and *Polyporus* spp. belonging to the Hymenomycetes which migrate independently through the soil from plant to plant or even from field to field by the active growth of their hyphal strands.

Armillaria mellea rhizomorphs spread from one root to the other in forest trees. Hyphal mats or strands grow out to short distances in the soil from the sclerotia of *Rhizoctonia solani* and *Sclerotium rolfsii*. The rate of such spread in *Phymatotrichum omnivorum,* responsible for Texas root rot of cotton crop, is estimated at 5 to 30 feet per season and 2 to 8 feet per month in alfalfa crop. It may be emphasized that in *Armillaria mellea*, the cause of the tree root disease, *Gaeumannomyces* (*Ophiobolus) graminis*, the cause of the take-all disease of wheat, *Phymatorichum omnivorum*, the cause of root rot of cotton, the pathogens do not grow and ramify through the soil as do the nonspecialized types, but require a food base such as the roots or crop debris of the host plants which they attack.

9.2.2 Wind Dispersal of Pathogens

This type of spread is termed by Gaümann (1950) as **anemochory**. Some plant pathogenic bacteria are carried to short distances by the wind. The seeds of some angiospermic plant parasites are effectively carried by the wind. All fungul pathogens which produce spores or conidia on the surface are disseminated by the wind, such as downy mildews, powdery mildews, rusts, smuts, leaf spot causing pathogens, sooty molds, etc. The pathogen must produce numerous spores which are successfully liberated, dispersed and deposited in a viable condition on susceptible plants under conditions conducive to infection; in other words, production, liberation, dispersal and deposition of spores are involved.

Many fungal pathogens are remarkably well adapted to wind dissemination for they produce and liberate into the air almost countless numbers of small and very light spores that are carried to long distances.

Spore production per unit area of infected leaf tissue may be increased by close aggregation of sporophores, branched or whorled sporophores, production of spores in chains or clusters, by a more or less continuous production of spores within the sorus or by successive crops of spores. The number of spores produced can be astronomical. Powdery mildews may produce several thousand conidia per square centimeter of infected leaf surface, a fairly, modest output compared with 100,000 or more in some downy mildews. A single smut sorus may contain millions of spores, a heavily infected barberry bush is said to produce up to about 70,000 million aeciospores of *Puccinia graminis* at one time (and there may be several crops of spores in the spring), and a relatively small apothecium of *Sclerotinia* can produce a bout of 30 million ascospores. The spores produced by large polypores, such as *Fomes*, are even more numerous. Heavily rusted wheat fields produce so many spores that spore clouds may be seen rising during harvesting and implements too may be layered with a fine reddish deposit of spores.

The duration and periodicity of sporulation is as important as the number of spores. Some pathogens produce a more or less simultaneous crop of propagules followed by other crops, as in the case of sporangiophores of *Peronospora.* The sporophores of others may bear several successive propagules, for example, those of *Phytophthora, Cercospora,* etc. The capacity to produce a steady stream of infective propagules over a prolonged period of time is likely to be advantageous to the pathogen. Some fungi show a diurnal periodicity in sporulation, as nocturnal production of sporangia by some species of *Sclerospora*. This is probably chiefly due to temperature-moisture effects, but might in some cases be related to periodicity in susceptibility of the host plant.

Apart from spores, bits of mycelium and nematode cysts are also sometimes disseminated by wind. In India, the molya disease of wheat and barley *(Heterodera major)* is believed to have been introduced into Haryana from Rajasthan in this manner.

Although fungus spores are generally more abundant in the air near the earth than at high altitudes, several investigators have encountered clouds of spores numerous bacteria, and other minute objects several thousand feet above the earth. Urediniospores of *Puccinia graminis tritici* have been caught as high as 14,000 feet above infected grain fields. Living spores of various fungi were caught from aeroplanes above the Carribean Sea 600 miles from their nearest possible source, and living spores of several common molds were caught in a spore trap released from the balloon Explorer II at 72,500 feet.

Long-distance dissemination of rusts is a well-established phenomenon. In the Indian subcontinent, rust spores are known to make big jumps from the source areas to the plains. Studies have shown that for South and Central India, the spores are wind-borne, rain-deposited and are transported from an altitude of around 700 mb (300 m). The urediniospores get transported by upper winds from the Nilgiri and Palney hills and are then washed down over Central India by rain. Such a spread occurs to a distance of 600 km or more without infecting the fields in between.

Turbulence redistributes spores and affects the progressive dilution of spores with increasing distance from their source. There is ample evidence that with normal wind and turbulence conditions, spores, especially small ones, can travel considerable distance. Those of *Phytophthora infestans* have been reported by various observers to travel distances of 200 m to more than 60 km depending upon the wind degree of turbulence.

An interesting example of wind dispersal is in case of fire blight bacterium (causing diseases in apples and pears). The pathogen produces fine strands of dried bacterial exudate containing bacteria, and these strands may be broken off and disseminated by wind.

9.2.3 Water Dispersal of Pathogens (Hydrochory)

Water, as an agent of dissemination, appears to be relatively less important as compared to wind. Although it sometimes facilitates dissemination in various ways, it does not furnish the motive power for long-distance dissemination, except as streams and rivers. Flood and irrigation water may carry inoculum, especially that in or near the soil. It is clear that water separates and distributes spores in a micro-environment. Spores that are extruded in gelatinous tendrils that may harden when dry, can be separated from each other and washed down trees and other kinds of plants during rain or heavy dews. The splashing and spattering of water during heavy rains may result in distributing inoculum to plant parts near the soil and may distribute bacteria and spores to different parts of the same plant or neighbouring plants.

Bacteria, nematodes, spores, and mycelial fragments of fungi present in the soil are disseminated by rain or irrigation water that moves on the surface or through soil.

In some places, rain seems to be more important than insects in disseminating the fire-blight bacteria, *Erwinia amylovora*, by its washing or spattering effect. The secondary spread from initial infection of *Xanthomonas campestris* causing black rot of cabbage and other crucifers of *X. translucens* which attacks cereals and grasses and of bacterial blight and canker of tomato, *Clavibacter michiganese* subsp. *michiganese*, results largely from splashing during rains.

Xanthomonas campestris pv. *malvacearum*, the bacterium that causes angular leaf spot of cotton, can be spread by wind-blown rain.

Surface water resulting from rains or irrigation is known to disseminate some pathogens and undoubtedly a large number of spores. The bacterium causing angular leaf spot or black arm of cotton is spread largely by irrigation water or surface water in Arizona, USA, and Uganda. *Plasmodiophora brassicae*, the organism causing club root of crucifers, is spread by surface water, and *Phytophthora infestans*, causing late blight of potatoes, has been observed to spread from plants in dry areas by means of irrigation waters.

Colletotrichum coffeanum, which produces spores within the acervuli in the maturing bark of coffee twigs and on diseased berries, is disseminated by water.

Water, then, is an important agent of local dissemination, but it is not as important as wind for long-distance dissemination.

Large drops of water in the form of rain or overhead irrigation, falling at their terminal velocity, are probably most active in splash dispersal, much more so than small drops. In a **splash dispersal**, the spores are carried and often deposited on the plant surface in a droplet of water which could provide the moisture required for their germination especially in humid weather. Splash dispersal, although usually considered a mechanism for liberating and dispersing slimy spores, may also play a part in the dispersal of roughened spores which adhere strongly to each other or the plant surfaces. In *Hemileia vastatrix* (coffee rust) the roughened urediniospores cling together and are broken up by water. They then float singly to the surface of the water drop, from which they are splash dispersed by further raindrops. The cells of many plant pathogenic bacteria and the slimy spores of such important genera of fungal plant pathogens as *Septoria, Fusarium*, and *Colletotrichum* are splash dispersed and it is surprising that relatively little information on splash dispersal is available.

The eelworm disease of potato and *Piper longum* have been known to be carried through surface water from field to field. The spread of infection in the red rot of sugar cane caused by *Colletotrichum falcatum* and the dissemination of the sclerotial bodies of *Sclerotium rolfsi* through soil are chiefly brought about by irrigation water. Water is also an important agency of dissemination for the seeds of flowering parasites such as *Orobanche,* dodder (*Cuscuta*), and *Striga* through the soil. Short-distance transmission of semi-aquatic pathogens from one leaf to another of adjacent plants occurs through splashing rains accompanied by wind such as for *Phytophthora arecae, P. parasitica, P. infestans*, and other downy mildew fungi. These are well adapted to wind as well as water dissemination due to the formation of deciduous conidia capable of germinating either by zoospores or germ tubes and thus develop into dangerous pathogens capable of widespread damage within a short period.

9.2.4 Dissemination by Insects, Mites, and Nematodes

1. Insects

Insects not only transmit and disseminate many plant pathogens but are sometimes agents of inoculation also. Moreover, some pathogens multiply within insects and some are present in them during periods when host plants are dead or dormant.

The largest number of insect vectors have sucking mouth parts which belong to the Homoptera. Within this order, the largest number of species which are known to transmit one virus or the other is contained within two families—Aphididae (aphids) and Cicadellidae (leaf hoppers). A few belong to the Aleyrodidae (white flies), the Cercopidae (Spittle insects), the Coccidae (mealy bugs), and the Membracidae (tree hoppers) families. Thrips transmit the spotted-wilt virus. Some viruses are transmitted by both chewing and sucking insects, while a few are transmitted only by the former. Squash mosaic, cowpea mosaic, and turnip yellow-mosaic viruses are transmitted by beetles (Coleoptera), while the last is transmitted also by grasshoppers (Orthoptera) and earwigs (Dermaptera).

It is of interest that practically all vectors of the so-called yellows group of viruses are leaf-hoppers while most vectors of the mosaic group are aphids. In other words, those viruses which are more readily transmitted mechanically are aphid-borne and those which do not yield to such transfers are leaf-hopper borne.

More than fifty species of aphids are known to transmit the viruses of plants. Some species can transmit only a few kinds and others transmit many; *Myzus persicae,* the green peach aphid, alone is the vector of fifty different viruses.

Some of the diseases transmitted by aphids are Katte or Marble disease of cardamom, mosaic and grassy shoot of sugar cane, mosaic streak of wheat, maize mosaic, barley mosaic, bean mosaic, cowpea mosaic, bunchy top of bananas, banana mosaic, papaya mosaic, potato necrosis (viral), potato leaf roll, tomato mosaic, and chilli mosaic.

Some of the diseases transmitted by white flies are tobacco leaf curl, mung (*Phaseolus aureus*), yellow mosaic, double bean (*Phaseolus lunatus*) yellow mosaic, and yellow mosaic of *Dolichos lablab*. Tungro of rice is transmitted by leaf hopper vectors, while bottlegourd mosaic is transmitted by the red pumpkin beetle.

Three important questions arise with respect to the transmission of viruses by aphids:

1. How long must the insect feed on infected plants before it can transmit the virus to healthy plants?
2. Is an incubation period in the insect necessary before the virus can be transmitted?
3. How long do the insects remain infective after having acquired the virus?

Some viruses are quickly acquired and lost, others are acquired more slowly and are quite persistent. Some require an incubation period and others do not. In general, the viruses that require no latent or incubation period in the insects are relatively nonpersistent and those that require one or more are persistent. Insects are important in the dissemination, inoculation, and persistence of certain bacterial pathogens of plants.

Erwinia tracheiphila, the cucurbit wilt organism, is completely dependent on cucumber beetles for its spread. *Erwinia* (*Xanthomonas*) *stewarti*, the corn-wilt pathogen, overwinters inside the corn flea beetle (*Chaetocnema pulicaria*). The cucumber beetles, *Diabrotica vittata* and *D. duodecimpunctata,* chew the plant parts affected by *E. tracheiphila* and carry the bacteria inside the body where they overwinter, passing on the host in the next season. It is believed that the relationship between the insects is symbiotic.

Insects are important agents of dissemination and inoculation of *Erwinia carotovora*, causing black leg of potato. The pathogen survives in the seed tubers, which can account for primary infection in the spring. It was originally thought that the organism did not survive in the soil and there was no secondary spread in the field. Circumstantial evidence later indicated that there must be a secondary spread, but the agents remained unknown until 1926 when it was proved that the seed-corn maggot, *Hylemya cilicrura*, is a very important agent of effective dissemination and inoculation.

The seed-corn maggot and closely related species are the pests of many kinds of plants, such as corn, cabbage, beets, beans, pear, radish, onion, tomato, etc. These insects and certain others contribute to the development of soft rot caused by *Erwinia carotovora* in some of these crops in much the same way as has been described for black leg.

Erwinia amylovora, the cause of fire blight of apples, pears, and certain other members of the family Rosaceae, was the first plant disease known to bacteria and transmitted by flies and ants. *Xanthomonas* campestris pv. *citri* which causes the well-known citrus canker is carried by the leaf miner from diseased to healthy plants.

Erwinia (*Xanthomonas*) *stewartii* and *X. campestris* pv. *vasculorum* which cause wilt of corn and gummosis of sugar cane respectively, are transmitted by insects and can persist inside the vectors. *Erwinia stewartii* is also transmitted on seed and *X. campestris* pv. *vasculorum* on sugar cane setts used for planting. The corn-wilt organism, however, is carried from plant to plant in the fields by flea beetles. *Chaetrocnema pulicaria* and *C. denticulata* by *Diabrotica duodecimpunctata,* mentioned in the discussion on cucurbit wilt and probably by other insects as well. Moreover, the bacteria can live during the winter inside the corn flea beetle, *C. pulicaria,* and can live for some time in the adult Diabrotica beetle.

The big-vein disease of lettuce has been known since the 1930s. The disease-causing agent was not identified at that time but was known to be soil-borne. In the 1950s, the chytridiomycetous fungus, *Olpidium brassicae*, was found to be intimately associated with this disease. Finally, it has been shown that the causal agent could be transmitted by grafting in the absence of *Olpidium*. Thus, this disease is now known to be caused by a virus. Later it has been shown that this virus is transmitted by the zoospores of *Olpidium* and survives within the resting spores of this fungus. *Olpidium* also transmits other viruses that are soil-borne. They are also known to be transmitted by various members of Chytridiomycota.

About 66 species of fungi causing plant diseases are transmitted by more than a 100 species of insects belonging to at least six orders. Transmission occurs in various ways with or without injury to the plant.

Spores of *Ustilago violaceae* (anther smut of Caryophyllaceae) and *Botrytis anthophila* (anther mold of clover) are thus carried from the infected anthers to uninfected flowers by pollinating insects. The conidia of *Claviceps* are spread from infected ovaries to healthy ones by insects which feed on the sugary honey dew on which the conidia are produced. Other fungi infecting the shoots and roots of plants are likely to be fortuitously disseminated by insects which move or are carried from plant to plant. Richardson and Saunders (1968), for example, attributed the partial control of potato blight by aphicide sprays to the destruction of aphids which carry mycelial fragments and sporangia of *Phytophthora infestans* on their bodies and leg bristles, and so spread the pathogen over short distances.

More frequently, however, transportation is accompanied by injury to the plant. The insect carries the fungus and places it in the wounds, thus facilitating infection. The injury may be caused by mechanical breaking, by other animals or by the insect itself. The spores of *Ceratocystis fagacearum* (oak wilt) are carried by various insects, chiefly Nitidulid and Scolytid beetles, to existing wounds caused by breaking, woodpeckers, squirrels, and other causes. Mycelial mat of the fungus develops beneath the bark, the crack is open and gives off a characteristic odour which apparently attracts insects. These feed on the fungus and also carry small spores which fertilize the receptive hyphae and so lead to the development of perithecia of the fungus. In some cases, the insect or the fungus is adapted to insect transmission. Female Siricid (Sirex) wood wasps carry hyphae and conidia of *Stereum sanguinolentum* (heart rot of conifers) within the intersegmental sacs at the anterior end of the long ovipositor. The eggs are contaminated with the fungus during laying and are deposited within the wood of the tree and the fungus is reported to precede the larva as the latter bores through the wood. The fungus is contained in the hypoplural organ modifications of the larval integument. *Ophiostoma* (*Ceratostomella) ulmi* (Dutch elm disease) grows within the egg galleries of bark beetles (*Scolytus*) and contaminates emerging adults, which also inject the slimy spores and thus spread the fungus to adjacent healthy trees. More sprecific are the diseases (stigmatomycoses) in which the fungus is injected into plant tissue by the insect vector. Several parasitic yeasts, notably *Nematospora*, are transmitted in this way, the acicular ascospores being sucked up the feeding stylet of stainer bugs (*Dysdercus, Nezara*) whence they lodge in the stylet pouches and are injected into young cotton bolls causing internal boll disease due to *Nematospora gossypi,* bean pods (*N. phaseoli*, yeast spots) or coffee berries (*N. coryli, N. gossypi*). It is thought that *Nematospora* is almost entirely dependent on plant bugs for dissemination and entry into the plant. In the case of chestnut blight caused by *Cryphonectria* (*Endothia) parasitica*, the insects responsible for the disease are the European bark beetle (*Scolytus multistriatus*) and the native elm bark beetle (*Hylurgopinus rufipes*).

Many insects attack the fruits of plants and fruit-rotting fungi often enter through the wounds. The spores of *Aspergillus niger*, which causes rot of fig fruits, are reported to be introduced by certain beetles and fruit flies, and *Sclerotinia* sp. (brown rot of various fruits) can enter through feeding and oviposition wounds made by insects. Some fungi causing stem rot (*e.g. Physalospora tucumanensis*, red rot of sugar cane) also invade through insect wounds and it is suspected that some leaf-attacking fungi behave in a similar way. *Pycnostysanus azaleae* (bud blast of *Rhododendrons*) is associated with the leaf hopper *Graphocephala coccinea* and the disease is very slight in the absence of the insect. Little is known of dissemination of the root pathogen by insects but it seems likely that this occurs in the soil and that insects feeding on roots also introduce fungi into them. The original insect damage may thus be greatly aggravated by subsequent fungal invasion. Insects which feed on cotton plants have been found to contain *F. oxysporum* f. sp. *vasinfectum* (cotton wilt), the fungus having been recovered in a viable condition from faecal pellets of various grasshoppers. It seems possible that such insects disseminate this normally slowly spreading fungus which might explain the occasional appearance of the disease in hitherto disease-free fields or parts of fields, but other explanations are possible, and the relationship between the soil-borne pathogens and insects needs further investigation.

Insects which act as diploidizing agents for some heterothallic fungi, transfer fertilizing spores (spermatia, oidia) between compatible strains and so bring about the formation of the perfect state. The best known example of this diploidization is that of receptive hyphae of rusts by insect-borne spermatia, but similar fertilization occurs in some Ascomycetes and other Basidiomycetes. This phenomenon has obvious similarities with insect pollination in higher plants.

2. Mites

These, although wingless, transmit some viruses (reversion disease of currants) which are perhaps carried by winged insects or wind. Transmission of fungal pathogens by mites has not been extensively studied.

3. Nematodes

Some viruses as well as some bacterial and fungal plant pathogens are transmitted by nematodes. *Corynebacterium fascians*, leafy gall of various herbaceous plants, is carried by ectoparasitic nematodes (*Aphelenchoides*) and typical cauliflower symptoms develop only when both nematode and bacterium are inoculated into the strawberry plant. Yellow ear rot of wheat caused by *Rathayibacter* (*Corynebacterium*) *tritici* is dispersed by the ear cockle nematode (*Anguina tritici*).

There is also evidence that some root-infecting pathogens, particularly fungi, may enter roots through wounds made by nematodes and this may perhaps be of more frequent occurrence then is at present realized. Cotton wilt caused by *Fusarium oxysporum* f. sp. *vasinfectum* is an example of this, the pathogen being associated with root knot nematodes and probably others. The nematodes may also affect the physiology of the root, making it a more suitable medium for the fungus. The reverse has also been reported; the fungus (*Pyrenochaeta lycopersici*) causing tomato brown root rot reduced the invasion of tomato roots by the potato root eelworm *Heterodera rostochiensis*. Bacterial wilt of carnations, caused by *Pseudomonas caryophylli*, is reported to be more severe in the presence of root-knot nematodes which probably facilitate the entry of the bacteria.

In 1958, it had been demonstrated that the grapevine fan-leaf virus is disseminated by the nematode *Xiphinema index*. Both polyhedral and tubular viruses have since been shown to be nematode transmitted, the former by species of *Xiphinema* and *Longidorus* and the latter by *Trichodorus* spp.

4. Animals

Grazing animals could easily transmit the disease mechanically from an infected plant to a healthy one. Domestic herbivores like goat, sheep, cows, and some other plant-eating animals including deer, feral pigs, rabbits, etc. tend to feed on different plant parts and thus transmit diseases, which is a mechanical mode of transmission. Sometimes as a result of their grazing, there is an increase in the population of insects like aphids due to change in the habit from long grasses to small grasses. These aphids are in turn responsible for the transmission of plant pathogens from one area to another.

9.3 BIOLOGICAL TRANSMISSION

Plant viruses could be transmitted from an infected to healthy plants through a bridging dodder, parasitizing both plants. An association of dodder and host cell is formed and transmission occurs through directional movement of nutrients.

Cochran (1946) has also found that transmission of tobacco mosaic virus, which occurs through dodder plants can be controlled by pruning the dodder and shading the healthy host while keeping the infected host in bright light.

Human Dispersal (Anthropochorus)

Humans are, to a large extent, responsible for the dissemination of plant pathogens which they do in two ways—through person and through the objects they transport from one place to another.

Local dissemination of pathogens on humans is illustrated by the angular leaf-spot disease of cucumber caused by the bacterium *Pseudomonas lachrymans*. When cucumbers are grown for pickles, the fruits are picked. When a person comes in contact with or picks an infected plant, his/her hands become contaminated, particularly when the plants are wet with dew. The pathogen is thus spread down the row of plants.

Another pathogen spread by humans is *Clavibacter* (*Corynebacterium*) *michiganense*, the cause of bacterial canker on tomatoes. This pathogen is seed-borne, and when it develops on tomato seedlings to be used as transplants, the workers handling the transplants may contact the bacterium on their hands. Since the handling of transplants often causes small injuries to the plant, the worker placing the transplants into the ground can easily inoculate healthy plants. In California, USA, this disease has become a rarity since most tomato fields are no longer planted with transplants but are direct seeded.

The viruses that are spread by animals can also be spread by humans. Tobacco mosiac virus is known to survive in dry tobacco leaves. Hence, workers who handle tobacco plants in the field are advised not to smoke while they are working and to thoroughly wash their hands before entering the field. The long-distance dissemination of plant pathogens by humans is usually accomplished by the transport of infected plant parts, such as seeds, nursery stock, or timber products. The cause of some of the worst plant-disease epidemics is the introduction of a pathogen into a new area. In many cases, the plants growing in these areas are very susceptible since they do not develop resistance through evolution in the presence of the pathogen.

Humans have repeatedly carried pathogens to areas which otherwise would have remained free of them. Almost all countries have suffered to such an extent from pathogens introduced by humans that the Food and Agricultural Organization of the USA and several international committees of plant pathologists are promoting studies and regulations designed to reduce the danger.

Humans are an agent of transportation of pathogens from continent to continent, from country to country and even from plant to plant on their own farms. Eternal vigilance is necessary to protect economic plants against the pathogens distributed by humans.

REVIEW QUESTIONS

1. What is the essential difference between distribution and dissemination?
2. Give some examples of viral diseases carried in the seeds.

3. Give examples of plant diseases in which dissemination of the plant pathogens is dependent upon special types of insects.
4. Which insects act as diploidizing agents for some heterothallic fungi?
5. Name the diseases which have been introduced in India from other countries.

MULTIPLE-CHOICE QUESTIONS

1. In anemochory, the dispersal of pathogens occur by
   (i) water (ii) wind
   (iii) animals (iv) human beings
2. The primary source of inoculum for *Phytophthora infestans* is
   (i) oospores (ii) zygospores
   (iii) dormant mycelium in tubers (iv) plant debris in soil
3. Which of the following is not correct for indirect transmission of pathogens?
   (i) Water (ii) Wind
   (iii) Nematodes (iv) Seeds
4. How does *Armillaria mellea* spread from one root to another in forest trees?
   (i) Seeds (ii) Birds
   (iii) Rhizomorph (iv) Animals
5. Which of the following fungus is transmitted with wind currents from one place to another?
   (i) *Synchytrium* (ii) *Olpidium*
   (iii) *Puccinia* (iv) *Plasmodiophora*
6. Which of the following fungus does not act as a vector for transmission of viruses?
   (i) *Olpidium* (ii) *Polymyxa*
   (iii) *Spongospora* (iv) *Agaricus*
7. Which of the following plant parasites is responsible for transmission of viruses?
   (i) *Cuscuta* (ii) *Orobranche*
   (iii) *Striga* (iv) *Loranthus*
8. Which of the following nematodes is responsible for transmission of plant viruses?
   (i) *Xiphinema* (ii) *Longidorus*
   (iii) *Trichodorus* (iv) All of these
9. Which of the following statements is correct regarding the dispersal of pathogens in or by the soil?
   (i) Contamination of soil (ii) Growth and spread of the pathogen in the soil
   (iii) Persistence of pathogens in soil (iv) All of these

Answers

1. (ii) 2. (iii) 3. (iv) 4. (iii) 5. (iii) 6. (iv) 7. (i) 8. (iv) 9. (iv)

10 PLANT-DISEASE EPIDEMIOLOGY AND PLANT-DISEASE FORECASTING

10.1 INTRODUCTION

The practical implications of forecasting have been incorporated in the definition proposed by Miller and O'Brien (1952) who state "Forecasting involves all the activity in ascertaining and notifying the growers of a community that conditions are sufficiently favourable for certain diseases, that application of control measures will result in economic gain, or on the other hand, and just as important that the amount of disease expected is unlikely to be enough to justify the expenditure of time, energy, and money for control".

Forecasting the incidence of plant disease must take into account not only the disease itself but the weather also, which is more or less predictable. Disease forecasting is made more reliable if the reasons for a particular disease developing and certain conditions and not under others are known.

Importance of Disease-Forecasting Services

Plant pathologists are interested in forecasting for three major seasons:

(i) disease forecasts are useful to farmers in the field in the practical management of disease of their crop,

(ii) research aimed at accurate forecasting helps us to identify or pinpoint the gaps existing in our knowledge, and

(iii) modern research on modeling on the complexities of epidemics is stimulating and rewarding in the long run.

Forecasting can be useful if the forecaster suggests that the risk of disease is great, there is a big advantage in spraying or using resistant varieties, may be, of moderate yield. Forecasting provides a complex tool for determining as to when and where a given management practice should be applied.

10.2 COMPUTER SIMULATION OF EPIDEMICS

The availability of computers has allowed plant pathologists to write programmes that allow simulation of epidemics of several plant diseases. In 1969, the first computer simulation programs called EPI-

DEM was developed and resulted from modeling each stage of the life cycle of a pathogen as a function of the environment. This (EPIDEM) was designed to simulate early blight of tomato and potato caused by *Alternaria solani*. Later on, computer simulators were written for Mycosphaerella blight of Chrysanthemums (MYCOS), for apple scab caused by *Venturia inaequalis* (EPIVEN), for downy mildew of grapes caused by *Plasmopara viticola* (PLASMO) and for southern corn leaf blight infected by *Helminthosporium maydis* (EPICORN).

BLIGHTCAST (Krause *et al.*, 1975) for late blight of potato caused by *P. infestans* is a computer programme written in Fortan IV developed in the USA. This was a synthesis of the Hyre and Wallin fungicide scheduling systems developed for northeastern and mid-western US. BLIGHTCAST is based on predicting periods when infection is likely to occur. Spray recommendation were issued in two different triggers: after 10 consecutive days when the cumulative rainfall exceeded 30 mm and the 5-day average temperature did not go beyond 25.5°C, or after an accumulation of 18-20 "severity units". Each day was scored on a 0-4 "severity" scale according to temperature and hours with relative humidity more than 90%.

TOM-CAST is a computer-based forecasting system for tomato early blight (FAST). TOM-CAST required direct measurement of wetness duration in the field which was read by wetness sensors and data-collecting systems such as those used with BLIGHTCAST.

PLASMO is a forecasting model developed in Northern Italy to control downy mildew of grapes caused by *Plasmopara viticola*. It is a very sophisticated computer programme which simulates the critical phases in the life cycle of the pathogen using rate factors determined by environmental parameters (temperature, relative humidity, rainfall, and leaf-wetness duration). Conventional control of the disease is done by regular spray of fungicides at 7 to 10 day intervals.

In addition to the above, some recent disease forecasting/warning simulation models have been developed which include *Pyricularia grisea* on rice and *Blumeria* (*Erysiphe*) *graminis* f. sp. *tritici* on wheat.

In a computer simulation of an epidemic, the computer is given data describing the various subcomponents of the epidemic and control practices. The computer then provides continuous information regarding not only the spread and severity of the disease over time, but also with regard to the final crop and economic losses likely to be caused by the disease under the condition of the epidemic as given to the computer. Computer simulation of epidemics is extremely useful as an educational exercise for students of plant pathology and also for farmers so that they can better understand and appreciate the effect of each epidemic subcomponent on the final size of their crop loss. The computer evaluates not only the current status of the disease but also the effectiveness of the applied management measures in controlling the epidemic. Computer simulation of epidemics is used fairly in plant-disease forecasting systems. Such systems allow farmers to take appropriate control measures against a disease as soon as the conditions that are likely to lead to disease development are present or are likely to be present soon.

10.3 SYSTEMS APPROACH IN EPIDEMIOLOGY

As explained earlier, weather is an important factor that influences disease development. Where the host and pathogen coincide, weather is the only variable that influences epidemic development. Therefore, by monitoring the relevant weather parameter and relating them to disease development, empirical predication models can be developed.

Everdigen's Method

Van Everdigen (1926) in the Netherlands apparently was the first to base his forecasting system for late blight of potato on scientific principles. His forecast was based on what became known as the **Four Dutch Rules**. These rules were based on four weather parameters:

(i) Minimum temperature (night temperature must not be lower than 10°C (50°F))
(ii) Dew (leaves must remain wet for at least 4 hours the day following night dew)
(iii) Cloudiness (the day following the night dew must be cloudy)
(iv) Rainfall (during such periods described above, there must be at least 0.1 mm of rainfall)

When all these conditions are satisfied under field conditions continuously for a few days then in Netherlands control measures are recommended.

Forecasting and warning systems involve considerable expense and should be economically justifiable, as in the following cases:

1. For destructive diseases of important case or food crops.
2. For diseases which can be reliably predicted early enough for effective and economical control measures to be applied.
3. For diseases against which economical control measures are available.

10.4 METHODS USED IN PLANT-DISEASE FORECASTING

Disease forecasting allows the prediction of probable outbreaks or increases in intensity of disease and, therefore, allows us to determine whether, when, and where a particular management practice should be applied. It is also useful to farmers in the practical management of crop disease. Most frequently, farmers need forecasts of plant disease development to decide whether to spray a crop right away or to wait for several more days before they spray, since, if they can wait, they can reduce the amounts of chemicals and labour used without increasing the risk of losing their crop. In developing a plant-disease forecast, one must take into account several factors of the particular pathogen, host, and environment.

The methods used in disease forecasting mostly fall into 5 main groups based on

1. Weather condition during the intercrop months, particularly those affecting the survival of inoculum
2. Weather condition during the crop
3. The amount of disease in the young crop
4. The number of propagules of the pathogen in the air, soil, or planting material
5. Change in the host physiology

10.4.1 Weather-based Predictions

When the importance of weather conditions in determining the host-pathogen interaction was realized, prediction based on weather conditions were frequently used in forecasting diseases. The severity and subsequent spread of tikka disease of groundnut caused by *Cercospora personata* (now *Cercosporidium personatum*) and *C. arachidicola* depend upon the prevalence of high humidity and low temperature during crop season. Similarly, the chances of rapid development of epidemic of Cercospora leaf spot of peanut are optimum when foliage remain wet for a period greater than 10 hours or above and minimum temperature is 21°C or higher for the next 24 hours.

Padmanabhan (1965) examined the meteorological factors associated with the epiphytotics of *Helminthosporium oryzae* involved in the 'Great Bengal Famine' of 1942. It was revealed that heavy rainfall in September accompanied by favourable temperature of 25°C to 30°C, followed by cloudy weather in subsequent months resulting in low solar radiation were the critical conditions for outbreak of brown leaf spot of rice. Padmanabhan (1965) used weather analysis for predicting outbreak of blast of rice (earlier *Pyricularia oryzae* now *P. grisea*). Nycto-temperature ranging from 20-24°C in association with 90% relative humidity or above lasting for a week during any susceptible phase of crop brought the epidemic of blast.

Nagarajan *et al.* (1980) correlated the epidemics of wheat rusts of India with the prevalence of certain environmental conditions at certain places in the country during crop season. According to them, epidemic of leaf rust (*Puccinia recondita*) of wheat is likely to develop if

1. Rust infections are observed in Uttar Pradesh and Bihar between 15th and 20th January at 5-6 different locations separated from each other by a minimum distance of 25 ± 5 km;
2. The number of rainy days during January to mid-April over North-West India are at least twice the normal number, i.e. 12 days, and
3. The weekly mean maximum temperature during March to mid-April over North-West India is with 1°C of the normal temperature (26°C).

If the first and the second condition exist fully then a severe epidemic of leaf rust is likely to occur. If they exist partially, isolated outbreaks of the disease are possible. If the mean maximum, temperatures are usually low, stripe rust is likely to become a serious problem throughout North-West India.

10.4.2 Inoculum-based Prediction

Forecasting of many diseases is based on the estimates of amount of initial inoculum likely to infect the host plant at the beginning of a crop season. In many cases, there is a positive correlation between the amount of primary inoculum in vicinity of host plant and severity of infection. To quote Shrum (1978) again, "the forecast based on inoculum potential contains a hidden assumption that the weather will continue to be favourable for disease". Plant-disease forecasting can be based on the amount of inoculum available in air, soil, or planting material. Utility of inoculum-potential based forecasting can however be found for a number of soil-borne diseases.

The severity of apple scab caused by *Venturia inaequalis* is related to the amount of perithecia surviving in dead leaves and the subsequent amount of ascospores liberated from them. The forecasting of apple-scab epidemic is mostly based on temperature and moisture conditions, affecting the discharge of ascospores from perithecia at full-bloom stage of the crop. The discharge of ascospores takes place when perithecia are thoroughly wetted and the infection takes place 30 hours after leaf wetness at 5°C, 14 hours at 10°C or 9.5 hours 15°C.

A system for predicting blister blight of tea caused by *Exobasidium vexans* has been developed which is based on the number of spores in the tea garden in the air and the duration of surface wetness on the leaves. It was found empirically, by observation, that the number of basidiospores of the pathogen in the air at a given time was related to the percentage of shoots which were infected. The duration of sunshine is negatively correlated with the duration of surface wetness of the leaves. Therefore, the duration of sunshine was used as a measure of surface wetness indirectly. A prediction equation for the development of epiphytotic has been developed.

10.4.3 Type of Reproduction of the Pathogen

Some of the microbes such as most fungi, bacteria, and viruses produce many offsprings while a few fungi, all nematodes, and all parasitic plants produce comparatively a small number of offspring. Some plant pathogenic fungi, bacteria, and viruses are polycyclic, i.e. they produce many generations of offspring in a single growing season. Most of the rusts, mildews, and leaf spots are polycyclic and are responsible for most plant-disease epidemics. Some of the soil fungi like *Fusarium, Verticillium*, and nematodes have one to four reproductive cycles per growing season. Such type of pathogens often cause localized and slower developing epidemics. Several monocyclic pathogens such as smuts require a whole year to complete a life cycle. In such cases (monocyclic diseases) the epidemics usually develop over several years and lead to long-term epidemics.

10.4.4 Ecology of the Pathogen

Some pathogens produce their inoculums on the surface of the aerial parts of the host and these can be dispersed easily over a range of long distances and can cause widespread epidemics. Some other pathogens such as bacteria, viruses, protozoa, and vascular fungi reproduce inside the host and spread of such pathogens are rare or impossible without the help of a vector. Therefore, such pathogens can cause epidemics only when vectors are in large number and active. Still some soil-borne bacteria, fungi and nematodes produce the inoculum on infected plant parts in the soil and here the inoculums disperses slowly and chances of sudden and widespread epidemics are little or minimum.

10.4.5 Mode of Spread of the Pathogen

The spores of rusts, mildews, and leaf spots are released into the air and can be dispersed by strong winds over long distances. Such type of fungi is responsible for widespread epidemics. The next most important group of pathogens include those whose inocula are carried by air-borne vectors. Some fungi (Dutch elm disease), bacteria (bacterial wilt of cucurbits), many of the viruses, mollicutes, and even nematodes are transmitted primarily by beetles, aphids, whiteflies, and leaf hoppers. Some of the pathogens such as anthracnoses and apple scab are transmitted by wind-blown rain and are responsible for severe but localized epidemics. Generally, soil-borne fungi often cause local disease but when such soil-borne fungi like *Sclerotinia sclerotiorum* on lettuce produce wind-disseminated ascospores, they can spread over a long distance and can cause epidemics in lettuce plants.

10.4.6 Host Physiology/Factors-based Prediction

According to Agrios (2005), the physiology of the host keeps changing in response to environmental conditions to which it is exposed during its growth. The study of physiology of a plant is important for forecasting plant-disease epidemics.

1. Resistance or Susceptibility of the Host

Host plants carrying vertical resistance will not allow a pathogen to become established and thus no epidemic can develop while plants carrying horizontal resistance allow a pathogen to become established, but the degree at which the disease and epidemics will develop depends on the environmental factors and the levels of resistance. Thus, susceptible host plants lacking genes of resistance against the pathogen provide the ideal substrate for establishment and development of new infections.

2. Genetic Uniformity of Host Plants

When genetically uniform host plants with regard to the genes associated with disease resistance are grown over longer areas, there is every possibility that a new pathogen race will appear that can attack the genome and result in an epidemic. The highest rates of development of epidemics generally occur in vegetatively propagated crops in comparison to self-pollinated or cross pollinated plants.

3. Type of Crop

In diseases of annual crops such as vegetables, rice, and cotton, epidemics generally develop usually in a few weeks in comparison to perennial woody plants. Some epidemics of fruit and forest trees take years to develop.

4. Age of the Host Plant

1. Pathogens attack their host plant at a certain period of its growth. For example, *Phytophthora infestans,* may cause severe infection of either sprouts on tubers or crop at its later stage of growth. Potato plants during intermediate stage of their growth do not show much susceptibility to the pathogen.
2. The host plant growing under normal conditions usually has sufficient nitrogen and mineral matter for requirement of the attacking fungus, but at certain stages of its growth, there is not sufficient spare carbohydrates for the fungal pathogens needed for their growth and reproduction. At this stage of its growth, the host plant generally shows less susceptibility to the attacking fungus.
3. Plants are more susceptible at that time when they have plenty of spare carbohydrates.
4. Leachates from various plant tissues are known to exert influence on growth, reproduction, and other activities of microorganisms in the environment of a plant. If the quality of leachates from roots and other plant tissues can be determined or regulated in relation to prevailing weather conditions, it may be possible to forecast the occurrence of many diseases caused by soil-borne and air-borne pathogens.

10.4.7 Effect of Human Cultural Practices

Human intervention has a direct or indirect effect on plant-disease epidemics as some of them favour and some reduce the rate of epidemics. Low-lying and poorly drained areas near the infected fields tend to favour the development of epidemics. Similarly, the use of infected seeds, nursery stock, and propagating material increases the initial inoculum within the crop and favours the development of epidemics. Another practice like monoculture, addition of large amount of fertilizers, dense planting, excess irrigation, and poor sanitation conditions increase the development of epidemics in a particular area. Generally fungicides, cultural practices, and biological control reduce the possibility of epidemics but continuous use of certain fungicides may lead to the development of resistant strains of the pathogen that lead to epidemics. The movement/transportation of infected seeds, tubers, and nursery stock from one place to another increase the possibility of introducing pathogens in new areas where the hosts are not resistant to these pathogens.

10.4.8 Molecular-tools-based Prediction

The most important molecular tools are the use of genetic (DNA) probes that allow the detection and identification of a particular plant pathogen in a mixture of other microorganisms in the vicinity of the host plant. The detection and identification of a particular pathogen by its genetic (DNA) probe are made through the use of PCR (Polymerase Chain Reaction) technique which amplifies a specific fragment of DNA present on a probe and produces many copies of it.

RAPD (Random Amplified Polymorphic DNA) markers is another technique used to detect genetic similarities among pathogenic strains known to show genetic heterogeneity can also be used for designing SCAR (Sequence Characterized Amplified Region) markers for detecting the pathogen in diseased plant tissue. The significance of those techniques lies in the fact that they detect the pathogen arrival much earlier, thereby allowing farmers to get ready and use whatever control measures are most effective against the pathogen. Even new mutant pathogens that could attack plant varieties or could not attack earlier or tolerate the fungicide and produce a new resistant pathogen can be detected. Such information is important in epidemiology of epidemics because such changes make revision of any previous predictions about the development of epidemics and recommendations for control of the disease.

10.4.9 GIS (Geographical Information System)

The GIS is a computer system that can be installed in a computer and is capable of collecting, storing and displaying data referenced by geographic coordinators. This system can be used from a single field to a large agricultural region. It is used in better understanding of environment and management of plant-disease epidemics which often go unnoticed without GIS. GIS techniques are also incorporated in disease-forecasting systems.

10.4.10 Global Positioning System (GPS)

GPS consists of a hand held device that is coordinated with man-made satellite and enables one to identify a particular tree or a specific area of a field infected by a pathogen. This particular area or tree could be treated with a particular fungicide without visiting the entire area. GPS can also be used to apply fungicide, plant nutrients, etc. in only areas that are infected with the pathogen or in areas deficient in particular micro or macro-nutrients. Elimination of a particular pathogen by early detection with the help of GPS is quite effective in not allowing the pathogen to cause an epidemic in a particular area.

10.4.11 Remote Sensing

It is an instrument used for measuring electromagnetic radiation reflected or emitted in the ultraviolet, infrared, or visible part of the spectrum from an object. The instruments used for remote sensing may be digital cameras, video systems, ground-based cameras, and radiometers that may be carried on balloons, aircraft, and satellites. With the help of this system, an epidemic of a particular tree disease or in a particular locality can be seen by aerial photography and that too without going to this particular field. The various remote-sensing instruments store data, analyze directly or transfer the data to a computer for further studies. Generally, infrared light or light of other wavelengths is used for the detection and progress of a disease in plants. Plants, when infected with various pathogens, change their colour like yellow or brown and have different reflectances from the healthy plant which are green. These different colours become more distinct when photographed with the different wavelengths which can be examined with specific equipment to distinguish disease-coloured plants and will provide information of newly diseased plants and rate of progress of a particular disease.

10.5 EXAMPLES OF PLANT-DISEASE FORECASTING SYSTEMS

Here, we shall discuss some Indian attempts regarding the plant-disease forecasting.

10.5.1 Predicting Rust Epidemic in India

Nagarajan and Singh (1975) formulated 'Indian Stem Rust Rules' for *Puccinia graminis tritici*. These rules explain satisfactorily if the primary inoculum from South India would travel tens of hundreds of kilometers before its deposition in central India or not. The stem-rust prediction methods proposed by them consist of the following steps.

1. Check if there were any rains during November in Central India coupled with southerly winds.
2. Check the urediniospore (uredospore) content of the rain.
3. Check if the Indian stem rust rules are satisfied. If a tropical cyclone occurred during November (Rule '1' of the Indian stem rust), the disease would appear near the area where the cyclone dissipated. If the inoculum arrives early, the disease can develop into severe proportions. If rainfall satisfying the weather condition of Rule 2 or 3 occurred during January or February, only traces of the disease would appear; that is too late in the crop season to be of serious threat.
4. Check the weather over the area of urediniospore (uredospore) deposition.
5. Check the viability of the transported spores, i.e. if the trajectory for 700/850 m bar passed near the south foci within 120 hours, spores would be viable.
6. Check if the ground-level conditions following the deposition by rain are favourable for infection or not.

If all the conditions are satisfied, the first detectable infection could occur in 25 to 30 days at locations close to where the cyclone dissipated. Once infections have occurred in Central India, their further spread can be predicted following the same rules with the deletion of steps 1 and 3. Further spread occurs to the east and north of India from these established foci. The more the number of foci, greater are the chances of epidemic development.

Based on this, Nagarajan *et al.* (1978) developed an empirical model, given below to predict whether brown rust of wheat can become severe or not.

1. At least five to six infection sites should be observed at a minimum distance of 25 ± 5 km around 15–20 January during the survey of wheat crop in Uttar Pradesh and North Bihar.
2. Amount and number of rainy days during January to mid-April over Northwest India should be two times the normal.
3. Over Northwest India, weekly mean maximum temperature during March and mid-April should be normal ±1°C. If the first and second steps are satisfied then severe epidemic would occur. But if all of them are partly satisfied then the chances are that isolated epidemic can occur. On the other hand, if the temperature is low then in addition to brown rust, yellow rust also would appear aggravating the situation.

10.5.2 Predicting Late Blight of Potato in India

According to Bhattacharyya *et al.* (1982), under Indian hill conditions, rainfall is the primary factor for the generation of conducive condition for late blight. Bhattacharyya further worked out a correlation between the weather data and the actual data of appearance of the disease under Simla (15 years), Shillong (7 years) and Ootacammund (5 years) conditions. The following was established:

1. If the day moving precipitation (30 mm for Shimla, 28.9 mm for Ootacammund, and 38.5 mm for Shillong which were observed to be the critical rainfall lines) associated with mean temperature of 23.9°C or less continues for 7 days consecutive days, the potato late blight would appear within 3 weeks.
2. Hourly temperature ranges within 18-20°C and associated with high RH (80% or more) for continuous 18 hours for 2 consecutive days, the potato late blight would appear within a week. Accordingly, a warning service for spraying fungicides on potato crop against late blight has been initiated in the Northern India hills and Nilgiris.

Establishment of disease-forecasting services in India is a must for important diseases of vegetables, fruits, and food crops. In India, priorities have to be decided on type of diseases that need to be considered for forecasting. For the development of reliable forecasting systems, extensive laboratory and field studies are required for understanding the behaviour of the pathogen and the host populations under varying environmental conditions. Forecasting of disease epidemics is still in its infant stage in India. There is a great scope and necessity to develop this branch which is, as pointed out in text, not only useful to farmers but also intellectually stimulating. No forecasting system at present in use is perfect.

REVIEW QUESTIONS

1. How does plant disease forecasting help control disease?
2. What are the environmental conditions necessary for the development of the epiphytotics of the following in India?
   (i) Potato blight (ii) Blast of rice
   (iii) Downy mildew of bajra (iv) Rusts of wheat
3. Give at least two examples of forecasts based on
   (i) amount of initial inoculum (ii) weather conditions
   (iii) host physiology
4. What are BLIGHTCAST, CERCOS, MYCOS, EPICORN, EPIVEN, and EPIDEMIC?
5. Discuss in brief the contributions of the following towards the epidemiology of plant diseases.
   (i) K C Mehta (ii) S Nagarajan
   (iii) Van Everdingen (iv) A Beumont
   (v) H T Cook (vi) S K Bhattacharya
   (vii) S Y Padmanabhan
6. What are GIS and GPS? How do they help in plant-disease forecasting?
7. How can molecular tools be utilized in plant-disease forecasting?

MULTIPLE-CHOICE QUESTIONS

1. Which of the following terms are correct for forecasting?
   (i) Prediction (ii) Prognosis
   (iii) Forwarning (iv) All of these
2. In 1969, the first computer simulation programme called EPIDEM was developed for
   (i) early blight of tomato (ii) early blight of potato
   (iii) late blight of potato (iv) both (i) and (ii)
3. For which of the following, forecasting and warning systems are economically justified?
   (i) For diseases of important crops or cash crops
   (ii) For diseases which can be predicted
   (iii) For diseases against which control measures are available
   (iv) All of these
4. Which of the following methods is used in plant-disease forecasting?
   (i) The amount of disease in the young crop
   (ii) Epidemiology
   (iii) Biological control
   (iv) Chemical control

5. On which hidden assumption is the forecast based on inoculum potential made?
   (i) Severity of infection
   (ii) Amount of initial inoculum
   (iii) Weather will be favourable for disease
   (iv) Soil-borne diseases
6. The highest rate of development of epidemics generally occur in
   (i) self-pollinated crops
   (ii) cross-pollinated crops
   (iii) vegetatively propagated crops
   (iv) both (i) and (ii)
7. Which of the following practices favour the development of epidemics?
   (i) Monoculture
   (ii) Dense planting
   (iii) Excess irrigation
   (iv) All of these

Answers

1. (iv) 2. (iv) 3. (iv) 4. (i) 5. (iii) 6. (iii) 7. (iv)

11 MANAGEMENT OF PLANT DISEASES

There are various methods of plant-disease control like cultural, biological, physical, and chemical depending on the kind of the pathogen, the host, the interaction of the two and many other variables.

The principal methods of controlling plant diseases are (1) avoidance, (2) exclusion, (3) eradication, (4) protection, and (5) immunization.

In order to eliminate all the impediments to maximum production of food, it is imperative to control the spread of plant pathogens by cultural methods, chemical methods, and by breeding disease-resistant varieties.

11.1 CULTURAL METHODS

Cultural measures involve agricultural cropping, harvesting and storage, tillage, crop rotation, soil management, growing of resistant varieties, planning of land use, and other related practices.

11.1.1 Avoidance of the Pathogen

Many plant diseases can be prevented by a proper selection of the land or field, choice of time of sowing, selection of varieties, seed and planting stock, and by modifications of cultural practices. The aim of these measures is to enable the host to avoid contact with the pathogen or to ensure that the susceptible stage of the plant and favourable conditions for the pathogen should not coincide.

1. Proper Selection of Geographical Area

Many fungal and bacterial diseases are more severe in wet areas than in dry areas. Crops which are susceptible to these diseases, if grown in wet areas, are likely to be affected by plant pathogens. *Moesziomyces penicillariae* (formely *Tolyposporium penicillariae*), the pathogen responsible for the smut disease of bajra and the ergot of bajra caused by *Claviceps fusiformis* (*C. microcephala*), are more severe in wet areas in regions where rain occurs for long durations during the flowering stage of the crop.

2. Selection of Field

The selection of a suitable area or field for cultivation is very important from the point of view of better yields as well as protection of the crop from the ravages of pathogens. In the case of many soil-borne pathogens, if the same field is chosen for cultivating a specific crop, there is every likelihood of the disease appearing in severe form due to the build-up of the inoculum potential of the pathogen. Hence, it is advisable not to grow the crop in the same field, where in the previous year; it was infected with a pathogen capable of surviving in the soil. *Colletotrichum falcatum* can persist in the soil for a couple of months. Even this short duration is sufficient for the development of disease in the case of sugar cane, if it is successively cultivated in the same field. It follows that a number of soil-borne diseases can be prevented if the same crop is not grown in the same field year after year. The examples are wilt of arhar, green-ear disease of bajra, late blight of potato, stem gall of coriander, powdery mildew of wheat, smut of bajra, ergot of bajra, ear cockle of wheat, root knot of vegetable crops, and bacterial wilt of solanaceous plants.

3. Choice of Time of Sowing

Pathogens are able to infect susceptible plants only under certain environmental conditions. For example, downy mildews require prolonged high moisture for infection. If the crop is grown in such a way that it is not attacked by pathogens during its most susceptible period, disease incidence can be considerably reduced. Rhizoctonia root rot of gram is severe if gram is sown immediately after the rains. This is due to the fact that the pathogen develops rapidly under high temperature and moisture conditions. A slight delay in sowing is helpful in this case. In the same way, *Puccinia graminis tritici* does more damage to a wheat crop sown late. Early seeding also induces losses due to bunt of wheat.

4. Disease-Escaping Varieties

Certain varieties of crops escape damage because of their growth characters. These varieties escape the onslaught of the pathogen and resist the attack due to their inherent characteristics, for example, early maturing varieties of wheat or pea escape damage due to *Puccinia graminis tritici* and *Blumeria (Erysiphe) polygoni*, respectively.

5. Selection of Seed and Planting Stock

In India, seed potatoes are invariably imported in the Northern plains from the Simla hills for control of viral and bacterial disease. In the case of foot rot and leaf rot of *Piper betle* caused by *Phytophthora parasitica* var. *piperina* and rhizome rot of ginger (*Zingiber officinale*) caused by *Pythium myriotylum*, it is advisable to obtain seeds from disease-free areas where the disease has not occurred due to certain environmental factors. Since many plants propagate by vegetative parts, the selection of disease-free planting material forms a very important control measure. The planting of disease-free fields is often an important control measure for certain diseases such as red rot of sugar cane, black scurf of potato, or various viral diseases of potato. Tuber indexing has been resorted to for obtaining virus-free seeds and protecting crops against the attack of various viral diseases of potato. Loose smut of wheat, bunchy top and Panama disease of banana, and whip smut of sugar cane can be effectively kept under check if seeds are properly selected before sowing or planting.

6. Modification of Cultural Practices

The distance between plants, time and amount of irrigation, quantity and quality of fertilizers or organic manures, time and method of planting, mixed cropping method, depth of sowing, and so on are some cultural practices which influence the incidence and severity of certain diseases. The necessary changes in these practices can help reduce disease incidence and loss of the crop.

Tillage is an effective method of control of pathogens harboured in crop residue, but it leads to loss of soil organic matter and increased soil erosion. Tillage provides a temporary improvement in aeration, accelerates drying, and redistributes substrates. These environmental and physical changes can predispose pathogens to antagonistic effects of the microbiota. Early and deep ploughing kills cotton roots and favours saprophytes rather that *Phymatotrichum omnivorum*. If the plant debris is buried deeper by ploughing, the pathogen is at a disadvantage as compared to other saprophytes in soil.

Faulty mechanical sowing makes plants susceptible to infection. This may also result in an unequal and uneven distribution of seeds. Overcrowding may result in damping-off of seedlings. The depth of sowing has an important effect on some pathogens notably those which attack seedlings. By delaying the emergence of seedlings, deep sowing may help increase the resistance of a susceptible crop to a pathogen. This is true in many diseases caused by *Fusarium* and *Rhizoctonia* species and in the smut diseases in which seedlings are attacked. The incidence of wheat bunt caused by *Tilletia foetida* and *T. caries* and flag smut *Urocystis agropyri* (formely *U. tritici*) of wheat is higher in crops sown deep.

Mixed cropping is of special significance in Asian countries. This practice helps to effectively check the spread of infectious diseases. Ploughing under the surface refuse before sowing controls *Sclerotium rolfsii* on peanuts. This deprives the fungus of a food base since it can grow only on or near the soil surface. The effect, therefore, is very much like that of crop rotation. Deep ploughing of soils that contain diseased potato tubers can effectively reduce the amount of inoculum of *Phytophthora infestans*.

The root rot of cotton caused by *Rhizoctonia bataticola* and the blight of pulse crops caused by *Phyllosticta phaseolina* have been successfully controlled in Punjab by mixed cropping. Mixed cultivation of cotton and 'moth' (*Phaseolus aconitifolius*) reduces the incidence of root rot of the former crop (caused by *Rhizoctonia solani* and *R. bataticola*). Mixed crops, such as wheat and barley, wheat and gram, arhar and jowar, and cotton and moth reduce the economic losses from diseases.

Management of soil water offers a means to manage the associated microbiota—directly, through water potential, and indirectly, through gas exchange of the soil.

7. Exclusion of Inoculum

These methods aim at preventing new pathogens and diseases from reaching an uninfected area and avoiding contact between the pathogen and the crop of field. Seed certification, crop inspection, growing crops in regions unfavourable for the pathogen, and quarantine measures are some means of preventing the spread of pathogens.

8. Seed Treatment

Seeds, tubers, grafts, bulbs, and other propagating material can be given heat, gas, or chemical treatment to keep them free of pathogens.

9. Inspection and Certification

The crops grown exclusively for seeds are inspected periodically for the presence of diseases that are disseminated by seeds and necessary precautions are taken to remove the diseased plants. The crop is then certified as disease-free. This practice helps to prevent the inter and intra-regional spread of seed-borne diseases.

10. Quarantine Regulation

Quarantine can be defined as a legal restriction on the movement of agricultural commodities for the purpose of exclusion, prevention, or delay in the spread of plant pests and diseases in uninfected areas. In most agriculturally advanced countries, plant quarantine legislation has been placed on the statute book to restrict the movement of diseased plant material or of fungi, bacteria, mycoplasma, viruses, or nematodes that cause diseases in plants. Quarantine measures are of three types: (1) domestic, (2) internal, and (3) total embargoes. The quarantine laws were first enacted in the USA in 1912, and are known as Federal Quarantine Acts. In India, the Destructive Insect and Pest Act was passed in 1914 and subsequently supplemented by other provisions to enforce regulatory measures on import of biologically active material. It has been followed by Seed Act (1966), New Policy on Seed Development (1988), National Seed Policy (2002), Insecticides Act (1968), Environment Protection Act (1986), Protection and Plant Varieties and Farmers Right (PPVFR) (2001) and Biodiversity Act (2002). Such quarantine laws were first enacted in France in 1660, and in Denmark in 1903. They aimed at a rapid destruction or eradication of barberry, which has been known since early times to harbour black rust.

The DIP Act in India was passed in 1914 and has been revised several times from 1930 to the present time, but it still does not fulfill the needs under the prevailing circumstances. Of late with the World Trade Organisation (WTO) coming in place, the country has a challenge to appropriately comply with governing rules of agreements under WTO. In India, there are 16 quarantine stations operating under the Directorate of Plant Protection and Quarantine—eight at seaports, six at airports, and two on land frontiers (including Hussainiwala in Ferozepur and Sukhiapokri in Darjeeling district). The enforcement of plant quarantine regulations is carried out by the technical officers of the Directorate of Plant Protection and Quarantine, Ministry of Agriculture, Government of India, under the overall supervision of the Plant Protection Advisor. The research material is examined by these agencies, The National Bureau of Plant Genetic Resources, New Delhi, for agricultural and horticultural crops; the Forest Research Institute, Dehradun; for forest plants and Botanical Survey of India, Kolkata for all other plants of general economic significance. NBPGR has established two field stations at Pantnagar (for tropical plants) and Ranichauri (for temperate plants) for post-entry quarantine checking of imported plant material (Chaube and Pundir, 2005). There are a number of lacunae in the DIP Act, and as a result many serious plant pathogens have already been introduced in the country. These are now established and are taking a heavy toll of crops in certain areas. Golden nematode (*Heterodera rostochiensis*) and wart (*Synchytrium endobioticum*) diseases of potato have been introduced in India from the European countries and are now well established in the Nilgiri Hills and Darjeeling. Fortunately, timely action had prevented the spread of wart disease to other parts of India. *Synchytrium endobioticum* was first described by Ganguly and Paul in 1953 and golden nematode of potato (*Heterodera rostochiensis*) was reported by Jones for the first time in India in 1961.

Bunchy top disease of banana was introduced into India around 1940 (Verghese, 1945). By 1943, it had spread to a few areas in Kottayam and it has now affected an area of about 3000 square miles in Kerala. The disease has also spread to Tamil Nadu, Orissa, West Bengal, and Assam.

11.1.2 Eradication of the Pathogen

These measures primarily aim at breaking the "infection chain" by removing the foci of infection and starving the pathogen. Knowledge of the life cycles of the pathogens, method of perennation, host range, and habit of growth are essential prerequisites for devising suitable measures to prevent the disease. Eradication involves eliminating the pathogen from infested areas, and the magnitude of the operation involved may vary considerably. Eradication is attempted through roguing, sanitation, crop rotation, eradication of alternate and collateral hosts, heat and chemical treatment of diseased plants, soil treatments, biological control, and so on.

1. Roguing

This practice entails careful removal and destruction of infected plants or plant parts from the fields at an early stage by removing the "foci" of infection and preventing wide dissemination of the pathogen. Roguing is employed in such diseases as loose smut of wheat, loose and covered smut of barley, sorghum and maize, red rot of sugar cane, green-ear disease of bajra, ergot of bajra, wilt of arhar, and many viral diseases, such as yellow vein mosaic of lady's finger, "Katte" disease of cardamom, Panama disease of banana, *Loranthus* growing on trees, etc.

2. Eradication of Alternate or Collateral Hosts

Many plant pathogens complete their life cycles on two hosts; one of them is generally a wild plant. The destruction of such a wild host plant breaks the chain in the completion of the life cycle of the pathogen. Barberry eradication in temperate countries helps control the black stem rust of wheat. Many plant pathogens have several alternate or collateral hosts on which they pass their life cycles. *Pyricularia grisea* (*P. oryzae*) can infect and survive on several wild species of rice and other grasses, such as *Panicum repens*, *Digitaria marginata* and *Setaria intermedia*. The destruction and removal of collateral hosts is an important control measure of the disease. The yellow vein mosaic of lady's finger (*Abelmoschus esculentus*) persists on wild hosts (*Hibiscus tetraphyllus*) in nature and the systematic removal of the wild plants helps control the disease.

3. Crop Rotation

Very often, continuous cultivation of the same crop or related crops leads to the perpetuation of a pathogenic soil-borne fungus and the gradual increase in the intensity of the disease. Soils generally suffer from what we call soil sickness. When the same crop is raised year after year in the same field, the soil-borne pathogens of that crop easily perennate in the soil and increase their population. The interposition of 2 or 3 nonsusceptible crops between the cultivation of the susceptible crops in the same field is effective in reducing the soil sickness and populations of soil-borne pathogens. Crop rotation helps to control many soil-borne diseases, such as wheat mosaic, wilt of arhar, wilt of linseed, finger and toe diseases of crucifers, wilt of pea, wilt of gram, red rot and wilt of sugar cane, ergot and smut of bajra, bunt of wheat, leaf smut and bunt of rice, flag smut of wheat, molya disease (a nematode disease) of wheat and barley and root rot of vegetable crops.

4. Sanitation

Like plant sanitation, field sanitation is also essential for the control of many plant diseases. Pathogens survive or perennate through dormant structures in plant organs in the soil. The collection and destruction of these structures form an important control measure. It is common practice to burn diseased stubble for plant disease control. In fact, sanitary conditions are helpful in almost all infectious diseases. Powdery mildew of wheat, barley, peas, downy mildew of peas and maize, red rot of sugar cane, Panama disease of banana, Verticillium wilt of cotton and wilt of arhar, Koleroga of Areca palms, foot rot and leaf rot of *Piper betle*, brown rot of stone fruits, late blight of potato, angular leaf spot of cotton, etc. are some examples where sanitary precautions help in minimizing losses. The spread of certain diseases, for example, tobacco mosaic virus, would not be rapid if the workers who smoke were to wash their hands before handling the plants.

5. Heat or Chemical Treatment of Diseased Plants

The pathogen present in the plant can be killed or inactivated by applying heat or chemicals. Hot-water treatments have generally been most successful and are used to eliminate *Ustilago segetum* (*nuda*) from wheat and barley seed, *Alternaria brassicae* from cabbage seeds, *Puccinia menthae* from mint runners, the nematode *Ditylenchus dipsaci* from Narcissus bulbs and ratoon stunt, and other diseases from sugar cane setts. The most important application of hot-air treatments is to eliminate viruses from vegetatively propagated material.

Thermotherapy is a time-old practice that is still in vogue in the control of several crop diseases. Thermal inactivation points of sugar cane and its pathogens are taken advantage of freeing infected stocks of many shoot, ratoon stunting smut diseases and red rot of sugar cane. Treating the seeds with fungicides, especially organomercurials, is a common practice to remove the superficially borne inoculum, and there are now systemic fungicides available which can eliminate deep-seated fungal infections. It has been found that soaking barley seeds for 48 hours in a 0.2% Chloranil (Spergon) solution at 22.2°C to 25°C satisfactorily controls the loose smut fungus, *Ustilago segetum* (*U. nuda*), on barley. Thiram soak has been successfully utilized to eliminate *Septoria apiicola* from celery seed, and *Mycosphaerella pinodes* and *Ascochyta pisi* from pea seed. The systemic fungicides, Vitavax and Plantavax, have been successfully used as seed dressing for the control of loose smut (*U. segetum* var. *tritici*) and rust (*Puccinia graminis tritici*) of wheat, respectively.

6. Soil Treatments

Soil treatment involves the use of chemicals, heat energy, flooding, fallowing, and so on. Plant pathogens can be eradicated from the soil by dry heat, steam or hot water, flooding and fumigation with chemicals. Heat and chemical treatments have been successfully used for the eradication of plant pathogens from greenhouse soils as well as soils in plant beds. The treatment of soils in large fields has not been so effective. Nematodes can be eradicated from the soil by fumigating it with DD mixture or with other nematicides. PCNB has been found to be effective in eradicating sclerotia-producing soil pathogens, carbon disulphide in eliminating *Armillaria mellea* and formaldehyde in controlling onion smut disease.

7. Physical Agents that Reduce or Eradicate the Inoculum

Many different types of control methods aim at eradicating or reducing the amount of pathogen present in an area or plant part.

(a) Soil Sterilization by Heat Soil can be sterilized by the heat carried in live or aerated steam or hot water. Soil sterilization is completed when the temperature of the soil has remained 82°C for at least 30 minutes, at which temperature most of the pathogens present in soil are killed. At about 50°C, some oomycetes, nematodes are killed, whereas most plant pathogenic fungi and bacteria are killed at temperature around 60 to 72°C. At about 82°C, some weeds, the remaining plant pathogenic bacteria, viruses and insects are killed. It is important to see that very high temperatures not only kill all microflora in the soil but also result in the release of toxic levels of some salts like manganese. In addition to this, by killing the nitrifying bacteria, there is accumulation of toxic level of ammonia which may kill the plants afterwards.

(b) Soil Solarization Solarization is just to capture the solar radiations/energy to heat up the soil. The soil is covered airtight with polyethylene mulches/sheeting during the period of intense solar radiation.

Katan *et al.* (1976) from Israel suggested for the first time, the actual technology of soil solarization, i.e. use of transparent polyethylene film to increase soil temperature to the levels to eliminate the propagules of *Fusarium oxysporum* f.sp. *vasinfectum* and *Verticillium dahliae* in soil. Soil solarization involves trapping of solar heat through polyethylene covering during sunny summer days, to raise the soil temperature at the top 5 cm of soil as high as 52°C, where it becomes lethal to temperature-sensitive soil microorganisms. The physical and chemical balance of soil is also altered by the action of trapped solar energy. Solarization is a complex process involving direct thermal inactivation/destruction of pathogen propagules, shifting in microbial balance populations and activity towards creating soil suppressiveness with simultaneous changes in physico-chemical properties of the soil.

(c) Hot-Water Treatment Hot-water treatment is used to kill any pathogen which may be present inside or outside of seeds, bulbs, and nursery stock. This treatment is quite effective for some diseases like loose smut of cereals in which the pathogen resides as mycelium inside the seed where it could not be treated by contact fungicides. Similarly, for nematodes that may be present in bulbs and nursery stock such as *Ditylenchus dipsaci* in bulbs of various ornamentals and *Radopholus similis* in citrus root stocks, hot-water treatment frees them from nematodes. In the loose smut of wheat, the seeds are kept in hot water at 52°C for 11 minutes, whereas bulbs treated for *D.dipsaci* are kept at 43°C for 3 hours because the dormant plant organ can tolerate higher temperature than their respective pathogens.

11.1.3 Disease Control by Radiation

Various radiations such as UV light, X rays, γ (gamma) rays, α and β particles have the ability to control post-harvest diseases of fruits and vegetables. Some pathogenic fungi like *Alternaria*, *Botrytis*, and *Stemphylium* sporulate only when they receive UV light below 360 nm. This can be done by constructing the greenhouse with a special UV absorbing vinyl film that blocks the transmission of light wavelength below 390 nm. Post-harvest diseases of strawberries, peaches, and tomato were controlled by γ (gamma) radiations (Agrios, 2005).

11.1.4 Disease Control by Drying Stored Grains and Fruits

All grains and legumes carry a large number of microorganisms that can cause decay and production of mycotoxins in the presence of sufficient moisture. Such decay can be controlled if such seeds are dried until the moisture content is reduced to about 12 percent. Similarly, many fruits like grapes, plums, dates, and figs can be stored dry for a long time and free of fungal infection if they are dried properly before storage and moisture is kept below a certain level during storage that will not allow fungal infestation.

11.1.5 Disease Control by Refrigeration

It is the most widely used and the most effective method of controlling post-harvest diseases of perishable fruits and vegetables. Low temperature inhibits the growth and inactivates the pathogens, thereby reducing the spread and initiation of new infection. The magnitude of disease control through this method and its value to consumers is immense.

11.1.6 Biological Eradication of Plant Pathogens

More than eighty-six years ago, a paper by Sanford (1926) on potato scab caused by the actinomycete *Streptomyces scabies*, brought the rather sudden realization that other soil microorganisms must exert a natural biological control even on root-disease pathogens. The term *biological control* clearly implies control of a disease through some biological agency, and the term *biological agency*, means a living microorganism or macroorganism other than the diseased or damaged plant acting as host and the pathogen or pest causing the disease or damage. According to Garrett (1965), "Biological control of plant disease may be defined as any condition or practice whereby survival or activity of a pathogen is reduced through the agency of any other organism (except man himself), with the result that there is a reduction in the incidence of the disease caused by the pathogen".

It was as early as 1932 that Weindling showed that *Trichoderma viride*, a common, saprophytic fungus, is able to parasitize the mycelia of other fungi. Later, he showed that the lethal action of *T. viride* is due to the secretion of an antibiotic substance called **gliotoxin**. Still later, Brian and McGowan (1945), isolated another antibiotic substance from *T. viride* which they named **viridin**. These discoveries raised hopes of controlling plant diseases by biological means.

In 1951, Bliss in the USA, who was working on root diseases of citrus caused by *Armillaria mellea* opened a new approach. He suggested that if the soil is fumigated with carbon disulphide, the disease-inciting fungus is killed not directly by the fumigant but indirectly by the dominance of *T. viride*, the population of which increases after fumigation.

The most conspicuously successful biological control was by Rishbeth in England. Rishbeth has been studying a root disease of pines (*Pinus sylvestris* and *P. nigra*) caused by basidiomycetous fungus *Heterobasidion (Fomes) annosus*. The disease spreads through infection by basidiospores produced in the large fructifications. The spores infect the leftover stumps when the trees are felled. Rishbeth showed that if paste of oidia of *Peniophora gigantea* (now renamed as *Phlebia gigantea)* is applied on the stumps, it parasitizes the cells rapidly and kills the tissues, making it easy for saprophytes to colonize and decompose the stump and making tissues unsuitable for an attack by *H.* (*Fomes*) *annosus* which is a specialized parasite.

Although a large number of microorganisms have been shown to be antagonistic to the growth of a large number of plant pathogens in the laboratory, only a few have been found effective and practicable in field conditions. According to Agrios (2005) around thirty microorganisms have been registered so far. Some of them are *Gliocladium virens* sold as GlioGard for control of seedling disease. *Trichoderma harzianum* sold as F-stop for control of several soil-borne fungi. *T. harzianum/T. polysporum* sold as BINAB-T for the control of wood decays, *Agrobacterium radiobacter* K-84, sold as Gallex or Galltrol for controlling crown gall disease caused by *Agrobacterium tumefaciens*, *Pseudomonas fluorescens* sold as Dagger G for use against *Pythium* and *Rhizoctonia* species, and *Bacillus subtilis* sold as Kodiak and utilized as seed treatment.

Pseudomonas rhizobacteria primarily of *P. fluorescens*, *P. putida*, *P. cepacia* and *P. aureofaciens* applied to seeds, seed pieces, and roots of plants have resulted in less damping off, less soft rot, and an increase in the growth of the crop.

Different mechanisms have been proposed to explain disease suppression by *Pseudomonas* spp. These include production of siderophores (iron chelating compounds), antibiotics, induction of resistance, and competitive colonization.

1. Mycorrhizal Fungi as Biocontrol Agents

Mycorrhizal fungi have received considerable attention in recent years. Mycorrhizal fungi grow better in infertile soils. The role of mycorrhizal fungi in the biological control of fungal root pathogens is well known. Several mechanisms have been postulated. These are (1) creating a mechanical barrier for the pathogen to penetrate, (2) inducing thickening of cell walls through lignification and production of other polysaccharides, (3) stimulation of host roots to accumulate metabolites, (4) stimulating flavanolic wall infusions, (5) increasing concentrations of dihydroxy phenols in roots, (6) producing antibacterial and antifungal antibiotics, and (7) stimulating microbial activity in the root zone.

Implementation of biological control is fraught with constraints, and successful application depends on a thorough knowledge of the ecology and biology of both the target pathogen and the biological control agent. Paramount to efficacious employment of biological control is knowledge of the specific mechanism of microbial antagonism involved. Documented biological control mechanisms for plant pathogens include competition, hyperparasitism, predation, and the production of extracellular metabolites such as antibiotics, hydrogen cyanide (HCN), and siderophores. Recently Miller *et al.* (1996) discovered a novel antimicrobial metabolite, a biosurfactant. Stanghellini and Miller (1997) have recently discussed the biosurfactants, their identity and potential applicability in the biological control of zoosporic plant pathogens.

2. Soil Amendments

Many root diseases such as potato scab have been controlled by ploughing organic material into the soil, which started the chain of these investigations. Several others have been successfully controlled in this way, such as Phymatotrichum root rot of cotton, take-all (*Gaeumannomyces graminis*), several Rhizoctonia and Fusarium diseases. In the case of Phymatotrichum root rot of cotton, it was shown that fresh organic materials stimulate the germination of sclerotia and this exhausts the inoculum.

In the case of *Gaeumannomyces* (*Ophiobolus*) *graminis*, it has been suggested that organic amendments, rich in carbon and deficient in nitrogen, control the take-all disease of wheat. This is explained as follows: As a result of organic amendments with a high C/N ratio, there is a tremendous increase in the microbial number and activity. There is considerable liberation of CO_2 by soil saprophytes, which suppresses the pathogenic activity of this fungus. In the process of survival also, low nitrogen content in the soil reduces the longevity of the fungus.

The organic amendments increase the activity of saprophytic organisms in a flush which results in the abundant liberation of CO_2 as a result of the respiratory activity of microorganisms. The pathogens sensitive to CO_2 are inhibited. Due to the rapid multiplication of microorganisms in the soil, the available nitrogen in the soil is rapidly utilized by the fast-growing saprophytes. This results in acute nitrogen scarcity which adversely affects the growth of the pathogens. It has been reported that soil amendment with either chitin or laminarin reduces incidence of bean root rot and also of radish wilt due to *Fusarium oxysporum* f. *conglutinans*. But none of these amendments can control diseases caused by pathogens, which do not contain chitin in their walls, such as *Pythium debaryanum* and *Agrobacterium tumefaciens*.

Phytophthora root rot of avocado was controlled by Zentmyer (1963) by amending the soils with alfalfa meal—a material of low C/N ratio. Similarly, Mehrotra and Tiwari (1976) could successfully control the root rot of *Piper betle* by organic amendments, and antagonists growing on organic amendment in the soil. Amendment with corn straw gave best control. Corn straw supplemented with nitrogen (NH_4NO_3) gave still better control. The effect of different antagonists growing on corn straw and till *Sesamum indicum* oil-cake amendments showed that *Trichoderma viride* gave best control. Mehrotra and Claudius (1972) tried to control the root rot and wilt diseases of *Lens culinaris* by organic amendments of the soil and also by introducing antagonists in the soil along with the organic amendments. The wilt of lentil caused by *Fusarium oxysporum* f. *lentis* was checked with amendments of wheat straw, wheat straw plus ammonium nitrate, corn straw, and sugar cane straw. The successful antagonists were *T. viride* and *Streptomyces gougeroti*.

3. Hyperparasitism

Certain fungi and bacteria are parasitic on plant pathogens. The genus *Cinccinobolus* includes species of fungi, which attack members of the family Erysiphaceae. *Cinccinobolus* seldom reduces the attacks of the powdery mildews in nature because it rarely occurs in large quantities. *Cinccinobolus* can be cultivated on artificial media, however, and some good results have been obtained by artificial inoculation. Several workers have tried to utilize *Eudarluca australis* (pycnidial stage known as *Darluca filum*) for controlling rusts. The fungus does control the disease under certain natural climatic conditions over which humans have little control. Extra inoculum of the hyperparasites will not be of much use as the weather, unless very favourable to the development of hyperparasite which is unlikely to permit any marked control of the disease. There are several other fungal species which attack rusts, such as *Tubercularia* spp. attack aecia and spermogonia of several rusts and *Verticillium hemileiae* on *Hemileia vastatrix*.

4. Entomogenous Fungi

There are several fungi, which are capable of growing in the bodies of insects, thus destroying them. One important fungus which has been utilized is *Entomophthora*. About 100 species are known, mostly as insect parasites. While some appear to be rather specific in their host range, others are capable of attacking a wide range of hosts. Some species have been grown in nature. One of them, *Entomophthora coronata*, a fungus which has been referred to by various names, for example, *Delocroixia coronata*, *Conidiobolus villosus*, *C. coronatus,* is a parasite of aphids and termites but is also a common saprophyte of plant detritus.

11.2 CHEMICAL METHODS

Plant diseases play an important role in determining the amount and cost of food. Plant pathology must alleviate the problem by devising new control measures and improving the older ones. The annual loss to world crops as a result of diseases has been estimated at 25,000 million dollars; of this, a major part is due to fungal pathogens. Surface deposits of fungicides can control many fungal diseases if properly applied, but it appears that more effective control can be achieved with systemic fungicides which can penetrate and move into the plant.

Before we discuss the various chemicals for plants disease control it becomes essential that we define a fungicide. A fungicide is a chemical agent that kills or inhibits the development of the fungus spore or mycelium. The fungicides used on plants may be classified as protectants, eradicants, and therapeutants on the basis of their uptake and mobility within plant.

1. Protectants

As the name suggests, protectant fungicides are prophylactic in their behavior. Protectant fungicides, which may be applied to seeds, plant surfaces, or the soil, cannot penetrate plant tissues in efficient amounts. Therefore, they act outside the plant prior to infection by the pathogen.

2. Eradicants

These chemicals eradicate the dormant or active pathogen from the host. They can remain effective on or in the host for some time and function as protectants also. Chemicals can be protectants as well as eradicants.

3. Therapeutants

A therapeutant is an agent that inhibits the development of a disease syndrome in a plant when applied subsequent to invasion by a pathogen. Therapy can be by physical means, such as solar-energy treatment or hot-water treatment of smutted wheat grains, but more often it is by chemical means and is then called **chemotherapy**. Usually the chemotherapeutants are systemic in their action, that is, they enter the plants and affect deep-seated infection.

To be successful, fungicidal sprays must have some constant characteristics. To be effective against rain and dew, they must be nearly insoluble in water, but at the same time, they must be soluble enough to function as fungicides. Tenacity, or adhesiveness, is another quality needed to enable them to stick well and last for a longer period. Further, they should not be toxic to the host parts (phytotoxic) so that they can be used safely. To meet all these requirements, various modifiers are mixed with the fungicides; these are known as stickers, spreaders, and safeners.

The fungicides are dealt with below under the following categories:

1. Inorganic copper compounds
2. Inorganic mercury compounds
3. Sulfur sprays
4. Organic sulfur compounds
5. Quinone and phenolic fungicides
6. Heterocyclic nitrogen compounds
7. Benzene compounds
8. Organomercurials
9. Systemic fungicides
10. Organophosphate fungicides
11. Strobilurins
12. Antibiotics
13. Soil fumigants
14. Oils

11.2.1 Inorganic Copper Compounds

1. Bordeaux Mixture

Millardet in collaboration with Gayon, a chemist, was able to announce in 1885 the successful use of a lime-copper sulfate mixture as a fungicide against *Plasmopara viticola.* This discovery has doubtless been the greatest boon to plant protection since about fifty million kg of copper sulfate is commonly used in the USA for this purpose every year.

The formula for Bordeaux mixture varies. The original mixture contained 15% copper sulfate and 8% lime (CaO). The proportion of these now varies according to the crop and the disease. Higher concentrations are useful in case of woody and tough tissues, while lower concentrations are used for succulent and softer parts. The following formula (5:5:50) is regarded as a good standard type.

Copper sulphate (blue stone)	5 lb (2.268 kg)
Stone or hydrated lime	5 lb (2.268 kg)
Water	50 gallons

The mixture is prepared by mixing $CuSO_4$ solution with limestone solution. The two solutions are prepared separately in wooden or earthen vessels and the diluted solution of $CuSO_4$ is slowly poured into a concentrated lime solution in a tub with constant stirring. The resulting mixture is neutral or slightly alkaline and is not injurious. The mixture is used almost immediately after preparation and deteriorates if left for some time. However, the mixture may be stabilized by adding sugar or jaggery at the rate of 0.5 lb in 50 gallons of the mixture. The mixture stabilized in this manner will not deteriorate for several days under normal conditions. Alternatively, a stock solution of the two ingredients can be prepared separately and stored in wooden vessels. These can be mixed when actually required.

The chemistry of Bordeaux mixture is complex and is not fully understood. The expected reaction is

$$CuSO_4 + Ca(OH)_2 = Cu(OH)_2 + CaSO_4$$

Cupric hydroxide is the active principle which is soluble and is toxic to spores and sporelings. The Bordeaux mixture is a suspension, which is fairly tenacious, nearly insoluble and spreads quite well. It may be detrimental or beneficial to different hosts. Beneficial effects have been observed on grape, potato, and other crops. Greener foliage is often the outstanding result and it also has insect-repellent properties. The Bordeaux mixture is sometimes phytotoxic to tender leaves and other parts and burns them as well as fruits and other aerial parts. Sometimes blemishes occur, as on apple fruit. The fungicidal action is due to the formation of soluble copper, which is toxic to spores or sporelings in drops of atmospheric water.

The Bordeaux mixture is specific against downy mildews, such as late blight of potatoes, koleroga of areca nuts, coffee rust, various leaf-spot diseases, blights, anthracnoses, and so on. It has remained the standard fungicide against downy mildews of grapes and other hosts since its inception in 1882.

2. Bordeaux Paste

Bordeaux paste consists of the same ingredients as those of Bordeaux mixture but it is in the form of a paste. It is mainly used for tree-wound dressings to prevent fungal pathogen attacks such as in the control of stem bleeding disease of coconut. It is generally prepared by adding 1 lb of copper sulfate and 1 lb of lime in 1 gallon of water.

Some other copper fungicides are Burgundy mixture, Cheshunt compound, and Chaubattia paste.

3. Inorganic Mercury Compounds

Many inorganic mercury compounds are highly effective as fungicides and bactericides. Mercuric chloride ($HgCl_2$) and mercurous chloride (Hg_2Cl_2) are used as 1:1000 dilutions for soaking the seed, rhizome and corms of vegetables and flowering plants to control certain bacterial diseases. Club-root disease of crucifers can be controlled by treatment of *Brassica* seedlings against the pathogen. The use of inorganic mercury compounds is limited because they are highly toxic to animals and humans.

4. Inorganic Sulfur Compounds

Sulfur is effective against powdery mildews on many plants, but is also effective against certain rusts, leaf blights, and fruit rots.

5. Self-boiled Lime Sulfur and Wettable Sulfurs

Lime sulfur has some eradicant properties also and is active against apple scab and powdery mildew. Wettable sulfurs are used against apple scabs, powdery mildews of grape, and other crops, ornamentals and vegetables. In greenhouses, it is used against rusts also. Strangely enough, the mode of action of sulfur as a fungicide is still unclear.

6. Organic Sulfur Compounds

It was in 1934 that the patenting of dithiocarbamates was announded by Tisdale and Williams who patented 'Thiram' and the use of this group of materials as insecticides and fungicides. Since then volumes of work have been done on the chemistry of thiocarbamates.

Carbamic acid is the starting point. It is derived from carbonic acid by the replacement of a hydroxyl group by an amino group.

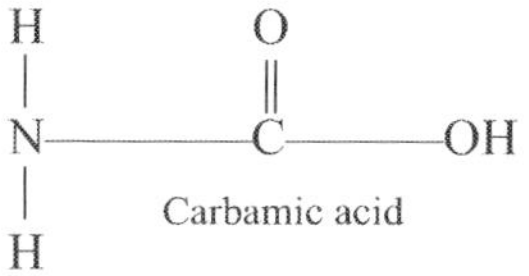

Carbamic acid

Dithiocarbamic acid is derived by the replacement of two oxygen atoms by sulfur, giving the hypothetical dithiocarbamic acid.

From dithiocarbamic acid, the following important fungicides have been derived.

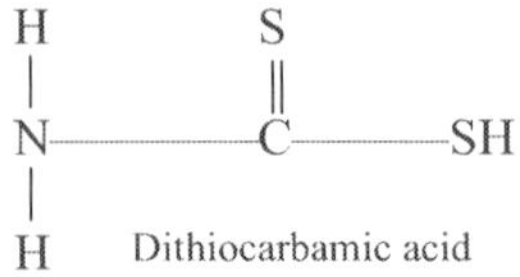

Dithiocarbamic acid

(a) Thiram (Tetramethyl Thiuram Disulphide) TMTD is a condensation product of dithiocarbamic acid (2 molecules in which the H atoms have been substituted by methyl groups). This was used as an activator in the production of synthetic rubber.

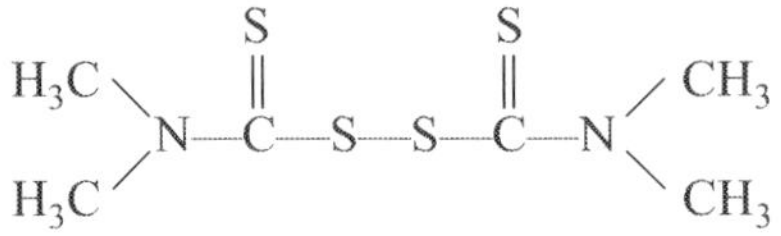

Thiram

Thiram is being sold under a variety of other tradenames, such as Arasan, Tundas, Terson, Tulisan, and so on.

It is a leading seed treatment for wheat, rice, gram, peas, mustard, linseed, sorghum, and onion. Thiram is also used in the control of foliage diseases and acts as an insecticide. Among the soil-borne pathogens controlled by it are *Pythium*, *Rhizoctonia*, *Fusarium*, and *Protomyces*.

As a foliage spray, Thiram was not as successful as its metallic compounds. For this reason, several compounds have been developed which are as follows.

(b) Ferbam (Ferric Dimethyl Dithiocarbmate) In Ferbam, three molecules of dithiocarbamic acid react with one atom of iron. Its trade names are Fermate, Ferbam, Karbam, or Coromet. Ferbam is compatible with most pesticides, but its use with copper, lime, or mercury compounds tends to reduce fungicidal efficiency. Ferbam has been successfully used as a protectant fungicide against a wide variety of fungus diseases of fruits and vegetables. As a foliage spray it is also used to correct iron deficiency.

(c) Ziram (Zinc Dimethyl Dithiocarbamate) The trade names are Cuman, Zerlate, Karbam white, Corozate, and so on. In India, this fungicide is available under the tradename of Cuman. It is used against many foliar diseases in field and orchard crops.

(d) Nabam (Disodium Ethylene Bisdithiocarbamate) This is sold as Dithane D-14, Parzate liquid, or Dithane A-40. Nabam has been found effective against many leaf blights, such as early and late blight of potatoes and seedling blights caused by *Pythium*, *Rhizoctonia*, and *Fusarium* species.

A similar compound of zinc, Zineb, was later developed.

(e) Zineb (Zinc Ethylene Bisdithiocarbamate) Zineb is derived by the substitution of sodium in Nabam with zinc. It is called Dithane Z-78, Parzate C, or Lonacol. Zineb is mainly used for foliar sprays against diseases, such as early and late blight of potatoes and tomatoes, blast of rice, leaf blight of rice, ripe rot of chillies, and downy mildew of maize.

(f) Maneb (Manganese Ethylene Bisdithiocarbamate) This is called Dithane M-45. Maneb is excellent for the control of foliage and fruit diseases of many vegetables, flowers, trees, turf, and some fruits. In many diseases, Maneb has been found more effective than Zineb. It is being used to control downy mildews, leaf spot, blight, and anthracnose diseases.

Two important clues to the toxic action of the dialkyldithiocarbamates and thiuram disulphides are their formation of metal chelates and their polymodal dosage response. The toxicity of ethylene bisdithiocarbamates is believed to be due to the production of isothiocynate which reacts with SH groups.

(g) Quinone and Phenolic Fungicides Though not as important as the carbamates, quinone and phenolic compounds are quite important as fungicides in certain respects.

(h) Chloranil or Spergon (Tetrachloro-p-benzoquinone) It was reported in 1940 by Cunningham and Sharvelle to give protection to lima bean seedlings from damping off. Chloranil is mainly used in seed treatment. It is sold as Spergon and is used in seed and bulb treatment of legume flowers and vegetables. It is also used as soil drench.

(i) Dichlone or Phygon (2,3 dichloro-1,4 naphthaquinone) This is a later development. Though it is used for seed treatment, it is also useful as a foliage spray against many plant diseases and is apt to cause dermatitis. There are two possible mechanisms by which chloranil and Dichlone act—bringing the quinone nucleus to –SH and NH_2 groups in the fungus cell and disturbances in the electron transport system.

7. Phenols

Phenols have been chiefly applied to textiles and woods as preservative fungicides, but have also been used as soil fungicides and in seed treatments. Derivatives of cresol and O-phenyl phenol have been mainly used as preservatives. The bisphenols are a particularly interesting group of phenolic fungicides.

8. Heterocyclic Nitrogen Compounds

Their principal uses are as foliage protectants and eradicants of fruits and vegetables. Glyodin (2-heptadecyl-2 imidazoline acetate) and Captan are the two important fungicides of this group.

(a) Captan Captan is the common name adopted for N-trichloromethyl thio-4-cyclohexene-1,2 dicarboximide. Kittleson (1952) prepared this compound and so it is sometimes called Kittleson's killer.

Captan is commercially sold under different names, such as Captan 50 W and Esso fungicide 406, but it is also claimed that it acts systemically. Although a fairly broad spectrum fungicide, it is not generally effective against rusts, powdery mildews, and downy mildews. Captan inhibits many oxidative enzymes, considering that its basic toxic action was an interaction of the intact Captan molecule with free sulfydryl groups. Captan is used as a spray, dust, or for seed dressing.

It has been used against diseases of different types of crops, but its use in fruit-disease control and as a seed-dressing fungicide has found popularity all over the world.

(b) Captafol or Difolatan [N-(1, 1, 2, 2-tetrachloro ethyl sulfenyl)-cis 4-cyclohexene 1, 2-dicarboximide] Captafol or Difolatan has properties similar to Captan. It is a good fungicide for the control of early and late blight of potatoes and tomatoes.

(c) Folpet (N-trichloromethylthio phthalimide) It is an analogue of Captan. Folpet is effective against *Sphaerotheca pannosa*. It is sold under the tradenames of Phaltan, Orthophaltan, and so.

(d) Glyodin (2-heptadecyl-2-imidazoline acetate) It has been suggested that the biological activity of Glyodin depended on the interference, in the biosynthesis, of those purines which are components of nucleic acids. It is effective against *Venturia inaequalis*, *Coccomyces hiemalis*, *Diplocarpon rosae*, and *Podosphaera leucotricha*.

9. Benzene Compounds

Many benzene compounds are toxic to microorganisms and several have been developed as fungicides. Dinitro-O-ceresol is contained in the formulation called Kerenite or Elgetol. It is used as a dormant spray for the control of many diseases of fruit trees and ornamental plants and for the treatment of wounds in trees.

(a) Chloroneb The fungicide Chloroneb, 1,4-dichloro-2,5-dimethoxy benzene, was introduced for the control of seedling diseases of cotton. It has been found active against *R. solani*, *S. rolfsii*, and *Phytophthora cinnamomi*. This is a systemic fungicide and is unusual in that its activity does not follow clear-cut taxonomic divisions. This fungicide has been used for the most part as a seed or as in-furrow soil treatment at planting time for systemic control of treatment of seedling diseases of cotton, beans, sugarbeets and soybeans.

(b) Pentachloronitrobenzene (PCNB) This is sold as Quintozene, PCNB, Terrachlor, Brassicol, and so on. The chlorinated nitrobenzenes were introduced as fungicides in 1935 by Brown. They control many soil-borne diseases. It is used primarily against *Rhizoctonia* and *Plasmodiophora*. PCNB also has nematicidal properties.

Diseases caused by *Rhizoctonia*, *Sclerotium*, and others are controlled by PCNB. It is not effective against pythiaceous fungi.

(c) Dexon This is p-(dimethylamino)-benzene–diazosodium sulfonate and is used against soil-borne phycomycetous fungi causing root rot and damping-off diseases of plants.

Dexon had a curative effect on avocado seedlings afflicted with *Phytophthora cinnamomi* when applied in the root region.

(d) Chlorothalonil (tetrachloro-isophthalonitrile) It is a broad-spectrum-contact fungicide belonging to the phthalimide group. It is used for the control of early and late blights, leaf spots, powdery mildews, downy mildews, apple scab, coffee-berry disease, rusts (yellow and black), sigatoka of banana, and leaf spots of groundnuts.

(e) Dinocap (Karathane) Dinocap, or Karathane, is a mixture of isomers of 2-(1-methyl-heptyl), 4-6 dinitrophenyl crotonate. Dinocap is an excellent substitute for sulfur for the control of powdery mildews and is highly specific against them. It is also effective against mites. Actually, it was developed as a miticide and is now extensively used against powdery mildews.

10. Non-aromatic Organic Compounds

This group includes Dodine (*n*-dodecylguanidine acetate) used for the control of apple scab and several other diseases caused chiefly by Ascomycetes and Fungi imperfecti. Due to its toxicity to some crop plants, its use is limited. The mechanism of action of Dodine is due to its binding with sterols in the cell membrane, with the result that the membrane structure is altered.

11. Organomercurials

Organomercurials are very widely used for seed treatment. Some of the important compounds are Ceresan M (ethyl mercury *p*-toluene sulphonanilidae), Panogen (methyl mercury dicyandiamide), new improved Ceresan (ethyl mercury phosphate), Ceresan wet, Agallol or Aretan (methoxy ethyl mercury chloride), Ceresan dry (phenyl mercury acetate), and Agrosan G N (phenyl mercury acetate and ethyl mercury chloride mixture). Many other tradenames are used for compounds which are used for compounds which are in wide use. The chemicals can be used either as dips or dusts. Organomercurial fungicides are mainly used to control externally seed-borne diseases since these compounds are able to eradicate the inoculum from the seed. Their use in the control of externally seed-borne cereal smuts has been widespread.

Tubers and bulbs are also similarly treated with dusts against adhering organisms, such as potato scab and black scurf. The hazards which organomercurials pose are too well known. Because of their persistence they get into the food chain. Several countries have now banned the use of these fungicides. India has also banned the use of organomercurials (Nene and Thapaliyal, 1993).

12. Systemic Fungicides

A new aspect was added to the control of fungal pathogens of plants when the oxathiin group of chemicals was developed in 1963. These compounds are systemic in their action and are receiving the active attention of pathologists. Most systemic compounds are absorbed by the roots and transported acropetally to the xylem. They are usually not transported downwards through the phloem tissue.

Major advances in plant-disease control will now be made from systemic fungicide compounds, which penetrate and move from their site of application to other parts of the plant. A systemic fungicide could eradicate established infection and protect the new parts of the plant as they are produced and it would not be subject to loss by weathering. The idea of controlling disease by systemic fungicides is not new. They can be applied in solution as spray on the foliage is less likely to be effective against the pathogens responsible for root diseases. Several systemic fungicides have been tested as seed-dressing to eliminate seed infection. Systemic fungicides have not been very successful in the case of trees and bushes. From the practical point of view, the systemics offer unique possibilities in the tropics, since the active spray residues are not subjected to tropical weathering, thus permitting fewer applications.

13. Action of Systemic Fungicides

Very little is known about the mechanism of systemic fungicides. The following are the possibilities:

(a) Inactivation of the enzymes and toxins of the pathogen.
(b) Selective accumulation of the fungicide due to greater permeability of the fungus cell wall.
(c) Damage to the membranes of the fungal hyphae and inhibition of structures, such as appressoria, cushion formation, emergence of germ tubes, and formation of haustoria.
(d) Inhibition of fungal enzymes or their destruction. Systemic fungicides are more specific in their action than nonsystemic fungicides.

The following are some of the systemic fungicides used in plant-disease control.

(a) Benomyl This is also marked as Benlate. This is methyl-N (butyl carbamoyl)-2 benzimidazole carbamate. The benzimidazoles have been very effective in the control of *Mycosphaerella* on banana and peanut. Benomyl hydrolyzes in acidic solution to Methyl Benzimidazole Carbamate (MBC). MBC has been shown to be the principal fungitoxicant of Benomyl, which is systemic in plant tissue.

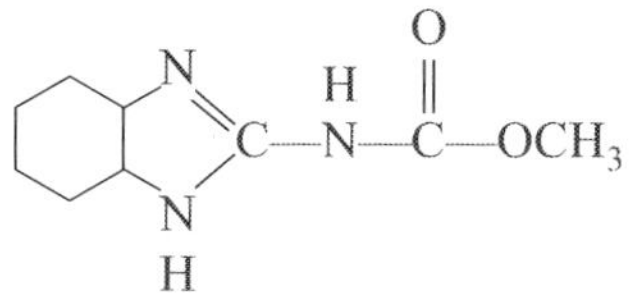

Methyl Benzimidazole Carbamate (MBC)

Benomyl has been found effective against a wide range of pathogens. It has a high degree of mite-ovicide activity. It is effective against *Cercospora* leaf spot of sugarbeet, rice blast, apple scab, powdery mildew on cucurbits, cereals, and legumes. It is ineffective against Oomycetous pathogens and dark-coloured fungi such as *Alternaria* and *Helminthosporium,* some basidiomycetes, and bacteria.

Application of Benomyl induced a remission of symptoms of spike disease of sandal caused by MLOs. Many soil-borne pathogens can be controlled, provided the roots are confined to the Benomyl-treated soil.

Other benzimidazole derivatives are Thiabendazole and Bavistin. It is effective against Ascomycetes and Fungi imperfecti which are considered hard to control. It is ineffective against Oomycetous fungi. Bavistin is translocated within the plant and has both prophylactic and curative actions. The action of Benomyl and MBC are usually the same with most fungi. DNA synthesis is severely affected before other systems are affected.

(b) Terrazole Terrazole-5-ethoxy-3-trichloromethyl-1,2,4-thiadiazole is toxic to Oomycetous fungi, such as *Pythium* and *Phytophthora*. It has a systemic action in plants. Little is known about its fungitoxic action.

(c) Thiophanates Thiophanates seem to occupy an exceptional position since they are not heterocyclic compounds, but instead disubstituted benzene derivatives. It was found that both methyl and ethyl thiophanate are broad-spectra fungicides which have each the same antifungal spectrum as Benomyl, though it is at a somewhat lower level of activity. Thiophanate (methoxy carbonyl thioureido benzene) is sold as Topsin M or Cercobin M.

The fungitoxicity, as well as systemic action of methyl and ethyl thiophanate, may at least in part be attributed to their easy transformation into MBC and its ethyl analogue, respectively.

(d) Oxathiins The oxathiins are systemic fungicides which are effective only against Basidiomycetous fungi, such as rusts, smuts, *Rhizoctonia solani* (*Corticium solani*). Vitavax (2,3 dihydro-5-carboxyanilido-6-methyl 1-1,4, oxathiin) has been used to free barley seed from *Ustilago nuda* (loose smut) and *Ustilago segetum* var. *tritici* (*U. tritici*) from infected wheat seeds. Plantavax, the sulfone derivative of Vitavax, is effective against rusts. Seed or soil treatment has given good control. The oxathiins are taken up by sensitive and insensitive fungi alike, with most of the fungicides accumulating in the ribosomal and soluble fractions of disrupted cells. Although synthesis of protein and nucleic acids is strongly interrupted, the site of action is the inhibition of succinate dehydrogenase in the mitochondrial fraction. Oxathiin compounds adversely affect nitrogen metabolism of the pathogens. The hexose monophosphate shunt appears to be an important respiration pathway in sensitive fungi.

(e) Calixin Among the other systemic fungicides is Calixin with excellent prophylactic and curative action against powdery mildew of cereals (*Blumeria graminis*), Sigatoka disease of banana (*Mycosphaerella musicola*), and other ascomycetous pathogens. Its chemical name is N-tridecyl-2,6-dimethyl morpholine, commonly known as Tridemorph.

14. Acyalalanines

A new class of chemicals which showed promise with regard to curative activity and systemicity against Oomycetous pathogens. The most important acylalanine fungicides are Metalaxyl. It is sold as Ridomil for use in the soil. It is also sold as Apron for use as seed dressing. Metalaxyl interferes with the development of the pathogens only after these pathogens penetrate their hosts. Metalaxyl interferes with RNA synthesis. It has also been reported that Metalaxyl fungicide elicits the production of phytoalexin in the plants which imparts resistance.

15. Organophosphate Fungicides

Historically, organophosphorus compounds are important because the first practically applied systemic fungicide, Triamiphos, is a member of the group. The fungicides in this group are Kitazin, Ediphenphos (Hinosan), Pyrazophos (Aufgon, Euramil), and Triamiphos (Wepsin).

(a) Kitazin The first systemic fungicide of this group was S-benzyl diethyl phosphorothioate (Kitazin). Later another compound, Kitazin P, was developed which is now used for the control of blast disease of rice. Kitazin inhibits the incorporation of C-glucosamine into the cell-wall fraction of *Pyricularia grisea*. Chitin biosynthesis is interfered with this systemic fungicide which was earlier developed as an insecticide.

(b) Triazoles Triazoles include several excellent fungicides such as RH-124 (INDAR), Triadimenol (BAYTON), Triadimefon (BAYLETON), Biloxazole (BAYCOR), Diclobutrazole (VIGIL), Tricyclazole, and several others. These fungicides show great protective and curative activity against a broad range of foliar, root and seeding diseases such as leaf spots, blights, powdery mildews, rusts, smuts, and others.

16. Morpholines

Morpholines are heterocyclic ring compounds, which have yielded derivatives showing systemic movements in plants. Two compounds have reached commercial use. Morpholines interfere with the function of cell membranes and also in some cases with respiration. Triademorph is marketed under the tradename Calixin. Triademorph is active against representatives of all taxonomic groups of fungi except Peronosporales.

17. Phosphites

Two phosphate compounds, Aluminium-tris (aluminium ethyl phosphite) and LS 73 1038 (sodium ethyl phosphite) have been developed as systemic fungicides based on organometallic compound effective against Oomycetous fungi. However, only Aluminium tris has reached commercial production. Aluminium tris is not toxic to fungi *in vitro* but gives excellent prevention and has curative activity in fields for the control of diseases caused by Oomycetous fungi.

Aluminium-tris (Aliette) This has been introduced under the tradename Aliette.

There are quite a number of other systemic fungicides such as fungicides belonging to Piperazines (heterocyclic nitrogenous compounds) such as Triforine (Saprol). Triforine controls cereal, vegetable, fruit, and flower diseases such as powdery mildews, rusts, scabs, rots, and post-harvest rots by foliar and soil applications.

18. Pyrimidine and Pyridine Derivatives such as Triarimol, Pyroxychlor

Triarimol is a broad-spectrum fungicide which is particularly effective for the control of scab and powdery mildew of apple and pear. Triarimol inhibits sterol synthesis in fungi. Pyroxychlor is a fungicide based on the pyridine nucleus. The compound is of great interest because it was the first compound effective against phycomycetous fungi and capable of controlling root diseases by foliar application because of its downward translocation. Then there are the Acetamides (acetic acid amides) such as DPX-3217 (Curzate) effective against Oomycetes, aliphatics such as Prothiocarbs (Previcur) effectively inhibiting Oomycetes such as *Pythium* and *Phytophthora*, *Peronospora*, and *Bremia* spp.

19. Strobilurins or QoI fungicides

It is one of the most important classes of agricultural fungicides. Strobilurins represented a major development in fungus-based fungicides. The first strobilurins fungicide was isolated from wood-rotting mushroom fungus called *Strobilurus tenacellus*. The discovery of strobilurin fungicide was inspired by a group of natural b-methoxycrylate, the simplest of which are strobilurin A and Oudemansin A. BASF announced the first development of strobilurins, Azoxystrobin and kresoxim-methyl respectively. Since that time, other products have been announced, namely Trifloxystrobin, Metominostrobin, Pyraclostrobin, Picooxystrobin, Fluoxastrobin, Oryzastrobin, and Dimoxystrobin. Strobilurins are mostly contact fungicides as they are absorbed into the cuticle and not transported any further. The strobilurins act by inhibiting mitochondrial respiration of fungi. Apart from this, they have suppressive effect, reducing competition for nutrients, disrupting metabolism and preventing growth of target fungi. Target diseases for nearly all Strobilurin fungicides include downy mildews, rusts, powdery mildews, and many leaf spots. One of the key reasons for the outstanding commercial success of Azoxystrobin is that it gives control of fungi from all four groups of plant pathogens, namely the Ascomycota, Basidiomycota, Deuteromycota, and Oomycota.

20. Antibiotics

Antibiotics are substances which are produced by microorganisms and which act against micro-organisms. Most antibiotics known until now are products of actinomycetes and some are from fungi and bacteria. The chemical nature of antibiotics is complex and they are not as a rule related to each other. A large number of antibiotics have been tried for plant-disease control but only a few have been successfully utilized. The important antibiotics for the control of plant diseases are Streptomycin, Tetracyclines, Griseofulvin, Cycloheximide and Aureofungin. The use of antibiotics is a comparatively new method of plant disease control, but antibiosis or antagonistic phenomena among microorganisms and their relation to plant diseases have been recognized for more than 80 years. Gliotoxin, the first antibiotic to be used for plant-disease control, was isolated and purified by a plant pathologist even before the discovery of Penicillin.

21. Bulbiformin

Bulbiformin is an antifungal polypeptide antibiotic produced by *Bacillus subtilis* and was reported by Vasudeva *et al.* in the year 1958. The production of this antibiotic was demonstrated in sterilized soil amended with molasses, groundnut cake, sweet clover roots, and so on. Inoculation of amended soil with *B. subtilis* controlled the wilt disease caused by *F. udum* on pigeon pea. The antibiotic was absorbed, and because it is systemic, considerably reduced the incidence of wilt of pigeon pea.

22. Soil Fumigants

The most promising method of controlling nematodes in the field has been through the use of chemicals called nematicides. Some of these, including chloropicrin, methyl bromide, Mylone, Vapam and Vorlex, give off gases after being applied to the soil. They are general-purpose preplant fumigants and are effective against a wide range of soil microorganisms, including, in addition to nematodes, many fungi, insects, and weeds. These are also highly volatile fumigants such as formalin and carbon disulphide.

(a) Formalin Various chemical treatments of the soil have been developed, one of the oldest being the formaldehyde drench. Formaldehyde corrodes the hands and is somewhat erratic in effectiveness. For the control of damping-off and seedling blights, a 37-40% solution in water is used.

(b) Carbon Disulphide (CS_2) Carbon disulphide is perhaps the earliest chemical used as a soil fumigant for checking nematodes. Carbon disulphide is used in citrus orchards against *Armillaria mellea*. It is used like formalin. This substance was originally employed against Phylloxera of the vine (*Phylloxera vitifoliae*). Carbon disulphide is cheap, volatile, and is an insecticide also. Some other soil fumigants are chloropicrin, vapam, methyl bromide, DD mixture, and Dazomet.

23. Oils

Various oils are capable of controlling several plant diseases. Sigatoka disease of bananas caused by *Mycosphaerella musicola* has been controlled by oils, which, because of their advantages have almost completely replaced conventional fungicides. Nearly 50 million liters are estimated to have been used in Latin America and the Caribbean in 1964 and perhaps another 7.6 million liters in other banana-producing areas.

The oils involved in disease control are mainly mineral oils from petroleum, and to a much lesser extent, glyceride oils from plants, and synthetic oils. It is not exactly known how mineral oils control plant diseases. Light viscosity oils atomized into very fine droplets of 50-1000 μm in diameter seem to be most effective and can apparently stop the growth of the fungi after it has become established inside the leaf.

The therapeutic action of oils is perhaps due to the oily physical barrier which interferes with gas exchanges in the leaf and alters its physiology so as to stop the development of the fungi. There is evidence that anti-transpirants may reduce disease by, for example, inhibiting stomatal penetration. It is also likely, but not certain, that oils exert their therapeutic action not directly on the pathogen but rather on the physiology of the host. *M. musicola* is encouraged by fairly high light intensities which no doubt favour photosynthesis and high-sugar concentration. Some high sugar pathogens, such as powdery mildews and rusts are apparently susceptible to oils.

11.2.2 Safeners, Spreaders, and Stickers

1. Safeners

A safener is defined as a chemical that reduces the phytotoxicity of another chemical. Copper sulfate is phytotoxic, but with the addition of lime its toxicity is reduced. Lime is, therefore, a safener. Lime is used universally with chemicals to prevent the formation of, or to neutralize arsenic acid, which is phytotoxic to sensitive foliage. Glyceride oils act as good safeners for copper sprays by providing a film around each particle, but they reduce the fungicidal value of the product.

2. Spreaders

Materials which are added to the spray mixture to establish improved contact between the fungicides and the sprayed surface are known as spreaders. These can be classified as water-soluble and water-insoluble. Mineral oils, glyceride oils, and terpene oils are water-insoluble spreaders. Soap has long been known as a spreader. Sulphated alcohols, sopamines, resinate, and petroleum sulphonic acids are some of the common spreaders available in the market. Spreaders reduce the surface tension and improve contact between the fluid and the sprayed surface so that the fluid does not collect into large drops and run off easily.

3. Stickers

Materials added to spray or dust, which improve its adherence to plant surfaces are known as stickers. Flour, starch, gum arabic, and dextrins are commonly used as stickers. Bentonite clays when added to dusts improve the latter's sticking quality. Oils, oil emulsions, and hydrocarbon oils have also been employed as stickers.

4. Carriers

When the fungicide is used as a dust, the active ingredient is generally diluted with a finely divided powder called a carrier.

11.3 DISEASE CONTROL BY IMMUNIZING THE HOST

Inoculation of plants with weak strain of pathogen leads to immunization of the plants to which the plants are generally susceptible. The resistance of plants can also be done by incorporating genes obtained from plants. Other pathogens code for the production of certain enzymes or toxins interfering with infection by the pathogen. Genetic-engineering technology has made possible the isolation of resistance (R) genes from resistant plants and the transfer of R genes into susceptible plant in which they induce the resistant response.

11.3.1 Cross Protection

This method applies to the protection given to a plant by infection with a mild strain of a pathogen from subsequent infection by a severe strain of the same pathogen that normally causes symptoms. Its application in controlling virus diseases has met with some success in cross-protecting citrus with mild strains of the Citrus tristeza virus, and tomatoes with mild strains of the tobacco mosaic virus. This method has some limitations because mild strains of viruses are often not available. In addition, there is a danger of mutation towards a new, more virulent strain and spread to other crops in which virulence might be higher.

11.3.2 Acquired Resistance to Fungicides

The term 'acquired resistance' is used when, in a population which is normally sensitive to a biocide, forms arise that is less sensitive to this biocide. This resistance may have a genetical basis and then as a rule be stable, or it may be caused by a physiological adaptation and disappear again rapidly when the organism is no longer exposed to the toxicant. This temporarily reduced sensitivity is called **tolerance** and the former type is called **resistance**.

The high specificity of action of a toxicant is both a strength and weakness. Many organisms are resistant to selective organic insecticides. The dithiocarbamates and inorganic compounds are nonspecific toxicants and do not penetrate the plant tissues. The development of resistance in microorganisms due to these toxicants is unusual. However, with the introduction of selective systemic fungicides, there was a significant development of resistance. In practice, sometimes resistance is so great that the use of the compound concerned has to be severely restricted or even abandoned in the treatment of certain diseases. Resistance to many of these new systemic fungicides could be readily induced in the laboratory by treatment with ultraviolet irradiation or other mutagenic agents.

Dimethrimol was introduced in Holland in 1968 to control powdery mildew in cucumbers. During the following two years, this control measure was found to be less effective. Even after the removal of the fungicide, tolerant fungi were reported in 1971. It has been reported that Benomyl was very effective in the control of heart rot in cyclamen. Within a short period, a strain of *Botrytis cinerea* developed that could tolerate 1000 ppm, while the wild type was controlled at 0.5 ppm. The tolerant strains showed cross resistance to Thiabendazole, Fuberidazole, and Thiophanate methyl. Benomyl protected sugarbeets in 1970 and 1971 at comparatively low levels. However, in July of 1972, resistance was noted and subsequently a large acreage of beets was lost. Finally, a number of ways to reduce fungicide tolerance were suggested. Among these is the avoidance of the use of one fungicide on an exclusive basis. Systemics with different modes of action would be alternated along with the protectant fungicide. An economically feasible level of fungicide dose should be used, phytotoxicity should be avoided, and toxic residues should be avoided. Resistance to the fungicide Metalaxyl has also been observed. Metalaxyl is used for controlling Oomycetous pathogens.

The continuous and prolonged use of a selective, systemic fungicide may result in loss of effectiveness, which may be incorrectly diagnosed as fungicide tolerance.

11.3.3 Plant Defense Activators

Resistance in plants to several diseases/pathogens such as TMV, *Peronospora tabacina*, and *Pseudomonas syringae* can be induced with salicylic acid, acetyl-salicylic acid (Aspirin), and

Dichloroisonicotinic acid (INA) when applied into the plant by injection or spraying. Such synthetic compounds have no antimicrobial action but mimic the biological induction of systemic acquired resistance in plants against many pathogens. For example, silicon seems to protect plants from powdery mildew infection by inducing localized defense reactions including papilla formation, callose production, and accumulation of phenolics along the cell wall.

11.4 CONTROL THROUGH USE OF TRANSGENIC PLANTS

DNA technology has made it possible to engineer transgenic plants that are transformed with gene coding for enzymes (chitinases and glucanases), for tolerance against a specific diseases or for tolerance of adverse environmental factors against a certain group of pathogens or with nucleic acid sequences that lead to gene silencing of pathogen.

11.4.1 Transgenic Plants that Tolerate Abiotic/Environmental Stress

Many plants have been transformed with genes that enable those plants to tolerate abiotic stresses beyond the normal range of these plants. For example, egg plant transformed with the bacterial gene coding for mannitol phosphodehydrogenase are tolerant against osmotic stress induced by salt, drought, and low temperature. Rice gene coding for glutamine S-transferase have been transformed in maize plant that enabled maize plants to tolerate low temperature and to germinate better submerged in water. Drought tolerance in tobacco plants was increased by introducing the yeast gene coding for trehalose phosphate synthase into the genome of tobacco chloroplasts (August 2005).

11.4.2 Transgenic Plants that Tolerate Specific Plant Genes for Resistance

There are many crops in which specific plant genes for specific pathogens have been isolated from resistant plants and transferred and expressed into susceptible plants. Such resistant plants are cloned and multiplied that is resistant to the specific pathogen. For example, hybrid rice transformed with the rice gene Xa21 coding for resistance of rice to the bacterial blight caused by *Xanthomonas oryzae* pv. *oryzae*. Similarly, when the resistance gene DRR 206 from pea was transferred into Canola, the transgenic Canola plants exhibited resistance to *Leptosphaeria maculans, Rhizoctonia solani*, and *Sclerotinia sclerotiorum*.

11.4.3 Transgenic Plants that Tolerate Gene Coding for Antipathogen Compounds

Gene coding for some enzymes like chitinase and some glucanase have been identified, isolated, cloned, and expressed in susceptible plants and thus providing resistance.

1. Origin of Fungicide-Resistant Strains

It is not clear how fungicide-resistant strains of fungi arise as little work has been done on their genetics. The disappearance of resistance in the absence of the fungicide could be due to sublethal concentrations of the latter. This results in the formation of a substance or substances conferring resistance which could otherwise not be formed. Perhaps, adaptive enzymes, the formation of which is induced by the fungicide, are involved.

Stable resistant strains perhaps arise by mutations, which occur either by chance, or in the case of certain fungicides, could be brought about by the fungicide itself. It is significant that the exposure of *Fusarium caeruleum* (potato dry rot) to tetrachloronitrobenzene (TCNB) resulted in the appearance of TCNB-resistant mutants and that variants of *Botrytis allii* (onion neck rot) were produced after the fungus was treated with the vapour of Pentachloronitrobenzene. Some aromatic hydrocarbons are said to produce gene mutation in fungi. The fungicide resistance of some fungi, including a few which attack plants, has been shown by genetical analysis to be of chromosomal origin, as in *Hypomyces solani* f .sp. *cucurbitae* (seedling blight and fruit rot of cucurbits), where resistance to TCNB is determined by three independently inherited genes. However, there is a little critical evidence that fungicides can induce mutations in fungi which bring about resistance to fungicides.

2. Mechanism of Resistance to Fungicides

The development of resistance to a fungicide may be due to changes in the fungal cell that inhibits the fungicide, to a greater or lesser extent, from reaching the site of action. Such changes include a decreased permeability of the protoplast membrane to the fungicide, or an increased detoxification before the site of action is reached. If the fungicide is able to reach this site, resistance may be due to a decreased affinity between the chemical and the reactive site, to circumvention of the blocked site by operation of an alternate pathway or to compensation for the effects of inhibition.

11.5 BREEDING FOR DISEASE RESISTANCE

The promise of resistance to pests and diseases has attracted the attention of plant breeders ever since the demonstration near the turn of the 20th century that resistance could be selected and that it is often simply inherited. The built-in protective resistance effective throughout a crop plant's life offers a compelling alternative to fungicides and insecticides. In the early years, minimizing the use of protective chemicals was desirable because it saved money but today there is an added advantage of reducing their hazard against those pests and disease such as cereal rusts, soil-borne smuts and certain nematode diseases.

The use of disease-resistant varieties for controlling plant diseases has been termed the "painless method" because it does not cost the farmer anything. In an underdeveloped country like India, it is all the more important since we cannot pay for the heavy costs of spraying and dusting crops on a large scale. The use of resistant cultivars and hybrids has several other advantages. It eliminates the hazard to human health and wildlife which is caused by large-scale use of dangerous fungicides and pesticides. It also reduces pollution which results from the use of poisonous chemicals and their residues. Resistant crop varieties check epidemics of pathogens and pests and thus help maintain the biological balance in the environment.

It has earlier been seen that resistance can be varied in different hosts and parasites. It can be mechanical, such as waxiness, hairiness, and toughness of the epidermis or it may be biochemical in which case it is the result of the interaction between the cytoplasm of the host and the parasite. The biochemical defense mechanism is due to the synthesis of polyphenols in the host that may be toxic to any organism present and to the plant itself. Susceptible and resistant reactions differ principally in the speed and scale of host responses which are more rapid in the latter. As a result, the development of the invading parasite is restricted.

Susceptibility and resistance to a parasite are largely inherited characteristics. The laws governing their inheritance are the same as those governing other characteristics of the host and parasite. It has been proved in several cases that the resistance or susceptibility to disease is located in the genes. The breeding of resistant varieties of crops is, therefore, based upon the laws of inheritance. The Darwinian principles of a variety's struggle for existence and survival of the fittest apply to all the wild varieties and their pathogens. Wild plants have been growing for centuries in the presence of pathogens and natural selection must have eliminated the most susceptible biotypes and left the most resistant ones. In this way, regional biological balances exist between the native plants and their pathogens and pests. The balance will tend to persist unless disturbed by new factors such as the introduction of new biotypes of pathogens or hosts. This is likely to occur if barriers such as oceans of high mountains protect the region. Often it is humans who disturb the natural balance by transporting plants and pathogens to new regions, or who, through crop improvement, produce new varieties of crop plants. Many of the serious outbreaks of plant diseases are traceable to the introduction of parasites to which nonresistant crop plants immediately succumb. The following are some of the notable examples:

1. A very superior variety of hops (*Humulus lupulus*) was cultivated in Western Europe. In 1917, the downy mildew *Pseudoperonospora humuli* was observed in England in a locality, which grew on seeds and plants imported from Japan. In 1922, the fungus attacked South-Eastern England. The pathogen soon spread to Belgium, France, Holland, Germany, and other European countries and had to be controlled at heavy costs by spraying.
2. *Cryphonectria (Endothia) parasitica*, the cause of chestnut canker, does not cause serious damage in the Eastern countries. When it was introduced in the USA in about 1904, it brought heavy losses to the chestnut forests (*Castanea dentata*) in a few years' time. It is estimated that within a period of about 10 years, it had caused losses of about 50 million dollars.
3. Taichung Native 1 (TN1) is a high yielding variety but it brought with it the bacterial blight a serious disease of this crop in 1966. TN1 is highly susceptible to bacterial blight.

There are so many other important diseases, which have been imported into countries which were earlier free from them. For example, *Phytophthora infestans* was brought in infected tubers from South America to England and other European countries in the 1830s and *Plasmopara viticola* from America to France in 1870.

Failures accompany successes while breeding resistant varieties. Plant breeders even now introduce genes for resistance, as they did in 1918. Borlaug (1965) has remarked that the average useful life of a rust-resistant variety of wheat is as short as five years in some parts of the world after which it succumbs to a new race of rust. Walker (1959) wrote, "By and large, the development of resistant varieties must be looked upon as a continuing programme. The potential variability of most pathogens (including viruses) will not permit any currently successful variety to remain so for an indefinite period". Several other examples can be cited where resistant varieties did not stay resistant for long, such as Hessian fly on wheat, powdery mildew on barley, *Cladosporium* leaf mould of tobacco, mosaic virus of tomato, golden nematode and leaf blight of potato. The question arises as to why these varieties were so short-lived and why the long and continued series of efforts failed to give the desired results. The reason is that while the breeder tries to make plants more resistant, nature tries to make pathogens more virulent, and ultimately nature gets the upper hand.

It was earlier stated that the characteristics of an organism are governed by the laws of genetics which are common to both the host and parasite. Resistance or susceptibility (both of which are relative expressions) is the result of interaction between these characteristics. It has now been shown that the genotypes of the host as well as the parasite can be varied by segregation—recombination of genes, mutation, heterokaryosis, somatic hybridization, etc. Genetic studies have now been extended to bacteria and viruses where new strains arise due to hybridization, transformation, and transduction. The previous conception that the pathogen is a fixed entity which damages the crops year after year without change is no longer acceptable. There is a great array of genotypes of the parasite as there is of the host.

11.5.1 Types of Resistance

Plant pathologists commonly divide resistance into monogenic, polygenic and oligogenic, according to whether resistance is governed by one gene, many genes or a few genes. Resistance can be described as genetic, mechanical, or epidemiological.

The general behaviour or the genetic behaviour of disease resistance is governed in many simple to complex ways.

Monogenic resistance (governed by a single gene) is governed by one gene with a 3:1 Mendelian ratio. It can be monohybrid dominant, such as the resistance of oats to *Helminthosporium victoriae,* sorghums to *Periconia circinata,* tomato to the pathogen *Fusarium oxysporum* f. sp. *lycopersici,* onion to the pathogen *Pyrenochaeta terrestris*, and to *Cercospora* leaf spot of cucumber. It can be monohybrid recessive, with resistance as a recessive character, such as powdery mildew of barley, sorghum smut, and yellow rust of wheat. *Oligogenic* resistance (governed by several genes) is determined by two to several genes.

Polygenic resistance (governed by many genes) involves many genes which are more difficult to analyze and which are known to be involved in a large number of diseases such as cotton wilt.

1. Extrachromosomal Inheritance

Extrachromosomal inheritance of disease resistance is rare but is of great importance. The widespread epiphytotic of southern leaf blight of corn in the USA was due to the sudden appearance and distribution of a new race identified as race T of *Helminthosporium maydis*. Resistance to the new race is mostly extra chromosomal. Corn with *cms*-T cytoplasm is more susceptible to yellow leaf blight caused by *Phyllosticta* sp. Symptom production in potato virus X in peppers is apparently determined by cytoplasm.

2. Gene Interaction

Disease resistance to pathogens is not always simply inherited. In many instances, the pattern of inheritance is quite complicated. The resistant phenotype is the ultimate expression of a chain of events in which genes for resistance perhaps control only a few links. Although genes may appear to act as autonomous units, in effect gene products interact and this becomes clear in genetic analysis by the appearance of complementation, modification, and epistasis.

3. Modifier Genes

A host may carry a gene for strong resistance but this is not expressed because of the presence of an inhibitor gene. In onions, resistance to smudge disease is controlled by a single dominated gene R. In the presence of a partially dominant inhibitor gene, I, the resistance is reduced or completely suppressed, depending on whether the gene I is present in the heterozygous or homozygous state. In wheat rust, resistance is completely suppressed by a dominant inhibitor gene. Inhibitor genes affecting recessive genes for rust resistance are also known to be present in maize.

4. Reversal of Dominance

A gene may be dominant for resistance to one race of the pathogen, but recessive for resistance to another race of the pathogen. This reversal of dominance may be due to dosage effect, that is, one dose of the gene in a heterozygote being sufficient to provide resistance against one race of the pathogen but not against the other, or due to closely linked genes for resistance one dominant and the other recessive.

5. Vertical and Horizontal Resistance

When a variety is more resistant to some races than to others, resistance is called "vertical", "perpendicular", or "specific". When resistance is such that it is uniformly spread against all races of the pathogen it is called "horizontal", "lateral", "generalized", "nonspecific" or "field resistance". The derivation of this term is mathematical and refers to the horizontal axis of the graph. Horizontal resistance can be demonstrated when a series of pathotypes is tested against a series of pathodemes. Horizontal resistance has been called field resistance against *Phytophthora infestans* and generalized resistance against *Puccinia graminis*. Vertical Resistance (VR) confers complete but impermanent protection, whereas Horizontal Resistance (HR) confers incomplete but permanent protection.

Vertical resistance is conditioned by oligogenes and is effective against only some races of pathogen, whereas HR, which is polygenic in inheritance, is effective against all races of a pathogen. The terminology is a bit confusing. All oligogenes are not necessarily major genes in the sense of being important genes. For example, the R1 gene governing resistance to potato blight merits being called a major gene. But the gene R4 is weak and of little importance to potato breeders, and hence, it would be an exaggeration to call the R4 gene a major gene. In the same way, all polygenes are not minor genes in the sense of being relatively less important.

The derivation of the term "vertical resistance" is mathematical and refers to the vertical axis of graph. Vertical resistance is proved by analysis of variance in which there is significant differential interaction between path types and pathodemes. Vertical resistance is generally (but not necessarily) inherited by means of major genes. It often (but not invariably) produces a quantitative presence of absence effect in that it operates completely or not at all. The epidemiological characteristic of vertical resistance is an apparent reduction in the initial inoculum; as a result, the epidemic is delayed (cf. horizontal resistance which slows down the epidemic).

11.6 INNOVATIVE METHODS OF PLANT-DISEASE CONTROL

In a multi-authored book entitled *Innovative Methods of Plant-Disease Control* edited by I Chet (1987) several aspects of innovative methods have been discussed. Mehrotra (1993) has discussed the innovative methods for plant-disease control. The first important innovative technique is soil solarization. There are others such as use of siderophores, mycovirus infection and hypovirulence-induced resistance, utilization of mycorrhizal fungi, genetic engineering for plant-disease resistance, pollen management, and biological control. Some of these have already been discussed earlier. Here, soil solarization and pollen management will be discussed in somewhat greater detail.

Soil Solarization

According to Katan *et al.* (1976), this technique for control of plant diseases refers to a new approach for soil disinfestations by means of solar energy. Solarization is done by covering the soil with transparent polythene sheets 100 μm thick for 6-8 weeks during summer before planting any crop. At ICRISAT, Pattancheru, Hyderabad, Chauhan *et al.* (1984) successfully controlled wilt of pigeonpea and chickpea with the help of soil solarization. Sharma *et al.* (1984) successfully demonstrated reduction in the viability of sclerotia of *Sclerotium oryzae* when soils were mulched with 0.5 mm thick transparent polythene sheets. Lodha (1989) showed the reduction of the population of *Macrophomina phaseolina* in arid soils of Rajasthan as a result of solarization. Dwivedi (1993, 1997) studied the soil mycoflora of sugar cane and sugarbeet crops in relation to soil solarization and reported encouraging results. Soil solarization has also been successfully employed to control soil nematodes in India.

11.6.1 Pollen Management

Thakur and Williams (1978, 1980) and Thakur *et al.* (1983) have reported excellent control of ergot in pearl millet through pollen management. Based on the findings that rapid pollination reduces ergot infection, experiments were conducted at ICRISAT Centre at Patancheru, A P, where an ergot less susceptible early flowering line was strategically planted as a pollen donor for susceptible hybrids, ergot severity was considerably reduced, and grain yields were increased in different hybrids.

Genetic manipulation (engineering) has been discussed in Chapter 29 where techniques and applications of biotechnology have been discussed in relation to biotechnology and plant-disease control.

11.6.2 Integrated Pest Management

IPM (Integrated Pest Management) is a paradigm that is widely used or adopted by all pest-control disciplines but whose early definitions and philosophical basis belong to entomologists (Jacobsen, 1997). In phytopathology, another term IDM (Integrated Disease Management) is being used and is a concept derived from integrated pest management. Integrated pest management aims at keeping the pest population below the level of economic injury using best combinations of various components of management. FAO has defined IPM as a pest-management system that, in the context of associated environment and the population dynamics of the pest species, utilized all suitable techniques and methods, in as compatible a manner as possible and maintained the pest population at levels below those causing injury. Srivastava (1999) has briefly summarized the concept of integrated pest management for sustainable agriculture. This new concept actually describes what the farmers were doing before chemical pesticides came into vogue. In IPM, the various components of management are so tailored so as to get cost-effective control without any adverse effect on the environment and ecosystems. The IPM or IDM relies on nontoxic means but does not intend to keep pesticides at bay from the management system, rather advocates minimum or requisite use of pesticides as may be appropriate to make disease management not only effective but nonhazardous, user- and environment friendly. IPM or IDM thus rules out total reliance on pesticides which by and large are the strongest weapons against pathogens or pests. The various components of IDM are host resistance, cultural practices (such as physical treatment of soil by baking the soil by direct exposure to the sun or through soil solarization, removal of infected plants before they contribute to disease severity such as removal of seedlings affected by downy mildew of pearl millet caused by *Sclerospora graminicola* within 3-4 weeks after sowing followed by gap filling by healthy seedlings). Crop rotation is immensely beneficial in reducing inoculum levels. In the case of flag smut of wheat caused by *Urocystis agropyri* (syn. *U. tritici*), manipulation in the dates of sowing is aimed at avoiding the susceptible age with congenial environmental period. For example blast of rice in Haryana can be minimized by planting Basmati varieties by the first week of July and ergot of bajra incidence can be minimized by sowing the crop by June end or the first week of July. Integrated disease management also encourages biological control of plant pathogens. This aspect has been discussed earlier. In India efficiency of several biological control agents has been reported from TANU, Coimbatore, PAU, Ludhiana, G B Pant University of Agriculture and Technology, Pantnagar, University of Agricultural Sciences, Bangalore and Kurukshetra University, Kurukshetra, and several potent biocontrol agents reported are species of *Trichoderma*, *Gliocladium*, *Chaetomium*, and certain *Pseudomonas*.

In recent years, several botanical pesticides have been reported to control plant pathogens. Neem products/neem extracts have become quite popular as insecticides. The efficacy of neem has also been reported against some fungal pathogens also. Though IPM or IDM relies on nontoxic chemicals for plant-disease control, it does not preclude the use of pesticides. It favours minimum and judicious use of chemicals. Of course growing resistant varieties is the best method of controlling any disease and will from an important strategy for managing plant pathogens.

This IPM or IDM is a system approach that combines a wide array of crop production and protection practices to minimize the economic losses caused by pests (insects, pests, disease, nematodes, weeds, rodents, birds, etc.). It emphasizes careful monitoring of pests and conservation of their natural enemies. The Indian Council of Agricultural Research (ICAR) established a National Centre of Integrated Pest Management (NCIPM) in 1988 to cater to the emerging plant-protection needs of agriculture in the country.

REVIEW QUESTIONS

1. What are the cultural methods of plant-disease control?
2. What are plant quarantines?
3. When and where was the first quarantine act enacted?
4. What is the DIP act? When was it first enacted?
5. Name some plant diseases, which have been introduced before and after the enforcement of plant quarantines in India.
6. Is quarantine necessary for all types of diseases?
7. Name some plant diseases which can be controlled by crop rotation.
8. What is seed certification?
9. What is biological control of plant pathogens? Give some suitable examples where it has been successfully used to control disease.
10. What are the different chemical systemic fungicides for plant-disease control?
11. How do the fungicides act on the pathogen?
12. What is chemotherapy?
13. What is the mechanism of the action of various groups of fungicides?
14. Name some important antibiotics used for plant-disease control.
15. Breeding for disease resistance is the best method of controlling plant diseases. Why?
16. What is integrated Pest/Disease Management (IPM or IDM)?
17. What is solarization?
18. What are the innovative methods of plant-disease control? Name a few.
19. Name some plant pathogens of plant quarantine importance to India.
20. What is resistance to fungicides by plant pathogens? What strategy will you employ to overcome this?

MULTIPLE-CHOICE QUESTIONS

1. The principal methods of controlling plant diseases are
   (i) avoidance (ii) eradication
   (iii) protection (iv) all of these
2. Which of the following is not a cultural method of plant-disease control?
   (i) Avoidance of pathogen (ii) Disease-escaping varieties
   (iii) Choice of time of sowing (iv) Bordeaux mixture
3. How many quarantine stations are operating under the Directorate of Plant Protection and Quarantine in India?
   (i) 16 (ii) 116
   (iii) 216 (iv) 6

4. Soil solarization is a method to control the diseases which are
   (i) soil-borne (ii) air-borne
   (iii) seed-borne (iv) all of these
5. Post-harvest diseases are effectively controlled by
   (i) radiations (ii) soil solarization
   (iii) mixed cropping (iv) crop rotation
6. Which of the following is used as biocontrol agent against pathogenic fungi?
   (i) *Trichoderma* (ii) *Polyporus*
   (iii) *Armillaria* (iv) *Ganoderma*
7. Which of the following mechanisms have been proposed to explain disease suppression in biological control of fungi?
   (i) Production of siderophores (ii) Antibiosis
   (iii) Competitive root colonization (iv) All of these
8. Which of the following is a dithiocarbamate group of fungicides?
   (i) Thiram (ii) Burgundy mixture
   (iii) Bordeaux mixture (iv) Inorganic sulfur compounds
9. Which of the following is a systemic fungicide?
   (i) Benomyl (ii) Pentachloronitrobenzene
   (iii) Cheshunt compound (iv) Captafol
10. Which of the following antibiotics inhibits cell-wall synthesis of bacteria?
    (i) Streptomycin (ii) Cycloheximide
    (iii) Griseofulvin (iv) Penicillin
11. Which of the following is not a soil fumigant?
    (i) Formalin (ii) Carbon disulfide
    (iii) Vapam (iv) Vitavax
12. Which of the following substance is used as a plant-defense activator?
    (i) Salicylic acid (ii) Bordeaux mixture
    (iii) Plantavax (iv) Terpene oil
13. Which of the following method is used to induce resistance in plants?
    (i) Extra chromosomal inheritance (ii) Gene interaction
    (iii) Reversal of dominance (iv) All of these
14. Which of the following are the Mexican draft varieties of wheat introduced in India?
    (i) Sonara 64 (ii) Lerma Rajo
    (iii) Sharbati sonara (iv) All of these
15. Name the Institute where the wild species of *Solanum* are maintained.
    (i) CPRI, Simla (ii) IARI, New Delhi
    (iii) CIMAP, Lucknow (iv) ICRISAT, Hyderabad
16. Which of the following is an innovative method of plant-disease control?
    (i) Pollen management (ii) Crop rotation
    (iii) Antibiotic method (iv) Chemical method
17. IPM and IDM are programmes used to minimize the economic losses caused by
    (i) insects (ii) disease
    (iii) nematodes (iv) all of these
18. The full form of DD mixture is
    (i) Dichloropropene and Dichloropropane
    (ii) Dimethylchloride and Dimethyloxide

(iii) Dichlorosulphate and Dichlorosulphite
(iv) none of these

19. Bulbiformin is an antifungal polypeptide antibiotic produced in
(i) *Bacillus subtilis* (ii) *Bacillus thuringenesis*
(iii) *Bacillus papillae* (iv) *Bacillus anthracis*

20. Which of the following is not a Tetracycline antibiotic?
(i) Terramycin (ii) Aureomycin
(iii) Gliotoxin (iv) Actinomycin

21. Which of the following chemicals are fungus-based fungicides?
(i) Ferbam (ii) Chloroneb
(iii) Strobilurins (iv) Plantavax

Answers

1. (iv)	2. (iv)	3. (i)	4. (i)	5. (i)	6. (i)	7. (iv)	8. (i)	9. (i)	10. (iv)
11. (iv)	12. (i)	13. (iv)	14. (iv)	15. (i)	16. (i)	17. (iv)	18. (i)	19. (i)	20. (iii)
21. (iii)									

12 PLANT DISEASES CAUSED BY FUNGI CHARACTERISTICS AND CLASSIFICATION OF PLANT-PATHOGENIC FUNGI

12.1 GENERAL FEATURES OF FUNGI

Although it is not easy to define fungi, biologists currently use the term 'fungus' to include nucleated, spore-bearing, achlorophyllous organisms with absorptive nutrition which generally reproduce sexually and asexually, and whose filamentous branched somatic structures are typically surrounded by cell walls containing cellulose or chitin or both.

Although fungi are a diverse group of organisms, a feature which they all have in common is their mode of nutrition. Like animals, fungi are heterotrophic organisms, which must consume preformed organic matter. They may live as saprophytes or parasites.

Fungi are of great economic importance to humans and play an important role in the disintegration of organic matter. They affect us directly by destroying food, fabric, leather, and other commercial goods. They are responsible for a large number of diseases of plants, animals, and man.

12.1.1 Cell Structure

Fungal cells may be minute and a single, uninucleate cell may constitute an entire organism. A single cell of *Olpidium* or yeast cell falls in this category. Alternatively, fungal cells may be elongated and strandlike and several may be joined to form a thread of cells—a hypha (pl. hyphae). A large number of hyphae collectively form the mycelium. The cell wall of fungi is generally made up of chitin or cellulose. The majority of the fungi have chitinous walls. Occasionally, both chitin and cellulose may occur simultaneously.

One cell in the hypha may be separated from another by a cross wall or a septum. Septa in different groups of fungi have important differences in structure. For example, the septa in the lower fungi (Chytridiomycota and Zygomycota) are pseudosepta, which are septa that are perforated by so many pores that they are sievelike. In Ascomycota and some Fungi Imperfecti, the septum is perforated by a single pore. In Basidiomycota and some members of Fungi Imperfecti, the septum is more complex than that found in the Ascomycota. The septum here is called a **dolipore septum**.

Hyphae characteristically grow at the tips and can usually produce lateral branches in their older parts. The mycelium may remain microscopic or develop easily visible organized structures, such as strand- or cord-like rhizomorph, compact resting bodies comprising fungal tissues made of prosenchyma or pseudoparenchyma.

All growing hyphae are filled up with cytoplasm, are frequently vacuolated and may be aseptate; hence, they form a single multinucleate cell (coenocyte) or are divided into segments by septa. Fungal nuclei are very small. The nuclei have well-developed nuclear membranes. The nuclear envelope consists of a double nuclear membrane and is marked by large gaps. During mitosis, the nuclear envelope does not disappear ordinarily, instead it constricts in a dumbell-like fashion and eventually separates into two daughter nuclei. Meiosis conforms more closely to the norm in other organisms than does mitosis. Ribosomes are of 80S and mitochondria occur in the cytoplasm. There is also the endoplasmic reticulum. Storage products occur in the fungal cell. These may be in the form of glycogen or lipids.

Golgi apparatus, similar to those in higher plants and animals, have been found in some fungi. A feature which is of widespread occurrence is the presence of **lomasomes**. These are made up of membrane-bounded tubules or vesicles. The function of lomasomes is not known.

12.1.2 Reproduction

Fungi reproduce both asexually as well as sexually. Asexual reproduction takes place by a variety of methods. This may be by (1) fragmentation of the soma, (2) fusion of the stomatic cells, (3) budding of the somatic cells, or (4) production of spores, such as conidia, sporangiospores—motile or nonmotile.

Sexual reproduction is of widespread occurrence in fungi. Fertilization can be brought about in a variety of ways. The most common methods are (1) planogametic copulation, (2) gametangial contact, (3) gametangial copulation, (4) spermatization, and (5) somatogamy.

12.1.3 Life Cycle

As in other organisms, in fungi too there is generally a cycle of haploid and diploid structures, corresponding to the gametophyte and sporophyte in the green plants. The diploid phase begins with karyogamy and ends with meiosis. In the majority of fungi there is no distinct alternation of generations. Raper (1954) recognized seven basic types of life cycles. However, Burnett (1976) recognizes five basic life cycles in fungi. These are

1. **Asexual**, in which sexual reproduction is lacking, namely, Deuteromycetes.
2. **Haploid**, in which meiosis immediately follows nuclear fusion. This is very commonly seen in fungi or some members of Ascomycota. Here the diploid phase is of minimum duration.
3. **Haploid-dikaryotic**, which is similar to the second type, excepting that paired, potentially conjugate nuclei persist in close physical association in the same hyphal segment (hence, dikaryon). The examples of this are many members of Ascomycota. Binucleate ascogenous hyphae develop just prior to ascus development and such a dikaryon cannot exist independently of the haploid phase. On the other hand, there may be a condition where the meiospores fuse to form a dikaryon so that the fungus cannot exist independently of the haploid phase. On the other hand, there may be a condition where the meiospores fuse to form a dikaryon so that the fungus is dikaryotic throughout its life cycle, excepting for the moment of fertilization and during the subsequent meiosis. This occurs in yeast (Endomycetales) and more commonly in smuts (Ustilaginales). In the Basidiomycota, the mycelium derived from the germination of a meiospore may persist in the haploid condition as a monokaryon, but once a dikaryon is formed, it shows unrestricted and independent growth and is the 'long-lived' phase of the life cycle.
4. **Haploid-diploid**, in which the haploid and diploid phases alternate regularly. This is restricted to the order Blastocladiales.

5. **Diploid**, in which the haploid phase is restricted to the gametes. The vegetative phase is diploid. Reduction takes place at the time of gametogenesis, such as *Pythium*, *Phytophthora*, *Saprolegnia* spp., in the Oomycetous fungi.

12.2 CLASSIFICATION OF FUNGI

Whittaker (1969) broke the tradition of a three-kingdom system of classification of all living organisms. In doing so, he recognized that the classification of all living organisms as prokaryotes, animals, and plants (including Fungi) did not reflect their relationships. Whittaker (1969) added two more kingdoms viz. Fungi and Protista.

In an attempt to recognize monophyletic groups, the organisms once classified as Fungi considered in three different groups—the monophyletic kingdom Fungi, Straminipila (various names have been applied to Straminipiles, also known as "Stramenopiles" or Stramenopila) or Chromista and four protist phyla. The kingdom Fungi includes six phyla—Chytridiomycota, Glomeromycota, Microsporidia, Zygomycota, Ascomycota, and Basidiomycota. Kingdom Stramenopila or Chromista includes the phyla Oomycota, Hyphochytriomycota, and Labyrinthulomycota. The four phyla of the Kingdom Protista or Protictista are Plasmodiophoromycota, Dictyosteliomycota, Acrasiomycota, and Myxomycota.

An outline classification of fungus and fungus-like organisms according to Alexopolous *et al.* (1996) is given below:

1. Kingdom Fungi

- Phylum Chytridiomycota
- Phylum Zygomycota
- Phylum Ascomycota
- Phylum Basidiomycota

2. Kingdom Straminipila or Stramenopila

- Phylum Oomycota
- Phylum Hyphochytriomycota
- Phylum Labyrinthulomycota

3. Kingdom Protista

- Phylum Plasmodiophoromycota
- Phylum Dictyosteliomycota
- Phylum Acrasiomycota
- Phylum Myxomycota

The classification scheme proposed by Blackwell and Spatafora (2004) is presented below:

Fungi	Straminipila	Slime Molds
Chytridiomycota	Oomycota	Plasmodiophorales
Zygomycota	Hyphochytriomycota	Dictyosteliomycota
Ascomycota	Labyrinthulomycota	Acrasiomycota
Basidiomycota	Thraustochytriales	Myxomycota

An outline classification followed in the 10th edition of the Dictionary of fungi by Kirk *et al.* (2008) is given below:

Table 12.1 Classification of Fungi (based on 10th edition of Ainsworth and Bisby's Dictionary of the fungi by P M Kirk, P F Cannon, D W Minter, and J A Stalpers, 2008, CAB International, UK).

Kingdoms	Phyla	Classes/Subphyla	
1. FUNGI	1. Ascomycota	14 Classes	1. Arthoniomycetes 2. Dothideomycetes 3. Eurotiomycetes 4. Laboulbeniomycetes 5. Lecanoromycetes 6. Neolectomycetes 7. Orbiliomycetes 8. Pezizomycetes 9. Pneumocystidomycetes 10. Saccharomycetes 11. Schizosaccharomycetes 12. Sordariomycetes 13. Taphrinomycetes 14. Leotiomycetes
	2. Basidiomycota	14 Classes	1. Agaricomycetes 2. Dacrymycetes 3. Tremellomycetes 4. Entrorrhizomycetes 5. Exobasidiomycetes 6. Ustilaginomycetes 7. Agaricostilbomycetes 8. Atractiellomycetes 9. Classiculomycetes 10. Cryptomycocolacomycetes 11. Cystobasidiomycetes 12. Microbotryomycetes 13. Mixiomycetes 14. Pucciniomycetes
	3. Chytridiomycota	2 Classes	1. Chytridiomycetes 2. Monoblepharidomycetes
	4. Glomeromycota	1 Class	1. Glomeromycetes
	5. Microsporidia (Microspora)	No Classes	
	6. Zygomycota	4 Subphyla	1. Entomophthoromycotina 2. Kickxellomycotina 3. Mucoromycotina 4. Zoopagomycotina
ii. CHROM-ISTA	1. Hyphochytriomycota 2. Labyrinthulomycota (=Labyrinthista) 3. Oomycota	1 Class 1 Class 1 Class	1. Hyphochytriomycetes 1. Labyrinthulea(=Labyrinthulomycetes) 1. Oomycetes
iii. PROTO-ZOA	1. Percolozoa (Acrasida)	1 Class	1. Acrasiomycetes
	2. Amoebozoa (Mycetozoa)	3 Classes	1. Dictyostelia (= Dictyosteliomycetes) 2. Myxogastria (=Myxomycetes) 3. Protostelia (=Protosteliomycetes)
	3.Cercozoa (Plasmodiophorida)	1 Class	1. Phytomyxea
	4. Choanozoa (Amoebidiales, Eccrinales)	No Classes	

12.2.1 Kingdom Fungi

As already indicated in Table 12.1, the kingdom Fungi is divided into six divisions or phyla according to the latest classification of Kirk *et al.* (2008). These are Chytridiomycota, Glomeromycota, Microsporidia or Microspora, Zygomycota, Ascomycota, and Basidiomycota.

1. Phylum Chytridiomycota

The phylum Chytridiomycota contains two classes Chytridiomycetes and Monoblephariomycetes. These are the only members of the kingdom Fungi that produce motile cells at some stage of the life history. The motile cells (both zoospores and gametes) of these organisms each possess a single, posterior flagellum. Other characteristics that these species of Chytridiomycota have in common but other fungi may also exhibit are 1. the coenocytic structure of the thallus, whether a globose or ovoid structure, an elongated simple hypha, or a well-developed mycelium, and 2. the conversion of the zygote into a resting spore or a resting sporangium, or in one order, its growth into a diploid thallus. Cell walls of these fungi contain chitin and glucan. Nuclear divisions in the group are intranuclear and centric.

Chytridiomycota are divided into 5 orders—Spizellomycetales, Neocallimasticales, Chytridiales, Blastocladiales, and Monoblepharidales. Ultrastructure of zoospore plays an important part in the classification of this division.

The order Spizellomycetales contains 2 families, 7 genera, and 27 species. They are a segregate from Chytridiales which they resemble except zoospores. Zoospores of most species have more than one lipid globule. *Olpidium* and *Rozella* are important. *O. brassicae* often attacks roots of cabbage. *O. brassicae* zoospores are vectors of lettuce big-vein virus, *O. uredinis* occurs as parasitic on the uredospores of *Hemileia canthii*. The order Neocallimastigales contains 16 species classified into 6 genera which contain both monocentric and polycentric chytrids that inhabit the rumen and caecum of herbivorous animals and are anaerobic. Their zoospores lack mitochondrium.

The genus *Synchytrium* with 121 species belonging to the order Chytridiales are plant parasites. *Synchytrium endobioticum* causes black-wart disease of potato. Members belonging to Blastocladiales which are important from the pathological point of view are *Coelomomyces* (Coelomomycetaceae), which are obligate parasites in the larvae of insects, usually the mosquito. The genus *Coelomomyces* shows the phenomena of heteroecism and is a potential candidate for the biological control of mosquitoes. Another genus *Physoderma* produces galls and spots. *P. maydis* is responsible for brown spot of maize. *Urophlyctis* species cause galls. Members of Monoblephariadales are not important from the plant pathological point of view.

2. Phylum Zygomycota

This division has been classified differently by different authors. Genera like *Rhizopus*, *Choanephora*, and *Blakeslea* are plant pathogens. Some are human and animal pathogens such as genera like *Absidia* and *Saksenaea*. Members of the order Entomophthorales are chiefly parasitic on insects (such as *Entomophthora*) and members of Zoopagales are mycoparasites.

3. Phylum Glomeromycota

This division contains 4 orders, 9 families, 12 genera, and 169 species. These are mycorrhizal fungi and are of great importance in agriculture and forestry. Some of the important genera are *Endogone*, *Glomus*, *Acaulospora*, *Scutellospora*, *Sclerocystis*, and *Gigaspora*.

4. Phylum Ascomycota

This division contains 14 classes, 68 orders, 327 families, and 64163 species (Kirk *et al*, 2008). They are saprobes, parasites (especially on plants) or lichen-forming group of fungi.

Leaf-curl diseases are caused by *Taphrina*, and *Protomyces* species cause galls. Powdery mildew diseases are caused by members of the family Erysiphaceae grouped into several main genera; these genera are distinguished from one another by the number of (one versus several) asci per cleistothecium. The main genera are *Blumeria*, *Erysiphe*, *Leveillula*, *Microsphaera*, *Oidium*, *Phyllactinia*, *Podosphaera*, *Sphaerotheca*, and *Uncinula*. Some of the important genera of this division (Ascomycota) are *Aspergillus* and *Penicillium* (anamorphic forms) of the order Eurotiales. These are powerful producers of antibiotics and mycotoxins and play an important role in spoilage of fruits and vegetables after harvesting. Filamentous pyrenomyceteous ascomycetes with perithecia are important parasites and symbionts of arthropods, endophytes, mycotoxin producers, mammalian parasites, plant parasites, and saprobes. Some of the important genera are *Hypocrea*, *Gibberella*, *Claviceps*, *Cordyceps*, *Epichloe*, *Balansia*, *Ophiostoma*, *Ceratocystis*, *Magnaporthe*, *Cryphonectria*, and *Rosellinia*. Ascomycetes with apothecia which are of pathogenic importance are *Rhytisma*, *Scelerotinia*, and *Monilia*. Then there are ascomycetes with ascostroma of plant pathological importance such as *Elsinoe*, *Guignardia*, *Mycosphaerella*, and *Venturia*.

5. Asexual Ascomycetes and Other Asexual Fungi (Deuteromycetes or Imperfect Fungi)

Classification here is mainly based on the mode of development of conidia. Some of the important genera are *Geotrichium, Aspergillus, Penicillium, Drechslera, Exserohilum, Botrytis, Monilia, Oidium, Chalara, Trichoderma, Gliocladium, Verticillium, Fusarium, Colletotrichum, Cercospora, Septoria, Phyllosticta, Alternaria, Rhizoctonia, Sclerotium.* The anamorphic stage, the mode of development of conidia and the presence of teleomorphic stage plays an important role in the classification of these genera.

6. Phylum Basidiomycota

According to Kirk *et al.* (2008), there are 3 subdivisions, 14 classes, 52 orders, 177 families, 1589 genera, and 31515 species in this division. Without going into the details of classification, it would be suffice here to mention some important orders, genera which are of plant pathological importance. The members belonging to Aphyllophorales or Polyporales are saprobic or parasites. Some of the important genera are *Schizophyllum* (not much of pathological importance but much important from the cytological and genetical point of view). *Poria*, *Fomes*, *Heterobasidion*, *Lenzites*, *Trametes*, *Irpex*, *Ganoderma*, *Polyporus* are lignicolous and are wood and timber destroyers or pathogenic on living trees. *Boletus* species are mycorrhizal. *Amanita* species are generally poisonous. Then there are edible mushrooms. *Armillaria* species produce rhizomorphs and parasites of trees. Species like *Psilocybe mexicana* are hallucinogenic. *Exobasidium vexans* is responsible for blister blight of tea.

Members belonging to rusts and smuts are well-known plant pathogens. Some of the important plant pathogens in this group are

Puccinia graminis tritici (Black stem rust of wheat), *Puccinia striiformis* (yellow or stripe rust of wheat), *P. triticina* (Syn. *P. recondita*) (brown rust of wheat), *Gymnosporangium juniperi virginianae* causing cedar-apple rust, *Hemileia vastrartix* (coffee rust), *Phragmidium rubi-idaei* (yellow rust of raspberry), *Uromyces pisi* (rust on pea), *Cronartium ribicola* (white pine blister rust), *Melampsora lini* (flax rust). The important smut genera are *Ustilago segetum* var. *tritici* (loose smut of wheat), *Tilletia indica* (Karnal bunt of wheat), *Sphacelotheca cruenta* and *S. reiliana* (Sorghum smuts), *Urocystis cepulae* (onion smut), *Neovossia barclayana* causing kernel smut of rice, *Entyloma oryzae* (leaf smut of rice), and several others.

12.2.2 The Four Groups of Protists Phyla

The 10th edition dictionary of fungi has named these (Protists) as Percolozoa (Acrasida), Amoebozoa (including Dictyostelia, Myxogastria, and Protostelia), Cercozoa (Plasmidiophorida), and Choanozoa (Amoebidiales, Eccrinales). These four groups are not closely related to fungi; however, mycologists often consider them along with of fungi because they occur in the usual habitats of fungi and have similar modes of nutrition.

REVIEW QUESTIONS

1. Fungi and fungal-like organisms are now placed in three kingdoms. What are these three Kingdoms? How are Oomycota different from other true fungi?
2. What are the criteria for classifying fungi and fungal-like organisms?
3. Give an outline classification of fungi and fungal-like organisms.
4. What are the salient differences in the classification of fungi as given in the *Dictionary of Fungi*, 10th edition, by Kirk *et al.* (2008) to that of Alexopoulos, Mims, and Blackwell (1996)?

MULTIPLE-CHOICE QUESTIONS

1. Which of the following division does not belong to the true fungi?
   (i) Oomycota (ii) Chytridiomycota
   (iii) Ascomycota (iv) Basidiomycota
2. Heterokontean biflagellates zoospores are with tinsel, and whiplash type are found in
   (i) Oomycota (ii) Chytridiomycota
   (iii) Basidiomycota (iv) Myxomycota
3. Phenomena of heteroecism is found in
   (i) Smuts (ii) *Coelomomyces*
   (iii) Zygomycota (iv) Ascomycota

Answers

1. (i) 2. (iv) 3. (ii)

13

DISEASES CAUSED BY CHYTRIDIOMYCOTA

Chytridiomycetes are also called **chytrids**. Chytridiomycetes are water or soil-borne fungi. They survive in soil as resting spores or in host plants as a spherical-shaped thallus. The resting spores germinate to produce one or many posteriorly uniflagellate zoospores which infect plant cells. The class Chytridiomycetes contains three important pathogenic genera: *Olpidium* which infects the roots of many plants; *Synchytrium* which causes black wart of potato, and the *Physoderma* species which causes brown spot of maize and crown rot of alfalfa. *Olpidium* can transmit viruses from the host in which it is produced to those it infects next. The fungus *Coelomomyces* belonging to the order Blastocladiales is not a plant pathogen but is of great biological importance as it is a heteroecious fungus and is a potential candidate for the biological control of mosquito larvae.

WART OR BLACK-WART DISEASE OF POTATO

It is found in most European countries, South Africa, India, Peru, Bolivia, and Mexico. In India, it is prevalent only in the North-East Himalayas.

The disease was reported for the first time in India from Rangbul in Darjeeling district (WB) in 1953. The disease is endemic in that area and about 60% of the potato area was affected.

1. Symptoms

All underground parts exhibit symptoms of the disease except the roots and basal part of the stem. The parts that show abnormal activity and the formation of wart symptom are buds of underground stems, stolons, and tuber but not on the roots. The size of the warts also varies depending on the disease intensity. Diseased tubers when lifted bear dark brown, warty, cauliflower-like excrescences. In severe infections, the galls may also be produced on the aerial shoots and they are then green with convoluted, leaflike masses of tissue. As the galls become confluent, enormous, warty outgrowths and tumours are formed. Tumours may turn brown to black with age. The yield of saleable potatoes from a heavily infected crop may be less than the actual weight of the seed potatoes planted (shown in figure).

Wart disease of potato (*Synchytrium endobioticum*) symptoms

2. Causal Organism

The pathogen responsible for wart disease of potato is *Synchytrium endobioticum* de Bary and Woronin which is a biotrophic parasite and has been successfully cultured outside living host cells. The pathogen produces thin-walled sporangia or summer spores (active phase) and the resting sporangia (dormant phase) are also known as winter sporangia. The resting sporangia are released by the decay of wastes and these may remain alive in the soil for many years (30–40 years). The resting sporangia function as a prosporangium on germination and the endospore balloons out to from a vesicle within which a single sporangium differentiates (Kole, 1965). Infection of the tubers takes place by uniflagellate zoospores. A zoospore on the surface of a host cell loses its flagellum and forms a wall around itself. The naked protoplast enters the host tissue through a pore developed on the surface. Once inside the host cell, it reaches the lower part of the cell. The host tissue surrounding the infected cell and the adjacent epidermal cells divide repeatedly to form a characteristic tumourous tissue. A golden-brown, double-layered wall is secreted around the thallus and at this stage, the thallus is termed a prosorus or summer spore. The summer spore or prosorus germinates to form a sorus of 5-9 sporangia and then ultimately forms zoospores. Zoospores are liberated outside in the soil and the cycle of infection can continue if environmental conditions remain suitable for infection.

3. Disease Cycle

The pathogen survives in the soil in the form of resting sporangia for many years (three to four decades). Viable resting spores were found in air-dry soil after 15 months and in dry waste material after 5 years. Contaminated manure and infested soil transported in various ways are means of spreading the pathogen. Short periods of saturation of the soil with water are essential for the germination of resting spores and sporangia, as well as for the liberation and movement of zoospores.

4. Disease Management

Various degrees of quarantine procedures have helped in keeping the disease under check. In India, the disease has been restricted to the North-Eastern Himalayan region (Darjeeling hills) by prohibiting the introduction of potatoes from that region (local quarantine under DIP Act 1914 II Section A) to other parts of the country. The following methods can be adopted.

- Practice long crop rotation (5 years or more).
- Rogue out plants of susceptible varieties by burning and destroying warty tubers.
- Diseased potatoes should not be used as seed potatoes.
- Diseased potatoes may be consumed or fed to cattle only after boiling.
- Intercropping with maize rotated with bean or radish was quite effective in the management of wart diseases.

The application of lime to the soil to raise the pH and the use of soil fungicides have met with partial success.

There are some other soil treatments, and amongst them were steam sterilization of the soil and application of mercuric chloride, ammonium sulphocyanate, copper sulphate, and formalin (5%). The only drawback with these treatments is that they are costly and cannot be applied on a large scale. Resistant varieties are being bred but the occurrence of virulent races and too many biotypes of the pathogen have complicated the whole programme. But there are varieties which are immune to wart disease and they are Kufri Jyoti, Kufri Sherpa, Kufri Kanchan (SC/1-1307), Pimpernel, Ackersegen, Voran, Sangdorze, and Ultimus.

REVIEW QUESTIONS

1. Name the important genera and diseases caused by Chytridiomycota.
2. Write symptoms and management of wart disease of potato.
3. Write the etiology and management of brown spot of maize.
4. What is heteroecism? Where does it occur in the fungus other than rusts?
5. What is a biotroph?

MULTIPLE-CHOICE QUESTIONS

1. Which of the following member of Chytridiomycota acts as a biocontrol agent against mosquito larvae?
   (i) *Spizellomyces* (ii) *Allomyces*
   (iii) *Olpidium* (iv) *Coelomomyces*
2. Which of the following part of the plant is not infected in wart disease of potato?
   (i) Underground stem (ii) Stolons
   (iii) Tubers (iv) Roots
3. The zoospore of *Synchytrium endobioticum* possesses
   (i) whiplash type of flagellum (ii) tinsel type of flagellum
   (iii) both (i) and (ii) (iv) single anterior flagellum
4. Which of the following varieties are immune to wart disease of potato?
   (i) Kufri Jyoti (ii) Ultimus
   (iii) Pimpernel (iv) All of these

5. In *Physoderma zeae maydis*, infection hyphae expand within the host cell to found several enlarged vegetative cells called
   (i) sammelzellen
   (ii) resting spores
   (iii) prosorous
   (iv) endosporangium

Answers

1. (iv) 2. (iv) 3. (i) 4. (iv) 5. (i)

14

DISEASES CAUSED BY ZYGOMYCOTA

Zygomycetes have coenocytic mycelia, and produce thick-walled zygospores and nonmotile spores in sporangia. They are terrestrial fungi, either weak parasites of plants or saprophytes or their spores are often found floating around in the air.

Three genera of zygomycota are known to cause diseases in plants or plant products.

1. *Choanephora* which attacks the floral parts of many plants and from there enters the fruit and causes a soft-rot disease primarily on summer squash, okra, pepper, and pumpkin.
2. *Rhizopus* causes post-harvest rot of many fruits and vegetables.
3. *Mucor*, both common bread fungi, which in addition cause soft rot of many fleshy fruits, vegetables, bulbs, flowers, and seeds.
4. *Saksenaea vasiformis* although is not a plant parasite is a serious human pathogen and causes zygomycosis in humans. Other genera are fungi that become associated with roots of higher plants and form endomycorrhiza, e.g. *Glomus*, *Sclerocystis*, *Gigaspora*, *Scutellospora*, *Acaulospora,* and *Entrophospora* are beneficial to plants. The pathogenic members secrete enzymes that diffuse into the living tissue and disrupt and kill the cells. The mycelium then grows into and colonizes the tissues it killed.

RHIZOPUS SOFT ROT OF FRUITS AND VEGETABLES

The disease is quite prevalent throughout the world on harvested fleshy vegetables and fruits during storage, transit, and marketing of these products.

1. Symptoms

Infected areas of fruits and vegetables appear water soaked at first and are soft. If the outer covering of the infected fruit and vegetables remains intact, the tissue loses moisture slowly until it shrivels into a mummy. More frequently, the mycelium grows into and outside the infected tissue producing tufts of sporangiophores and sporangia. Later, these infected tissues give off a typical smell but soon yeast and bacteria move in and a sour odour develops. When the infected organs of a fruit or vegetable lose moisture quickly, it finally dries up and mummifies.

2. Causal Organism

The pathogen responsible in the several species of *Rhizopus*.

3. Disease Cycle

The pathogen is air-borne and if sporangiospores tend on wounds of fleshy fruits and vegetables, they germinate and cause soft rotting by producing pectolytic and cellulolytic enzymes.

4. Disease Management

Avoid wounding fleshy fruits and vegetables during harvest, handling, and transportation. Discard the wounded, rotted, fruits, and vegetables separately from healthy ones. Clean and disinfect storage containers and warehouses with copper sulphate solution, formaldehyde, and chloropicrin. Keep fruits and vegetables at temperatures around 25–30°C because at this temperature, cut surfaces cork over and subsequent penetration by fungi do not exist. *Candida* and *Pichia* are potential biocontrol agents against Rhizopus soft rot of fruits and vegetables.

REVIEW QUESTIONS

1. Describe the symptoms and causal agent of Rhizopus soft rot of fruits and vegetables.
2. Describe the following briefly:
   (i) Management of Rhizopus soft rot of fruits and vegetables
   (ii) Difference between *Rhizopus* and *Mucor*
   (iii) Why are endomycorrhizal fungi beneficial to plants?
   (iv) Endomycorrhizal fungi

MULTIPLE-CHOICE QUESTIONS

1. Hyphae of *Rhizopus* are
   (i) unbranched, tubular and unicellular
   (ii) branched, tubular and coenocytic
   (iii) unbranched, tubular and coenocytic
   (iv) branched, tubular and unicellular
2. During asexual reproduction, *Rhizopus* produces many aerial hyphal branches. The branches are called
   (i) sporangiophores (ii) antheridiophores
   (iii) conidiophores (iv) all of these
3. In *Mucor*, sexual fusion takes place between
   (i) two sporangia (ii) two conidia
   (iii) two hyphae (iv) two gametangia
4. Zygospore of zygomycota is
   (i) haploid (ii) polyploid
   (iii) diploid (iv) aneuploid

5. Under certain conditions, *Rhizopus* causes soft rot of fruits and vegetables. This degree of parasitism is known as
   (i) biotrophs (ii) symbionts
   (iii) facultative parasites (iv) all of these
6. Which of the following fungus is a serious human pathogen of Zygomycota?
   (i) *Choanephora* sp. (ii) *Rhizopus* sp.
   (iii) *Mucor* sp. (iv) *Saksenea vasiformis*
7. Which of the following fungus shows symbiotic association with the roots of higher plants?
   (i) *Rhizopus* sp. (ii) *Mucor* sp.
   (iii) *Choanephora* sp. (iv) *Glomus* sp.
8. Which of the following fungus of Zygomycota helps in uptake of nutrients from soil?
   (i) *Glomus* sp. (ii) *Sclerocystis* sp.
   (iii) *Gigaspora* sp. (iv) All of these
9. Which of the following fungi are potential biocontrol agents against Rhizopus soft rot of fruits and vegetables?
   (i) *Candida* sp. (ii) *Pichia* sp.
   (iii) Both 'a' and 'b' (iv) *Cercospora* sp.

Answers

1. (ii) 2. (i) 3. (iv) 4. (iii) 5. (iii) 6. (iv) 7. (iv) 8. (iv) 9. (iii)

15

DISEASES CAUSED BY ASCOMYCOTA

15.1 DISEASES CAUSED BY ASCOMYCOTA

The members belonging to the order Erysiphales are responsible for diseases which are popularly called **powdery mildews** because of the powdery appearance of the conidia on infected plants. They are all obligate parasites of angiosperms, often showing great physiological specialization that is having strains or races confined to a narrow range of host plants.

The taxonomy of powdery mildew fungi (order *Erysiphales*) recently underwent extensive revision based on DNA sequence data. According to Heffer *et al.* (2006), powdery mildew genera are now grouped into five tribes. Currently, these genera are identified using the following five features: (i) location of mycelium in the host tissue (epiphytic or endophytic); (ii) conidial production (singly or in chains—pseudo chains or true chains; (iii) Number of asci in the chasmothecium (cleistothecium)—one or several; (iv) type of appendages (mycelial, bulbous base, circinately coiled, or dichotomously branched; and (v) host specificity.

Erysiphaceae are parasitic on some 40,000 plant species representing well over 40 orders of angiosperms (Hirata, 1986). Examples of economically important hosts include cucurbits, apples, grapes, cereals, and various ornamental plants such as roses, etc. Some powdery mildew diseases are extremely destructive. Some of the important and destructive powdery mildew diseases are as follows:

1. *Uncinula necator* → Powdery mildew of grapevine
2. *Sphaerotheca humuli* → Powdery mildew of hops
3. *Sphaerotheca pannosa* var. *rosae* → Powdery mildew of roses
4. *S. fuliginiea* → Powdery mildew of cucurbits
5. *S. mors-uvae* → Powdery mildew of gooseberries
6. *Podosphaera leucotricha* → Powdery mildew of apples
7. *Blumeria graminis* → Powdery mildew of cereals and grasses
8. *Erysiphe cichoracearum* → Powdery mildew of cucurbits and many other plants
9. *E. polygoni* → Powdery mildew of peas
10. *E. cichoracearum* → Powdery mildew of cucurbits
11. *Microsphaera alni* → Powdery mildew of lilac

12. *Leveillula taurica* → Powdery mildew of *Medicago sativa*
13. *Phyllactinia dalbargiae* → Powdery mildew of *Dalbergia sissoo*

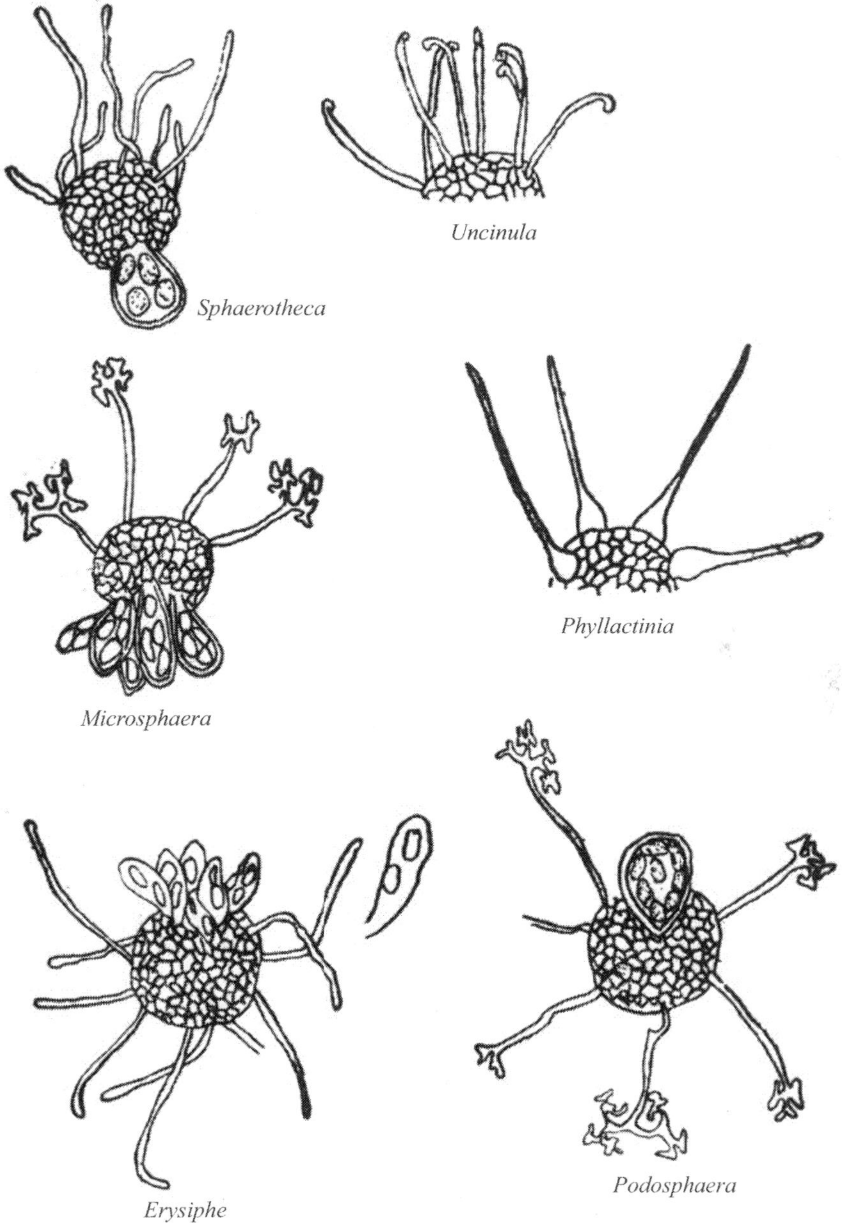

Fig. 15.1 Some important genera of Erysiphaceae

The Erysiphaceae have dark, spherical perithecia as well as colourless, septate, uninucleate hyphae, and conidiophores and conidia, which appear white on plant surfaces. They occur primarily on leaves, buds, flowers, and fruits. Some of their other characteristics are the occurrence of haustoria in the epidermal cells of their hosts; rapid development in rain-free seasons; high water content of their large, turgid, passively liberated, air-borne conidia; diurnal periodicity with respect to several characters; compatible association with their natural hosts, macroscopically visible hypersensitive reaction shown by some incompatible host reaction; slow injury to their hosts; and vulnerability to control by sulfur fungicides. The powdery mildews are favoured by cool, shaded environments and by succulent host tissues. The appearance of powdery mildew is due to the conidiophores and conidia. Three types of conidiophores can be recognized within the family.

1. The Oidium type, which consists of a short strip of one or more cells, a cell which gives rise to the conidia, that is a generative or spore mother cell, and chain of maturing conidia. This is found in species of *Erysiphe*.
2. The Oidiopsis type in which the conidiophores are branched and typically arise from a stroma, as in *Leveillula*.
3. The Ovulariopsis type with clavate conidia, and conidia are not in chains as in *Phyllactinia* species.

Disease Management

The powdery mildews are effectively controlled by sulfur dust. On certain plants, such as cucurbits and some apple varieties, sulfur powder cannot be used as it causes injury. Such plants are called sulfur-shy, and for them there are various new alternatives, the most important of which is Dinocap [2-(1-methyl-n-heptyle)-4, 6-dinitrophenyl crotonate] known commercially as 'Karathane'. This is used as a wettable powder or dust. Dinocap is both protective as well as therapeutic. In the 1960s, 2-aminopyrimidine derivatives—dimethrimol and ethrimol—were introduced as effective seed dressing fungicides. Calixin or Tridemorph, a morpholine derivative, is quite effective against barley mildew and rose powdery mildew. Triazole fungicides such as Triadimefon (Bayleton), Triadimenol (Bayton), Etaconazole (Vangard), Flusilazole, Myclobutanil, and Penconazole are currently being used for controlling powdery mildews.

15.2 POWDERY MILDEW OF PEAS (*PISUM SATIVUM*)

This is distributed throughout the world. The disease appears in epidemic form almost every year when the plants are in the pod stage towards the end of January and in February. Unlike the downy mildew which is favoured by moist weather, this mildew disease is at its worst when conditions are dry. Several other leguminous crops, such as *Medicago, Vicia, Lupinus, Lens esculenta, Trifolium dubium,* etc. are attacked. Losses depend on the stage of maturity of the pea crop when infection occurs.

1. Symptoms

In the early stages, small irregular powdery spots appear on the upper surface of leaves. When plants are in the flowering and pod stage (in January and February), the disease assumes epiphytotic proportions. The powdery whitish spots completely cover the leaves, petioles, stem and even the pods. The plant assumes a greyish white appearance, and the leaves turn yellow and are finally shed. The yield from the crop may be greatly reduced (Fig. 15.2).

Histological changes in the host include collapse of the necrotic epidermal cells in case of sub-infection and collapse of tissues below the penetrated epidermal cells.

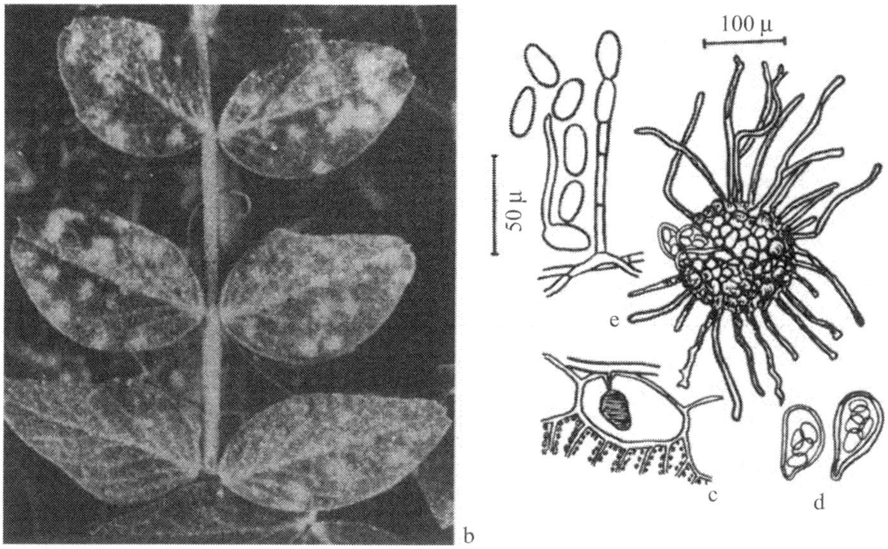

Fig. 15.2 *Erysiphe polygoni* on pea: (a) Symptoms (b) Haustorium (c) Cleistothecium with mycelioid appendages (d) Asci with ascospores (e) Conidiophores and conidia

2. Causal Organism

The organism responsible for the powdery mildew of peas is *Erysiphe polygoni* DC.

3. Disease Cycle

The fungus is an obligate parasite and the disease perennates through cleistothecia in the soil. Ascospores from the perennating cleistothecia first infect the lowermost leaves near the soil. The conidia formed abundantly during primary infection are responsible for the secondary spread of the disease. This is also suspected to be seed-borne.

The primary inoculum in the form of conidia comes from these places by wind currents to infect the pea crop in the plains. The conidia can germinate at temperature from 20°–24°C, with relative humidity up to 70 percent.

4. Disease Management

Field sanitation is important. Diseased plant refuse should be collected and burnt in the field. For vegetable crops, early varieties are preferred. Several workers recommended sulfur dust. Cosan, Elosal, Karathane WD, and Thiovit sprays controlled the disease in Rajasthan.

The other most effective fungicides were Calixin and Karathane followed by Milstem, and Sultaf Aureofungin and Morocide. The use of seed- and disease-free areas is recommended. Seed treatment with organosynthetic fungicides, especially Captan, Cerenox, and Thiram, also gives good control. On the basis of power or efficacy, one spray of Calixin (0.1% to 0.05%) was better than Bavistin or Sulfex. One spray of Bitertanol (Baycor) checked the disease intensity up to 20%. Triadimenol (Bayton) and Dichlobutrazol reduced the disease up to 50 percent.

Bitertanol, Triarimenol, and Dichlobutazole are some recent fungicides used. They are not only protective but curative also. Several fungal biocontrol agents such as *Ampelomyces, Cladosporium, Tilletiopsis*, and *Verticillium* are natural control agents of powdery mildew of pea. Singh *et al.* (2003)

have reported that *Oidium* and *Phyllactinia* which are nonpathogenic to pea, induce resistance to pea against powdery mildew. The pea cultivars which are resistant to powdery mildew in Himachal Pradesh are DPP-54, DPP-26, Pant-8, Sugar Giant, and JP-71.

Control of powdery mildew of pea by some novel chemicals, plant products, and biological agents has been tried with varying success. The inhibitory effect of bergenin (2 β-D-glucopyramosyl 4-0-methyl gallic acid–0-lactose), a naturally occurring isocoumarin isolated from *Flueggea macrocarpa* against *E. pisi in vitro* and also in glasshouses, has been seen. Control of powdery mildew of pea by seed bacterization of *Pseudomonas fluorescens* and *P. aeruginosa* alone and the two bacterial species or Neemazal (a natural product of "neem") has been found to be quite efficacious.

Out of the 670 genotypes tested, P185, P6583, P6587, P6588, T10, T56, and P388 proved resistant. It was interesting to note that all the strains which exhibited resistance to powdery mildew belonged to the table pea group (*Pisum sativum*). The germ plasm lines could be utilized for breeding pea varieties resistant to powdery mildew. In Rajasthan, Mathur *et al.* (1991) have reported resistance in P6583, P6587, P6588, P383, P388, P185, DPS3, DRS6, HUP2, DMR9, and Rachna.

15.3 POWDERY MILDEW OF APPLE

This disease is not confined to any specific region. It is common in the apple orchards of northern India. Nursery plants are damaged more than the adult plants. Peaches (*Prunus persica)*, quince (*Cydonia vulgaris*), and *Photinia* spp. are also attacked (Hirata, 1986).

1. Symptoms

The disease appears soon after the buds develop into new leaves and shoots. Sometimes buds are so heavily infected in the previous season that they are killed before they develop into leaves. The affected leaves are longer and narrower than normal leaves and are covered with a whitish growth (Fig. 15.3). Nursery plants are damaged more than adult plants. In nursery plants, the disease prevents the formation of wood in the stem. Fruit buds suffer more damage than vegetative buds.

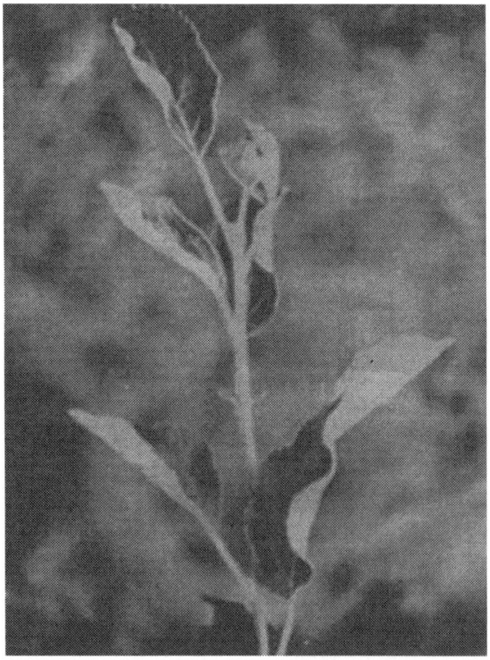

Fig. 15.3 Powdery mildew of apple (*Podosphaera leucotricha)*: symptoms

2. Causal Organism

The disease is caused by *Podosphaera leucotricha* (Ellis and Ever.) Salm.

3. Disease Cycle

Cleistothecia play a very little part, if any, in the perennation of the fungus. In most places, the fungus survives in the form of resting mycelium or encapsulated haustoria in the buds (dormant terminal, lateral shoot buds, and in blossom buds). Secondary spread is by wind-borne conidia.

4. Disease Management

Effective fungicides include microfine sulfur, lime sulfur, wettable sulfur and Karathane. Lime sulfur sprays should be undertaken according to the following schedule:

1. 1:15 dilution when the buds are green
2. 1:35 dilution at the open-cluster stage
3. 1:60 dilution at the blossoming or full pink stage
4. 1:100 dilution when about half the petals have fallen

Breeding resistant varieties appears to be the most promising control measure, some good sources of resistance for breeding varieties are known. Susceptibility to mildew in apple varieties is governed by one completely dominant gene and is related to weather conditions.

Four sprays of Bavistin (0.05%) or Morocide (0.1%) starting at bud-swell stage and repeated at 15 days' intervals give control of the disease. Among newer fungicides, (0.05%) Triadimefon (Bayleton) gives best control by persisting on the host surface for up to 15 days. Maximum reductions of germination of spores and germ-tube growth by Triforine followed by Tridemorph, Bitertanol, and Carbendazim have also been reported.

Strobilurin or Trifloxystrobin, a formulation of Strobilurin, are highly effective against powdery mildew. The fungicide inhibits mitochondrial respiration and inhibits conidia and conidiophore formation. Bupirimate (Nimrod) and Triadimefon (Bayleton) are reported to give better control. Podem–TM, a disease-forecasting model developed in UK, is effective in optimizing the timing of sprays.

Plant oil of sunflower, olive, maize, rapeseed, and soybean have provided effective control against powdery mildew as these oils, especially sunflower oil, inhibits conidial germination and mycelial growth of the pathogen.

In biological control, suppression of apple powdery mildew by the mycoparasites *Ampelomyces quisqualis* is reported. The cultivars Red Delicious and Golden Delicious are slightly resistant but apple cultivar with Vf resistance to apple scab are resistant to powdery mildew of apple.

15.4 LEAF SPOT OF TURMERIC (*CURCUMA LONGA*)

The disease is common along the coastal areas of Orissa and Andhra Pradesh and in the turmeric growing areas of Gujarat and Tamil Nadu where it causes considerable damage. The disease is also quite common in northern India where it causes considerable damage and in certain areas almost all the leaves on the plant are infected.

1. Symptoms

Spots are produced on both surfaces of the leaves but are more prominent on the upper surface. Individual spots are comparatively small and are light yellow to dark yellow. Spots, however, form larger patches causing the drying up of the entire lamina (Fig. 15.4a). Attacked plants are not killed, but due to excessive spotting and destruction of the green tissue of the leaf, there is a heavy reduction in yields. It may be emphasized that the pathogen does not stimulate the hyperplasia and hypertrophy of tissues in contrast to peach leaf curl caused by *Taphrina deformans*.

2. Causal Organism

Taphrina maculans Butler is the causal organism (Fig. 15.4 a,b).

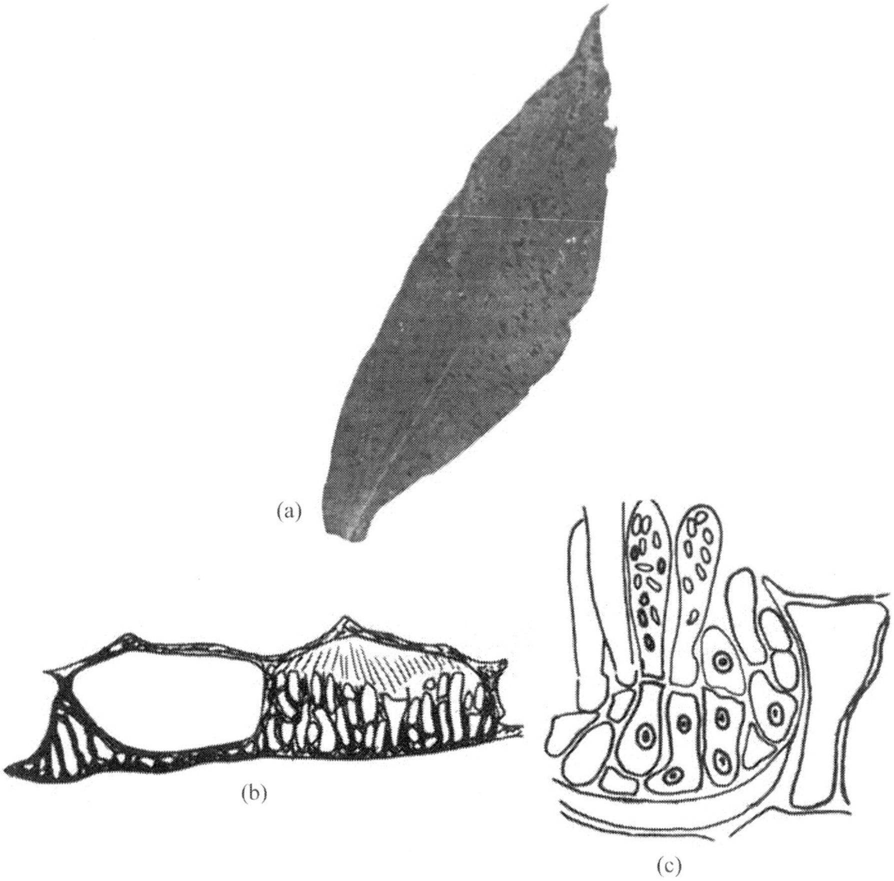

Fig. 15.4 Leaf spot of turmeric caused by *Taphrina maculans*: (a) Symptoms (b) Mycelium in the walls of epidermal cells (c) Asci and ascospores

3. Disease Cycle

The exact mode of transmission of the fungus is not yet known. How long the ascospores remain viable in plant debris has not yet been ascertained. The possibility that the mycelium reaches the rhizomes to remain there in a dormant state and then gives rise to the disease when the rhizomes are sown the following season, is remote. It is just possible that the inoculum might be surviving on the dried infected leaves in the soil in the fields from which the asci might be the chief source of the subsequent year's infection.

4. Disease Management

Some of the cultural methods like removal of diseased leaves should be adopted. Timely spray of Bordeaux mixture (5:5:50), Perenox, Fytolan, Blitox 50, and Dithane Z-78 checks the spread of the disease. Some resistant varieties like 'China' and 'Jaweli' had been reported. Cultivars Ca 69 and Shillong are resistant to leaf spot of turmeric.

15.5 BLAST DISEASE OF RICE

Blast disease has been reported from almost 70 rice-growing countries of the world. It has, for a long time, been recognized as a major problem of rice production in Japan, Taiwan, the USA, and many other countries. Losses due to blast may range up to 90%, depending upon the part of the plant infected. It is a major disease of rice and is most destructive in South India particularly from Andhra Pradesh, Madras, Kerala, Maharashtra, Gujarat, Orissa, and Jammu and Kashmir.

1. Symptoms

The disease produces brownish lesions and spots on leaves, leaf sheaths, culms, and panicles and finally the foliage may be completely blasted. Lesions on the neck of the culm and panicle branch bases result in typical "rotten necks". There are also nodal infections (Fig. 15.5a).

(a) (b) (c) (d)

Fig. 15.5 Blast of rice (*Pyricularia grisea*): (a) Symptoms on the inflorescence and leaf (b) Spindle-shaped spots on the leaves (c) Node blast (d) Neck blast

The leaf spots are typically elliptical with more or less pointed ends. The colour of the spots is usually gray or whitish, and the margin is usually brown or reddish brown. However, both the shape and colour of the spots vary, depending upon environmental conditions, the age of the spots, and the degree of susceptibility of the rice variety. The spots are usually small, water soaked, whitish, grayish, or bluish dots. They enlarge quickly under moist conditions, are 1–15 cm long, 0.3–0.5 cm broad, and usually develop a brown margin. Spots on susceptible varieties growing under moist, shaded conditions show very little brown margins but instead sometimes have a yellow halo around the spot. On highly resistant varieties, only minute, brown specks, each of the size of a pinhead, may be observed. Varieties with intermediate reaction show small, round, or short elliptic lesions, a few millimeters long, with a brown margin.

Brown to black spots or rings are formed on the rachis of the maturing inflorescence. Ears may also show similar small spots. The most characteristic symptoms appear on the culm. The neck becomes shriveled and covered with a gray fluffy mycelium. The affected plants can very easily be identified by examining the bluish patches on the neck or the stem. If the infection has occurred before grain formation, the latter are not formed and the panicle hangs down. However, due to necrosis of neck tissues, the ear tends to break and fall off. This stage of the disease causes maximum damage.

2. Causal Organism

The causal organism of blast disease is *Pyricularia grisea* (Cooke) Sacc. (*Pyricularia oryzae* Cav.). The perfect stage of the fungus is *Magnaporthe grisea* (Herbert) Yaegashi, and Udagawa. The pathogen produces two toxic substances, piricularin and α-picolinic acid—in culture filtrates as well as from severely diseased plant tissues. Piricularin is stable in water, especially on the acidic side. The toxin inhibits the growth of the rice seedlings as well as germination of the conidia of the fungus.

3. Disease Cycle and Predisposing Factors

The fungus survives through hot dry months in the tropics, and through the cold winter in the subtropics and in temperate rice areas by very different means. One common method of survival is through the infection of collateral hosts, such as sugar cane, *Setaria intermedia, Digitaria marginata, Dinebra retroflexa, Panicum repens, P. proliferum, Brachiaria mutica, Arundo donax, Leersia hexandra, Echinochloa crus-gali, Digitaria sanguinalis, Stenotaphrum secondatum, Eremochloa ophiuroides.*

4. Epidemiology

The carryover of the blast inoculum from one season to the next has not been fully understood, except perhaps in the northern hilly regions where the pathogen can survive the winter in infected plant parts under snow. However, it is doubtful whether the pathogen can survive in the infected plant debris or in the soil through the hot and dry summer months in the principal rice-growing regions in the plains of India.

According to Padmanabhan (1974), the survival of the pathogen through summer is most probably through an ascomycetous stage, with the ascospore discharge taking place under conditions favourable to infection just after the cessation of the monsoons.

5. Disease Management

(a) Culture Methods Field sanitation and destruction of weeds are precautionary methods that should be followed. Apart from this, early planting usually shows less disease than late planted crops. Application of lesser amount of nitrogenous fertilizer and in split doses generally reduces the disease severity.

(b) Chemical Methods Among the fungicides recommended for the control of blast disease are the copper fungicides (Blitox, Bordeaux) and organomercurials (Ceresan, Agrosan). Seed treatment with Agrosan GN (organomercurial) is effective in externally eliminating seed-borne inoculum. Copper fungicides are not effective in severe epidemics because it is phytotoxic to rice and reduces yield. But according to Thirumalachar (1967) when a mixture of 200 ppm copper sulfate is mixed with 20 ppm of Aureofungin, it is effective in controlling seed-borne inoculum.

Among the other fungicides and antibiotics found highly effective against rice blast were Kitazin and Hinosan (1.5 ml/litre water), Brestanol (0.44 g/liter water), Pyroquilon (Coratop 2.5 G), Hexaconazole (Contaf 5EC) at 2 ml/litre, Triflumizole (Rilan 70 WP) at 1 g/litre, Fungorene (8 g/kg seed), Kasumin 2% WP (1 g/litre water, Du-ter (2.5 g/litre).

There are some systemic fungicides which had strong systemic action against the rice-blast disease. Seed treatment with Benomyl (1:400 W/W) gives protection to seedlings because it inhibits spore germination and appresorium formation of the pathogen. Benlate, Bavistin, MBC (all 0.1%), and Dithane M45 (0.25%)—a nonsystemic fungicide—were also effective against the disease.

Apart from these fungicides, some antibiotics developed and used in Japan were Cephalothecin, Antiblastin, Antimycin A, Blasmycin, Blasticidin A, and Blasticidin–S but none could become popular due to chemical instability and toxicity to fish.

(c) Resistant Varieties In India the work had been summarized by Padmanabhan (1963). Taichung Native 1 and other similar varieties introduced in India for their high yield have been found to show susceptible reaction to blast. Studies at CRRI, Cuttack, indicate that the source of resistance to the disease had been identified. Three varieties—Tetep, Tadukan and Zenith—had been found to be resistant to all the physiological races, of the pathogen occurring in the country. Varieties T-603, T-141 (in Orissa) and A-67, A-90, A-200, and A-249 (in Maharashtra), GS-397, GS-480, ADT-20, CH-20, Suchi, Alkulu, Kukulu, Gennibera, Kamala (in Bihar), Co-4, TKM-1 (in Tamil Nadu), IR-579, Bala Lata-6 (in Himachal Pradesh), Jaya, Pankaj, Jayanti (in Raipur, MP) have been found resistant to blast of rice.

15.6 STEM GALL OF CORIANDER (*CORIANDRUM SATIVUM*)

This is a common and widespread disease of coriander causing heavy damage, especially when associated with wilt. Losses up to 23% have been reported by some workers.

1. Symptoms

The disease appears in the form of tumorlike swellings on leaf veins, leaf stalks, peduncles, stems, and fruits. The swellings on the veins give a hanging appearance to the leaves. The tumours are at first glossy but later rupture and become rough. They are about 5 mm in diameter (Fig. 15.5a).

2. Causal Organism

The disease is caused by *Protomyces macrosporus* Unger. The mycelium of the fungus is found only in the tumours. Attempts to trace the mycelium in tumour-free regions have failed to show the mycelium. It is believed that the infection is caused by the resting spores of the fungus. The infection becomes systemic. The hyphae are intercellular and closely septate. Branching is irregular. Scattered cells in the hyphae swell to form ellipsoidal or globose bodies which later develop into chlamydospores (Fig. 15.7).

3. Disease Cycle and Predisposing Factors

The pathogen survives in the soil in the form of chlamydospores and this seems to act as the primary source of infection. However, there are some who believe that the disease is both soil- and seed-borne.

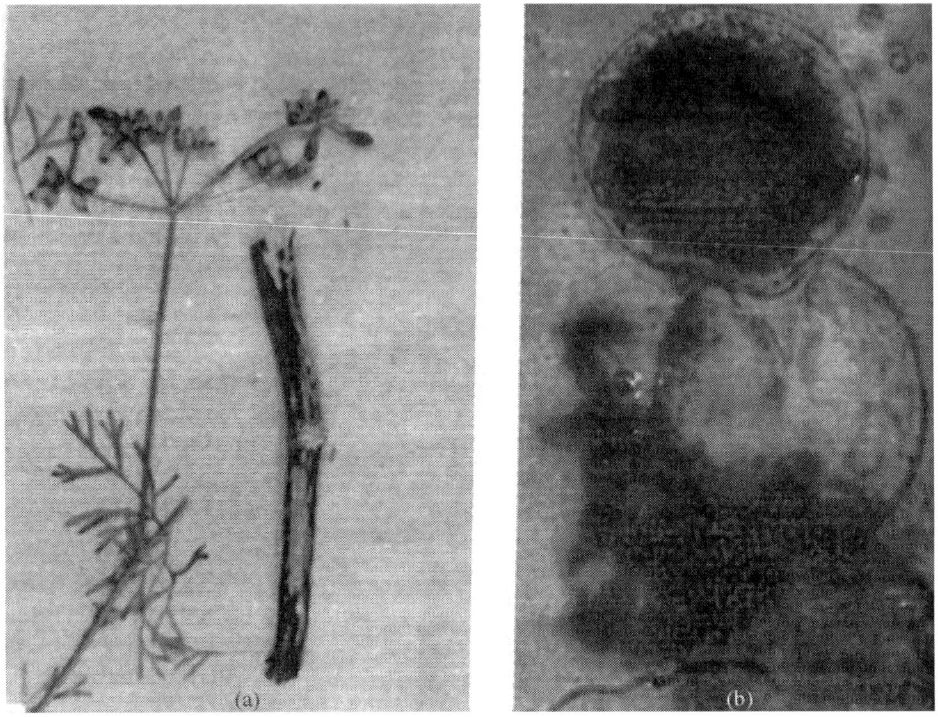

Fig. 15.6 Stem gall of coriander (*Protomyces macrosporus*): (a) Symptoms (b) A germinating chlamydospore

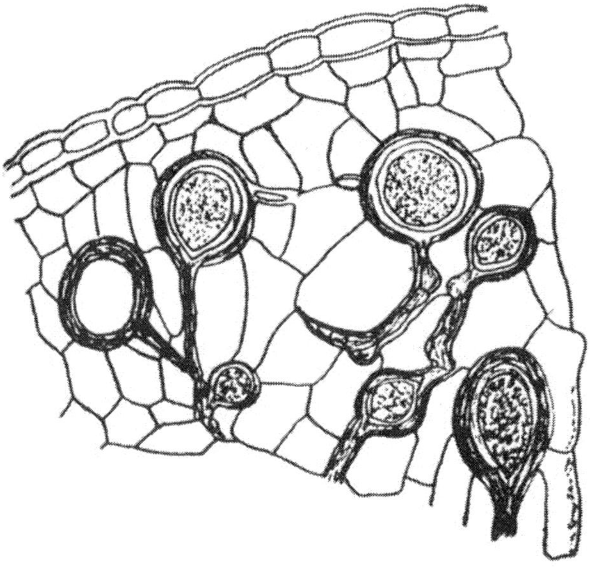

Fig. 15.7 Stem gall of coriander (*Protomyces macrosporus*): a section through the gall showing intercellular septate mycelium with terminal and intercalary chlamydospores

4. Disease Management

(a) Cultural Methods The following are some cultural practices that can avoid the disease incidence.

- Use of healthy and clean seeds
- Proper field sanitation
- Destruction of the diseased plants
- Rotation of crop

(b) Chemical Methods Nene *et al.* (1966) have reported very effective control by combined seed and soil treatment with Thiram. For seed treatment, the dose recommended is 250 g/100 kg seed and for soil treatment, 20 kg/ha.

15.7 LEAF CURL OF PEACH (*PRUNUS PERSICA*)

This disease occurs throughout the world wherever peaches are grown. In India, the disease is prevalent in the peach orchards of Kumaon, Kulu, and other parts the sub-Himalayan range.

1. Symptoms

The disease first becomes manifest in early spring. The leaf blades become thickened and puckered along the midrib and the leaves curl downwards. The leaves are pale green or yellowish, but later they become reddish, thick, and fleshy. The upper surface is covered with a whitish bloom as the fungus begins to produce spores. Affected leaves die and drop prematurely. The dormant buds become active to produce new leaves. Twigs may also be affected. Young shoots when attacked, become swollen and distorted. Even flowers and fruits may be infected and they drop quickly (Fig. 15.8).

Fig. 15.8 Peach leaf curl (*Taphrina deformans*): symptoms

2. Causal Organism

Taphrina deformans (Berk.) Tul. is the causal organism. The mycelium of the pathogen forms a compact, subcuticular layer of hyphae that fragment into individual ascogenous cells, each of which develops into an ascus. Asci are formed in a palisade-like layer above the epidermis. Each ascus contains ascospores but these may bud and give rise to secondary spores which are often referred to as conidia.

3. Disease Cycle and Predisposing Factors

The pathogen overwinters in the form of ascospores and sprout conidia which lodge on the surface of trees and cause infection in the following spring season. If they are washed down and lodged on the young unfolding leaves and buds, they infect the young meristematic tissues in the early spring when the buds are opening.

Mycelium of the pathogen can also persist in the cortex of the infected twigs. Towards the end of winter, the mycelium penetrates the buds beneath the infected area.

4. Disease Management

(a) Cultural Methods

- All fallen and diseased leaves should be burnt away from the orchard.
- Orchard and tree sanitation is important.

(b) Chemical Methods Peach leaf curl is controlled by spraying with a suitable fungicide, such as Bordeaux mixture (6:6:50), Perenox, Fytolan (0.3%), or Blitox 50 (0.25%). The fungicide should be sprayed before bud-break in spring.

It has been recommended that Bavistin (0.05%) or Dithane M-45 (0.2%) should be sprayed at the dormant stage (January) rather than the flowering stage (February). Mehdi and Shah (1994) have reported good control of the disease by fungicides such as Hexacap (Captan), Dithane M-45, Bavistin Topsin-M, Baycor, Bayleton, and Calixin; and out of these fungicides tested, Captan (1500 ppm) gave 90% control followed by Dithane M-45 (2500 ppm) giving 87.9% control, and Calixin giving 71.6% control. Chlorothalonil (Bravo or Daconil) controls the disease if applied twice in late fall and in early spring before the leaf bud swells. Two sprays of Captafol and Ziram given prior to green-tip stage provided best control of leaf curl followed by Captan and Chlorothalonil. Difenoconazole—a sterol inhibitor fungicide—when applied twice during bloom showed eradicational activity and was as effective as Chlorothalonil.

5. Resistant Varieties

Late-blooming varieties are generally resistant to the disease. Tolerant cultivars like Bed Will's Early and July Elberta have been found.

15.8 APPLE SCAB

Apple scab is one of the most serious diseases that exists worldwide but is more severe in areas with cool, moist springs and summers. In India, it was first reported in Kashmir in 1935. In addition to the Kashmir Valley, the disease is also reported from Shimla, Kulu (Himachal Pradesh), Kumaon Hills (Uttarakhand), Sikkim, Arunachal Pradesh, and Conoor and the Kodaikanal hills of Tamil Nadu. Gupta (1992) had reviewed the work done on this disease till then.

Its primary effect is reduction of quality of infected fruit, reduced fruit size, resulting in premature fruit drop, defoliation and poor fruit bud development for the next year. Losses from apple scab may be up to 16-34% in Himachal Pradesh percent.

1. Symptoms

The first symptom is light, irregular olive-coloured spots appearing on young leaves or lower surface of the sepals of the flower bud. Soon after, the lesions become olive green to gray with a velvety surface. Finally, the lesions appear black in colour and may be slightly raised. Infected leaves remain small and curled and may fall off. Sometimes twigs are also infected (Fig. 15.9).

On fruit, circular scab lesions, initially olive green in colour appear but later become darker and sometimes cracked. The cuticle of the infected fruit is ruptured at the margin of the lesion. The infected and cracked fruits drop prematurely and decrease the yield. Mature fruits with few small lesions may develop in dark scab during storage.

2. Causal Organism

It is caused by *Venturia inaequalis* (Cke.)Wmt. The unequal size of 2 cells of ascospore gives the species its name, *inaequalis*.

Fig. 15.9 Apple scab symptoms on mature fruit and leaves

3. Disease Cycle

The pathogen survives in the off season on dead leaves on the ground as immature pseudothecia and conidial stage which is parasitic on leaves, flower buds, and shoots on the tree during spring and summer. The pseudothecia, which is saprophytic or over wintered stage, is considered the major source of survival of the pathogen and provides the inoculum in the form of ascospores for primary infection of new leaves and flower buds in spring. This starts the parasitic stage by formation of lesions on leaves in which many conidia are formed. These conidia are responsible for repeated summer cycles of the pathogen and spread of the disease. When these infected leaves fall in the winter season, the pathogen enters its saprophytic stage on dead fallen leaves. The presence of scab lesions on old shoots/twigs may be another source of survival of the pathogen in winter. The shoot lesion as a primary inoculum may be of importance in those areas where fallen leaves decompose and eliminate the pseudothecial inoculum (Fig. 15.10).

4. Plant Disease Forecasting

Since 1990, considerable progress has been made in developing simple or computerized apple-scab prediction systems of spore release and infection for scheduling fungicide applications for scab control. All the predictive systems are based on the interaction among temperature, amount and duration of rainfall, and duration of leaf wetness, on the one hand, and the period required for the pathogen to initiate infection on the other. The first computerized simulation of apple scab was EPIVEN (Agrios, 2005).

5. Management of Disease

(a) Elimination or Reduction of Perithecial Production Fallen infected leaves are the main source of survival of the pathogen. It has been reported that 5% urea after fruit harvest at the pre-leaf fall stage and 2% spray of urea in spring on overwintered fallen leaves not only reduced ascospore number but also fastened decomposition of leaves, thus destroying the perithecia. Another explanation is that it also increases the population of fungal antagonists like *Trichoderma* and *Cladosporium* spp. Post-harvest spray of fungicides such as Benlate (0.4%) also reduces perithecia formation in dead fallen leaves. Some of the microbial antagonists such as *Trichoderma* sp., and species of *Microsphaeropsis*, *Athelia*, and *Diplodia* applied on fallen leaves can reduce the amount of ascospore formation.

(b) Reduction or Elimination of Primary and Secondary Infection A large number of fungicides such as Mancozeb, Cuman, Difolatan, Captan, Benlate, Bavistin, Topsin-M, Saprol have been found effective against the apple scab. Among the protectant fungicides, Difolatan, Mancozeb, and

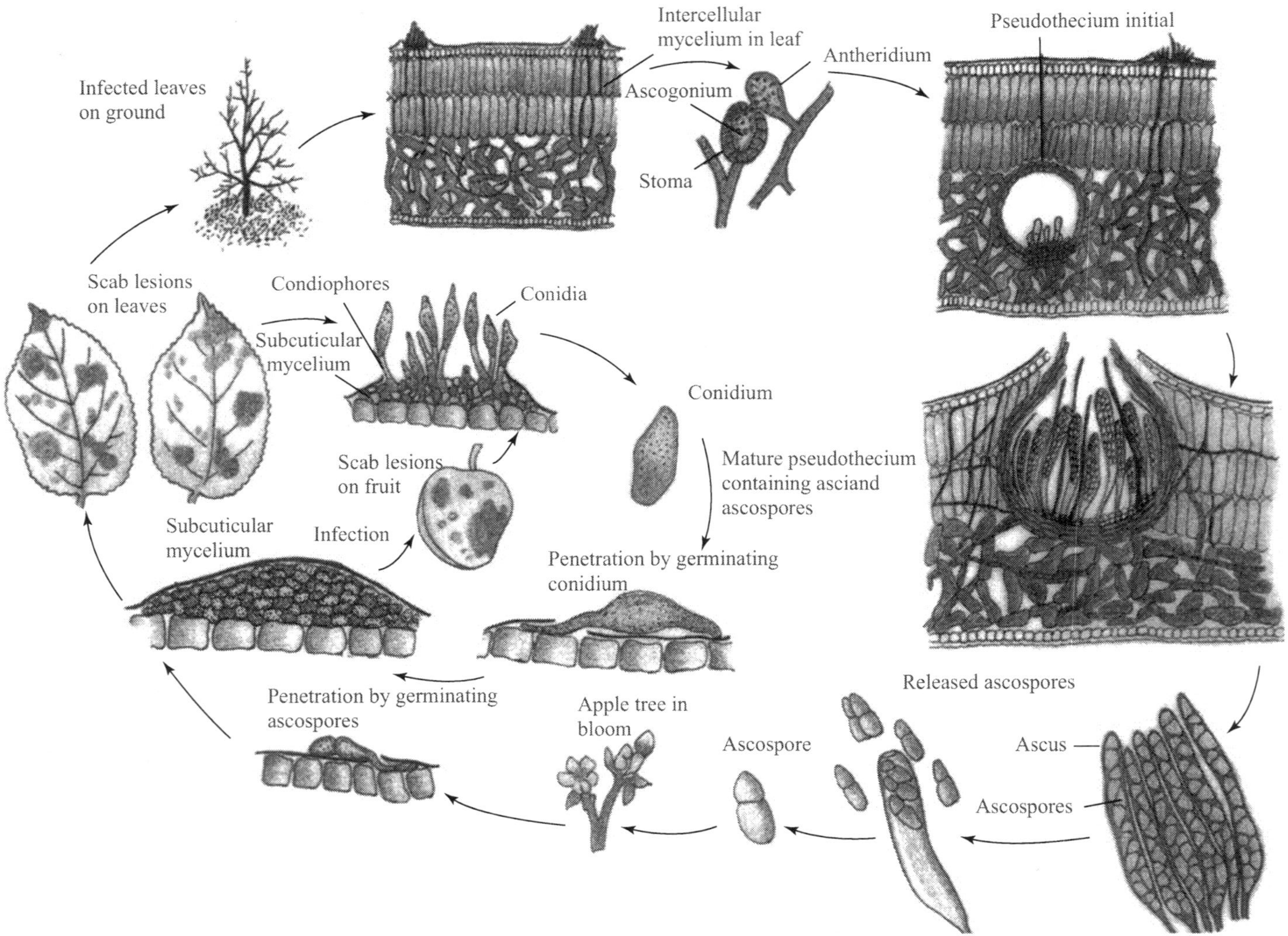

Fig. 15.10 Disease cycle of apple scab caused *Venturia inaequalis* (Redrawn from G N Agrios (2005), *Plant Pathology*, Academic Press, USA. pp.506)

Captan had been extensively used. The systemic fungicides Benlate and Bavistin were found to be excellent as they have preventive as well as curative action. The spray of these fungicides prevents the formation of perithecia and ascospore formation on dead fallen leaves, thus reducing the primary inoculum. Attempts have been made to use Sterol Biosynthesis Inhibiting (SBI) fungicides which have better protective, curative, and eradicating properties. The choice of SBI fungicides such as Triazole fungicide, Difenoconazole, Fenarimol, Flusilazol, Myclobutanil, Hexaconazole, and Bitertanol are highly effective in controlling the disease as they reduce fungal infection stages that are formed after penetration in the leaf. Since these SBI fungicides have been used very frequently, the problem of resistance to SBI fungicides in *V. inaequalis* is also very common. The new generation of microbial fungicides (the Strobilurin fungicides) produced by the symbiotic fungus *Strobilurus tenacellus* have a broad and balanced spectrum of action against the pathogen. Several fungicides of this group, viz. Kresoxim-methyl, Azoxystrobin, Trifloxystrobin, and CGA 279202, provide a long-lasting protection and no resistance to these fungicides have developed so far.

To overcome the problem to develop resistance to a particular systemic fungicides, the mixture of nonsystemic (protectant) and systemic fungicides are preferred. Combination of nonsystemic and SBI fungicides have also been suggested because they have better chances of preventing resistance development.

In Himachal Pradesh, the chemical control of apple scab was recommended as such:

- One spray of 5% urea on trees before leaf fall.
- One spray of Difolatan (330 g/100 L in water) or Bavistin (50–100 g/100 L water) at silver-tip to green-tip stage.
- One spray of Dithane M-45 or Captan (200 g/100 L water) at the petal-fall stage
- Again, one to three sprays of Dithane M-45 (250 g/100 L water) at the post-blossom stage at an interval of 10–15 days are quite effective in controlling apple scab

Similarly, in the Kashmir Valley, the following schedule is recommended to the growers.

- First spray of Mancozeb or Ziram (0.3%) at the silver-tip to green-tip stage
- Second spray of Captan or Dodine or Captafol at the pink-bud stage. This spray should be done after an interval of 10–12 days of first spray.
- After 12–15 days' interval, third spray of Bavistin or Topsin-M (50 g/100 L H_2O).
- After 12–15 days' interval after the third spray, the fourth spray of Mancozeb or Captan or Ziram is recommended.
- In the fifth spray after a gap of 15–20 days, any Dithiocarbamate group of fungicide spray is recommended.
- In the sixth spray after a gap of 25–40 days, spray of Bavistin or Topsin-M is given.
- If necessary depending on the disease, a pre-harvest spray of Ziram or Mancozeb can be given to avoid scab development during storage or transportation.

(c) Biological Control *Chaetomium globosum* and *Athelia bombacina* are promising fungal bio-control agents against scab pathogen. Spray of water extract of anaerobically fermented spent of mush-room compost during green-tip to petal-fall stage is quite effective. The mechanism of action is that microorganisms are complete for nutrients and the space of an infection site.

(d) Resistant Varieties/Cultivars The apple-scab resistant cultivars in India are Prima, Sir Prize, Macfree, Priscilla, Red free, Liberty, and Freedom.

Introducing endochitinase genes from fungi into apple increased the resistance of apple to scab but it also reduced vigour of the plant.

15.9 SOME OTHER DISEASES CAUSED BY ASCOMYCOTA

Name of Disease	Causal Organism	Symptoms	Disease Cycle	Management
Powdery mildew of wheat	*Blumeria* (*Erysiphe*) *graminis*	Grayish white powdery growth on leaf sheath and floral parts; leaves get reduced, weak, twisted, and deformed in infected plants	Soil-borne	Systemic fungicides like Benomyl, Milstem and nonsystemic fungicide like sulfur dust are effective. CPAN 1922 is a resistant variety to all races.
Powdery mildew of grapes	*Uncinula necator*	White patches on leaves, young blossoms, and young berries; diseased vines have a wilted and dwarfed appearance.	Diseased buds, fallen berries or over-wintered stems act as primary source of inoculum (soil-borne)	Pruning of diseased parts, Bordeaux mixture, Calixin, Topsin M, Bayleton, Bayton, Azoxystrobin, Kresoxin-methyl, Trifloxystrobin give effective control. *Ampelomyces quisqualis* is a mycoparasites on pathogen. Black prince, Sultana Red, Isabella, and Champion are resistant varieties.
Powdery mildew of cucurbits	*Erysiphe cichoracearum* and *Sphaerotheca fuliginea*	Tiny, white, superficial spots on leaves and stems which become powdery and cover the entire host.	Soil-borne and in some cases conidial stage survives on wild host.	Karathane, Benomyl, Bavistin, Azoxystrobin, Kresoxim-methyl, and Trifloxystrobin are effective in controlling disease. *Ampelomyces quisqualis* and *Sporotrix flocculosa* act as biocontrol agents.
Powdery mildew of rose	*Sphaerotheca pannosa* f. sp. *rosae*	White mycelial growth on the surface of the plant tissues.	Fungus overwinters mostly as mycelium in the buds.	Sulfur fungicides can control the disease.
Powdery mildew of Mango	*Oidium mangiferae*	Pathogen attacks all parts except roots; wefts of white mycelium appear on infected plants and later get covered with white powdery mass	Soil-borne	Several fungicides like Karathane (0.06%), Benlate (0.2%), Wettable sulfur (0.2%), Calixin (0.1%), Saprol (0.2%), and Anvil (0.05%) are effective in controlling disease.
Dutch elm disease	*Ophiostoma* (*Ceratocystis*) *ulmi*	Sudden prolonged wilting of the leaves, infected branches lose their turgor and wilt; the syndrome is called flagging. Diagnostic feature is brown discolouration of tissue in vascular region.	Spread by bark-boring beetle (*Scolytus* sp. and *Hylurgopinus* sp.)	Eradication and sanitation of diseased leaves accompanied by insecticide to eradicate the insect vector. *Pseudomonas* or non-aggressive strain of *Ophiostoma* or *Verticillium* induce resistance in plants.

(Contd.)

Name of Disease	Causal Organism	Symptoms	Disease Cycle	Management
Stem rot of rice	*Sclerotium oryzae* (perfect stage *Magnoporthe salvinii)*	Sparse growth of plant and leaves which become small and pale, culm is covered with black discolouration near the soil, excessive tillering, young leaves are darker green and stiff, small grain size	Soil-borne	Burning of rice stubble, soil solarization; Bavistin and Topsin M are effective in reducing the disease. MPN 234, IET 5633, ARC 12751, and Caloro are resistant varieties.
Ascochyta blight of chickpea	*Ascochyta rabiei*	All parts above ground, are infected, dark spots on stem, leaves and pods, lesions girdle the stem, the portion above the point of attack rapidly dies	Seed borne	Seed treatment with organomercurials; Thiram, Benomyl, Bavistin is reported to reduce seed-borne inoculum.
False smut of paddy	*Ustilaginoidea virens* (perfect stage *Claviceps oryzae sativae*)	Symptoms are confined to ears , due to development of fructification of pathogen; seeds are converted into large velvety green masses, size of transformed kernels are several times the size of normal kernel	Over-wintered sclerotia in soil or seed	Clean cultivation, crop rotation, dipping the seeds in brine solution or mercuric chloride, Bordeaux mixture, or Blitox are effective in controlling disease. IR22, IR 26, IR 28, Surya, and Vijaya are resistant varieties.
Ergot of bajra (pearl millet)	*Claviceps fusiformis*	Small droplets of pinkish or light honey-coloured fluid on infected spikelets which become dark, coalesce and cover large areas of cob. Alkaloids like Ergotoxin, Ergotamine, and Ergomatrine are present in sclerotia and responsible for poisioning of cattle.	Overwintered sclerotia in soil	Long crop rotation, use of sclerotia-free seeds, deep ploughing in summer, mixed cropping are effective in cultural practices. Spraying Ziram, Zineb, Bavistin, Benlate, and Brestan have provided partial disease control. *Fusarium roseum*, *F. sambucinum,* and *Dactylium fusarioides* are used as mycoparasites of sclerotia.
Black leg of crucifers	*Leptosphaeria maculans* (Anamorph stage *Phoma lingam*)	Infection either in the seed bed or at any stage but oval or linear light brown canker on stem and circular, light brown with ash-gray centre on leaves; infected seeds are shrivelled.	Seed-borne	Crop rotation, hot-water seed treatment destroys the seed-borne inoculum. Fungicides like Difenoconazole, Difenoconazole + Carbendazim are quite effective.
Ergot of cereals	*Claviceps purpurea*	Honey-dew stage typically occurs on the inflorescence	Soil-borne or seed-borne	Wild host as alternate (collateral) host should not be allowed to grow; clean seeds, rotation of crops are effective in controlling the disease. *Fusarium roseum* is a highly potential biocontrol agent.

(*Contd.*)

Name of Disease	Causal Organism	Symptoms	Disease Cycle	Management
Sclerotinia rot and wilt of various crops	*Sclerotinia sclerotiorum*	The pathogen attacks 383 plant species in 225 plant genera of 64 families, on eggplant and tomato; rotting occurs. On pea pods, white mycelial growth; on sunflower, and safflower rotting of flower heads; partial or complete wilting in tomato, potato, pea and chickpea; in Brassicas, the fungus causes stalk rot.	Soil-borne or seed- borne	Seed treatment with Captan, PCNB + Thiobendazole, Thiram, Benomyl, Trichoderma + Mancozeb is effective. For soil-borne inoculum, Carbendazim+ Thiram, burning of crop refuse, deep ploughing, flooding of soil, soil solarization, crop rotation and organic amendment of soil is helping in controlling the disease. *Pythium aphanidermatum* and different species of *Trichoderma* act as biocontrol agents.

REVIEW QUESTIONS

1. What are powdery mildews and how do they differ from downy mildews?
2. How different genera of powdery mildew can be identified on the basis of the oidial stage? How do these oidial stages differ from each other?
3. What do you understand by parasitic or physiological specialization? Discuss the phenomenon giving suitable examples from powdery mildews?
4. What group or groups of fungicides are used for controlling powdery mildews?
5. How will you distinguish the following pairs of plant pathogens:
   (a) *Sphaerotheca and Erysiphe*
   (b) *Uncinula and Phyllactinia*
   (c) *Podosphaera and Microsphaera*
   (d) *Blumeria and Erysiphe*
6. Name the causal organism responsible for the following diseases along with the major methods of disease control in each case:
   (a) Powdery mildew of peas
   (b) Blast of rice
   (c) Apple scab
   (d) Leaf curl of peach
   (e) Stem gall of coriander
7. Mention the seed-borne pathogens of the following:
   (a) Stem gall of coriander
   (b) Blight of check pea
   (c) Black leg of crucifers
   (d) Ergot of cereals.
8. What is soil solarization? Mention many important disease of paddy which can be controlled by soil solarization.

9. How does apple scab pathogen survives during spring and summer season? What control measures can be applied to control it?
10. What is false smut of paddy? Why is it called a false smut?

MULTIPLE-CHOICE QUESTIONS

1. Which of the following is not responsible for powdery mildew diseases?
   (i) *Blumeria* (ii) *Uncinula*
   (iii) *Phyllactinia* (iv) *Peronospora*
2. Which type of ascocarp is generally seen in Erysiphales?
   (i) Apothecium (ii) Perithecium
   (iii) Ascostroma (iv) Cleistothecium
3. Which of the following fungus acts as a biocontrol agent against powdery mildew of pea?
   (i) *Ampelomyces* (ii) *Cladosporium*
   (iii) *Verticillium* (iv) All of these
4. Which of the following fungus causes powdery mildew of apple?
   (i) *Podosphaera anserina* (ii) *Erysiphe graminis*
   (iii) *Blumeria sp.* (iv) *Uncinula necator*
5. Which of the following forecasting models is used against powdery mildew of apple?
   (i) PODEM-TM (ii) BLIGHTCAST
   (iii) EPIVEN (iv) TOMCAST
6. Which of the following fungus does not stimulate the hyperplasia and hypertrophy tissue?
   (i) *Taphrina deformans* (ii) *Taphrina maculans*
   (iii) *Plasmodiophora brassicae* (iv) All of these
7. Which of the following forecasting model is used against apple scab?
   (i) BLIGHTCAST (ii) EPIVEN
   (iii) TOMCAST (iv) All of these
8. Which part of the coriander plant is not affected by *Protomyces macrosporus*?
   (i) Stems (ii) Fruits
   (iii) Roots (iv) Leaf stalks
9. Which type of ascocarps is produced by *Taphrina deformans*?
   (i) Cleistothecium (ii) Perithecium
   (iii) Naked asci (iv) Apothecium
10. Which of the following is a resistant variety of the apple scab caused by *Venturia inaequalis*?
    (i) Sir-Prize (ii) Mac Free
    (iii) Liberty (iv) All of these
11. The pathogen responsible for powdery mildew of cucurbits is
    (i) *Erysiphe cichoracearum* (ii) *Sphaerotheca fuliginea*
    (iii) Both (i) and (ii) (iv) *Uncinula nector*
12. Which of the following alkaloids are present in ergot of bajra caused by *Claviceps fusiforms*?
    (i) Ergotoxin (ii) Ergotamine
    (iii) Ergometrine (iv) All of these

13. Which of the following method is effective to control soil-borne pathogen?
    (i) Crop rotation
    (ii) Dipping the seeds in mercuric chloride solution
    (c) Both (i) and (ii)
    (d) None of these
14. False smut of paddy is caused by
    (i) *Ustilaginoidea virens*
    (ii) *Ustilago tritici*
    (iii) *Urocystis tritici*
    (iv) *Ustilago hordei*

Answers

1. (iv) 2. (iv) 3. (iv) 4. (i) 5. (i) 6. (ii) 7. (ii) 8. (iii) 9. (iii) 10. (iv)
11. (iii) 12. (iv) 13. (i) 14. (i)

16 DISEASES CAUSED BY BASIDIOMYCOTA

16.1 SMUTS: USTILAGINALES

In the 10th edition of the *Dictionary of the Fungi* by Kirk *et al.* (2008), the Phylum or Division Basidiomycota has been divided into three subphyla, i.e. (1) Agaricomycotina (2) Pucciniomycotina and (3) Ustilaginomycotina. The subdivision Ustilaginomycotina contains two orders, i.e. (1) Urocystidiales, and (2) Ustilaginales. The members belonging to these orders are smuts and bunts.

The Ustilaginales are the 'smuts' and 'bunts' numbering approximately 1200 species in more than 50 genera. These fungi generally infect the ovaries and sometimes other meristematic growing tissues of the flowering plants, especially Gramineae and Cyperaceae, and develop in them the characteristic thick-walled brown or black spores—the **brand spores**—which consequently impart a sooty appearance. The origin of the term "bunt" is unclear. According to Duran and Fischer (1961), it may have been derived from a local pronunciation of "bunt" as "buhnt". They spend a part of their life cycle (sporidia) saprophytically and a part (mycelium and smut spores) parasitically within the plant tissues. They are, thus, facultative parasites. Many smuts have been grown in culture and a few have been induced to complete their entire life cycle—from teliospore (teleutospore) to teliospore—on artificial media. The smuts resemble the rusts in several ways and the two orders appear to be closely related. The main features of rusts and smuts may be compared and contrasted as follows:

	Smuts	Rusts
Perfect spore	intercalary	terminal
Sporidial number	indefinite	definite (4)
Mechanism of discharge	passive	active
Basidiocarps	rare	absent
Parasitic mycelium	intercellular with haustoria	intercellular with haustoria
Alternation of generations	indistinct	distinct
Parasitism	facultative	obligate- (Biotroph)

(*Contd.*)

	Smuts	Rusts
Clamp connection	common	rare
Sex organs	absent	specialized
Heteroecism	absent	common
Polymorphism	absent	distinct

Smuts most commonly attack the reproductive structures of the plant but may also be found on the stems, leaves, and occasionally the roots. The smut may be confined to a very small part of the plant, such as the ovaries or anthers. Economically, the smuts are very important causing enormous monetary losses by damaging crops. Smut spores may also accumulate in threshing machines that may explode violently, thus destroying the machine and the grain. Allergies may also be caused by the inhalation of smut spores by human beings.

The smut life cycle is considerably less complicated than that of the rusts. The smut mycelium which initiates the infection is usually dikaryotic, although in *Ustilago maydis,* the infecting mycelium is haploid in the beginning but the dikaryotic condition is soon established by mycelial fusion. Usually, there are no specialized haustoria, but short branches of the mycelium may enter host cells. The slender, septate mycelium develops intercellularly. Clamp connections and haustoria may be present. The mycelium in the sorus breaks apart into dikaryotic cells, which differentiate into teliospores. The teliospores have dark, thick walls, which may be smooth, spiny, or reticulated. In some cases, the spores are borne in spore balls. The two nuclei in the chlamydospore fuse, and at maturity, the spore contains a single diploid nucleus. Unlike its regularly triseptate counterpart in the rusts, the smut chlamydospore or brand spore or teliospore, on germination, produces a metabasidium which may be non-septate or, if septate, any number of septa may be formed. The majority of septate species form a three- or four-celled metabasidium. Meiosis occurs, and the metabasidium usually gives rise to uninucleate haploid basidiospores. Depending upon whether the basidium is septate or not the basidiospores are formed from each cell or from the apex of the metabasidium. The basidiospores may give rise to additional spores and in culture will form a yeast like colony of spores. In this yeastlike phase, many smuts are capable of prolonged saprophytic growth, but parasitic growth by the haploid phase does not usually occur.

Sori may develop in various parts of the inflorescence, including the ovaries, ovules, anthers, petals, etc. or they may form in leaves, fruits, stems, axillary buds, rhizomes, and even roots. In *Entyloma, Burrillia, Doassansia, Tracya*, and *Tuburcinia,* and sori are almost invariably formed in the leaves where they tend to remain embedded, even after maturity. However, the sori of other leaf smuts, when mature, may erupt to various degrees, as in some species of *Urocystis, Ustilago, Schizonella, Tolyposporella*, and *Tilletia.* Some smuts, however, sporulate on most host organs, as does *Mundkurella heptapleuri* which forms sori in fruits, stems, petioles, and leaves. *Melanotaenium* spp. similarly sporulate in most parts of the host which are above the ground and in the roots as well.

In the 10th edition of the *Dictionary of the Fungi* by Kirk *et al.* (2008), smuts and bunts have been placed in the class Ustilaginomycetes but for the sake of convenience and till the new classification becomes accepted by various authors, the present authors have followed the old classifications to avoid any confusion.

Some important genera of smuts of two families are as follows:

1. Family Ustilagianaceae

Promycelium septate, basidiospores produced laterally from each cell of the promycelium. Important genera are *Ustilago, Sphacelotheca, Sorosporium, Tolyposporium,* etc.

Ustilago —Sori almost never enveloped in a fungus peridium nor formed around columella of host tissue.
Sphacelotheca —Sori enveloped in a peridium of fungus cells, forming around a central columella of host tissue, at times with the vestiges of vascular elements remaining.
Sorosporium —Spore balls evanescent or tending to fragment into single spores.
Tolyposporium —Spores in the balls held together by interconnected thickenings of the exospores walls; germination as in *Ustilago*.
Melanopsichium—Sori forming in hard, gall-like masses, mostly in the inflorescence.

2. Family Tilletiaceae

Promycelium non-septate, basidiospores in a terminal cluster. Important genera—*Tilletia, Neovossia, Entyloma* and *Urocystis*.

Tilletia —Spores large, mostly 16–54 μm in diameter, exospores reticulate, smooth, and verrucose, or occasionally in the leaves, probably limited to Gramineae.
Neovossia—Sori less dusty at maturity, spores with a large hyaline appendage; on Gramineae.
Entyloma —Sori mostly forming in most parts other than roots. Sori tending to remain embedded in the host tissue after maturity, forming more or less distinct spots in leaves or other parts above the ground.
Urocystis —Spore balls formed in saccate cavities in galls, the balls consisting of one to several fertile spores surrounded by bladder-like sterile cells; spore germination Tilletiaceous.

16.2 MODES OF INFECTION

There are three ways in which smut fungi infect their hosts. These are (1) embryo infection, (2) seedling infection, and (3) shoot infection.

16.2.1 Embryo Infection

The best example is loose smut of wheat caused by *U. segetum* var. *tritici* (*U. tritici*). Here infection takes place at the time of blossoming. Penetration usually takes place through the ovary wall and not via the stigma and style, as was previously supposed. The hyphae cross the cells of the pericarp (fruit coat), enter testa (seed coat) and then move towards the ovule. After reaching the ovary, the hyphae turn sharply from the testa, enter the scutellum and pass into the growing point of the embryo. Within the pericarp and testa, the mycelium is intercellular but becomes intracellular within the tissues of the embryo. The mycelium of the pathogen remains dormant until the seed is sown and begins to germinate after which the pathogen remains dormant until the seed is sown and begins to germinate after which it becomes active. It passes into the crown node of seedlings and permeates the tissues. When the tillers lengthen due to the growth of the lowest internodes, the pathogen is carried upon the inflorescence primordia.

16.2.2 Seedling Infection

In many smuts, the infection takes place at the seedling stage of the plants. However, the exact mode of infection varies with the species involved. In the case of loose smut of oats, caused by *U. segetum* var. *avenae,* the spores are dispersed at flowering, similar to *U. segetum* var. *tritici* (earlier called *U. nuda*). On germination, the mycelium becomes established in the glumes and outer parts of the seed, that is, in the pericarp; it does not invade the embryo. When such seeds are sown, this dormant mycelium becomes active and invades the young seedling. In the case of *T. caries* responsible for bunt of wheat (also called stinking smut), the spores *en masse* have an odour of bad fish due to the presence of trimethylamine. The bunt balls often remain intact until threshing, when they are broken and the spores released in this manner contaminate the healthy seeds. When such contaminated seeds are sown, the spores on the seed coat germinate and the binucleate hyphae, which result from the fusion of sporidial cells, infect the young coleoptile. Further development of the pathogen is similar to that of *U. segetum* in wheat or barley. The inflorescence promordia are colonized at an early date.

Seedling infection from spores and/or mycelium also occurs with stripe smut of grasses caused by *U. striiformis* and flag smut of wheat and grasses caused by *Urocystis agropyri* (earlier called *U. tritici*).

In certain other smuts where seedling infection takes place, seed-borne inoculum is relatively unimportant. In the case of dwarf bunt of wheat (*Tilletia contraversa*) and onion smut (*Urocystis cepulae*), infections arise mainly from the soil-borne spores which may persist for many years. Infection takes place during the first 15–20 days after germination after which the cotyledons become resistant and provide a barrier to infection.

16.2.3 Shoot Infection

In a few smut fungi, the pathogen gains entry through the young shoots, particularly the young buds. Fisher and Holton (1957) described this for the first time in *U. violacea* on *Melandrium album.* In the case of *Ustilago scitaminea* on sugar cane, the entry is again through buds on the young offshoots or setts. Shoot infection also occurs in long smut of sorghum caused by *Tolyposporium ehrenbergii.*

In the smuts mentioned so far, the infection of the hosts is 'systemic' but with some smuts, infection is localized and sporulation is restricted to the part of the plant it has penetrated. One good example is common smut of corn caused by *U. maydis* (Syn. *U. zeae*). Another example is bunt of rice caused by *Neovossia horrida.* In the case of leaf smut of rice caused by *Entyloma oryzae,* there is localized infection on rice leaves.

16.2.4 Effect on Host

In the smut of cereals, particularly loose smut of wheat and barley, there is a direct loss in yield through the replacement of the flowering parts. There are very few indications of any adverse effect on the host growth. Excessive tillering and dwarfing may also take place. Of the two species, *T. caries* causes more dwarfing than *T. foetida. T. contraversa* causes a high degree of dwarfing.

U. maydis induces large, blisterlike galls on corn, and *Sporisporium* (*Sphacelotheca*) *reiliana* on the same host results in leaflike structures in the staminate and pistillate flowers which is a phenomenon known as phyllody. When ovulate plants of *Melandrium* are infected by *U. violacea,* the normally vestigial stamens develop and at maturity, the anthers contain smut spores instead of pollen; this may be accompanied by some atrophy of the gynoecium. A similar induced hermaphroditism in pistillate flowers of *Knautia arvensis* infected with *Ustilago scabiosae* has been reported. Thus, the pathogen changes the sex of the female plants. The nature of this morphogenetic stimulus is not known. According to the effect on the host, the smuts may be of two types: 1. **Ovariculous**, when the inflorescence develops normally, and ovary also functions in the normal manner but the contents of the ovary are ultimately destroyed and replaced by smut spores, and 2. **Culmiculous**, where the full development of the inflorescence is suppressed.

16.3 LOOSE SMUT OF WHEAT

Loose smut of wheat has worldwide distribution but it is a serious problem in the humid and semihumid wheat-growing areas. Smutted plants produce less tillers, shorter peduncles, reduction in height of plants.

1. Symptoms

The disease manifests itself only when the plants are in ears. Diseased ears emerge out of the boot leaf a little earlier than the healthy ones and a black powdery mass of spores replaces the flowers. The pathogen is internally seed-borne. The growth of the plant and its general appearance is not affected. Smutted heads consist of deformed spikelets filled with a black, dry, powdery mass of spores which entirely replace all the floral parts and glumes; only the ends of the glumes escape transformation (Fig. 16.1). These spores easily separate from the diseased host and are blown off by wind leaving a bare rachis behind. In many varieties, partially smutted ears or only few tillers show smutted ears while others are free from the disease.

Fig. 16.1 Wheat smut (*Ustilago segetum* var. *tritici*): smutted ear heads

2. Causal Organism

The pathogen responsible for loose smut of wheat is *Ustilago segetum* (Pers.) Roussel var. *tritici* Jensen which was earlier known as *Ustilago tritici* (Pers.) Rostr.

3. Disease Cycle

The fungus resides in the kernel as a dormant mycelium and from superficial examination at harvest time, infected grains cannot be distinguished from healthy ones. The fungus survives when the infected grains are sown and the hyphae just behind the growing point keeps pace with the apical meristem. At the time of ear formation, the fungus enters the reproductive phase (Fig. 16.2).

Although the pathogen is internally seed-borne, its active phase is visible only in the floral parts.

4. Disease Management

In 1908, Jensen, a Danish investigator, found that if the dormant mycelium is induced to germinate, it is vulnerable to heat, by the application of which it can be killed. The seeds are soaked in water at a temperature of 26–30°C for 4–5 hours which induces the dormant mycelium to germinate. The seeds are then quickly transferred to hot water at 54°C for ten minutes which helps kill the intraseminal mycelium. The thermal death points of the fungus and the embryo are very near each other, and therefore, extensive care has to be taken in this treatment. There are several other modifications of this procedure. The seeds are then finally dried. It is claimed that the efficacy of the hot-water treatment

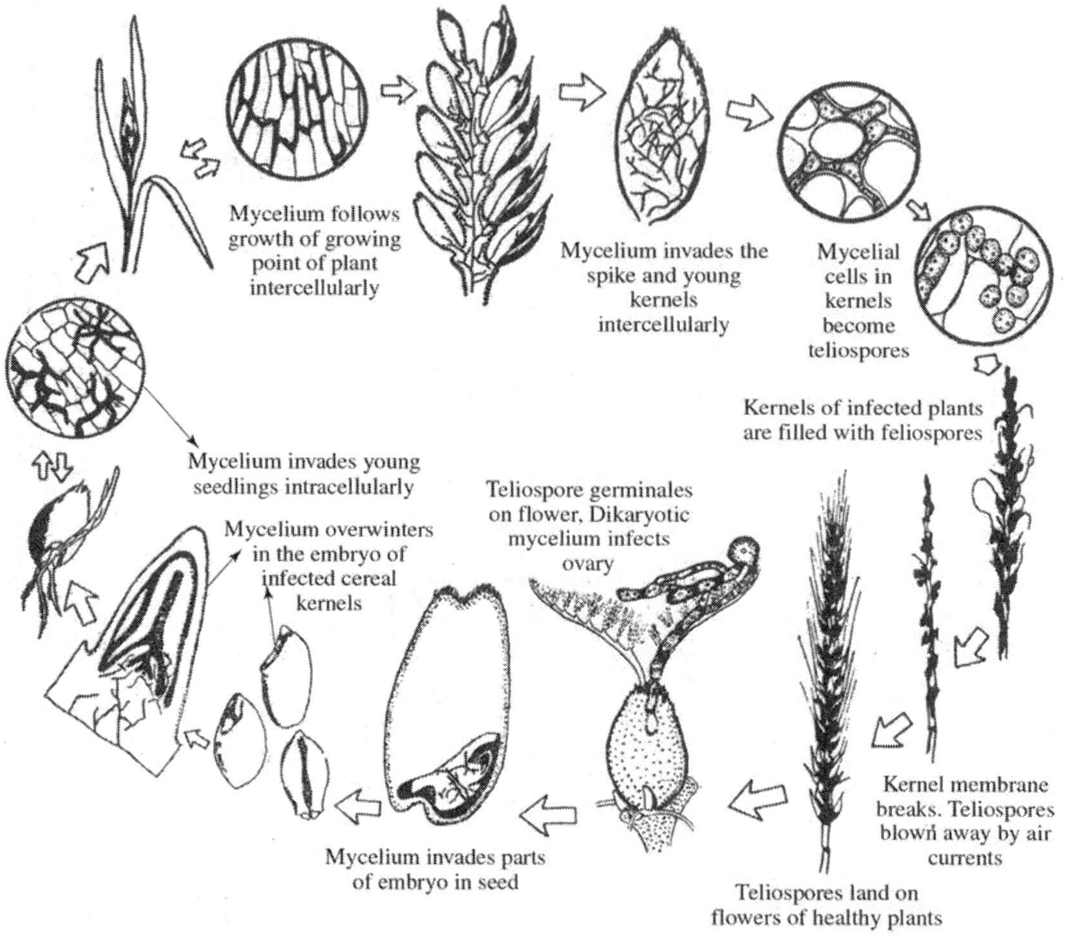

Fig. 16.2 Life cycle of loose smut of wheat (*Ustilago segetum* var. *tritici*). (Redrawn from G.N. Agrios, *Plant Pathology*, Academic Press, New York, 1969)

is largely due to the exclusion of oxygen, inducing anaerobic respiration. Alcohol and other catabolic products are set free which are lethal to the fungus.

In the plains of northern India, the summer temperatures are very high (42-44°C). Luthra and Sattar (1934) suggested the following simple and economical procedure: the seeds are soaked in water for four hours in the forenoon on a bright summer day followed by drying for four hours in the sun. This is called **solar energy treatment**. It has been found that even drying for one hour is sufficient to kill the intraseminal mycelium under conditions prevalent in Punjab.

Weibel (1958) reported a good method of controlling loose smut which entails the immersion of seeds in water soak to which Ceresan or sodium hypochlorite has been added, or by treatment in a water soak, followed by storage in airtight containers. Gera *et al.* (1963) showed that the fungus could be eliminated by soaking the infected seeds in water for 23–29 hours at 33°C or for 41–48 hours at 25°C.

It is believed that the soaking of the seeds results in the formation of certain quinones inside the seeds due to anaerobic respiration. It is because of this that the fungus is killed inside the seed (Tyner,

1957). A more recent successful method in the USA could also be adopted in India. This consists of soaking the seeds for 2–4 hours in water between 60–70°F and keeping the moist seeds in the airtight containers for 65–70 hours and thereafter drying them. This method is called **anaerobic seed treatment**. Soaking of seeds in 3 and 4% extract of onion or garlic at 30°C for 29 hours brought down infection from 12.5 to 1.4%.

Excellent control of loose smut of wheat is now obtained by seed dressing with systemic fungicides, such as Carboxin (Vitavax) and Benomyl at 0.2–0.25%. Bavistin SD (25% a.i.) has been found as effective as Bavistin WP (50% a.i.) and is also cheaper. The combination of Vitavax and Thiram or copper quinolate has been found equally effective and reduces the cost.

Varietal differences in the reaction to loose smut of wheat have been found in different countries. The high-yielding varieties, such as Kalyan 227, PV 18, WG 307, and C 302 have been found to be resistant to this disease and with the widespread cultivation of Kalyansona, there has been a considerable reduction in the loose smut of wheat.

Trichoderma viride is a promising biocontrol agent of several plant diseases. A strain of this antagonist, viz. TV-5, showed potentiality against loose smut of wheat. However, strain TV-5 was incompatible to the commonly used fungicide, Carboxin. Studies on development of mutants which would be compatible with commonly used fungicides is in progress.

The advent of some new systemic fungicides like Azoxystrobin and Picoxystrobin (Strobilurin fungicides) when applied as spray, reduce disease incidence and increase grain yield (Ruske *et al.* 2003).

It has been reported that when seeds of cereals (barley, oat, wheat, and rye) were treated with bacterial cultural of *Pseudomonas chlororaphis* (*Pseudomonas aureofaciens*) control seed-borne diseases caused by *Dreschlera* spp. and *Ustilago* spp. effectively. Singh and Maheshwari (2001) have reported that biocontrol agents *Trichoderma viride*, *T. harzianum*, *Pseudomonas fluorescens* and *Gliocladium virens* in combination with half dose of Vitavax were active against loose smut of wheat. However, Beniwal *et al.*, (2002) have reported that biocontrol agents like *Trichoderma* spp. and *Bacillus subtilis* control the disease but not better than Vitavax. Sinha *et al.* (2000) claimed that varieties PBW65, HDR 70, VL421, and WL 410 were resistant to loose smut of wheat.

16.4 KARNAL BUNT OF WHEAT

Karnal bunt of wheat was first reported by Mitra (1931) from Karnal (Haryana), India. This bunt was formerly restricted but of late it has become more widespread and common in Punjab, Haryana, Delhi, Himachal Pradesh, and the western districts of UP. It is sporadic in its occurrence, but in certain years it assumes serious proportions.

1. Symptoms

The symptoms of Karnal bunt are entirely different from other bunts. Here, all the ears are not infected in a stool and in the ears all the grains are not infected. The disease is evident when the grains are formed; only then can it be known whether the grains have been partially or wholly converted into black powdery masses enclosed by the pericarp. The smutted grains are very irregularly distributed, suggesting airborne, localized infection. The transformation of the kernel into small sori is not even complete in many cases, only a part being destroyed and the rest remaining healthy (Fig. 16.3a and d).

2. Causal Organism

Karnal bunt, or partial bunt of wheat, caused by *Tilletia indica* Mitra (*Neovossia indica* (Mitra) Mundkar) was first reported by Mitra in 1931 on *Triticum aestivum* from Karnal town of India; hence the name Karnal bunt.

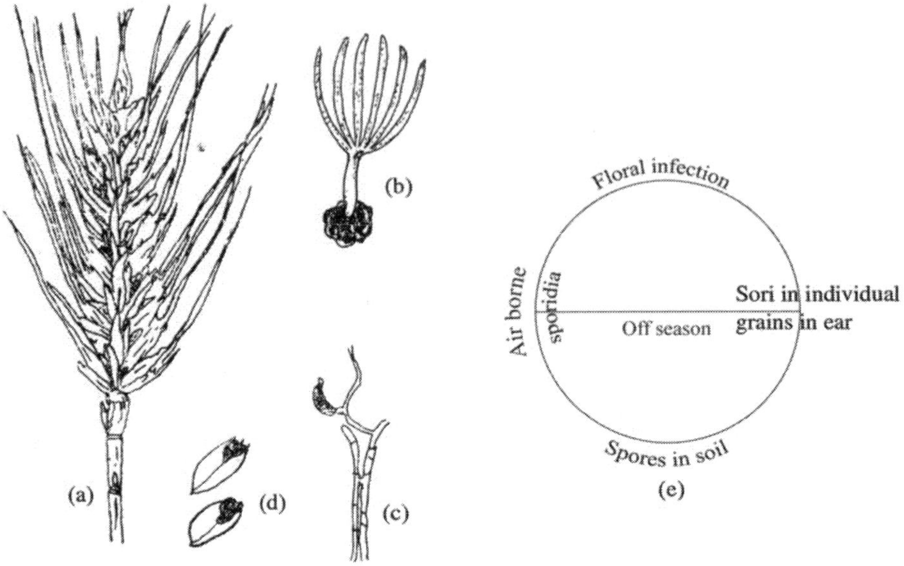

Fig. 16.3 Karnal bunt of wheat *Tilletia indica* (*Neovossia indica*): (a) Infected ear (b) Germinating spore with sporidia (c) Paired sporidia germinating to form secondary sporidium (d) Partially infected grains (e) Disease cycle of Karnal bunt of wheat

3. Disease Cycle and Predisposing Factors

Mitra (1937) reported that Karnal bunt is a soil-borne disease, but later it was shown to be air-borne. Disease cycle of the pathogen is depicted in Fig. 16.3. Higher dozes of nitrogen fertilizer, heavy manuring, and late sowing favour increase in the incidence of the disease.

4. Disease Management

The pathogen is seed and soil-borne and at the time of flowering, becomes air-borne as well. Therefore, an integrated approach including strict quarantine measures, cultural practices, chemical application, use of disease-free seed and resistant cultivars can check the disease to a satisfactory level.

5. Cultural Practices

Crop rotation, use of disease-free seed, judicious use of nitrogenous fertilizer and adjustment of date of sowing, irrigation management, and intercropping with gram or lentil have been worked out to minimize the disease incidence. Deep ploughing fallow after harvest, reduced plant density and planting of wheat on light soils have generally been advocated which are merely based on observation instead of long-term experiments and, therefore, have limited practical utility.

6. Chemical Control

Different chemicals have been tested against seed-borne inoculum. Chemicals like Butrizol, copper carbonate, Formalin, Thiram, Trifenilentin acetate, and Triphenile hydroxide have been found to reduce the teliospore germination. The fungicides as seed dressers have limited practical utility because no chemical eradicates the seed-borne inoculum completely. Thus, seed treatments with systemic fungicides are moderately effective. Fungicides like Bitertanol, Bentizol, Carbendazim, Carboxin, copper oxychloride, Fentinhydroxide, Mancozeb, Pentachloronitrobenzene, and Propiconazole have been found effective in controlling the air-borne sporidial inoculum when applied at the time of heading stage of the wheat crop. Tilt (Propiconazole), Baycor (Bitretanol), Bavistin (Carbendazim), and Dithane M-45 (Mancozeb) have also been found effective against the disease. Folicur 250 EW has also been found effective.

7. Host-Plant Resistance

Cultivation of Karnal-bunt tolerant varieties like WL 1562, PBW 54, HD 2281, several durums, and efforts to diversify the cultivars grown by farmers have contributed in reducing the disease incidence over years. Several wild species of *Triticum* have been found to possess resistance to the pathogen. Durum and aestivum wheats show promising resistance while *T. vulgare* is susceptible. In screening done in Mexico, no bread wheat were found resistant but 41 of the 448 durum wheat and 151 of 710 triticale lines were bunt free. In varietal trials at IARI, New Delhi, the varieties HD 1907, HI 358, HP-743, L176, L191, M-137A, and MW 59-4 X6A had shown resistance to *N. indica*. At Pantnagar, UP270, UP368, HD 2222, 2227, and 2235 remained free from bunt.

16.5 LOOSE SMUT OF SORGHUM

It is distributed in Africa, Asia, Europe and North, Central, and South America. It is less common than the grain smut. In India, it occurs in the states of Andhra Pradesh, Maharashtra, Karnataka, Madhya Pradesh and Tamil Nadu.

1. Symptoms

The affected plants can be detected before the ears come out. They are shorter than the healthy plants by about a foot, with thinner stalks and marked tillering. The ears come out much earlier and are more loose than the healthy ones. The glumes are hypertrophied (2.5 mm in length). All the spikelets are infected and sori may be borne on glumes and pedicels, besides the essential organs of the spikelets. The sorus may include the transformed pistil and the stamens. The size of the sorus varies with the varieties of the host (Fig. 16.4).

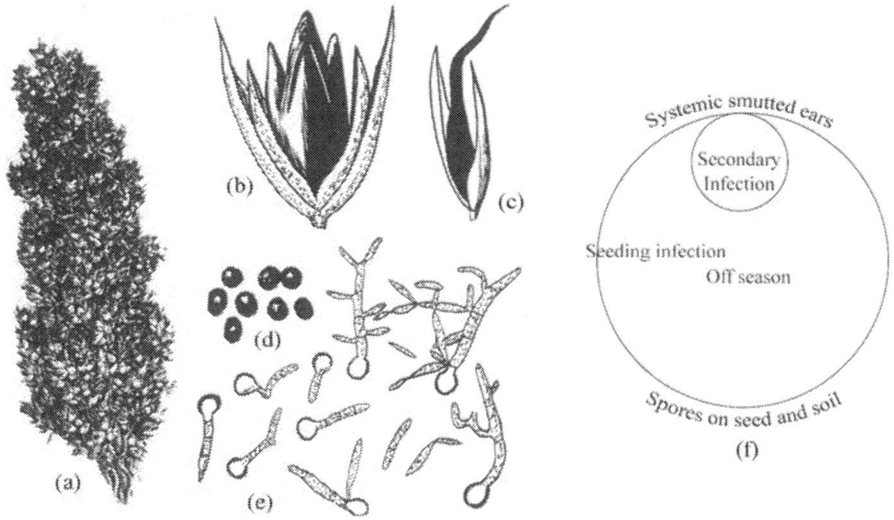

Fig. 16.4 Loose smut of Jowar *Sporisporium (Sphacelotheca) cruentum*: (a) Smutted ear (b) Sorus involving stamens (c) Exposed columella after the spores have fallen (d and e) Spores and their germination (f) Disease cycle of loose smut of sorghum caused by *Sporisporium (Sphacelotheca) cruenta*

2. Causal Organism

The pathogen responsible for loose smut of sorghum is *Sphacelotheca cruenta* (Kuhn) Potter. It has been renamed *Sporisorium cruentum* (Vanky,1994).

3. Disease Cycle and Predisposing Factors

As in grain smut, *S. cruentum* is mainly externally seed-borne. It is stated to be soil-borne to some extent in dry soils. Spread within the crop by floral infection by air-borne spores has been observed (Fig. 16.4).

4. Disease Management

Control measures are the same as those for grain smut of sorghum. The disease can be controlled by sanitation and crop rotation where the spores survive in the soil.

16.6 HEAD SMUT OF SORGHUM AND MAIZE

In India, it is prevalent in Andhra Pradesh, Karnataka, Maharashtra, Madhya Pradesh, Uttar Pradesh, Punjab, and Tamil Nadu. In Kashmir, it causes significant damage to the maize crop.

1. Symptoms

The disease is not noticeable by a difference in either the size or the growth of the plant. The entire inflorescence is converted into a big sorus, about 10–13 cm in length and 4.6 cm in breadth. The large smut sori replace the tassel and ear in corn. The sorus consists of the conductive elements of the suscept surrounded by the spore mass and the fragile exterior fungal membrane (Fig. 16.5).

Fig. 16.5 Head smut of jowar *Sporisorium (Sphacelotheca) reilianum*: (a) Smutted earhead covered by a membrane (b) The same with membrane ruptured showing long, thin, dark-coloured filaments (vascular bundles)

2. Causal Organism

The pathogen responsible for heat smut of maize or sorghum is *Sporisorium reilianum* [(Syn. *Sphacelotheca reiliana* (Kuhn) Clinton) and *Sorosporium reilianum*].

3. Disease Cycle and Predisposing Factors

Spores present in the soil cause infection. Only young plants are susceptible. The fungus becomes systemic (Fig. 16.6).

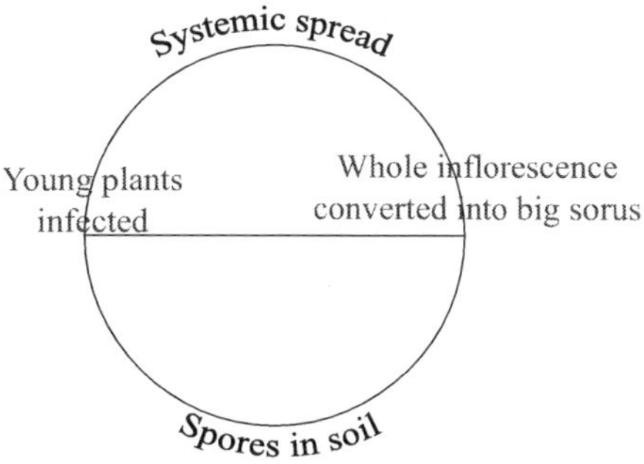

Fig. 16.6 Disease cycle of head smut of sorghum caused by *Sporisporium* (*Sphacelotheca*) *reilianum*

4. Disease Management

Crop sanitation to reduce the inoculum potential, and crop rotation are of primary importance. Resistant varieties are available and most of these are also resistant to covered smut. Application of urea, ammonium sulphate and triple super-phosphate reduces disease considerably.

Disinfection of seeds has been found partially effective in disease control in USA. Growing barley for 2 years has reduced infection of maize grown subsequently. Some insects like *Phalacrus, Brachytansus*, and *Lystronynchus* feed on the spores of *S. reilianum*. *Bacillus megaterium* is an effective biocontrol agent against smuts as it produces antibiotics.

16.7 SMUT OF BAJRA (*PENNISETUM TYPHOIDES*) OR PEARL MILLET

Bajra smut occurs in Africa, India, Pakistan, and Mesopotamia. In India, it is more prevalent in Tamil Nadu, Andhra Pradesh, and Maharashtra.

1. Symptoms

Only individual grains, irregularly distributed in an ear, are infected. Sori are oval or top-shaped, generally two to three times the size of the grains, 3–5 mm long and limited by a dark brown membrane of the host tissue which disintegrates at maturity. They are bright green to dirty black in color (Fig. 16.7).

2. Causal Organism

The pathogen responsible for this disease is *Moesziomyces bullatus* Syn. *Moesziomyces penicillariae* (Bref.) Vanky (earlier known as *Tolyposporium penicillariae* Bref.).

3. Disease Cycle and Predisposing Factors

The smut balls remain in the soil to germinate during the next season, producing masses of sporidia which become air-borne and infect the ear (Fig. 16.8).

4. Disease Management

According to Ramakrishnan (1963), seed treatment has given conflicting results. There is no effective method for the control of bajra smut and all the agronomically important varieties are susceptible. Attempts had been made to control this disease by the cultivation of resistant host varieties.

Other control measures suggested are the removal of diseased ears, clean seed, hot-weather deep ploughing, field sanitation, and crop rotation. Use of resistant varieties is recommended. Intercropping of Mung bean with pearl millet reduces smut incidence. Spray of fungicides such as Carboxin, Captafol, and Carbendazin at boot stage has been recommended.

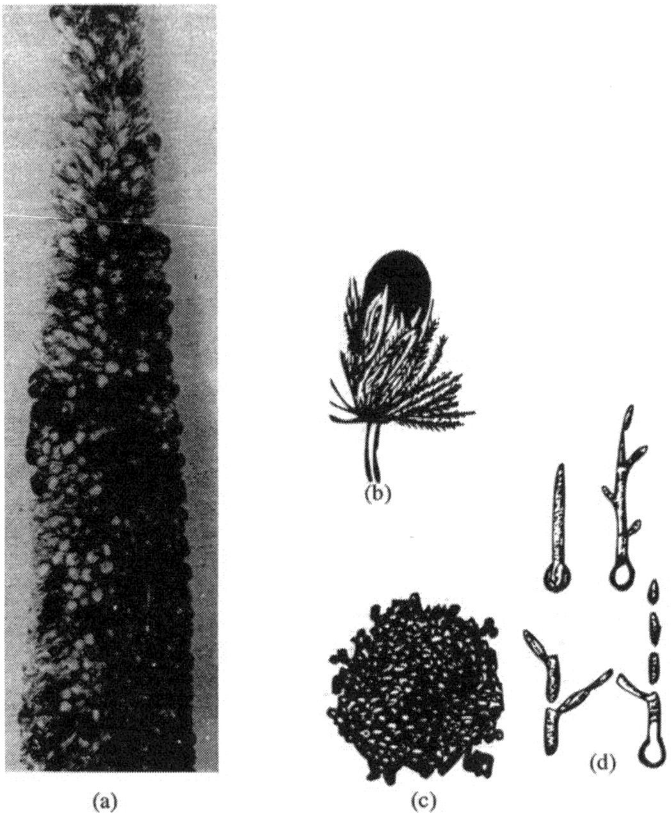

Fig. 16.7 Smut of bajra *Moesziomyces* (*Tolyposporium*) *penicillariae:* (a) Portion of the affected ears showing scattered sori (b) Single sorus (c) Spore wall (d) Germinating spores

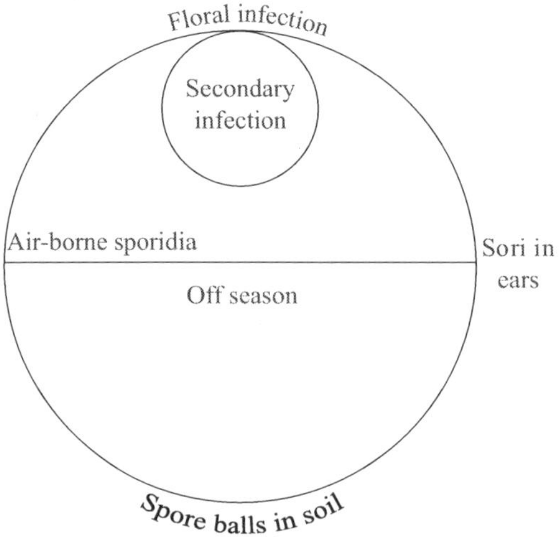

Fig. 16.8 Disease cycle of smut of bajra caused by *Moesziomyces* (*Tolyposporium*) *penicillariae*

16.8 PLANT RUSTS: UREDINALES

There are about 8016 species of rust fungi distributed among 190 genera in use today. In nature, all are obligate parasites (biotrophs), although there has been a breakthrough in artificial culture. They parasitize green plants—from the ferns and conifers to most of the more highly evolved families of both dicotyledons and monocotyledons. Earlier attempts to grow them in pure culture or in tissue culture met with limited success. It was, therefore, surprising when it was shown that *Puccinia graminis tritici* could be grown on Czapek-Dox, yeast extract agar (Williams *et al.*, 1966). Several rusts have been cultured axenically such as *Gymnosporangium juniperi virginianae*, *Melampsora lini*, *Uromyces dianthi*, and several others. They grow internally with intercellular mycelium, which rarely bears clamp connections. Sporulation is by means of sporogenous hyphae emerging through stomata by breaking through the surface of the host (erumpent), or in some cases, by internal production of spores, released only when the host decays. They are highly destructive and are responsible for widespread epiphytotic all over the world. The cereal rusts; the white-pine blister rust, cedar rust of apple, and rust of coffee and linseed are known for their destructive nature. Parasitism and specialization are highly developed among these fungi. According to the 10th edition of the *Dictionary of the Fungi* by Kirk *et al.* (2008), plant rusts have been kept in the order Pucciniales.

16.9 LIFE CYCLES

The rusts have a pleomorphic life cycle. Although morphologically simple, some rusts have the most complex life cycle found in the fungi, in addition to a high degree of specificity for hosts of a single species and sometimes of a particular genetic strain. Many rusts are long-cycled, passing through regular alteration of generations and a succession of five spore stages to complete a life cycle. The spore stages may be indicated as follows.

Stage 0—Pycnium or Spermatium

A pycnium (or spermatium) bearing pycniospores or spermatiophores. Pycnia are haploid structures, often flask-shaped. The pycniospores are haploid, uninucleate structures, which function as spermatia when transferred to another pycnium.

Stage 1—Aecium or Aecidium

An aecium or aecidium bearing aeciospores or aecidiospores. Aecia are frequently open, cupulate sori which bear dikaryotic aeciospores in chains. The aeciospores are the first dikaryotic spores. Aecia can be of the *Caeoma*, *Aecidium*, *Roestelia*, and *Peridermium* types. The form genus *Caeoma* includes all aecial stages in which a peridium is lacking. In the Aecidium type, the aeciospores are produced in a structure shaped like a cup with a short straight lip. In *Roestelia*, the lip protrudes much beyond the surface of the host and soon becomes shredded, often bending back and forming an ornamental fringe around the mass of spore chains. In the *Peridermium* type, the aecial lip is large and conspicuous. It is generally white and appears as a thin structure.

Stage 2—Uredinium or Uredium

A uredinium or uredium bears urediniospores or urediospores. The uredium is usually a sorus and the urediospores are dikaryotic. The stage is repetitive as a urediospores may germinate, forming a mycelium, which again gives rise to uredia.

Stage 3—Telium

A telium is containing teliospores or teleutospores. A telium is usually a sorus. The teliospores function as probasidia and are the site of karyogamy and meiosis.

Stage 4—Basidiospore (sporidia)

The haploid basidiospores produced from the teliospores are capable of reinfecting the host and initiating the life cycle again.

Normally, a rust fungus produces five different spore forms in a definite sequence, which is fixed and is not reversible. However, one or more forms may be omitted from the life cycle, thus indicating the extreme plasticity and flexibility of these fungi.

16.10 HETEROECISM AND AUTOECISM

Heteroecism has been demonstrated in a large number of rust fungi, notably cereal rusts, white pine, blister rust, cedar rust of the apples, bamboo rust in India (*Dasturella divina* and *Uromyces rumicis*). Autoecious rusts include *Uromyces appendiculatus*, *Melampsora lini*, *Puccinia helianthi*, *Uromyces hobsoni*, and *Puccinia asparagi*.

16.11 RUSTS OF WHEAT

Wheat occupies nearly 200 million hectares which are estimated to produce nearly 250 million tons in the world. In India, wheat is grown on over 24 million hectares which nearly yields 57.0 million tons. Wheat rusts are regarded as the most destructive pathogens on the crop in many countries. Wheat rust epidemic in India have been recorded as early as 1827 in the Central Provinces (Sleeman, 1839) and in 1907 by Milligan. In 1947, the black-rust epidemic in Madhya Pradesh (then known as Central Provinces) almost caused a famine. There were also reports of a black rust epidemic from Bihar in 1957. In 1971–72, a brown-rust epidemic was reported from the north-western part of the country and according to Joshi (1976), it caused an estimated loss of 1.5 million tons. Wheat rusts created an economic disaster in 1916 in the USA and Canada. However, an even worse disaster struck in 1935, when about 100 million bushels of wheat were lost. Another wheat rust epiphytotic swept through the wheat belt of the USA in 1937.

In our country, the major area is under bread wheat (*Triticum aestivum* L.) followed by *T. durum*, the macaroni wheat. The increase in area as well as in productivity in the past three decades has helped the country not only to achieve self-sufficiency but also to build up buffer stocks of wheat.

16.12 BLACK STEM RUST OF WHEAT

In India, black rust occurs in almost all the wheat-growing areas. In northern India, this rust does not appear before March and by this time, the crop reaches the 'dough' stage. Therefore, the damage to wheat crop by the rust is usually not severe. On the other hand, in the peninsular and southern parts of our country, it may appear as early as the fourth week of November and hence losses to the crop are generally great. Severe infection results in premature ripening and shrinking of the grain and sometimes as much as 90% reduction in grain yield may occur.

1. Symptoms

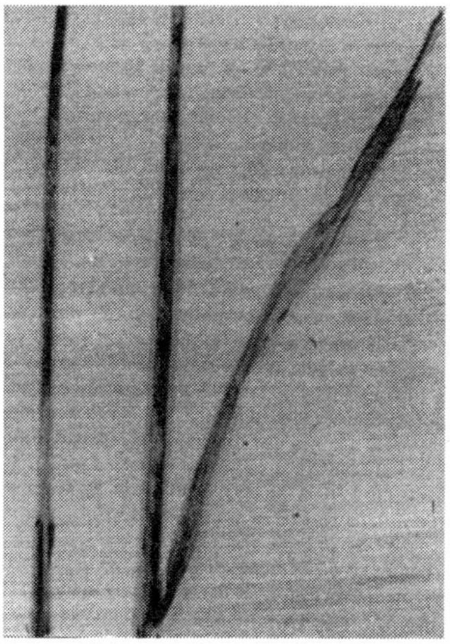

Fig. 16.9 Black stem rust of wheat (*Puccinia graminis tritici*): symptoms

In black rust of wheat, the stems are most severely attacked, then the leaf sheaths, leaves, and ears. Uredinia are the first to appear and are very prominent on culms, leaf sheaths, and also leaves. The urediosori are large, elongated, coalescing and dehiscing early, and breaking up large fragments of the epidermis.

Telia arise later in the season. Teleutosori are like urediniosori but black in color. These burst rather early and are found on all green parts of the plant, but least on the leaf blades (Fig. 16.9).

Apart from these physical changes on the plant, in case of severe attack, plants fail to produce normal ears. The grains are shrivelled and lighter in weight.

An increase in respiration rate accompanies infection of plant tissue by rust pathogens. The rate of photosynthesis shows an initial increase in early stages of infection which is transitory and goes down finally.

2. Causal Organism

Puccinia graminis tritici (Pers.) Erikss. and Henn. (Fig. 16.10) is heteroecious rust, and its pycnial and aecial stages are found on species of *Berberis* and Mahonia. These form the alternate hosts for this rust. *Berberis vulgaris* is the most common alternate host. Wind-borne sporidia infect the young tissues of these plants, where the binucleate stage is initiated by sexual fusion. This phase is followed by the formation of aeciospores.

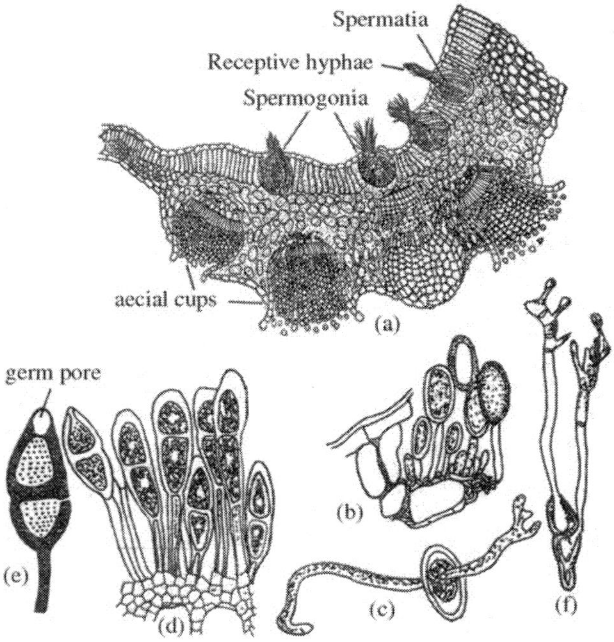

Fig. 16.10 *Puccinia graminis tritici*: (a) Pycnial and aecidial stage in section of barberry leaf (b) Uredospores on wheat (c) A uredospore germinating (d) Part of a teleutosorus (e) A teleutospore enlarge showing the germ pores (f) A germinating teleutospore

Pycnia are flask-shaped, and consist of pycniosporophores and receptive hyphae with a pycnial chamber in the centre in each.

Aecia are cup-shaped and are enclosed by a peridium. The aeciospores are echinulate and provided with six germ pores and arise in chains from the base of the aecial cup. Aeciospores are capable of infecting the grass host, that is, wheat.

Uredinia and telia are produced on the wheat plant. The urediniospores are brown in colour, oval in shape, 25–30 $\times$17 µm and provided with a single thick wall with tiny spines, and four equatorially placed germ pores.

The teleutospores are two-celled, chestnut brown, 40–50 µm long and 15–20 µm broad, and have a thick smooth wall. The apex is rounded or sometime pointed, with a germ pore through which the promycelium extrudes. On germination they form long, four-celled promycelium and from each promycelial cell, a strigma develops, which bears sporidia.

3. Epidemiology of Black Stem Rust of Wheat in India

In India, *Berberis* do not occur in the plains. In the hills, moderately susceptible species of *Berberis* are found; highly susceptible species of *Berberis*, that is, *B. vulgaris* does not occur. However, the condition prevalent at the time of formation of teliospores (April, May) and their subsequent exposure to warm weather, as well as the time of appearance of the new leaves on *Berberis*, when temperature and humidity are unfavorable for the germination of teliospores and infection of leaves, exclude the possibility of natural infection of *Berberis*. So far, in spite of extensive search, not a singe case of *Berberis* infection observed in nature has been connected with wheat. Such aecia have been found to belong to *P. graminis* var. *agropyri.*

We owe whatever information we now possess on the physiology and epidemiology of wheat and barley rusts in India, largely to the pioneering work carried out by Mehta (1929, 1940, 1952).

In the plains of India, the cereals are sown in October-November and harvested in March-April, or May. Experiments carried out here and other countries have shown that weather conditions as well as the age of host are quite favorable for infection of crops. However, initial outbreaks of rusts are delayed by 2-3 months from the time of sowing of the wheat crop in the plains.

There is hardly any chance of the survival of urediospores of any of the three rusts of wheat from the previous crop in the plains of India due to the intense summer heat. As already pointed out, the role of *Berberis* has been fully discussed by Mehta (1940). He (1933, 1940) supplied detailed information regarding the survival of the uredo stage of all the three rusts of wheat on "out-of-season" wheat during summer in the hills. He has also dissertated on the outbreak of rust on new crops sown earlier in the year in the foothills. A detailed comparison shows that there is heavier infection in the foothills that in the neighbouring plains and also that the periods of rust appearance in the foothills and the plains in different parts of the country are not the same.

Mehta (1933, 1940, 1952) safely concluded that the rust inoculums are reintroduced into the plains of India year after year from some other source, and this source, in all probability, lies in the hills where all the three rusts are able to oversummer in the uredo stage (altitudes 4000–8000 ft.). On account of milder summer and requisite humidity, the rust oversummer in the uredo stage on volunteer wheat, stubbles, and so on and on the regular crops (summer and winter) in the Nilgiris, Pulney Hills, and the submountainuous regions of the Himalayas in the north. These were regarded as the chief foci of infection, but regular surveys conducted in 1953–54 and 1954–55 have shown that the Panchgani-Mahabaleshwar range in the Western Ghats is also a focus of infection.

4. Role of Collateral Hosts

Although several grass hosts, such as *Bromus patulus, Brachypodium sylveticum, Hordeum* spp., Aegilops spp., and *Lolium* spp. are infected with rust, their growing period coincides with that of wheat and hence, they are not of importance in the annual recurrence of the rust in the plains.

As discussed earlier, the *Berberis* species seem to play no functional part in the annual recurrence of the rust in India and not a single case of infection on *Berberis* spp. in the hills can be traced to *P. graminis* tritici infection. Hence, the question of appearance of new races by hybridization is almost completely ruled out.

5. Disease Management

In a country like India, where rusts are perpetuated annually through the agency of urediniospores in the hills, Mehta (1929) had suggested that rust epidemic can be controlled to a considerable extent by reducing the initial inoculum at the source. To achieve this, he emphasized the importance and urgency of breeding rust-resistant wheat varieties and gave priority to the hills which act as foci of infection. Till such time as these suitable varieties are available and the hilly areas are saturated with them, he advocated the cleaning up of the volunteer wheat and barley plants, the suspension of summer crop in the Nilgiri and Pulney Hills, and the replacement of wheat and barley cultivation by that of oats in the hills.

Adjustment of the time of sowing, fertilization, destruction of volunteer wheat plants and application of fungicides are but adjuncts to the fundamental problem of breeding cereal varieties that are resistant to rust which is the most certain, effective, and economical means of controlling the rusts.

Some of the varieties of the series, NP 700 and NP 800, are resistant to one or the other of the three rusts.

Breeding for rust resistance is complicated by the appearance of newer races. Breeding for resistance is, therefore, a continuous process. The newer Mexican dwarf varieties introduced into India seem to be resistant; Lerma Rojo is reported to be highly resistant to all the three rusts in the field in North India, though it is susceptible to some races occurring in South India, particularly in the seedling stage; Safed Lerma and Sonalika are also highly resistant to all the three rusts; Sonora 64 is susceptible to yellow rust but is moderately resistant to black and brown rusts; and Choti Lerma is highly resistant to all the three rusts in the field.

Mixed cropping of wheat and barley with suitable crops gives good crop insurance even if the main crop fails. The susceptible surface area is reduced in mixed cropping and this reduces the inoculum for secondary infection and spread of the disease.

Nitrogenous fertilizers tend to increase the susceptibility of wheat crop to rusts. While potassium has the opposite effect, reduction of N in the NPK ratio helps in reducing the incidence of rust in a susceptible variety of wheat.

According to Forysth and Peruson (1958), effective control is obtained with 4-5 application of Nabam and zinc sulphate.

In Uttar Pradesh, experiments have shown that sulfur dusting at the rate of 13.6 kg percent acre at intervals of four days during the period when there are chances of the appearance of rust, is effective in controlling cereal rusts. In India, Grewal, and Dharm Vir (1959) have reported that four sprayings of Parzate liquid + zinc sulfate at fortnightly intervals is a good control measure. Wallen (1955, 1958) has controlled rust through the antibiotic cycloheximide derivatives of Actidione at 50 ppm (182 litres/ acre) as a systemic fungicide. Greenhouse experiments at the University of Minnesota have demonstrated that certain sulfa compounds, including sodium salts of sulphadiazine and sulphapyrazine, are effective as post- infection sprays.

Chemical control of rust has been receiving much more attention in recent years than in the past. A number of chemicals, such as RH-124 and Plantavax, have given quite encouraging results.

The effectiveness of Vitavax and Plantavax against rust pathogens have been reported. These Oxathiin fungicides effective as soil, seed, and spray treatments. Plantavax was found more effective than Vitavax.

Triadimefon and Fenapenil were reported as good fungicides for the control of stem rust. Propiconazole (Tilt) at 0.1% was quite effective in controlling the three wheat rusts because it persists for 12 days and completely inhibits urediniospore germination.

6. Genetic Resistance

Genetic resistance is the most commonly used and most effective means to control stem rust. For reasons that we do not fully understand, a few genes of race-specific (vertical) resistance to stem rust in wheat remained highly effective for many years. The most successful of these was *Sr31*, a gene that occurs on a segment of a chromosome from rye that was transferred into wheat by a complicated process of interspecfic hybridization. Wheat with *Sr*31 quickly became popular worldwide, because, in addition of *Sr31* the rye chromosome segment also carried genes for increased grain yield as well as addition of genes for resistance to other rust diseases. Since the 1980s, wheat varieties with *Sr31* were widely grown in nearly every major wheat-producing region throughout the world other than Australia. The effectiveness of *Sr31* was so great that wheat-stem rust declined to almost insignificant levels nearly everywhere in the world by the mid-1990s.

Recently, the resistance of *Sr31* was finally overcome. A new race of wheat-stem-rust fungus highly virulent to wheat varieties with *Sr31* was found in Uganda in 1999. The new race, tentatively designated **Ug 99**, rapidly dominated the fungus population in Uganda and spread to Kenya and Ethiopia where it caused major epidemics. Within a few years, Ug 99 was found in South Africa and in Yemen, from which it has spread to the North and East as far as Iran. It seems inevitable that Ug 99 will soon invade one of the world's richest wheat-producing areas in the Punjab of India. Previous examples of long-distance dispersal of wheat fungi include spread of a unique race of wheat-stem rust from South Africa to Australia, spread of coffee rust from Africa to South America, and spread of Southern corn rust from Central America to Africa. To make matters worse, the Ug 99 lineage of the stem-rust fungus has expanded its virulence through mutations that allow it to overcome the resistance of at least two other vertical resistance genes that wheat breeders have relied on for protection from stem rust in North America and many other parts of the world.

For the long term, however, wheat breeders may rely more on minor genes with additional effects of partial resistance expressed primarily in adult plants (i.e. horizontal resistance). A number of high-yielding wheat lines with moderate levels of horizontal resistance have been developed at the International Centre of Wheat and Maize Research (CIMMYT). These advanced lines are being intercrossed to produce wheat varieties combining as many as four to five horizontal resistance genes to effectively suppress stem-rust epidemics. Jain *et al.* (2011) have reported a new pathotype of *Puccinia graminis tritici* in India and has discussed its implication in the breeding programme. According to Jain *et al.* (2011), new sources of resistance and *Sr* genes in combination, should be used in breeding programmes to address the challenge posed by this new variant of stem rust.

16.13 BROWN RUST OF WHEAT

This rust attacks leaves almost exclusively, rarely leaf sheaths, and very rarely stems. This rust, although more general in occurrence, is more common in the northern and eastern parts of India than in peninsular India. It is perhaps the earliest rust appearing in certain parts of Bihar, Punjab, and Uttar Pradesh, and together with the yellow rust, causes great loss and damage. Epidemic of brown rust occurred in the north-western part of the country in 1971–72 and 1972–73 causing a loss of 0.8 to 1 millions tons and 1.5 million tons, respectively.

1. Symptoms

As a rule, the urediopustules develop on the leaves, being rare on the sheath and stalks. They burst on the upper surface as points of a bright orange color. They are never in rows like those of yellow rust but they may be gathered in small clusters or may be irregularly scattered. They are bigger in size than those of yellow rust. When old, the urediopustules of orange rust cannot generally be distinguished by colour from the yellow rust but for their irregular arrangement. The sori burst early and shed the urediniospores.

Telia are formed only rarely and in some ears may not be present. They are small, oval to linear, black, and covered by the epidermis. Paraphyses, which are abundantly present, divide the sorus into numerous compartments (Fig. 16.11).

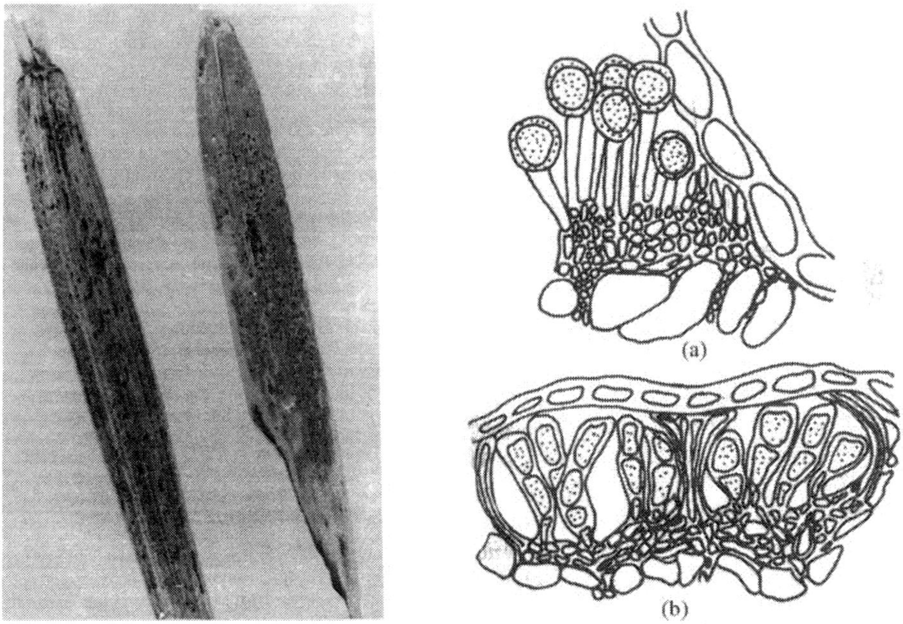

Fig. 16.11 Brown rust of wheat (*Puccinia recondita*): symptoms (a) Section through uredosorus (b) Section through teleutosorus

2. Causal Organism

The pathogen responsible for brown rust of wheat is *Puccinia recondita* Rob. ex. Desm. The pathogen is heteroecious. The aecial and pycnial stages are produced on *Thalictrum*.

The urediniospores of *P. recondita* are brown and spherical and 16–28 μm in diameter. The wall is minutely echinulate and furnished with 7–10 germ pores. Infection by germ tubes from the uredinio-spores occurs through the stomata on either side of the leaf. The process of penetration is of the same type as in blank rust.

The teliospores are oblong to cuneiform, slightly constricted at the septum and the apex is rounded with prominent thickening (Fig. 16.11a and b).

3. Disease Cycle

In the case of brown rust of wheat, the only known functional host of the rust, *Isopyrum fumarioides,* does not occur in India, though susceptible species of *Thalictrum* are common in the hills. All the workers except D'Oliveira have always regarded *Thalictrum* as a nonfunctional host of *P. recondita.*

Brown rust has two foci of infection. If we take the whole of India, brown rust of wheat is the most widespread and important. It has been observed that in early January and sometimes even earlier, the rust gets established in the foothills of the Himalayas in the north and also in the plains of Tamil Nadu and Karnataka in the south. The first build-up of inoculum takes place in the plains of Karnataka in the south. It moves northwards to Maharashtra and Madhya Pradesh. At the same time the inoculum of the foothills of Bihar and UP moves to the northern plains. Therefore, the brown rust appears slightly later in the western foothills of north India. The rust population of the north and south move in opposite directions, finally merging into each other, and cause epidemics in any part of the country under favorable climatic conditions.

Brown rust can stand warm weather better than the yellow rust and has been found to survive during summer at altitudes of 1450 m to 2050 m.

4. Disease Management

As in the case of black rust, growing of resistant varieties forms the best control measure. Chemical control measures applied for the black rust can also be used here. Reports on the field control of this rust by fungicides, such as some of the Dithiocarbamates and RH-124. 1 or 2 foliar sprays, have shown good control.

16.14 YELLOW RUST OF WHEAT

This is also known as **stripe rust** of wheat. The yellow rust is usually seen first in the northern foothills and adjoining places, such as Gurdaspur and Ropar (Punjab), and Pantnagar in the Tarai region (Uttarakhand). Stripe rust, or yellow rust, of wheat is as damaging as stem rust. The extensive cultivation of Kalyan Sona over a large area of north-western India has aggravated the stripe-rust problem during 1968–69 and its breakdown during 1970–71 was due to the appearance of a new virulence of stripe-rust pathogen.

1. Symptoms

In mild attacks, the cereal pustules are formed chiefly on the leaves but in more severe attacks, they appear on the sheath, stalk, and glumes also. The green colour of the leaves fades in long streaks on which rows of small uredosori appear. Each row consists of a series of oval, lemon-yellow pustules, arranged end to end and each distinct from the one above and below. In severe attack, this serial arrangement seems to be lost. The uredosori do not break through the epidemism as quickly as in other rusts, but they do so eventually (Fig. 16.12).

2. Causal Organism

Yellow rust of wheat is caused by *P. striiformis* West. [=*P. glumarum* (Schmidt) Erikss. and Henn.] (Fig. 16.13). The urediniospores are almost globose and not oval as in black rust. Their epispore is covered with fine spines. There are 6–11 germ pores.

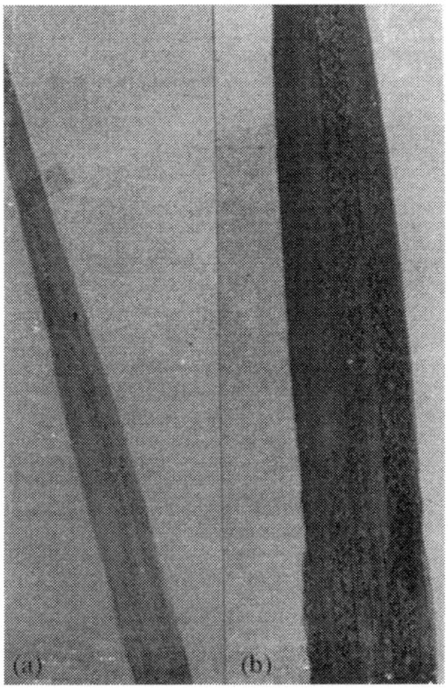

Fig. 16.12 Yellow rust of wheat (*P. striiformis*): symptoms (a) uredosori (b) teleutosori on leaf

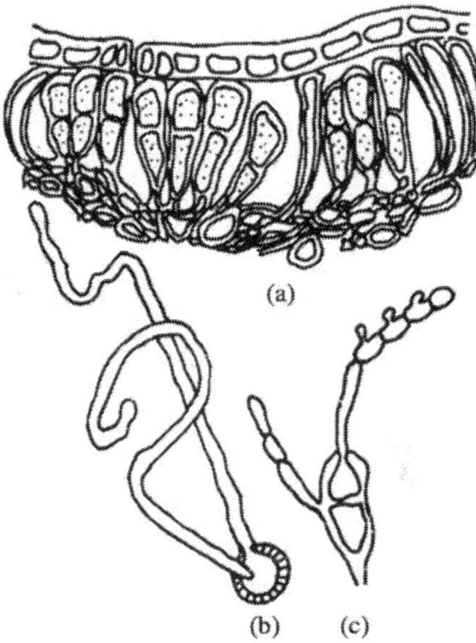

Fig. 16.13 Yellow rust of wheat (*P. striiformis*): (a) section through teleutosorus (b) a germinating urediniospore (c) a germinating teleutospore

The telia are hypophyllous or culmicolous and are long, narrow lines, like the uredia. They are covered for a long time by the epidermis with brown paraphyses surrounding the spores and intermingling with them along the edges of the telium.

The teliospores are oblong to cuneiform, slightly constricted at the septum and the apex is less thickened and pointed than in *P. graminis*. The teliospores are capable of immediate germination when mature.

3. Disease Cycle

The rust is heteroecious but the alternate host is not yet known. The inoculum responsible for starting annual epidemics comes from the hills. The black rust is able to oversummer at an elevation of 1375 m above sea level; *P. striiformis* requires still cooler temperature, and therefore, oversummers at 1900 m and above. The further sequence of events with regard to spread is the same as in black rust.

A grass, *Muehlenbergia hugelli*, growing wild in the Simla hills, has been found to harbour yellow rust, but its role in the annual recurrence is yet to be determined. Some 90 weeds (e.g. *Agropyron semicostatum*, *Bromus catharticus*, *B. japonicus*, *Hordeum murinum*, etc.), are known to serve as collateral hosts for the rust. There are 23 races and biotypes of yellow rust in our country reported so far, namely: 13, 14, 14A, 19, 20A, 24, 31, 38, 38A, 57, A, D, E, F, G, H, I, J, K, L, M, N.

4. Disease Management

Cultivation of resistant varieties is the best way of controlling this rust. The chemical-control measures used for the black and brown rust can also be used here for controlling the rust.

16.15 RUST OF LINSEED (*LINUM USITATISSIMUM*)

The disease causes reduction in seed yield and the weakening and disfiguring of fibre in flax. The pathogen occurs in Africa, Asia, Australasia, Europe, and North and South America. Saharan (1988) has reported 16 to 100% loss in yield.

1. Symptoms

Diseased plants become very conspicuous in the field because of the bright orange colour of the affected parts. This is due to the uredinia which occur in large number on both the surface of the leaves and other aerial parts of the plant. The leaves die prematurely and hence, telia, which appear late, are formed mostly on the stems. The brown to black telia appear as crusts covered by the epidermis (Fig. 16.14).

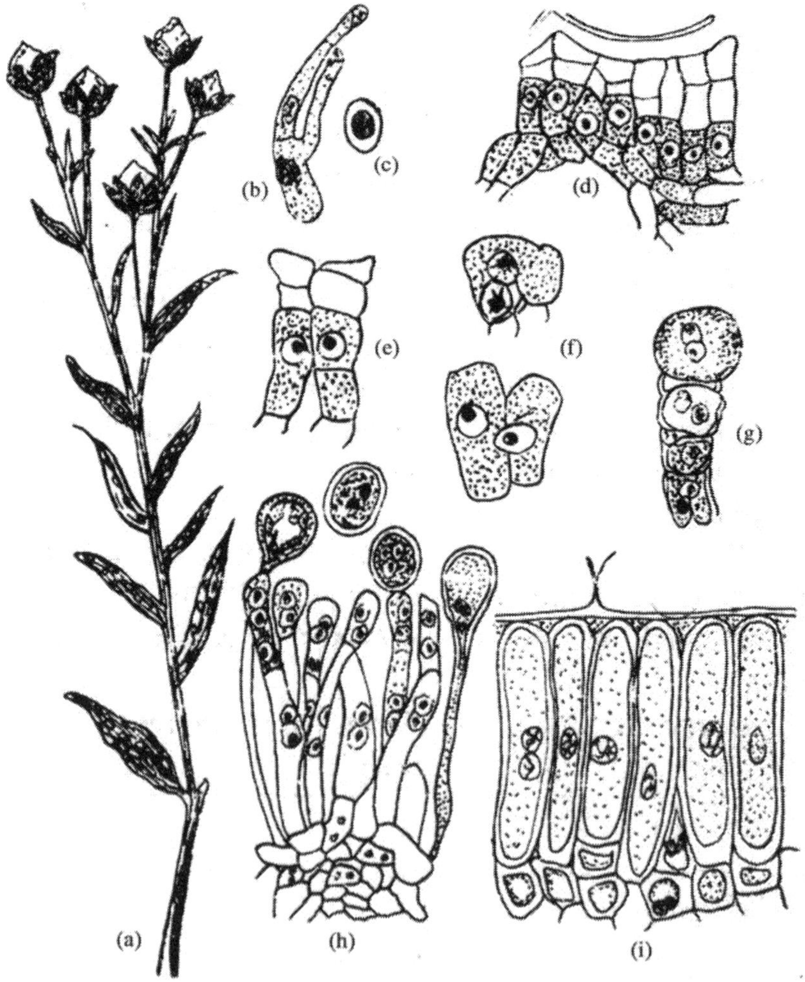

Fig. 16.14 Linseed rust (*Melampsora lini*): (a) Symptoms (b) Pycniophore (c) Pycniospore (d) Aecium (e) Paired cells at the base of aecium (f) Fusion of cells (g) A chain of binuclear aeciospores in formation (h) Uredinium and urediniospores (section) (i) Telium and teliospores (section)

2. Causal Organism

The pathogen responsible for linseed rust is *Melampsora lini* (Ehrenb.) Lev. It is an autoecious rust. Pycnia are amphigenous and about 5–10 are grouped closely together. They are usually flask-shaped but occasionally diffused, spermatiophores, without a definite structure, appear. Aecia are caemoid (i.e. lack peridium), hypophyllous, associated with and often surrounding the pycnia, concentrically grouped and often joining together to form a ring. Aeciospores are ellipsoidal, 15–25 μm in diameter, uredinia are amphigenous and caulicolous, irregularly scattered, 0.5–1 μm diameter. Paraphyses intermixed with urediniospores, capitate, up to 80 μm long, head 20–30 μm in diameter, hyaline. Urediniospores ellipsoidal or obovoidal, 15–25 × 15–22 μm, wall hyaline, finally echinulate, 1.5–2 μm thick. Telia amphigenous and culmiculus circular or elongated along with stems, often fused and covering large areas, sub-epidermal, black. Teliospores cylindrical, 45–80 μm long × 8–12 μm broad, wall luteous, 1 μm thick at the side, sienna to chestnut, 2–4 μm thick above.

3. Physiological Specialization

Flor (1956), by simultaneous studies on the inheritance of virulence in races of flax, put forward his concept of gene relationships between the host and pathogen. Person (1959) elaborated it further, showing that more than one gene for rust resistance may exist in some of the differential flax varieties (Finchan and Day,1971).

4. Disease Cycle

In temperate countries, soil-borne teliospores on crop refuge are the source of primary inoculum. These germinate to form the sporidia which infect young leaves. Prasada (1940) found that nearly 60% of the telia of the fungus collected in March and stored at 5–10° C during summer, germinate after eight months. They die in the field at summer temperature in the Indian plains. However, in the hills (1080–1380 m), variable teleutospores have been recovered from plant debris collected with the seed and there is evidence of the active role of such teleutospores in initiating fresh outbreaks of the disease. Volunteer flax plants, including some wild species, may also serve as important sources of infection (Laundon and Waterston, 1965) and enable the overwintering of the uredinial stage in New Zealand. *Linum mysorens*, a wild linseed, has been reported to harbour the fungus in India. However, it is not definite if the rust race on this plant is the same as on linseed.

5. Disease Management

The most promising method of control of linseed rust lies in the cultivation of resistant varieties. Another important control measure for this rust is the avoidance of excessively nitrogenous manures which favour lush growth.

Heggeness (1942) had reported that flax rust could be completely controlled by application of Borax at the rate 60 kg/ha. Since the source of primary inoculum lies in the hills, seed treatment to inactivate the teliospores and cultivation of resistant cultivars in the hills will go a long way in controlling the rust of linseed. Some cultural practices such as destruction of diseased crop debris and weed host have also been recommended. The cultivars (RR) 9, 10, 56, Jawahar- 7, 17, 552, LC-54, K-2, LC-216, and LC-255 have been cited as resistant to the disease.

16.16 RUST OF COFFEE

Coffee-leaf rust is not only one of the most dangerous diseases of coffee but ranks as one of the classical plant diseases of international importance. It appeared in Sri Lanka in 1868 and by 1875, the disease had assumed disastrous proportions and was responsible for the abandonment of the coffee plantations in Sri Lanka. The disease changed the entire economy of Sri Lanka. The disease also spread and caused devastation in South India, entered South Africa in 1878, Central Africa in 1913.

The main cultivated species of coffee are attacked in the field, but all possess strains resistant to the different races. In general, varieties of *C. arabica* appeared to be the most susceptible, while those of *C. liberica* are least susceptible. Varieties of *C. canephora* fall in the intermediate position.

1. Symptoms

Coffee-leaf rust produces yellowish orange, powdery, rounded blotches on the lower surface of the leaves, which may coalesce with others to form an irregularly shaped lesion. The disease is restricted to the leaves, but on rare occasions has also been recorded on berries and young shoots (Fig. 16.15).

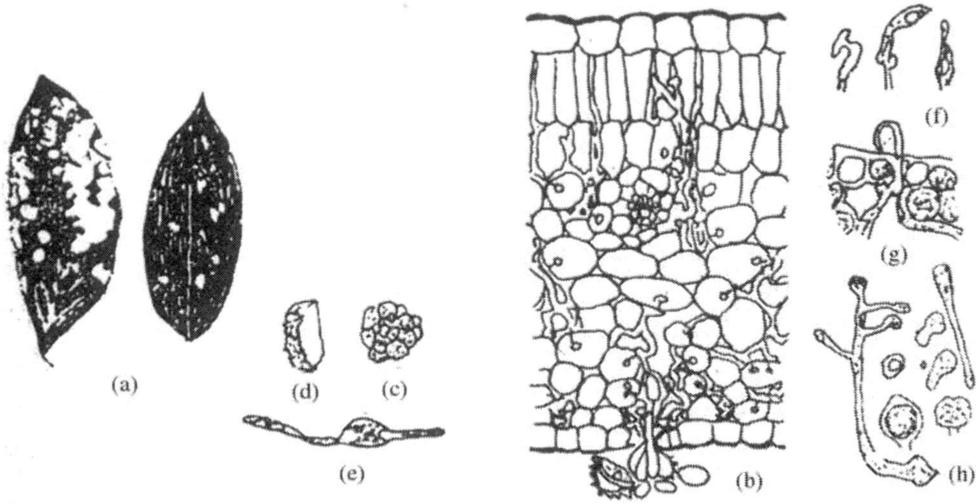

Fig. 16.15 Coffee rust (*Hemileia vestatrix*): (a) Symptoms (b) Section through an affected leaf showing mycelium, haustoria, and spore cluster (c) A spore cluster seen from below (d and e) Urediniospores showing germination (f) Appressoria (g) Infection through stoma of leaf (h) Teliospore and their germination

2. Causal Organism

The pathogen responsible for coffee rust is *Hemileia vastatrix* Berk and Br. The rust is hypophyllous, densely scattered and gives a powdery appearance or yellowish orange, rounded blotches, about 3–25 μm or more in diameter. The pathogen consists of numerous narrow interwoven feeder hyphae and rounded cells below the stomata, bearing clavate filaments which emerge through the stomata and whose tip bear numerous small pedicels on which the spores are borne. New urediniospores arise from the same stalk lower down, each stalk thus producing several urediniospores and becoming something like a swollen basal cell. Urediniospores are not shed immediately, but remain attached to the top of the pedicels by short strigmata. In mature uredium, the pedicel loses its individuality and forms a compound column at the ends and sides of which a densely crowded mass of strigmata carries the spores. Urediniospores are reniform, 28–36 × 18–28 μm; wall hyaline, strongly warted on the convex surface, smooth on the straight or concave face, 1 μm thick. Teliospores arise in the same sorus. They were once thought to be rare but after a careful search have been collected abundantly from Karnataka and Sri Lanka. They are mixed up with the urediniospores. Teliospores are spherical or turnip shaped, 20–28 μm in diameter; wall hyaline, smooth, 1 μm thick. These germinate without a rest period and *in situ* (Fig. 16.15).

3. Disease Cycle

It has been found that shelter from wind, intermittent rain, dew or moist, ample light, very light overhead shade, and moderately high temperature predisposes the plants to severe attacks. Earlier workers attributed spore dispersal to wind but more recent work has demonstrated the role played by rain splash over short distances. Air current may be responsible for long range intercontinental dispersal. Two species of thrips have been found feeding on and aiding the dispersal of urediniospores in India. Crowe (1963) has shown that two species of hymenopterous parasite of larvae of cecidomyid midges were found feeding on urediniospores in Kenya.

4. Disease Management

(a) Efforts are now being made in Karnataka and Coorge to secure varieties of coffee resistant to leaf rust. As a rule, *Coffea robusta* appears to be less susceptible than *C. arabica*, but the former species does not yield good-quality coffee. A variety, Coorg, was quite resistant.

(b) Control by spraying with copper fungicides has given good results when applied at the proper time. In South India, two sprays of Bordeaux (2:2:40), one during April–June and the other during September–November, has been recommended. Muthappa and Kumari (1979, 1981) have reported that coffee rust could be controlled by the application of Carboxin 20EC and Oxycarboxin 20EC. The Triazole fungicide, Triadimenol (Bayton) applied to coffee seedlings disrupts the rust hyphae in the host tissue.

(c) Among the cultural practices, sanitation is essential—fallen leaves should be destroyed.

(d) Possibilities of biological control of the disease have been reported by using hyperparasites.

16.17 RUST OF GROUNDNUT

Groundnut rust is now of economic importance in most areas. In India losses in pod yield to the tune of 14–32% have been observed. Early and late spots caused by *Cercospora arachidicola* and *Cercosporidium personatum* coincide with the appearance of the rust and hence the magnitude of loss is quite great.

1. Symptoms

Orange-coloured pustules (uredinial) appear on the lower surface of the leaves (leaflets). Later on pustules may appear on the upper surface of the leaf and on other aerial parts except flowers. Infected leaves become necrotic, dry up but tend to remain attached to the infected plants.

2. Causal Organism

Rust of groundnut is caused by *Puccinia arachidis* Speg. and only uredinial and telial stages are known. Spermogonia and aecial stages are not known.

3. Disease Cycle

Urediospores, or urediniospores, are short-lived in infected crop debris in the tropic and the fungus is unlikely to survive from season to season under post-harvested conditions that include a break of more than one month or more. The practice of continuous cultivation of groundnut without any significant break in parts of India or other parts of the world may be instrumental in the perpetuation of the rust. It has been suggested that groundnut crops in the southern part of India may act as a reservoir for infection from which the urediniospores are carried northward with the wind during the monsoon period. The pathogen may also survive from season to season on ground keepers and volunteer groundnut plants.

4. Disease Management

Based on several years of trials, it was noted that spraying of a mixture of Carbendazin (0.5%) and Mancozeb (0.25%) at 2–3 weeks interval starting from 4–5 weeks after planting of the crop was quite efficacious in controlling the rust disease.

Of course breeding for resistance is the best method of controlling the disease. At ICRISAT, some progress has been made in this direction.

Mycoparasites such as *Verticillium lecanii, Penicillium islandicum* and several others have been reported. Mycophagous insects feeding on urediniospores of groundnut rust have been reported but no attempts have been made for the biological control of groundnut rust disease in India on the field scale so far.

16.18 RED RUST OF TEA

The disease is widely distributed in the zones of north-east India, Sri Lanka, and Bangladesh. The disease appears sporadically every year in most of the tea gardens causing different degree of losses. In Bangladesh, about 8–25 tea estates are affected by this disease.

1. Symptoms

Red rust is a predominant disease in tea gardens caused by an alga *Cephaleuros parasiticus*. When tea plants are affected by red rust disease, the leaves become variegated. In extreme cases, it causes severe damage to young tea plants by killing stem tissues in patches. The disease is most common in young tea and old tea bushes in seed berries. The alga can infect the branches at any stage of growth. Once the branch is infected, the alga remains latent for a year. In the following year depending on the first rain, fruiting bodies of the pathogen are produced.

2. Causal Organism

The pathogen responsible for red rust of tea is *Cephaleuros parasiticus* Karst. The predisposing factors of the disease are mainly poor fertility, alkalinity and lack of aeration of the soil, inadequate or complete absence of shade, drought, and waterlogging.

3. Management

It has been observed that healthy leaves of tea plants contain more nitrogen and potash in comparison to infected ones. The effective treatment for controlling the disease was application of potash fertilizer which was followed by foliar spray with copper hydroxide and Carbendazim. Carbendazim, copper oxychloride, Hexaconazole, Propiconazole-based fungicides were approved for controlling the disease. Recently, Mancozeb (Sedomil 72 WP), and Mancozeb and Metalaxyl (Recozeb 80WP) were found quite effective in controlling red rust disease of tea.

16.19 SOME OTHER DISEASES CAUSED BY BASIDIOMYCOTA

Flag smut of wheat	*Urocystis agropyri* (*U. tritici*)	Stem, culm and leaves are affected, gray or grayish black sori on the leaf blades, leaves get twisted, began to drop and then wither, if grains formed are shriveled and do not germinate.	Soil- and seed-borne	Seed treatment with Benlate, Bavistin, Vitavax, Fenfuran, Triadimefon, and Triadimenol give good control. Cultural practices such as early sowing, burning of stubbles, crop rotation and resistant varieties like Kalyan Sona, Lerma Rajo and Sonara 64 have been reported.
Hill bunt or stinking smuts of wheat	*Tilletia caries* and *Tilletia foetida*	All the ears in the diseased plant and all the grains in the ear are turned into small balls, foul smell of trimethylamine is a good indication of infection.	Soil- and seed-borne	Seed treatment with copper fungicides, mercurial fungicides, Benlate (0.3%), Vitavax (0.2%), Bavistin (0.1%) were quite effective. *Pseudomonas fluorescens* inhibit teliospore germination.
Bunt of rice (syn. black smut, kernel smut, rice smut)	*Tilletia barclayana* (*Neovossia horrida*)	Only a few grains in the panicle are infected and most of the sori are hidden by the glumes.	Soil- and externally seed-borne. The sporidia are wind-borne to the flowers.	Field sanitation, crop rotation and cultivation of resistant varieties are recommended.
Leaf smut of rice	*Entyloma oryzae*	Linear, rectangular, or angular black-colored spots on leaves.	Soil-borne	Clean cultivation and cultivation of resistant varieties are recommended.
Long smut of sorghum	*Sporisporium* (*Tolyposporium*) *ehrenbergii*	Only few smut sori develop and scattered irregularly in the ear.	Air-borne	Early sowing, crop rotation and field sanitation should be practiced to control the diseases.
Covered smut, kernel smut, short smut, grain smut of jowar	*Sporisporium* (*Sphacelotheca*) *sorghi*	In place of the normal grain, an oval, cylindrical gray sac; the sorus is covered by a tough membrane, interior of the grain sac is completely filled with the spore powder.	Seed-borne	Steeping the grain in 3% solution of copper sulphate or 0.5% formaldehyde or copper carbonate are effective. Other fungicides such as Agrosan, Ceresan, Captan, Carboxin, and Bavistin are also effective in controlling diseases.
Common smut of maize	*Ustilago maydis*	*U. maydis* produce galls on the ears, axillary buds, tassels, stalks and on the leaves which are light coloured initially but turn into a black mass of spores which results in dwarfing or death of the plant shoot.	Seed-borne	Crop rotation, field sanitation, and growing resistant varieties are the most important measures.

(*Contd.*)

(*Contd.*)

Whip smut of sugar cane	*Ustilago scitaminea*	Floral shoot gets transformed into a long whip-like structure which is covered in its earlier stages which soon flakes away.	Smutted setts	Control can be achieved by removal of infected or smutted whips, discouraging ratooning, avoiding planting of setts from diseased canes and chemical treatment of setts before planting with Vitavax, Benlate, and Bavistin.
Covered smut of barley	*Ustilago segetum* (*U. hordei*)	Every ear in the diseased shoot and every grain in the diseased ear is affected. The black spore masses are held together by the persistent membrane of the grain.	Externally seed-borne	Seed treatment with Agrosan G N, Ceresan, and Vitavax are effective for the control of diseases.
Loose smut of barley	*Ustilago segetum tritici* (*U. nuda*)	Sori are formed in spikelets, in later stages they rupture and spore mass is shed leaving behind the rachis.	Seed-borne	Use of resistant varieties (C4, C50, CN292), hot water or solar treatment are effective in controlling diseases.
Onion smut	*Urocystis cepulae*	Black powdery mass streaks on the leaves, scales, and bulbs.	Soil-borne	Rotation with non-allium crop, shallow sowing, seed treatment with Thiram or Captan reduces seedling infection.
Rust of beans	*Uromyces phaseoli typhica* on beans and *U. phaseoli vignae* on cowpea	Rust sori on leaves are reddish brown and circular shaped, severe infection cause defoliation and loss of crop.	Soil-borne	Removal and burning of crop debris, long crop rotation and removal of weeds are cultural practices. Application of Mancozeb, Maneb, Zineb, and Triadimefon can control the disease.
Rust of pea	*Uromyces fabae*	All the green parts of the host including the pods show uredinial sori.	Soil-borne	Chemicals which give control include Bordeaux, Zineb, Ziram, Thiram, Chloranil, and Agrosan.
Rust of gram	*Uromyces ciceris arietini*	The initial symptoms are small, round cinnamon-brown pustules on leaves petioles, stems and also on pods.	*Trigonella polycerata* a collateral host in the hills	Varieties like NRC-34, NEC-249, JM-583, and HPC-63 have been found resistant under natural conditions.
Rust of bajra	*Puccinia penniseti*	Hypertrophied patches, yellowish green, circular in shape develop on leaves, premature drying of leaves occur due to severe infection of leaves.	Survive on alternate host like brinjal	The only effective method of controlling disease is by breeding resistant varieties.
Rust of jowar	*Puccinia purpurea*	Rust pustules which are reddish brown appear on leaves.	Survive on alternate host, *Oxalis corniculata*	The development of resistant varieties form the best control measures for this rust.

(*Contd.*)

(*Contd.*)

Blister blight of tea	*Exobasidium vexans*	Initially the lesions on leaves are translucent, concentric, and slightly depressed but later rapidly enlarge and develop into typical blister lesions on the underside of the leaf and produce a large number of basidiospores.	The disease spread through air-borne basidiospores.	Application of the protectant-eradicant copper and nickel fungicides at short intervals is necessary to control the disease.

REVIEW QUESTIONS

1. In spite of the fact that the inoculum of wheat rusts does not survive in the plains of India, the damage from wheat rust is very high. Discuss this statement critically.
2. What new information do we have on the epidemiology of black-stem rust of wheat in India?
3. What are the alternate hosts for the following rusts?
   (i) Black stem rust of wheat
   (ii) Brown rust of wheat
   (iii) Rust of bajra
   (iv) Rust of jawar
   (v) Rust of gram
4. Give at least three examples of autoecious and heteroecious rusts.
5. Who cultured *Puccinia graminis tritici* for the first time in axenic culture?
6. How will you control the wheat rusts?
7. What are the important rust diseases in your locality?
8. How will you differentiate between the following?
   (i) *Ustilago* and *Tolyposporium*
   (ii) *Ustilago* and *Sphacelotheca*
   (iii) *Urocystis* and *Tolyposporium*
   (iv) *Neovossia* and *Ustilago*
   (v) *Entyloma* and *Tilletia*
9. What is meant by systemic infection? Describe the life cycle of *Ustilago segetum tritici* along with the various control measures of the disease it causes.
10. What is the major difference between smuts and bunts?
11. Name the causal organisms responsible for the following diseases along with the major methods of disease control in each case:
    (i) Loose smut of wheat
    (ii) Karnal bunt of wheat
    (iii) Flag smut of wheat
    (iv) Whip smut of sugarcane
    (v) Smut of bajra
    (vi) Common smut of maize
12. Name at least three types of smuts found in *Sorghum,* indicating their causal organism and the symptoms produced. What is the major method of control in each case?
13. Who discovered solar heat treatment? Mention its special features.
14. Would you advise seed treatment with chemicals in all seed-borne diseases?

MULTIPLE-CHOICE QUESTIONS

1. The discharge mechanism of smut spores are
   (i) passive (ii) active
   (iii) both (i) and (ii) (iv) none of these
2. Which of the following angiospermic families are generally infected by smut fungi?
   (i) Gramineae (ii) Cyperaceae
   (iii) Asteraceae (iv) Both (i) and (ii)
3. Which part of the plant is most commonly attacked by smut fungi?
   (i) Stem (ii) Inflorescence
   (iii) Roots (iv) All of these
4. Which of the following smut fungi attacks the leaves of plants?
   (i) *Entyloma* (ii) *Tolyposporium*
   (iii) *Ustilago* (iv) *Tilletia*
5. Which of the following genus belongs to the family Ustilaginaceae?
   (i) *Tolyposporium* (ii) *Tilletia*
   (iii) *Neovossia* (iv) *Urocystis*
6. Which one of the following feature is an important symptom of bunt of wheat?
   (i) Typical smell of trimethylamine (ii) Typical smell of ammonia
   (iii) Typical smell of H_2S gas (iv) All of these
7. Which of the following fungus results in leaflike structure in the flower which is called phyllody?
   (i) *Tolyposporium* (ii) *Ustilago*
   (iii) *Urocystis* (iv) *Sphacelotheca*
8. Which of the following fungicide is effective in controlling loose smut of wheat?
   (i) Mercurial fungicides (ii) Bordeaux
   (iii) Vitavax (iv) All of these
9. The pathogen responsible for loose smut of wheat (*Ustilago segetum* var. *tritici*) is
   (i) air-borne (ii) internally seed-borne
   (iii) externally seed-borne (iv) soil borne
10. Luthra and Sattar (1934) suggested which of the following simple and economical method to control loose smut of wheat?
    (i) Solar energy treatment (ii) Soil treatment
    (iii) Seed treatment with Vitavax (iv) All of these
11. The pathogen responsible for Karnal bunt of wheat is
    (i) *Tilletia indica* (ii) *Tilletia foetida*
    (iii) *Tilletia caries* (iv) all of these
12. Which of the following factor favours increase in the incidence of Karnal bunt of wheat?
    (i) Higher doses of N_2 fertilizers (ii) Heavy manuring
    (iii) Late sowing (iv) All of these
13. Which of the following approaches are effective for control of Karnal bunt of wheat as the pathogen is soil-, seed- and air-borne?
    (i) Cultural practices (ii) Quarantine measures
    (iii) Chemical application (iv) Integrated disease-control approach

14. The new name of the fungus responsible for smut of bajra is
    (i) *Moesziomyces bullatus* (ii) *Tolyposporium penicillariae*
    (iii) *Tolyposporium ehrenbergii* (iv) None of these
15. The main source of recurrence of black-stem rust of wheat in the northern part of India is mainly due to
    (i) survival of uredo stage in soil
    (ii) *Berberis* plant as an alternate host
    (iii) survival of uredo stage in Niligiri and Pulney Hills
    (iv) seed-borne inoculum
16. The effective method to control black-stem rust in India is
    (i) replacement of wheat and barley cultivation in Niligiri and Pulney Hills
    (ii) eradication of *Berberis* plant
    (iii) seed treatment
    (iv) all of these
17. Which of the following plant acts as an alternate host of brown rust of wheat?
    (i) *Berberis* sp. (ii) *Thalictrum* sp.
    (iii) *Hordeum* sp. (iv) *Lolium* sp.
18. The alternate host of yellow rust of wheat is
    (i) *Berberis* sp. (ii) *Isopyrum* sp.
    (iii) Both (i) and (ii) (iv) *Muehlenbergia*
19. *Melampsora lini* responsible for rust of linseed is a
    (i) heterocious rust (ii) autoecious rust
    (iii) Both (i) and (ii) (iv) none of these
20. The pathogen responsible for rust of groundnut is
    (i) *Cercospora arachidicola* (ii) *Cercosporidium personatum*
    (iii) *Puccinia arachidis* (iv) *Urocystis tritici*
21. *Puccinia graminis tritici* is rust, that is
    (i) macrocyclic, heteroecious (ii) macrocyclic, autoecious
    (iii) microcyclic, heteroecious (iv) microcyclic, autoecious
22. Macrocyclic fungi is the name given to rust fungi
    (i) which produces bigger teleutospores
    (ii) when all the five spore stages are produced
    (iii) which complete its life cycle on single host
    (iv) all of these
23. Which of the following workers is famous for his work on the epidemiology of Wheat rust?
    (i) A F Blakeslee (ii) K C Metha
    (iii) E A Bessey (iv) H C Bold
24. The basidiospores are
    (i) exogenous spores (ii) endogenous spores
    (iii) Both (i) and (ii) (iv) none of these
25. Clamp connections are very common in
    (i) Ascomycota (ii) Zygomycota
    (iii) Basidiomycota (iv) Oomycota

Answers

1. (i)	2. (iv)	3. (ii)	4. (i)	5. (i)	6. (i)	7. (iv)	8. (iii)	9. (ii)	10. (i)
11. (i)	12. (iv)	13. (iv)	14. (i)	15. (iii)	16. (i)	17. (ii)	18. (iv)	19. (ii)	20. (iii)
21. (i)	22. (ii)	23. (ii)	24. (i)	25. (iii)					

17

DISEASES CAUSED BY ANAMORPHIC FUNGI (DEUTEROMYCETES)

Typically, the wilt diseases are associated with plants that contain vessels in the xylem, that is, the angiosperms. While wilt, or loss of turgidity, is a common symptoms of disease, it may result from one of many possible causes. The vascular wilts are set off as a group because of the association of symptoms with the presence of the pathogen in the vascular system of the host, usually in the xylem tissue. The most common pathogens are vascular Fusaria (life cycle of a typical vascular pathogen is depicted in Fig. 17.1), *Verticillium* spp., and bacteria, for example, the pathogens of cucurbit, bacterial wilt, and southern bacterial wilt. The wilt syndrome includes vein clearing, epinasty, chlorosis, vascular discolouration, stunting, temporary, permanent loss of turgidity, and death. Apart from wilting, the syndrome of this group of disease is not unique. The browning reaction occurs both in the local lesions, and wilt diseases as a result of similar sequences of the biochemical reactions. Respiration is accelerated, and auxins, phenols, and other aromatic products accumulate in a similar manner in the two cases.

Water relations are drastically changed in infected plants. The water content in diseased plants diminishes, and eventually dries up altogether. Stress initially occurs in the tissue of individual leaves, and then throughout the plant. The water in the leaf blades is affected before mycelium enters them. In infected plants, leaves lose turgor, and eventually die and become desiccated. The mechanism by which wilt occurs is debatable. The debate enters around two theories put forward to explain the phenomenon of wilting. The first theory is called the **plugging theory** or **vascular occlusion theory**—when there is very high resistance to the flow of water through the xylem, the leaves receive less water than they transpire, and in consequence, they wilt. This is the basis of the plugging theory. The second theory is called **toxin theory**, and is based on the fact that the permeability of parenchymatous cells in the leaves increases so that they lose their osmotic efficiency, and this results in wilting.

17.1 VASCULAR OCCLUSION, OR PLUGGING THEORY

Probably no single factor is responsible for vascular dysfunction in the wilt disease. High resistance to the flow of water through infected plants is probably the result of several factors. Some originate from the host as a result of response to the pathogens, and others due to the pathogens themselves. The infected plants may contribute to the altered resistance to flow by virtue of the changes in the vessel size that accompany infection, and development of tyloses in conductive elements. These are

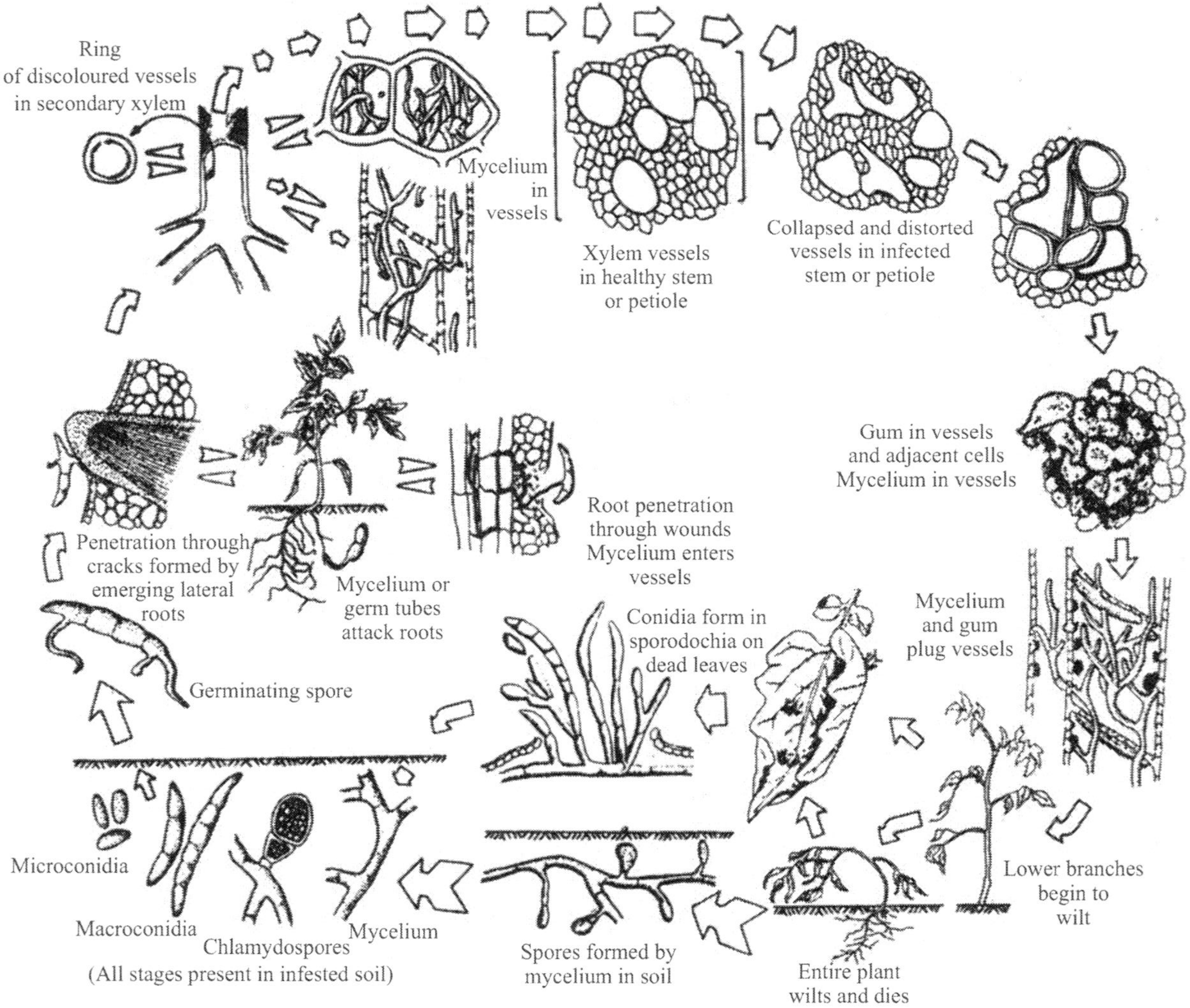

Fig. 17.1 Life cycle of a typical vascular wilt pathogen *Fusarium oxysporum* f. *lycopersici* (Redrawn from G N Agrios, *Plant Pathology*, Academic Press, London, p. 291)

(1) collapse of vessels, (2) formation of tyloses, (3) the physical presence of the mycelium, production of bud cells, microconidia and endoconidia within the vessels, (4) altered viscosity of tracheal fluids, and (5) plugging by compounds of high molecular weight, such as polysaccharides, polymerized melanoid pigments, gums, gels, and pectins.

1. Collapse of Vessels

Chambers and Corden (1963) have described the frequency with which vessels collapse, especially in petiolar bundles, and in tomato plants infected with *F. oxysporum f. lycopersici*. IAA or culture filtrates of the pathogen also result in the collapse of vessels as in the infected plants. If a vessel's diameter has been reduced to half as a result of the collapse of the vessel, it will conduct the water at only one-eighth of its original rate.

2. Tyloses

Tyloses are found more frequently in infected than in healthy plants. Development of tyloses in conductive elements reduces the effective radius available for transport, especially when tyloses are abundant. Some species of plants are more prone to form tyloses than others. New conducting elements may be formed in infected plants, and partly compensate for the reduced efficiency of the blocked vessels so that the extent of wilting caused by tyloses is uncertain. Beckman (1964) suggested that tylose and gel formation are basically a defensive reaction of the infected plants, and that in susceptible plants, tylose formation is inhibited or delayed. This inhibition is perhaps due to the pathogen. Metabolites, when present in sufficient quantities, reduce the plant's capacity to produce the growth-regulating substances which bring about the tyloses, and it may be that growth-promoting substances produced by the pathogen are involved. The pathogen moves through the xylem vessels by transported propagules, and that rapid tylose formations reduce such transport. The rapidity with which tyloses are formed determines whether they will confer resistance or susceptibility to the host. If tyloses are formed rapidly well before the spread of the pathogen, they will confer resistance to the host, but if they are formed after the colonization of the pathogen, they will themselves be instrumental in bringing about more wilt symptoms.

The importance of the tyloses in reducing the efficiency of the transport system as a whole depends upon a number of factors. When no alternative unoccluded pathway of flow exists, tyloses may block the flow to distal tissues. The extent of interference with flow will also depend on the production of vessels containing tyloses, the extent of occlusion of the vessels, and even the area of development of tyloses along the vessels.

3. Mycelium

Physical blockage in xylem vessels by the growth of fungal mycelium and/or spores produced in those vessels is perhaps the most obvious explanation of vascular occlusion. Plugging of tracheal elements by mycelium has frequently been considered a cause of vascular dysfunction. Individual vessels may become densely packed with mycelium but in general mycelium is not sufficiently abundant through all the vessels to be the sole cause.

As is well known, the vascular fungi commonly produce bud cells, microconidia, or endoconidia that break free from the mycelium, and are transported through elements. Conidia are carried upwards in the vascular stream, and many accumulate at the end walls or the opening of the pits, and thus cause impediments in the flow of water. The movement of conidia through xylem vessels leads to a more rapid colonization of the host, and to the production of large quantities of the metabolic products of the pathogens, resulting in more severe symptoms.

4. Altered Viscosity of Tracheal Fluids

In some bacterial wilt diseases, the tracheal fluids become highly viscous as the plant is dying. When the stem is cut open, and separated, white milky strands extend from one segment to the other. Although in moribund tissues, viscosity of tracheal fluids may be high, this can be questioned as it could be a post-mortem change made possible after the flow has already ceased. Husain and Kelman (1958) demonstrated that the pathogen *Ralstonia solanacearum* provides an extracellular polysaccharide, and that the pathogenicity of different strains of organisms is correlated with their ability to produce this polysaccharide. However, in Fusarium wilt of tomato, the viscosity of tracheal fluids is the same as that in healthy plants (Waggoner and Dimond, 1954). Polysaccarides are known to be produced by *X. campestris* pv. *phaseoli, Ralstonia solanacearum* (earlier called *Pseudomonas*), *Ophiostoma ulmi* (*Ceratocystis*), *Ceratocystis fagacearum, V. albo-altrum,* species of *Fusarium,* and *Cephalosporium.* At least three species *Corynebacterium* produce glycopeptides, and their presence in the diseased plants has been demonstrated. When polysaccharides and glycoproteins are introduced into vascular tissue they introduce wilting. Wilting has been ascribed to increased viscosity of tracheal fluids but this cannot be the sole cause.

17.1.1 Plugging by Compounds of High Molecular Weight—Polysaccharides, Pectins, Gels, Gums

Polymerized melanoid pigments are all compounds of high molecular weight. The walls of conducting elements behave as ultra filters, and remove materials of high molecular weight from the transpiration stream. In the process, the channels become plugged, and this impedes the flow of water. Polysaccharides and glycopeptides can affect the water economy of plants in two ways. A portion of these may accumulate in pit membrane of vessels which act as ultra filters. Such entrapments will reduce the lateral and longitudinal flow of water through the system. Secondly, smaller molecules transported to leaves disrupt water transfers from veins, and accumulate in pit membranes of parenchymatous cells where they cause damage which appears as marginal drying.

Although vascular browning is a regular symptom of wilt disease, it has never been regarded seriously as a cause of wilting. The browning originates in vascular parenchyma when cells are adjacent to invaded vessels. It is now established that vascular browning is the result of the oxidation and polymerization of phenolic compounds through the action of phenolases. Studies of the biochemical reactions involved in vascular browning point to the formation of melanoid pigments of high molecular weight. These result from the oxidation of phenols, polymerization of these products, and formation of products between oxidation products and proteins.

In *Fusarium*-infected tomato plants, especially in xylem parenchyma, phenols accumulate, and phenolases become active. The phenols are oxidized to fungitoxic quinones, and their polymerization leads to fungitoxic products in these cells. Polymerization to dark-coloured melanins may then occur in tomato wilt (*F. oxysporum f. sp. lycopersici*), conjugated phenols, such as glucosides, tannins, and lignin may be hydrolyzed by enzymes to polyphenols, which are in turn oxidized by PPO, possibly an induced enzyme, to toxic quinones which then polymerize to cause the characteristic vascular discoloration.

Winstead and Walker (1954) emphasized the role of pectic enzymes in vascular browning. Culture filtrates of several pathogens induced browning in the cuttings of a variety of plants. Browning occurred when both PG and PMG enzymes were present. This may mean that maceration caused by pectic enzymes results in the disorganization of host cells, and thus, phenolic oxidation can occur.

With the progress of wilt disease, the parenchyma cells are damaged, and pigments leak into the vessels. When free in vessels, the large pigment molecules can cause plugging of the type observed by Chambers and Corden (1963), or they can become trapped in gels when the vessel is already plugged.

Gels and gums are also thought to be formed as a result of infection by several wilt pathogens. Beckman (1964) has discussed the frequency with which plants respond to infection and wounding by forming gums and gels. Several workers have clearly indicated the formation of such materials in banana plants infected with *F. oxysporum f. cubens.* When liberated as colloidal solutions into the transpiration stream, they can cause plugging at cross walls. It is doubtful whether these materials are the same as those responsible for plugs in vessels in moribund plants, and have been attributed to the accumulation of partially hydrolyzed wall material resulting from the action of pectolytic enzymes. The occluding plugs in elm infected by *Ophiostoma ulmi* contain both pectin and lignin. This material in tomatoes infected by the pathogen *Fusarium* cannot be removed by phenol. In banana roots, the materials consist of a mixture of pectin, calcium pectate, and hemicellulose.

17.1.2 Blockage Caused by the Action of Cell-Wall-Degrading Enzymes

The wilt pathogens, in common with many microorganisms, excrete extracellular pectolytic and cellulolytic enzymes. The vascular wilt pathogens known to produce these enzymes are *R. solanacearum, V. albo-atrum, V. dahliae, Ophiostoma ulmi* (*Ceratocystis*), *Ceratocystis fagacearum, F. oxysporum* f.sp. *lycopersici, F. oxysporum* f. sp. *cubense, F. oxysporum* f. sp. *vasinfectum*, and *F. oxysporum* f. sp. *niveum.*

The evidence of the role of these enzymes in pathogenesis is based upon their presence in diseased plants, the ability of the pathogen to produce them on host tissues, chemical evidence of disappearance of their substrates in diseased plants, microscopic evidence of degradation of middle lamellae and cell walls, and the ability of such products to produce vascular dysfunction.

Another piece of evidence implicates pectolytic enzymes in the pathogenesis of wilt diseases.

17.1.3 Toxin Theory

Toxins have been extensively investigated, especially in tomato wilt caused by *F. oxysporum* f. sp. *lycopersici.* The first wilt toxin to be studied was lycomarasmin which was isolated in pure-form culture filtrates. Another important toxin thought to be involved in vascular wilt is fusaric acid. Fusaric acid was first isolated as the metabolic product of *Fusarium heterosporum,* a nonspecific parasite. It was recognized as an important toxin of *Fusarium lycopersici, F. vasinfectum, Gibberella fujikuroi.* It is now known to play a pathogenic role in the wilt of tomato, and cotton, in the bakanae disease of rice, and in the Panama disease of banana caused by *Fusarium oxysporum* f. sp. *cubense.*

Fusaric acid is called a wilt toxin as it is known to play a distinct role in the wilt of tomato and cotton plants. Even when the pathogen is restricted to the roots, symptoms like vein clearing and epinasty are seen in the leaves, and there is reduction in respiration and growth, necrosis of cortical tissues overlying the vascular bundles (leading to the furrowing of stem), and epinasty of petioles. This seems to be due to the toxic metabolites translocated to distant parts of the plant from the focus of infection. Pure-toxin fusaric acid stimulates some of the symptoms caused by pathogens.

Fusaric acid, 5-*n* butylpyridine-2-carboxylic acid has the empirical formula $C_{10}H_{13}O_2N$ with molecular weight 179. It has the following structural formula.

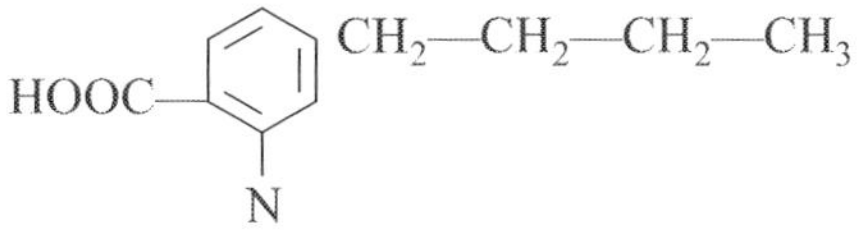

Fusaric acid

The existence of dehydrofusaric acid in preparation of fusaric acid was also demonstrated. This acid appears after the appearance of saturated fusaric acid. The toxin fusaric acid impairs the permeability of plasma membranes. Fusaric acid also impairs the energy metabolism at higher and lower concentrations.

Fusaric acid forms metal chelates. This toxin interferes with metal containing enzymes, especially Fe-porphyrin oxidase, and consequently depresses respiration. On the basis of antagonism reported between *ferrioxamine* B, and fusaric acid, it can be conjectured that fusaric acid specially interferes with the incorporation of Fe in the porphyrin skeleton.

Sanwal (1961) described the physiological effects of fusaric acid and suggested that it could cause a partial uncoupling of oxidative phosphorylation. Its role in pathogenesis remains problematical, and Deverall (1964) concluded that it plays no part in the development of wilt symptoms although it might act in conjugation with other metabolites. However the Sadasivan school placed great importance on the toxin being responsible for the wilt disease in cotton.

The toxin, lycomarasmin, is a polypeptide with a molecular weight of 277.3 and empirical formula $C_9H_{15}O_7N_3$. The following structural formula has been tentatively assigned to this toxin:

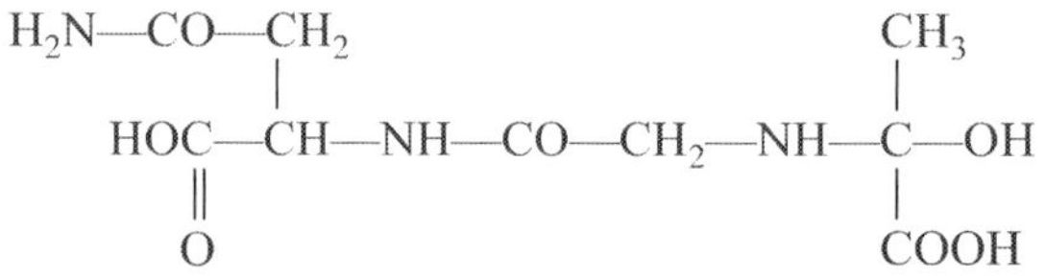

Lycomarasmin

The toxicity of this substance seems to arise from the fact that it can form metal chelates (such as with iron). The pathogenic effect of this toxin is partly due to the fact that its molecules, on their way through the infected plant shoots chelate iron in the host which otherwise would not be able to migrate, and transport themselves into the leaves in the form of a lycomarasmin-iron complex. Here, part of the lycomarasmin molecules are regenerated as chelating agents, and cause local deficiencies, while at other places there is excess of iron (i.e. iron plethora). In addition to fusaric acids and lycomarasmin toxins, vasinfuscarin and phytonivein have been implicated in vascular wilts. **Vasinfuscarin** is apparently a protein with enzymic properties, and it is reported to produce the vascular browning typical of vascular wilts. **Phytonivein** is present in culture filtrates of *F. oxysporum* f. sp. *niveum* and involved in the wilt of watermelons.

Diffusible toxins may be implicated in several other wilt diseases, including the sleepy disease of tomato, wilt of Michaelmas daisy, and pea wilt.

It is, however, the complexity of host-pathogen interaction which is of interest. Physical blockages occur in the vascular system of infected plants as a result of integrated metabolic reactions between fungus and plants. In addition to the production of toxin, and cell-wall-degrading enzymes mentioned earlier, the invading pathogen causes the plants to produce other additional compounds during infection. Ethylene, and IAA are produced by both the pathogen, and the susceptible host plants. Both these factors stimulate the production of tyloses in host plants leading to increased phenol-oxidizing and pectolytic enzymes. These phenolic compounds which are reported in the vascular elements IAA and ethylene also induce the formation of gels which accumulate at perforation plates and contribute to vascular blockage.

17.2 WILT OF COTTON (*GOSSYPIUM ARBOREUM*, AND *G. HERBACEUM*)

Wilt is one of the major diseases of cotton and is found wherever this crop is grown. The disease has spread to almost all the black-cotton-soil tracts of the country (Maharashtra, Madhya Pradesh, Karnataka, and South Gujarat). The losses in the wilt are very high to the extent of 40 to 60% because of reduced stand, stunted growth, small bolls, and poor lint quality.

1. Symptoms

The plants are affected during all stages of growth. The first symptoms appear on the young seedlings. These are vein clearing of the leaves followed by interveinal tissue necrosis. There is yellowing, and browning of the cotyledons, and the formation of brown rings on the petiole. The seedling soon wilts and dies. If the plants are affected a little later, the leaves become flaccid which result in the drooping and wilting, starting from the oldest leaves at the base, and proceeding upwards finally involving the branches and the whole plant. Often there may be complete defoliation, leaving the stem alone standing in the field. Discoloration of the stem may also be seen. The basal portion of the stem in young plants may turn black. In adult plants, the discolouration may be partial. Growth is retarded, and the plant eventually wilts (Fig. 17.2).

Fig. 17.2 Wilt of cotton (*Fusarium oxysporum* f. sp. *vasinfectum*): wilted cotton plants

2. Causal Organism

Fusarium oxysporum f. sp. *vasinfectum* (Atk.) Snyder and Hansen is the pathogen responsible for vascular wilt of cotton.

3. Disease Cycle

The pathogen is soil-borne and survives on the stubble of the diseased plants. There is some evidence that it may be seed-borne also. The wilting occurs primarily due to the excretory toxic substances produced by the fungus in the vascular system which are translocated throughout the plant. Fusaric acid is the wilt toxin which has been demonstrated to be produced in the plant tissues in the rhizosphere as well as in the soil around the roots. Wilt of cotton has an important relationship with the nematode injury to roots which helps in the entry of the pathogen into the host roots.

4. Disease Management

(a) Seed Treatment Since the pathogen is seed-borne also, treatment of the seeds with fungicides such as Bavistin, Topsin M, Thiram, and mercurials can effectively control the seed-borne infection. Treatment of the seeds with Chlorothalonil, Thiabendazole, and Carboxin also reduces the incidence of wilt by resistance to infection by inducing increased phenolic enzyme activity.

(b) Chemical Treatment Soil drenching with Benlate and Bavistin can control wilt to some extent but it is not economical because of the cost of chemicals. A systemic fungicide (Miejuncuzhangi) from China was reported to reduce disease infection up to 89% when used in combination with Carbendazim, as a foliar spray (Singh, R S 2007). High potash application in combination with nitrogen- and phosphorus-containing salts give reduction in cotton wilt. Trace elements like zinc, molybdenum, lithium, aluminium, nickel, boron, and cobalt were inhibitory to conidial germination of the pathogen. The presences of these trace elements were beneficial for plant growth. Inoculations with nonpathogenic strains of *Fusarium* or dried biomass of *Penicillum chrysogenum* induces cross protection and systemic resistance in cotton seedlings against the pathogen.

(c) Agronomic Practices The incidence of the wilt can be reduced to some extent by manipulating the sowing date, application of potash, and amendments of zinc.

(d) Resistance Varieties In India, tetraploid cotton varieties are immune to *F. oxysporum* f. sp. *vasinfectum*. A number of varieties/strains which are either immune or resistant to wilt have been developed. These are C-C-1-35, JLA -101, AKH-590, SM-143, LD327, and LD-254. These are being used in hybridization programmes. All American cotton varieties, even those highly susceptible, were resistant to wilt in India, and vice versa. The differences may be due to different races and their origin.

17.3 FUSARIUM HEAD BLIGHT OF WHEAT

Fusarium head blight of wheat, also called **scab of wheat,** is a destructive disease in humid and semi-humid wheat-growing areas of the world. Scab incidence epidemic ranged from 25 to 75%. There is a growing concern over the potential of the disease in causing crop loss in parts of Punjab, the wheat bowl of the country. Since the pathogen produces mycotoxins, storability of the grains and usage as feed is affected. This disease occurred in some areas of Amritsar and Gurdaspur districts of Punjab.

1. Symptoms

The first visible symptoms of infection usually occur on the first florets of the inflorescence, generally near the middle of the head. Under conditions highly favorable for infection, such as continuous wetness at 25°C, visible lesions develop within 2 to 4 days. Lesions may be brown, purplish brown, or brown with a bleached center. Under moist conditions, lesions may be water-soaked initially, later turning purplish brown. The presence of pink fungal spore masses is a reliable diagnostic feature of Fusarium head blight. In some cases, the first indication of infection may be the fungus itself, giving a pinkish tint to partially exposed anthers, and slightly pinkish mycelium on floret surfaces. In more advanced infections, the fungus can produce macroconidia in association with a pink colour on the margin, surfaces of florets, and glumes. As lesions enlarge, and coalesce, entire florets become bleached or brown. As the disease progresses in wheat, floret discoloration spreads apically and basally in the head from individual infected spikelets (Fig. 17.3).

(a)

(b)

Fig. 17.3 Symptoms of Fusarium head blight of wheat: (a) Diseased field (b) Individual infected spikelet

2. Causal Organism

Many species of *Fusarium* cause wheat scab, and the symptoms caused by different species are almost the same. *Fusarium graminearum* Schw. is the principal pathogen responsible for head blight of wheat. Other related species such as *F. culmorum*, *F. avenaceum*, *F. moniliforme*, *F. oxysporum*, *F. poae*, and *Microdochium nivale* may also contribute to the head-blight complex but are generally less important than *F. graminearum*. The perfect stage of *F. graminearum* is *Gibberella zeae* Schw. and Petch. *F. graminearum* can survive as mycelium, ascospores, macroconidia, and chlamydospores. Microconidia are not produced.

3. Disease Cycle

F. graminearum survives between wheat crops in living or dead host tissues. Ascospores, macroconidia, chlamydospores, and hyphal fragments all can serve as inoculum. Ascospores and macroconidia may be the principal inoculum that initiates an epidemic of head scab, because aerial dispersal is necessary for the fungus to reach the infection site.

4. Effect on Grain Quality

In addition to yield and quality loss, scab may also reduce the value of a crop because of the patent mycotoxins produced by the fungus. The three most important mycotoxins produced by *F. graminearum* are the estrogenic toxin Zearalenone (ZEA), and the trichothecenes deoxynivalenol (DON) and nivalenol (NIV). These mycotoxins have been linked with livestock toxicoses or feed refusal, and thus reduce the value of the harvest either as grain or for feed. DON-contaminated grain becomes less palatable to livestock, and may cause emesis, depressed feed intake, and feed refusal in pigs. Humans consuming flour of scab by wheat containing DON have been reported to suffer from nausea, and headaches. *F. graminearum* mycotoxins are thus toxic to humans and animals (Agrios, 2005).

5. Management of Disease

(a) Cultural Methods Because of the ubiquitous nature and wide host range of *F. graminearum*, and other *Fusarium* spp. capable of causing scab, adequate control by crop rotation alone is not effective. However crop rotation, coupled with deep ploughing to bury infected crop residue and weed hosts can be effective. Appropriate methods of soil preparation, rotation with noncereal crops, timely harvest, and proper storage all help reduce the disease by reducing primary inoculum and the likelihood of extensive infection. The adoption of minimum tillage for soil conservation greatly reduces options for this disease-management technique.

(b) Biological Control Antibiotic substances produced by *Bacillus subtilis* and *Pseudomonas putida* have inhibited spore germination, and reduced the *in vitro* growth of *F. graminearum*. The activity of the gliding bacterium *Lysobacter* sp. against *Fusarium* head blight is believed to be due to induced resistance as well as chitinolysis.

(c) Chemical Control Fungicide applications should not only reduce the disease to an acceptable level, but also control toxin contamination of the grain. A large number of fungicides like Benomyl, Mancozeb, Thiobendazole, Thiophanate-methyl, and Triadimefon are quite effective for blight control. Folicur (Tebuconazole) application increased grain yield by 21%, and decreased mycotoxin (deoxynivalenol—a trichothecene metabolite) contamination in grains. Benlate (50WP, Benomyl) increased yield by 10 percent. Benomyl, and Benomyl-Mancozeb mixture were nearly as effective as Tebuconazole. Bavistin (4L/ha) and Libero (1.5L/ha) also gave the best control with an efficacy of 60, and 61 percent respectively. The systemic Triazole fungicides and the Strobilurins do not translocate from leaves to head, except in trace amounts. Therefore, an early application will protect only leaves whereas the later emerging heads are unprotected.

Control of wheat-head blight by fungicide application or other practices is neither practical nor sustainable. It is said that the only effective method of controlling wheat scab is to grow resistant varieties, and this is still true today. Chen *et al.* (1999) explored an alternative strategy for scab control in wheat through the deployment of pathogenesis related (PR) protein genes. A rice Thaumatin-Like Protein (TLP) gene (*tlp*) and a rice chitinase gene (*Chi 11*) were introduced into the spring wheat cultivar Bobwhite by co-transformation of the plasmids PGL2 ubi-tlp and pAHG11. The transgenes tlp and bar, under the control of the ubiquitin promoter-intron were expressed. After inoculation with conidia with *F. graminearum*, the symptoms of scab development were significantly slower in transgenic plants of the T_1, T_2, and T_3 generations expressing the tlp gene than in nontransformed control plants.

17.4 ROOT ROTS

Root rots are essentially different from wilt diseases. In vascular wilts, the pathogen grows mainly in the vascular region of the plant, and produces toxins and enzymes which bring about the wilt; while in the case of root rot, the pathogen invades the parenchymatous tissue of the root, cortex, and pith, and there is little invasion of the vascular tissues. Root rots are caused by a number of pathogens belonging to genera such as *Aphanomyces, Pythium, Phytophthora, Rhizoctonia, Sclerotium, Phymatotrichum, Thielavia, Helicobasidium, Ophiobolus, Armillaria*, and so on.

17.5 RHIZOCTONIA ROOT ROT OF COTTON

Root rot is one of the severe diseases of cotton in India. The disease occurs in severe form in the provinces of Punjab, Haryana, Rajasthan, and Gujarat in India, but it is of less importance in Karnataka, Tamil Nadu, Madhya Pradesh, and Western Uttar Pradesh. In contrast to wilt which is mainly found in black cotton soils, root rot is confined to sandy soils. In Rajasthan, this disease causes major losses—up to 90%.

1. Symptoms

This disease is easily distinguished from the damping-off disease. The stem of the infected plants remain erect, and the diseased parts are not water-soaked. The disease spreads in the fields in concentric circles. The pathogen attacks the seedlings as well as mature plants. If the affected plants are pulled out and examined, the entire root system may be found rotting. The lateral and thinner roots get completely rotted. The tap root is intact in the beginning while the lateral and thinner roots get shredded with yellow slime on the wood. Minute fungal sclerotia appear as minute black dots on the surface of woody tissues and on the rotting bark.

When a young seedling is infected, the cotyledonary leaves may be affected, and they assume a 'pinched' look. Root rot of cotton is essentially different from the wilt of cotton. In the root rot there is necrosis of the parenchymatous region of the root, while the wilt infection is mainly confined to the vascular tract. The mechanism of wilt injury is also different to the root rot. *Rhizoctonia* root rot is also called the **dry root rot**.

2. Causal Organism

Root rot of cotton is caused by *Rhizoctonia bataticola* (Taub.) Butler, *R. solani* Kuhn in Punjab and Haryana, and only by *R. bataticola* in Gujrat. The perfect stage of *R. bataticola* is *Thanatephorus cucumeris* (Frank) Donk which is also known as *Corticium vagum* var. *solani* Bunt ex Rolfs or *C. solani* (Prill, and Del) Sourd. Gaiz, or *Pellicularia filamentosa* (Pat) Rogens. *R. bataticola* produces pcynidia, and this stage is known as *Macrophomina phaseolina* (Maubl) Ashby. The sclerotia produced by *R. bataticola* has its pcynidial stage in the fungus *Macrophomina phaseolina* while the perfect stage of *R. solani* is *Pellicularia filamentosa.*

3. Disease Cycle

Both *R. solani* and *R. bataticola* are capable of living saprophytically in the soil on organic matter. Both these pathogens can also infect several other crops, such as groundnut, cowpea, and several leguminous plants, oilseeds, and vegetable crops on which they are also pathogenic.

4. Disease Management

Since the disease is found both in Indian and American cottons, attempts to evolve resistant varieties have not been successful. Varietal differences in susceptibility to root rot have been observed. Among *arboretum* race *indicum* cottons, the perennial type—Rohi of Gujarat—is known to be more or less immune to root rot (Balasubrahmanyan, 1950). Large number of cotton strains received from the USA and other countries have been tested under field, and screenhouse conditions. Three lines, viz. KH-33-146, 15-Kw-2(MB), and 9- Kw-2(MB) have been found to be resistant to both *R. solani* and *R. bataticola.*

(a) Seed Treatment Treatment of seeds with Quintozene, Carbendazim, and Oxathiin are quite effective in reducing the seed-borne, and soil borne infection.

(b) Soil Drenching Pre- and post-sowing drenching of soil with Quintozene and Carbendazim provided maximum control. Seed treatment plus pre-sowing soil drench is most effective.

(c) Agronomic Practices Certain cultural practices have been recommended for reducing the incidence of root rot of cotton such as adjusting the time of sowing of the crop when soil temperatures are not very high. *Phaseolus aconitifolius* and Sorghum are commonly grown in mixed cropping.

(d) Biological Control Kalyan and Verma (2002) identified *Penicillium petuletum*, *Aspergillus terreus*, *A. fumigatus*, and *Chaetomium* sp. as antagonistic to *R. solani*. Seed coating with *Trichoderma* spp. reduced disease incidence up to 83%. The effectiveness of fluorescent Pseudomonads/PGPR against seedling infection caused by *R. solani* has also been reported. The secondary metabolites of fluorescent Pseudomonads have received considerable attention because of their ability to produce a range of secondary metabolites that are inhibitory to the pathogen. Siderophores, metabolites with antimicrobial properties, and other substances like cyanide are the main groups of metabolites known to be involved in pathogen suppression. The biocontrol activity of a genetically manipulated strain of *Trichoderma virens* was enhanced against cotton-seedling disease caused by *R. solani* due to the over-expression of a chitinase gene (Back *et al.*, 1999). A strain of *Pseudomonas fluorescens* is generally found in the rhizosphere of cotton seedlings. Seed treatment with this bacterium increased seedling survival. The bacterium produces an antibiotic identified as Pyrrolnitrin. Seed treatment with the antibiotic also increased seedling survival.

17.6 ROOT AND STEM ROT OF JUTE

Root and stem rot of jute caused by *Macrophomina phaseolina* is the most prevalent disease of jute in Assam, Bihar, West Bengal, Orissa, and the eastern districts of Uttar Pradesh.

1. Symptoms

The disease occurs during almost all stages of plant growth. On young plants, infection starts at the collar region, and the cotyledons are also infected. The affected seedling shows damping-off symptoms. Dark, thin streaks develop at the collar region which rapidly increase under humid conditions and rapidly kill the plant. On older plants, the leaves are attacked. Lesions appear on the margins and apex of the leaves which fall down leaving the jute plants with bare branches. The stems rot and the roots get shredded. Pycnidia are produced on the infected stem and the rotting tissue. When the pathogen infects the capsules, they become black, and the seeds become discoloured and small. The sclerotia of the pathogen are often seen on the capsules.

2. Causal Organism

The pathogen responsible for the root rot of jute is *Macrophomina phaseolina* (Tassi) Goid. The sclerotial stage of the pathogen is *Rhizoctonia bataticola* (Taub.) Butler.

3. Disease Cycle

The pathogen is both soil and seed-borne. Rajan and Patel (1946) established the seed transmission of the pathogen. The fungus is capable of growing saprophytically on dead organic matter in the soil. It has a very wide host range including cotton, tobacco, sesamum, potato, eggplant, mulberry, and so on.

4. Disease Management

The macronutrients play an important role in predisposing the plants to various diseases. It has been found that the application of potash was effective in checking the stem-rot incidence in Formosa. The application of nitrogen alone incited the incidence of stem rot and root rot. It has been observed that low K_2O/CaO ratio in plants, potash deficiency in soil, and application of nitrogen alone (about 60 kg/ha) predispose the crop to severe stem-rot incidences, and it was found that stems rotting on jute caused by *Macrophomina phaseolina* increased with increasing levels of nitrogen and was maximum at 80 kgN/ha.

Since the disease is seed-borne (both externally as well as internally), treatment of seeds with suitable systemic fungicides is recommended. Seed treatment with Bavistin at the rate of 2 g/kg seed has been recommended.

Fungi like *Aspergillus aculeatus*, *Trichoderma lignorum*, *T.viride*, *Streptomyces albus*, *Bacillus cereus*, and *B. subtilis* have been found to inhibit growth of *M. phaseolina in vitro*. Cultural practices like more number of irrigation to reduce temperature, and slower growth of fungus may be helpful. *M. phaseolina* is inhibited *in vitro* by Aureofungin at concentration below 1.5 μg/ml.

The Jute Research Station, Kendrapara (Orissa) has reported five Capsularis Jutes, namely Halmhera, C-58-9433, Bangkph, JRC-9826, and Patchy albino, and one Olitorius Jute (JRO-333I) to be field resistant. There is no absolute resistance available in existing jute germplasm. There was a positive decrease in the incidence of disease due to the application of micronutrients in combination with nitrogen, phosphorus, and potash or alone. It was observed that the incidence of stem rot increased in plots where no micronutrients or NPK were added. The application of zinc, iron, and boron has shown superiority in reducing the incidence of stem rot.

17.7 LEAF SPOTS AND LEAF BLIGHTS

'Blight' implies sudden and extensive damage to leaves, but in many leaf diseases, the area of necrosis is limited which gives the appearance of spot. These spots may be big or small, depending upon the organism responsible for them, and the environmental conditions. Leaf spots can also be due to unfavourable water relationships or temperatures, mineral deficiencies or excesses, or by insects. Viruses can also cause leaf spots, particularly "ring spots".

Species of *Erwinia, Pseudomonas,* and *Xanthomonas* may induce leaf spots and blights of plants. Fungal leaf infections are of two types, depending upon the activity of the pathogen. Some leaf-infecting pathogens are actually saprophytic, killing suscept cells by enzymatic, and toxic action well in advance of themselves, while others are truly parasitic, obtaining food from living tissues, which die as a result of attack by the pathogens.

Fungal leaf spots are caused by a large number of fungi mainly belonging to Fungi Imperfecti, such as species of *Gloeosporium, Colletotrichum, Sphaceloma, Phomopsis, Phoma, Phyllosticta, Ascochyta, Diplodia, Botryodiplodia, Septoria, Alternaria, Helminthosporium, Cercospora,* etc. Leaf spots and leaf blights resemble each other considerably in the expression of symptoms as well as mechanism of pathogenesis. In the leaf blight, a larger area of leaf or the whole leaf is involved. In certain cases, it may involve the whole plant. The limited development of lesions caused by many leaf-spot pathogens has been attributed to several causes, hypersensitivity being a common cause. Akai (1951) showed the formation of abscission cells around necrotic areas in peach-leaf spots induced by *Xanthomonas campestris* pv. *pruni.* Infection by *Cladosporium carpophilum* leads to perforation of spots in the leaves of *Prunus amygdalus.* Formation of an effective cork barrier around the lesion prevents its further development, or limitation of the lesion may mark the point at which the defensive reactions of the parasitized tissue are able to halt the pathogen. In unspecialized leaf-spot fungal pathogens, the fungus generally produces its reproductive structures in the center of the lesion in the dead areas, while in specialized ones, such as *Phytophthora infestans,* the sporangia are produced mostly at the margins of the necrotic spots. Bacterial and viral leaf-spot pathogens have been discussed under bacterial and viral diseases. Some of the fungal leaf spots or blights have been discussed under downy mildews (Oomycota), while some others are discussed in this chapter.

17.8 EARLY BLIGHT OF POTATO

The disease is common wherever potato is grown, and under favorable conditions may cause serious damage to the crop. In those areas where both late and early blight diseases occur, it is the early blight that comes in the season earlier than the late blight. While late blight epidemics occur in cooler areas, the early blight is free from inhibitions caused by weather conditions, and occurs in cool as well as warm areas. The disease also occurs on tomato, chillies, eggplant, and related wild hosts. Among the fungal diseases, early blight is the most destructive.

1. Symptoms

The disease first become visible as small, isolated, scattered, pale brown spots on the leaflets. These spots become covered with a deep growth of the fungus. The lowest leaves are attacked first, and the disease progresses upwards. In the necrotic spots, concentric rings appear on the older leaves, and darkened areas on the stem. This gives a target boardlike effect (Fig. 17.4 A). There is usually a narrow chlorotic zone around the spots, which fades into normal green, and increases with an increase in the size of the spots. There may be a few spots or a large number of them may appear, occupying major portions of the leaf. When the leaf lesions involve larger veins, chlorosis commonly extends well beyond the necrotic spots. This is due to the toxin-alternaric acid produced by fungus, and translocated through the veins. In dry weather, the spots become hard and the leaves curl. In humid weather, the affected area increases, and large rotting patches may appear. In severe attacks, leaves shrivel and fall down. In severe infections, all the aerial portions of the plant have a blighted appearance.

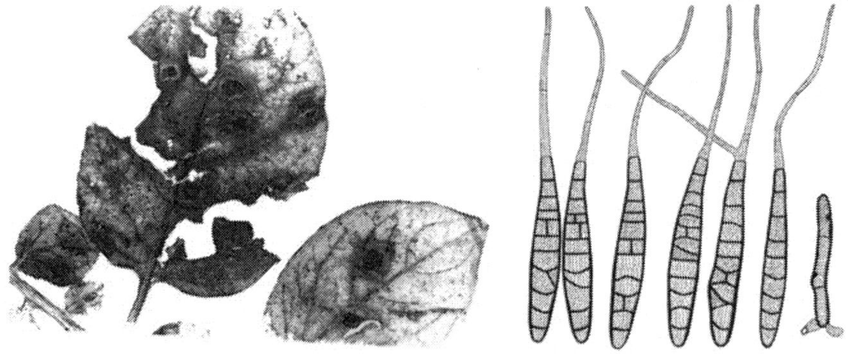

(a) (b)

Fig. 17.4 Early blight of potato caused by *Alternaria solani*: (a) Symptoms (b) Conidia with long beaks

2. Causal Organism

The causal organism for the early blight of potatoes is *Alternaria solani* (Ell. and Martin) Jones and Grout (Fig. 17.4B). There are both transverse and longitudinal septa in the mature conidia. *Alternaria solani* produces a nonspecific toxin known as alternaric acid. The role of this acid in disease is doubtful.

3. Disease Cycle

The pathogen can survive in soil, and plant debris particularly in temperate climate. The primary infection may be through tubers. The mycelium of the fungus remains dormant in dry infected leaves for a year or more.

4. Disease Management

Since the disease is soil-borne, crop rotation is helpful. Dead haulms should be raked together and burnt. Use only disease-free tubers for raising the crop. The crop must be given only balanced doses of fertilizers, especially nitrogen as the incidence, and severity of early blight is generally high in the crop receiving imbalanced doses of NPK, particularly low nitrogen with significant yield loses. The crops, if sprayed with 1% urea at 45 days, and given one more subsequent spray after 8-10 days easily escape the severe onslaught of early blight. Copper oxychloride (0.30%) sprays have been recommended for the control of early blight of potato. Fungicidal sprays, preferably with copper fungicides, are quite effective. Zineb, Brestan 60 or Dithane M-45, given at 15-day intervals, effectively controls the disease. Application of Difenoconazole (Score), a protectant, curative, and eradicant fungicide for the control of early blight has been recommended. Some of the moderately resistant varieties against early blight are Kufri Naveen, Kufri Sindhuri, and Kufri Jeevan.

17.9 LEAF BLIGHT OF WHEAT

It is a common and destructive disease especially in the northern parts of India. Although the disease was noticed as early as 1924, it did not cause much concern till 1966 when it became widespread. Many of the wheat varieties, which are resistant to rusts, have been found highly susceptible to this disease; the younger seedlings are not affected.

1. Symptoms

The disease first appears as small, discolored lesions which are irregularly scattered on the leaves. These spots become irregular in shape, and then increase in size, and appear brown to gray in color (Fig. 17.5). A light-yellow zone surrounds the spots. Several lesions coalesce to cover large areas of the leaf, and in severe infections the whole leaf dies. Black powdery mass of conidia are produced on the surface of the leaves.

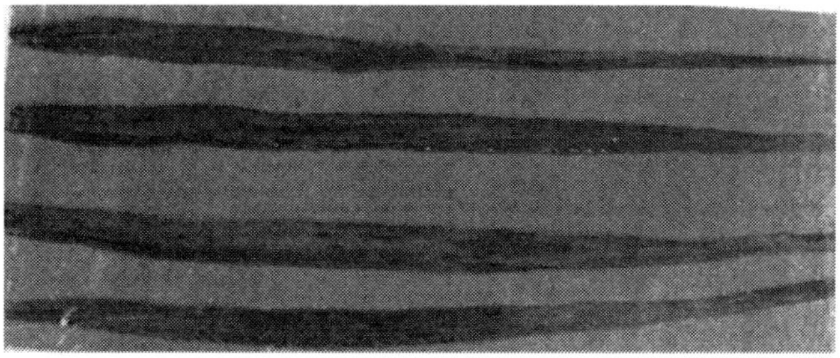

Fig. 17.5 Leaf blight of wheat caused by *Alternaria triticina*: Symptoms

2. Causal Organism

The disease it caused by the fungus *Alternaria triticina* Prasada and Prabhu. This was described by Prasada and Prabhu in 1962 for the first time.

3. Disease Cycle

The pathogen is both externally as well as internally seed-borne. It also survives in diseased plant debris in the soil. The possibility of infected seeds being a potential carrier of the disease was reported.

4. Disease Management

As the disease is deep-seated, seed-borne infection cannot be removed by ordinary organomercurial seed dressing (Prabhu and Prasada, 1967). Pre-soaking the seed in water for four hours followed by a ten-minute dip in hot water at 52°C gives satisfactory control. Since the pathogen can also survive on diseased plant debris in the soil, it is essential that suitable sanitary precautions should be observed. Sankhla *et al.* (1972) reported that Cuman was the most effective chemical for checking the vegetative growth of the pathogen, and reducing disease incidence. Dithane M-45 and Bisdithane could control the disease significantly at 1% level, and Miltox and Vitavax at 5% level. Singh and Singh (1971) have reported effective control by six sprayings of Dithane Z-78 or Dithane M-45, each at 0.25% concentration. Basant Ram and Joshi (1977) reported that Thiram, Dithane M-45, and Zineb proved effective in checking the disease as well as in reducing the loss in grain weight. Tilt 25 EC @ 500 ml/ha has been found very effective in reducing the blight incidence when applied as foliar spray at fortnightly intervals twice in the crop season (Goel and Sinha, 1996).

Some of the resistant varieties reported by many workers are NP4, NP52, NP200, NP809, Arnautka, E6160, K7340, HP1163, HD1941, Janak, M-134, K8904, K8908, and UP262.

17.10 BROWN LEAF SPOT (HELMINTHOSPORIOSIS) DISEASE OF RICE

Like blast, brown leaf spot (also known as sesame spot, helminthosporiosis, seedling, and leaf blight) of rice is a widespread disease in almost all the rice-growing areas of the world. In India, the disease is prevalent in all rice-growing areas, especially in heavy-monsoon areas in West Bengal, eastern part of UP, Assam, Tamil Nadu, and parts of Kerala. The disease is seed-borne, and rice seeds have been indiscriminately taken in the past from India to other countries without any precautions.

The disease is of great importance in several countries, and has been reported to cause enormous losses in the leaf-spotting phase when it can assume epiphytotic proportions. Two major epidemics have been recorded in India. The most notorius of these outbreaks was in Bengal in 1942 when loss in yield up to 90% was recorded. Nothing as devastating as the Bengal epiphytotic of 1942 has been recorded in plant pathological literature. The great Bengal famine has been discussed by Padmanabhan (1973).

1. Symptoms

All the aerial parts show symptoms of the disease; irregular, rectangular brown spots with a dark center appear on the coleoptiles, leaf blades, leaf sheaths, and glumes but are most commonly seen in the leaves. Spots first appear as minute brown dots, later becoming ellipsoidal or oval to circular, and when fully developed, have a light brown, gray centre, and a dark or reddish-brown margin (Fig. 17.6 A). Seedlings are often heavily attacked with numerous spots, about 2.5 mm in diameter; in such cases leaves may dry out and die. Badly affected nurseries can be very easily distinguished from a distance by their brownish scorched appearance, but seedlings are not usually killed. Spots also appear on the older leaves where they become larger than on the seedlings. These may be as much as 1 cm in length but are usually smaller. On the coleoptiles, the spots are brown and small. On the glumes, the disease is indicated by dark brown or black oval spots, or when the attack is more severe, the whole surface of the grain may be blackened, and the seeds may be shriveled and discolored. Thus, there are three types of infections, resulting in losses. The first is due to poor germination of infected seeds, the second is leaf infection resulting in the reduction of the effective leaf surface, and the third is the attack on the grains themselves.

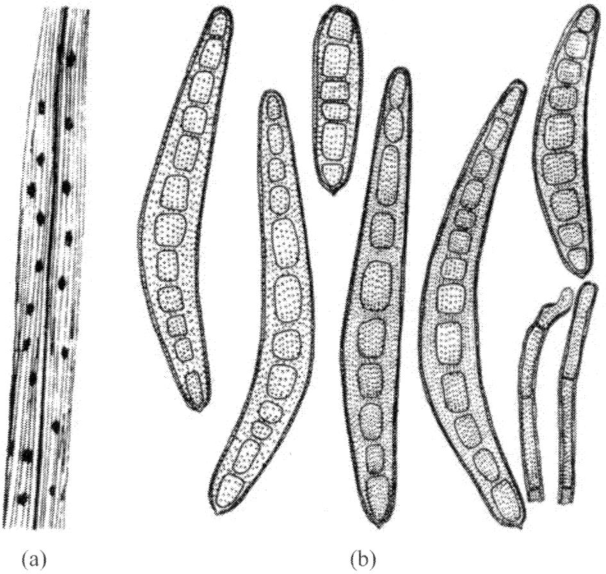

Fig. 17.6 Leaf spot of rice caused by *Drechslera oryzae*: (a) Symptoms (b) Conidiophores and conidia

2. Causal Organism

The pathogen responsible for the brown-spot disease of rice is *Drechslera oryzae* (Breda de Haan) Subramanian and Jain, earlier called *Helminthosporium oryzae* Breda de Haan [=*Cochliobolus miyabeanus* (Ito and Kuribay) Drechsler ex Dastur]. Conidiophores are short, erect, branched only at bases, short, segment, dark brown to olivaceous at the base, and somewhat paler at the growing tip. The points of attachment of successive conidia are marked by scars at regular intervals in the conidiophores. Conidia are olivaceous brown, slightly curved; tapering towards the rounded ends, and vary greatly both in size as well as in the number of septa (Fig. 17.6b).

The perfect stage of the fungus is *Cochliobolus miyabeanus*. The fungus produces two toxins, viz. Cochliobolin and Ergosterol, the former (Cochlibolin toxin) which is toxic to rice seedlings. The toxin also affects respiration of the plant by disturbing the equilibrium of the protoplasm. The fungus also produces proteolytic enzymes which are responsible for the physical and chemical breakdown of protein fragments of the cell walls.

3. Disease Cycle

The disease is soil- and seed-borne. The fungus may live for two to three years in infected plant parts, particularly the seeds, and the disease often starts from infected seeds.

4. Disease Management

(a) Cultural Methods

- After the harvesting, the stubbles should be burnt so that there is no soil-borne inoculum present in the soil.
- All the alternate hosts like *Setaria, Leersia, Echinocloa*, etc. that harbour the pathogen in off season should be removed.
- During irrigation, the water from infested fields should not pass into healthy fields.

- Leaching down of the important elements like K, Fe, and Mn during rains should be checked as leaching of these elements from the soil makes the plants more susceptible to the disease.
- Dry conditions of the soil during plant growth should be avoided. It has been observed that flooding the field when soil temperature is 20-28°C will prevent conidial germination.
- Severity of the disease decreases with increase in the potassium level of the soil during growth period of the plant.

(b) Chemical Control Since the disease is seed-borne in many regions, in such conditions the seeds should be disinfected with a suitable fungicide, such as Ceresan or Agrosan GN. In Japan, seed treatment with mercuric chloride, silver nitrate, copper sulfate, calcium hypochlorite, formaldehyde, and phenol had been found quite effective in controlling the seed-borne inoculum. Seed treatment may ward off the seedling blight phase. The disease can be controlled by spraying copper fungicides or Dithane M-45 (0.3%) at regular intervals. Several other fungicides like Chlorothalonil, Iprodione, Vinclozolin, and Maneb can be applied at the onset of the disease, and then repeatedly at 10–15 days intervals; this helps in controlling the disease considerably. Some antibiotics like Aureofungin (200 ppm) in combination with copper sulfate gave good control of the disease. Apart from antibiotics, some sulfa drugs like sulfanilamide, and antibiotics like Nystatin and Griseofulvin have also been used for seed treatment.

Hot-water treatment at 55°C for 10 minutes gives better control than chemical seed treatment.

Secondary air-borne spread of the disease can be checked by chemical treatment. Two to three sprays with Bordeaux mixture (5:5:50) or Dithane Z-78 (0.2%), Hinosan (0.1%), Aureofungin (25 ppm), Blitox (0.3%), Benlate (0.2%) had been found effective in controlling the disease. Some systemic fungicides like Brestanol, Brestan, Tricyclazole (0.08%) have been found to give good control of the disease.

5. Resistant Varieties

Growing of resistant varieties is the best way to control the disease under fungicidal umbrella. Some varieties have been bred in different centers in the country. Varieties Padma and IR-24 are comparatively resistant, and should be grown in areas where epiphytotics occur.

17.11 TIKKA DISEASE OF GROUNDNUTS

One or both of the imperfect stages, *Cercospora arachidicola* and *Cercosporidium personatum,* of these two species occur in all the main groundnut-growing areas of the world depending on the type of the groundnuts grown, and the weather conditions. The perfect stages, *Mycosphaerella arachidicola* and *M. berkeleyii,* which play an important part in the survival of the pathogen, and in establishing primary infection in the North American crop, are not found in other groundnut-growing regions.

Losses from the disease have been reported from 20 to 50% but can be up to 70% if the tikka disease is associated with rust (*Puccinia arachidis*).

1. Symptoms

Symptoms appear when the plants are one or two months old. Due to excessive spotting on the leaves, there is gradual weakening of the foliage, which results in defoliation. Consequently, fewer and smaller nuts are formed. According to Sundaram (1965), severe intensity of infection leads up to 22% reduction in yield. Woodruff (1933) designated the disease caused by *Cercospora arachidicola* as early spot, and that caused by *Cercosporidium personatum* as late spot, based on the relative time of their appearance in the USA. The first symptom of both leaf spots is the appearance of pale areas on the upper surface of older leaves. As the lesions develop, the two species can be distinguished. *Cercospora arachidicola* forms circular or irregular reddish brown to brown lesions, which are bigger on the lower surface. The lesions are approximately 1 mm–10 mm in diameter, and are surrounded on the upper-leaf surface by a narrow yellow halo. Haloes are indistinct on the lower surface. Their presence or absence may be due to varietal or nutritional effects. *Cercosporidium personatum* forms darker brown to black leaf spots which are smaller (1–6 mm), almost circular, and with a less diffuse margin than those of *Cercospora arachidicola.* They rarely have haloes, and are abundant on leaves, stipules, petioles, stems, and pegs. Ramakrishnan and Appa Rao (1968) studied the symptoms caused by two species on groundnut, and they found that neither the shape of the lesion nor the period of appearance of the halo is a reliable criterion for distinguishing the spots produced by *Cercosporidium personatum* and *Cercospora arachidicola.* A more reliable criterion which serves to easily identify the spots produced by the two pathogens is the colour of the spot in the lower surface which is light brown in the case of *Cercospora arachidicola* and carbon black in the case of *Cercosporidium personatum* (Fig. 17.7).

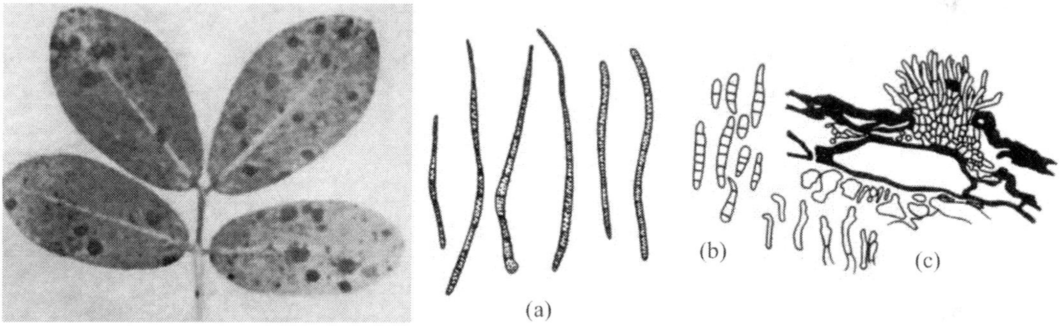

Fig. 17.7 Tikka disease of groundnut caused by *Cercospora arachidicola* and *Cercosporidium personatum*: symptoms (a) Conidia of *C. arachidicola* (b) conidia of *C. personatum* (c) conidiophores of *Cercosporidium personatum*

2. Causal Organisms

The causal organisms for the tikka disease of groundnut are *Cercosporidium personatum* (Berk. and Curt.) Deighton, earlier known as *Cercospora personata* (perfect stage *Mycosphaerella berkeleyi),* and *Cercospora arachidicola* perfect stage (*Mycosphaerella arachidicola* W A Zenkins) (Fig. 17.7 a, b, c). In 1983, J A von Arx proposed the new name *Phaeoisariopsis personata* (Berk. and Curt.) V. Arx on the basis of the formation of small synnemata or long conidiophores, and less thickened, darkened but bulging scars to the pathogen *Cercosporidium personatum.* Species of *Cercospora* produce a phytotoxin, cercosporin. Rajgopal and Vittal Rai (1983, 1984) have isolated two water-soluble, nonspecific phytotoxic glycopeptides from *in vitro*, and *in vivo* conditions. The toxins inhibit germination of seeds of several crop plants.

3. Disease Cycle

Both the disease caused by *Cercospora arachidicola,* and *Cercosporidium personatum* are disseminated by wind which blows the spores from leaf to leaf. Attacks on the new crop are apparently initiated by the spores, which lie in the soil. The spores remain viable in the soil for a long time, and attack the succeeding crop, as the conditions become favourable. There is evidence that they are seedborne, especially in the shells. The disease is perpetuated through conidia lying in the soil on diseased plant debris, and through conidia carried in the shell.

4. Disease Management

(a) Cultural Methods Crop residues should be removed as they may harbor infection, and provide a congenial medium for the growth of other diseases such as *Sclerotium rolfsii.* Burning and burying crops together with crop rotation is recommended for the control of the disease. Early planting helps to reduce infection. Mineral deficiencies such as deficiency of magnesium should be corrected by soil or spray application as these deficiencies may favor the development of the disease. Change in the date of sowing can help in avoiding the incidence of the disease. Early maturing varieties usually escape the losses caused by the disease. Varieties that have more foliage are less liable to the damage than erect and less foliaged varieties.

(b) Chemical Control Chemical spraying or dusting is used in some areas as a routine practice. Some of the protectant fungicides used are Bordeaux mixture (4:4:50), Dithane Z-78 (0.2%), Dithane M-45 (0.2%), Fycol 8E, Cosan, Brestan (0.1%), and copper sulfate mixture (15-25 kg/ha). Apart from these, some systemic fungicides like Benomyl, Bavistin, Brestanol, and Cercobin are highly effective against Cercospora leaf spot of groundnut. Dubey *et al.* (1995) reported that Bavistin (Cabendazin) is better than copper fungicides in controlling the disease. Among protectant fungicides, Du-Ter gave better result than dithiocarbamates. Other recent fungicides like Tilt 25EC (Propiconazole), SAN-619 (Cyproconazole), both at the rate of 1.2 g/lit., and Hexaconazole (0.2%) are reported to be effective against leaf spot of groundnut. Finally, breeding of resistant varieties forms an important control measure.

17.12 ANTHRACNOSES

The term 'anthracnose' literally means *like coal*, and was first used by Fabre and Dunal to describe a disease of grapes in which blackening of tissue was a characteristic feature. The term is now commonly used for plant diseases characterised by black lesions, usually sunken, caused by certain imperfect fungi that produce conidia in acervuli, that is *Colletotrichum, Gloeosporium,* and some closely related *Sphaceloma* spp.

A feature that distinguishes the fungi is that their conidia are borne on acervuli, which are erumpent, cushionlike masses of conidiophores. The conidia are hyaline, one-celled, ovoid to oblong. There are three important genera—*Gloeosporium Colletotrichum,* and *Sphaceloma.* The distinction between *Gloeosporium,* and *Colletotrichum* is that whereas the latter has long sterile hairs or setae, *Gloeosporium* does not possess setate in the acervulus. However, this distinction is not clear because many *Colletotrichum* spp. either produce setae rarely or only under certain conditions (Frost, 1964). The distinction between these two genera and *Sphaceloma* is also confusing. The conidiophores arise from a stroma in *Sphaceloma,* whereas in *Gloeosporium* and *Colletotrichum,* they do not. However, in natural infections, gradation between the stromatic, and nonstromatic type is not uncommon.

17.13 RED ROT OF SUGAR CANE

Red rot was first described in Java 1893 under the name red smut, and it soon became a major cause of the decline of several varieties of sugar cane in the USA, Australia, India, and Hawaii. In 1906, Butler reported that the disease cause severe damage in certain cane varieties in India, and renamed it as red rot. It was later reported by Butler and Khan (1913) at Pusa. Epiphytotics of red rot disease have been reported in 1939 and 1942, when a virulent form appeared in North Bihar, and the Eastern United Provinces of India (now Uttar Pradesh) that threatened the very existence of the sugar industry in India.

1. Symptoms

It is difficult to recognize the disease in its early stages in the field. The first symptoms of the disease are seen after the rainy season when the plant stops growing, and sucrose starts forming. The first external symptom of the disease is that the upper leaves of a shoot, usually the one approaching maturity, begin to lose color, and droop slightly. The entire tip withers, and the withering progresses down the margins. At this stage, there is no visible alteration in the cane stem, except that the withered and green shoots may occur in the same stool. This condition may be assumed to be due to drought. In later stages, the canes themselves show the effects. They become shriveled, the rind shrinks and becomes longitudinally wrinkled. On the midrib of the leaves, infection originates as a dark reddish area which elongates rapidly, forming blood-red lesions whose margin becomes darker (Fig. 17.8). The center becomes straw coloured in older lesions. When the fructifications of the pathogen are formed, the lesions are covered with a powdery mass of conidia.

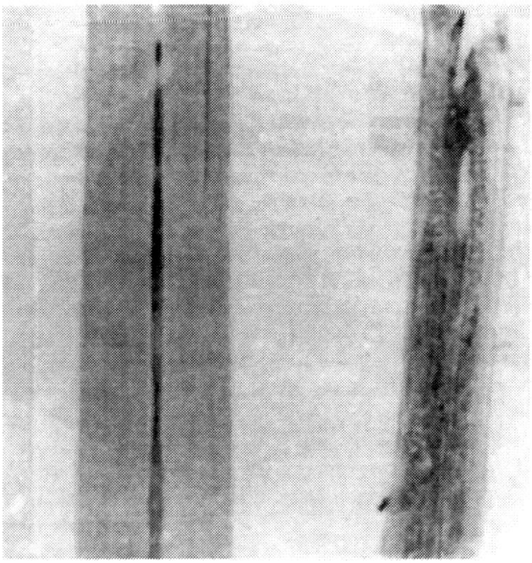

Fig. 17.8 Red rot of sugarcane (*Colletotrichum falcatum*): Spots on the midrib, and reddening of the pith with white transverse bands

If the affected canes are split open, the tissue of the internodes which are normally white or yellowish white will be found to be longitudinally reddened in one or more internodes (Fig. 17.8), usually towards the base. The reddening is most conspicuous in the vascular bundles, and progresses towards the pith. The red colour is interrupted transversely, and this symptom is very characteristic of the disease. It may be emphasized that in sugar cane, any injury leads to the reddening of tissues, and the characteristic cross bars in the reddened areas give an indication that the disease is due to red rot pathogen.

2. Causal Organism

The pathogen responsible for red rot is *Colletotrichum falcatum* Went. Its teleomorph is *Glomerella tucumanensis* (Speg.) von Arx and Muller. The pathogen produces its conidia from conidiophores arranged in acervuli. These appear as minute, black, dotlike bodies on the surface of the host under favorable environmental conditions. Acervuli are rounded or elongated and 70–300 μm across. Brown-colored setae are interspersed between conidiophores, and are about 150 μm long, 4–6 μm wide, 1–4 septate, and swollen at the base with a tapered to rounded tip. Conidia are borne on small conidiophores, and are falcate (sickle-shaped), hyaline, and aseptate; Conidia are 20–38 × 5–7 μm in size, and possess a large oil globule in the center.

3. Disease Cycle

It has been reported that the pathogen is capable of growing saprophytically in soil and producing acervuli. The fungus, however, does not survive in the active stage beyond 3–4 months. The pathogen also produced chlamydospores and these can survive for a long period in the soil. Diseased seed setts are the chief means of survival, and spread of the pathogen. When such diseased seed setts are sown, they give rise to infected shoots. When such infected shoots appear in the field, secondary infection is caused by conidia, which are produced in acervuli, and transmitted through insects, wind, and water.

4. Disease Management

(a) Cultural Methods Rotation of crops after two to three years is also useful. Field sanitation is one of the important means of checking inoculum build-up. In the field, fallen diseased leaves should be collected and burnt. Ratooning should be discouraged. Cultivation of crop in low-lying areas should be avoided, and finally same variety should not be cultivated for several years in a particular field. With a view to eliminating the inoculum present in the seed material, heat therapy is practiced. Divergent views have been put forward on the success of this therapy. Good control of the disease was claimed by Joshi (1954) when setts were treated with hot water at 52°C for 18 minutes. Of late, aerated steam and moist hot air have been observed to be effective in eradicating red-rot inoculum for seed material. By aerated steam, canes are heated at 52°C for 4–5 hours, and by hot air at 54°C for 2 hours.

(b) Chemical Control Seed treatment (seed setts) with organo-mercurials such as Aretan, Agallol (0.25%) for 5–10 minute dips helps only in eradication of superficial inoculum, and not deep-seated mycelium. Waraitch (1983) and Singh, K., *et al.* (1984) reported that Bavistin (0.5%) and Thiram (0.1%) could reduce the incidence of infection. Apart from chemical treatment, to eliminate seed (setts) borne infection, hot air treatment is recommended by various workers.

5. Resistant Varieties

The best means to control the disease is to use resistant varieties. *Saccharum spontaneum* is the chief source of resistance, and several disease-resistant varieties have been bred at Coimbatore from time to time. Some of the resistant varieties recommended are Co244, 62399, 62101, 1336, 1261, 1148, 1007, 975, 951, 846, 568, 562, 561, 513, 508, 475, 449, 393, 385, 370, 356, 349, 301, CoK32, 30, CoS510, 109, 443, BO32, 24, 3, 7, 10, 11, and 22.

17.14 GRAY BLIGHT OF TEA

Gray blight disease of tea has been reported from all tea-growing countries of the world.

1. Symptoms

The small, oval, pale yellow-green spots first appear on young leaves. Often the spots are surrounded by a narrow yellow zone. As the spots grow, brown or gray concentric rings with scattered tiny black dots become visible, and eventually the dried leaves fall, leading to defoliation. The entry of the fungus is associated with either seen scorch injury or scorch caused by chemical sprays. The typical gray patches are either seen at the tip, margin, or the centre of the leaf. Leaves of any age can be affected.

2. Causal Organism

It is caused by *Pestalotiopsis theae*. Rain splash transports the spores from one plant to another.

3. Management

Six systemic fungicides—Folicur, Calixin 80EC, Azoxystrobin, Contaf 5EC (Hexaconazole), Procure 30 WP (Triflumizole), and Baycor 25 WP (Betertanal)—could suppress the pathogen completely at 150 ppm. Neem-based botanicals (Nimbicidine) at 5% concentration were quite effective in controlling the disease. Among the cultural practices, adequate spacing of tea bushes to permit air circulation and reduced humidity are effective in reducing the disease incidence.

It has been reported that beneficial biocontrol strains, *Pseudomonas* and *Bacillus* sp., showed higher antagonistic effect against the gray blight pathogen *in vitro*. This indicated that they were capable of producing an array of lytic enzymes which is the main mechanism of bacterial biocontrol agents. Moreover, the lytic enzymes would suppress the fungal growth. These bacterial strains proved tolerant with selected fungicides (Hexaconazole, Carbendazim) *in vitro* (Vidhya Pallavi *et al*., 2012).

17.15 SOME OTHER DISEASES CAUSED BY DEUTEROMYCETES

Wilt of pigeon pea	*Fusarium udum*	Wilting of seedlings and adult plants which is characterized by sudden yellowing, withering, and drying of leaves; base of the stem is blackened.	Soil-borne	Crop rotations, hot-weather cultivation with deep ploughing, soil solarization, mixed cropping are effective cultural practices. Green manuring increase the antagonistic population in rhizosphere, and produce bulbiformin by *Bacillus subtilis* that inhibited the growth of *F. udum*. ICP 4769, 7118, 7182, 8862, and 9168 showed resistance.
Wilt of linseed	*Fusarium oxysporum* f.sp.*lini*	Plants become yellowish before they wilt, roots of young seedlings may rot, and plants may drop off.	Soil-borne	Crop rotations, hot-weather cultivation with deep ploughing, soil solarization, mixed cropping are effective cultural practices. Zinc application is reported to reduce fusaric acid production by pathogen. Some wilt resistant lines are NP 12, 21, 124, RPSB, PR80, B5128.
Panama disease of banana	*Fusarium oxysporum* f.sp. *cubense*	Sudden wilting of the plants, followed by unfurling; leaf goes yellow and dies; pseudostem splits, and the vascular symptoms extend up the pseudostem.	Soil-borne	Flooding the soil, soil fumigation, Carbendazim (2%) and Bavistin (0.1%) are effective control measures. Resistant varieties are Cavendish, Poovan, Moongal, Peyladen, and Rajabale.
Black scurf of potato	*Rhizoctonia solani* (perfect stage *Thanatephorus cucumeris)*	In stem canker, growing tip of plant may be killed but in black scurf of tubers, black sclerotial bodies are formed on diseased tubers, and infected plants are killed, and the tubers start rotting in the field.	Soil- as well as tuber-borne	Seed treatment with organo-mercurials (Agallol), soil application with Brassicol, soil solarization is effective. *Trichoderma viride* is a biocontrol agent.
Stripe disease of barley	*Drechslera graminea* earlier called *Helminthosporium gramineum* (perfect stage *Pyrenophora graminea*)	Long parallel, brownish streaks on leaf sheath, and leaves, and finally the tissues dry out, and leaf blade shredded.	Seed-borne	Field sanitation along with seed treatment with organo-mercurials is quite effective. Some resistant varieties are Morocco, Algerian, K12, CN 292, VL10, and HBL107.

(*Contd.*)

(*Contd.*)

Ripe fruit rot, and die-back of chillies	*Colletotrichum capsici*	**Fruit rot:** Only ripe fruits are affected; black, circular spot, generally sharply defined or at times diffused, diseased fruits lose red color, and numerous black colored acervuli are produced. **Die Back:** The plant dies back as the attack spreads downwards, as the disease progresses; the infected part of the stem takes on an enamel white colour, and the white of the diseased portion is punctuated by scattered black minute elevations called acervuli of the pathogen.	Soil- and seed-borne	Seeds should be taken from healthy fruits; Thiram, Brassicol, and Bisdithane eliminate the seed-borne inoculum. Zineb and Mancozeb are effective against die-back, and fruit rot. Bengal green, Lorai, Perennial are some resistant varieties.
Mango anthracnose	*Colletotrichum gloeosporioides* (perfect stage *Glomerella cingulata*)	On the leaves—dark brown necrotic areas, on the twigs—black necrotic patches, on the main and secondary rachis of the panicle, small dark spots, and black spots on the young fruits. The affected fruits rot during storage.	Pathogen survives in diseased twigs and fallen leaves	Pruning, burning of diseased parts, and overcrowding should be avoided. Bordeaux mixture, Zineb, Captan, Bavistin are quite effective in controlling the disease.
Anthracnose of grapes	*Gloeosporium ampelophagum* (Perfect stage *Elsinoe ampelina*)	Infected young shoot growth is arrested, and dry up, spots are formed on the petiole and fruits, infection on berries cause maximum damage to yield.	Pathogen is deep seated in the infected parts of the plant.	Pruning of diseased parts, Blitox, Ziram, Aureofungin, Bavistin, Topsin-M kills the deep-seated mycelium in the shoots.
Black spot of crucifers	*Alternaria brassicicola*	Small, dark coloured concentric rings in the leaves.	Seed-borne	Hot-water treatment, seed treatment with Arasan or Spergon, Iprodione, Fenpropimorph is effective. *Streptomyces arabicus*, *Trichoderma harzianum* act as biocontrol agents. Mustard cultivars such as RW 351, Pusa Kranti, and Seeta have been found resistant.
Purple blotch of onion	*Alternaria porri*	Small, water-soaked purplish colored area develops on the leaf or the seed stalk, and immediately turns brown, bulbs of affected plants show decay.	Over-seasons in crop debris, and symptomless bulbs	Long crop rotation, well-drained fields, deep ploughing, Mancozeb, Difolatan, Zineb, Daconil gave maximum yield. *Aureobasidium pullulans*, *Sporobolomyces roseum*, and *Cladosporium herbarum* reduce the infection of onion leaves.

(*Contd.*)

(*Contd.*)

Wilt disease of sugar cane	*Cephalosporium sacchari*	Stunting of plant growth followed by yellowing, and/or withering of the plant, brown patches on the rind, and gradually the pith becomes hollow and sometimes discoloured.	Soil- or seed-borne	Destruction of crop debris, use of healthy setts, proper maintenance of soil, 0.1 % Agallol treatment to setts reduce wilt incidence. Many species of *Streptomyces, Bacillus, Trichoderma,* and *Pseudomonas* are reported to suppress pathogen. Some resistant varieties are Co87010, Co91010, Co92003, Co92023, etc.
Verticillium wilt of tomato and eggplant	*Verticillium albo-atrum* (for potato and tomato wilt) *V. dahliae* (for tomato, eggplant, and pepper)	On tomato, yellow leaves, premature fall of leaves, necrosis of leaf tissues may lead to wilting of the plant, and reduction in size of fruits. On eggplant, unilateral wilting, chlorosis, and defoliation are seen.	Soil-borne	Soil fumigation with Metham sodium, and Chloropicrin+ Methyl bromide, soil solarization, crop rotation with rice, hot-weather cultivation is reported to reduce soil-borne inoculum. Biological control of *V. dahilae* with *Talaromyces flavus*. *Trichoderma*, *Gliocladium*, *Mortirella*, *Glomus mosseae* suppresses the pathogen.
Fusarium wilt of tomato	*Fusarium oxysporum* f.sp. *lycopersici*	Main symptoms are clearing of veinlets, drooping of the petioles, and finally entire plant wilts. Various toxins (Fusaric acid and Lycomarasmin) are also produced by the fungus which helps in development of wilt symptoms.	Soil-borne, and seed-borne	Seed treatment with Carbendazim (2.5 g/kg seed) is recommended for seed-borne inoculum. Deep ploughing in summer, burning of diseased crop debris, long crop rotation are helpful. Biological control by nonpathogenic strains of *Fusarium oxysporum*, *Trichoderma* spp, *Gliocladium virens*, *Penicillium oxalicum*, *Pseudomonas fluorescens*, *Burkholderia cepacia* are effective.
Anthracnose of bean	*Colletotrichum lindemuthianum* (perfect stage *Glomerella cingulata* f.sp. *phaseoli)*	On cotyledons, brown to black spots, young plants suffer from necrosis of veins, leaves show small, angular brown spots, on pods black sunken with gray central area, and on stem, the spots are eye shaped.	Soil and seed-borne	Control strategies include disease-free seed, seed treatment (Thiram 75 WP, Carbendazim, Mancozeb, Triadimenol, Carbendazim+ Thiram), foliar spray of fungicides, crop rotation, cultivation of race-specific resistant varieties.
Charcoal rot of soybean and other crops	*Macrophomina phaseolina.* It's synonyms are *M. phaseoli, Rhizoctonia solani, Sclerotium bataticola*	Light brown discolouration in the tap roots and parts of stem, in advanced stages, leaves turn yellow and wilt.	Soil and seed-borne	Avoid crowding of seedling, proper nutrition of the crop, long crop rotation, organic, and amendments with neem seed cake significantly reduce charcoal rot of cowpea. *Trichoderma viride*, *T. harzianum*, fluorescent *Pseudomonas*, *Paecilomyces lilacinus* act as biocontrol agents.

(*Contd.*)

(*Contd.*)

Sheath blight of rice	*Rhizoctonia solani*	Lesions on the leaf sheath, and culm are ellipsoidal or ovoid with centre of the spot becoming white with brown or purplish margins, and finally killing the entire leaves.	Seed-borne and some wild plants act as hosts in off season	Rotation with noncereal crop and silicon fertilization is a promising method of disease control. Application of PCNB, Bavistin, Topsin-M, Propiconazole, Ediphenphos, and Iprodione are effective fungicides. Seed treatment with *Trichoderma harzianum*, *T. viride* are quite effective in disease control.

REVIEW QUESTIONS

1. How do wilts differ from root rots? Discuss in brief the physiology of hydromycotic wilting in plants.
2. Are toxins alone involved in vascular wilts? Discuss in brief.
3. How does browning occur in vascular wilts?
4. Do enzymes play a role in vascular wilts? Discuss critically.
5. Are vascular wilt pathogens primitive or advanced? Give reasons.
6. Name some of the important genera involved in vascular wilts.
7. What is vascular occlusion?
8. How will you control vascular wilts?
9. American cottons are resistant to wilt when grown in India and Indian cotton varieties are resistant to wilt when grown in America. Why is this so?
10. Why is *Rhizoctonia* more common in sandy soils and Fusarial wilt in black cotton soils in India?
11. What is soil solarization? How is this used in controlling stem rot of paddy?
12. Name any *Helminthosporium* disease, which is internally seed-borne.
13. How will you differentiate the following:
    (i) *Alternaria* and *Helminthosporium*
    (ii) *Cercospora* and *Pyricularia*
    (iii) *Colletotrichum* and *Gloeosporium*
14. How will you control the following?
    (i) Leaf blight of wheat
    (ii) Tikka disease of groundnut
    (iii) Anthracnose of grapes
    (iv) Ripe fruit rot and die back of chillies
15. What are anthracnoses?

MULTIPLE-CHOICE QUESTIONS

1. Which of the following diseases is caused by members of Deuteromycetes?
   (i) Wilts (ii) Powdery mildew
   (iii) Downy mildew (iv) Rust

2. Which of the following symptoms are caused by anamorphic fungi during vascular wilting?
   (i) Vascular discoloration (ii) Vein clearing
   (iii) Wilting (iv) All of the above
3. Which of the following biochemical changes occur as a result of wilting in plants?
   (i) Water content decreases (ii) Nutrient uptake increases
   (iii) Both (i) and (ii) (iv) None of these
4. Vascular dysfunction in the wilt diseases is due to
   (i) collapse of vessels (ii) formation of tyloses
   (iii) physical blockage in xylem vessels (iv) all of these
5. Which of the following pathogen is responsible for wilt diseases?
   (i) *Fusarium* sp. (ii) *Ceratocystis* sp.
   (iii) *Verticillium* (iv) all of these
6. Which of the following toxin is responsible for wilt diseases?
   (i) Fusaric acid (ii) Dehydrofusaric acid
   (iii) Lycomarasmin (iv) All of these
7. Which of the following agronomic practices can reduce wilt of cotton to some extent?
   (i) Manipulating sowing date (i) Seed treatment with Bavistin
   (iii) Application of potash (iv) Both (i) and (iii)
8. Which of the following toxins is produced by *Fusarium graminearum* causing FHB of wheat?
   (i) Deoxynivalenol (ii) Nivalenol
   (iii) Both (i) and (ii) (iv) Alternaric acid
9. Root rot disease is caused by which of the following pathogen/s?
   (i) *Rhizoctonia* (ii) *Sclerotium*
   (iii) *Ophiobolus* (iv) All of these
10. The perfect stage of *Rhizoctonia bataticola* is
    (i) *Thanatephorus cucumeris* (ii) *Gibberella fujikuroi*
    (iii) *Alternaria solani* (iv) *Phytophthora infestans*
11. The pathogen responsible for root and stem rot of jute is
    (i) *Macrophomina phaseolina* (ii) *Fusarium oxysporum*
    (iii) *Fusarium culmorum* (iv) *Verticillium albo-altrum*
12. Leaf spots and leaf blights disease is caused by which of the following pathogen/s?
    (i) *Colletotrichum* sp. (ii) *Helminthosporium* sp.
    (iii) *Cercospora* sp. (iv) All of these
13. Alternaric acid produced by *Alternaria solani* is a
    (i) host-specific toxin (ii) nonspecific toxin
    (iii) Both (i) and (ii) (iv) an enzyme
14. Which of the following fungus is responsible for the Bengal famine in 1942?
    (i) *Helminthosporium oryzae* (ii) *Alternaria solani*
    (iii) *Rhizoctonia solani* (iv) All of these
15. Tikka disease of groundnut is caused by
    (i) *Cercospora arachidicola* (ii) *Cercosporidium personatum*
    (iii) *Puccinia arachidis* (iv) Both (i) and (ii)
16. The anthracnoses disease is generally caused by
    (i) *Colletotrichum* (ii) *Gloeosporium*
    (iii) *Cercospora* (iv) Both (i) and (ii)

17. The perfect stage of red rot of sugar cane caused by *Colletotrichum falcatum* is
    (i) *Glomerella tucumanensis*
    (ii) *Mycosphaerella berkeleyi*
    (iii) *M. Arachidicola*
    (iv) *Glomerella cingulata*
18. Black scurf of potato is caused by
    (i) *Rhizoctonia solani*
    (ii) *Helminthosporium graminearum*
    (iii) *Colletotrichum capsici*
    (iv) *Fusarium oxysporum*
19. Wilt disease of sugar cane is caused by
    (i) *Fusarium oxysporum* f. sp. *lycopersici*
    (ii) *Fusarium oxysporum* f. sp. *cubense*
    (iii) *Fusarium oxysporum* f. sp. *lini*
    (iv) *Cephalosporium sacchari*
20. Sheath blight of rice is caused by
    (i) *Helminthosporium oryzae*
    (ii) *Drechslera oryzae*
    (iii) *Rhizoctonia solani*
    (iv) *Sclerotium bataticola*

Answers

1. (i)	2. (iv)	3. (i)	4. (iv)	5. (iv)	6. (iv)	7. (iv)	8. (iii)	9. (iv)	10. (i)
11. (i)	12. (iv)	13. (ii)	14. (i)	15. (iv)	16. (iv)	17. (i)	18. (i)	19. (iv)	20. (iii)

18

DISEASES CAUSED BY OOMYCOTA

Damping off fungi, white rusts, and downy mildews all belong to the order Peronosporales. According to the 10th edition of the *Dictionary of the Fungi* as already mentioned earlier in Chapter 12, the Phylum or Division Oomycota of the Kingdom Chromista (Straminopila) has been divided into 8 orders: (1) Albuginales, (2) Leptomitales, (3) Myzocytiopsidales, (4) Olpidiopsidales, (5) Pernosporales, (6) Pythiales, (7) Rhipidiales, and (8) Saprolegniales. The genus *Pythium* has been separated from *Phytophthora*, the former has been placed in Pythiales, and *Phytophthora* has been placed in Peronosporaceae. However, in this edition of book, the authors have followed the old classification followed in the 9th edition of the *Dictionary of the Fungi*, till some other mycologists/taxonomists accept this revised classification. Many fungi in this group are parasites of higher plants, and may cause diseases of economic importance, such as late blight of potato, downy mildew of grapevine, and white rusts of crucifers. Some of them, particularly the members of the Peronosporaceae and Albuginaceae, appear to be biotrophs (obligate parasites), and do not survive saprophytically in soil or decaying vegetable matter.

Peronosporales are coenocytic mycelial fungi which are parasitic in plants or saprobic in water or soil. The cell walls are mainly made up of a glucan-cellulose complex in addition to protein containing hydroxyproline and lipid. The parasites form haustoria.

Asexual reproduction occurs by zoosporangia or conidia which vary greatly in shape. The sporangia possess one or few pores with or without a papilla, and produce biflagellate zoospores or a plasma, either in a vesicle or naked, or a germ tube. Sexual reproduction is oogamous.

18.1 KEY TO FAMILIES OF THE ORDER PERONOSPORALES

Obligate parasites of plants with unbranched, club-shaped sporangiophores. Each bearing a basipetal chain of deciduous sporangia in dense sub-epidermal clusters on the host, white or creamish sori; oogonial periplasm persistent, and conspicuous haustoria are knoblike *Albuginaceae. (Albugo)-A. candida* grows worldwide on crucifers, *A. tragopogonis* is common on Compositae; *A. bliti* on Amaranthaceae, and *A.ipomoeae-panduranae* on Convolvulaceae—***Albuginaceae***.

Strict parasites of plants, some are biotrophs with branched treelike sporangiophores or conidiophores of determinate growth, emerging singly or in tufts usually via stomata, producing sporangia or conidia singly at the branch tips, periplasm persistent, and conspicuous, haustoria varied usually branched—***Peronosporaceae***.

Non-obligate parasites or saprophytes, sporangiophores, or condiophores usually undifferentiated from the mycelium, branched, indeterminate, if determinate then the sporangiophore similar to the vegetative mycelium. Periplasm is a thin layer or absent, haustoria absent or branched—***Pythiaceae***.

18.1.1 Key to the Important Genera of Peronosporaceae

Sporangiophore unbranched, apices with short sterigmata bearing papillate sporangia, germinating by zoospores, oospores aplerotic—***Basidiophora***.

Sporangiophore 2-3 times dichotomously branched in the upper part, sporangia usually nonpapillate, germinating zoosperes or germ tube, oospore aplerotic—***Sclerospora***.

S. graminicola on sorghums, millets, maize, and grasses, *S. sacchari* and *S. spontanae* on sugar cane, *S. maydis, S. philippinensis* and *S. sorghi* on maize and *Sorghum*.

Sporangiophores dichotomously branched at acute angles, and taper to gracefully curved pointed tips on which sporangia are borne. Sporangia always germinate by germ tube—*Peronospora. P. parasitica* on crucifers, *P. destrustor* on leeks, and other alliums. *P. tabacina* on tobacco, *P. viciae* on peas, etc. *P. farinosa* on spinach.

Sporangiophores dichotomously branched at acute angles. The tips of the branches are expanded into saucer-shaped structures—***Bremia. B***. *lactucae* on lettuce.

18.1.2 Key to the Important Genera of Pythiaceae

Sporangiophores micronemeous (undifferentiated), Sporangia ovoid or obyriform or filamentous, protoplasm emerging almost or completely undifferentiated into a spherical vesicle usually at the tip of a short discharge tube. Zoospores formed in a vesicle, antheridia, always paragynous—***Pythium***.

P. ultimum, and *P. irregulare* (damping-off and root rots), *P. aphanidermatum* (common in warmer areas as root rot pathogen), *P. debaryanum* is rare.

Sporangiophores macronemous, or sympodially branched, sporangia pyriform or globose. Sporangia do not mature synchronously, zoospores differentiated inside the sporangia. Antheridia mostly amphigynous (in some paragynous), oogonia smooth or occasionally ornamented, oospore aplerotic, oospore germination direct by a germ tube—*Phytophthora.*

P. infestans (late blight of potato and tomato), *P. palmivora* (Cocao black pod, and diseases of rubber, and other tropical crops); *P. megasperma, P. cactorum, P. cinnamomi,* and *P. nicotianae,* and its variety *parasitica* (on a large number of hosts such as tobacco, castor, and *Piper betle).*

Sporangiophores determinate (growth definite), not extensively branched; conidia/sporangia mature synchronously, oogonial walls thick, confluent with that of the oospore, oospore germination direct by a germ tube—***Sclerophthora***.

18.2 DAMPING-OFF DISEASE OF VEGETABLE SEEDLINGS

Damping-off of seedlings is very common all over the world. It occurs in agricultural and forest soils, in tropical and temperate climates, and in almost every greenhouse. Damping-off can be of two types, that is, pre-emergence damping-off in which the seed and radicle rot before the seedlings emerge out of the soil, and the post-emergence damping-off in which the newly emerged seedlings are killed at ground level after they emerge from the soil, causing them to collapse or topple over. This is a common symptom of post-emergence damping-off. This disease is most noticeable in nursery beds, greenhouse flats, and row crops because symptoms develop suddenly, killing large numbers of seedlings in local areas. Among the pathogens in this class are the species of *Pythium, Phytophthora, Corticium,* and *Fusarium;* the former two will form the subject of our present discussion. Most species of *Pythium* mainly affect the juvenile or succulent tissues. This restricts their parasitism to seedlings, the feeder roots or root tips of older plants, and to watery fruits or stem tissues. They also cause fruit rots of crops, such as beans, squash, and watermelons. At a later stage, when the cells of stems and main roots have developed secondary thickenings, infection is restricted to feeder roots. This causes seedlings to become stunted, and chlorotic. Peach decline, which kills thousands of trees each year, has been shown to be a feeder-root problem caused by a combination of factors, including *Pythium* spp. Pectolytic, and cellulolytic enzymes produced by *Pythium* and *Phytophthora* spp., playing an important role in tissue disintegration and maceration. There are two recent books on *Phytophthora* diseases, i.e. *Phytophthora* diseases worldwide by Erwin and Ribeiro (1996), and *Phytophthora* diseases in India by Mehrotra and Aggarwal (2001).

Pythium debaryanum Hesse, and several other species of *Pythium* such as *P. aphanidermatum* *Pythium ultimum*, and *P. arrhenomanes* are the pathogens.

1. Symptoms

As has already been stated, damping-off of vegetable seedling occurs in two stages, that is the pre-emergence and the post-emergence phase. In the **pre-emergence phase**, the young seedlings are killed before they reach the soil surface. The young radicle and plumule are killed, and there is complete rotting of the seedling. The **post-emergence phase** is characterized by infection of the young, juvenile tissues of the collar at the ground level. The infected tissues become soft and water-soaked. The seedlings topple over or collapse. Such symptoms are very common in seed beds in nurseries.

2. Causal Organisms

The pathogen responsible for damping-off is *Pythium debaryanum.* Among other common species of *Pythium* are *P. arrhenomanes, P. aphanidermatum,* and *P.ultimum.*

3. Disease Cycle and Predisposing Factors

Pythium spp. are natural inhabitants of the soil, where they are probably low-grade parasites on fibrous roots. As a plant parasite, the pathogen can infect succulent organs of plants but tissues that have matured, and developed secondary thickening become resistant to its attack. The oospores serve as the overwintering or oversummering organs.

The disease cycle of a typical *Pythium* species causing damping-off and seed decay is given in Fig. 18.1.

4. Disease Management

Most *Pythium* spp. produce oospores and chlamydospores, and persist for many years. Direct control of *Pythium* on field scale is difficult and expensive. On a small scale, *Pythium* spp. can be eliminated from the soil for greenhouse use by heating. The common method is by steam treatment or pasteurization. The latter method is preferred as it does not change the microflora drastically.

Fumigation with Chloropicrin or Methyl bromide is a standard practice in many nursery, and horticultural operations.

Very often the most effective measure against damping-off of fungi involves the use of seed protectants to keep away the pathogens in the pre-emergence phase, and to adopt sanitary precautions in the nursery to check post-emergence damping-off. The most commonly used seed-protectant fungicides now are Thiram, Captan, Zineb, and some systemic fungicides such as Metalaxyl. Drenching the soil

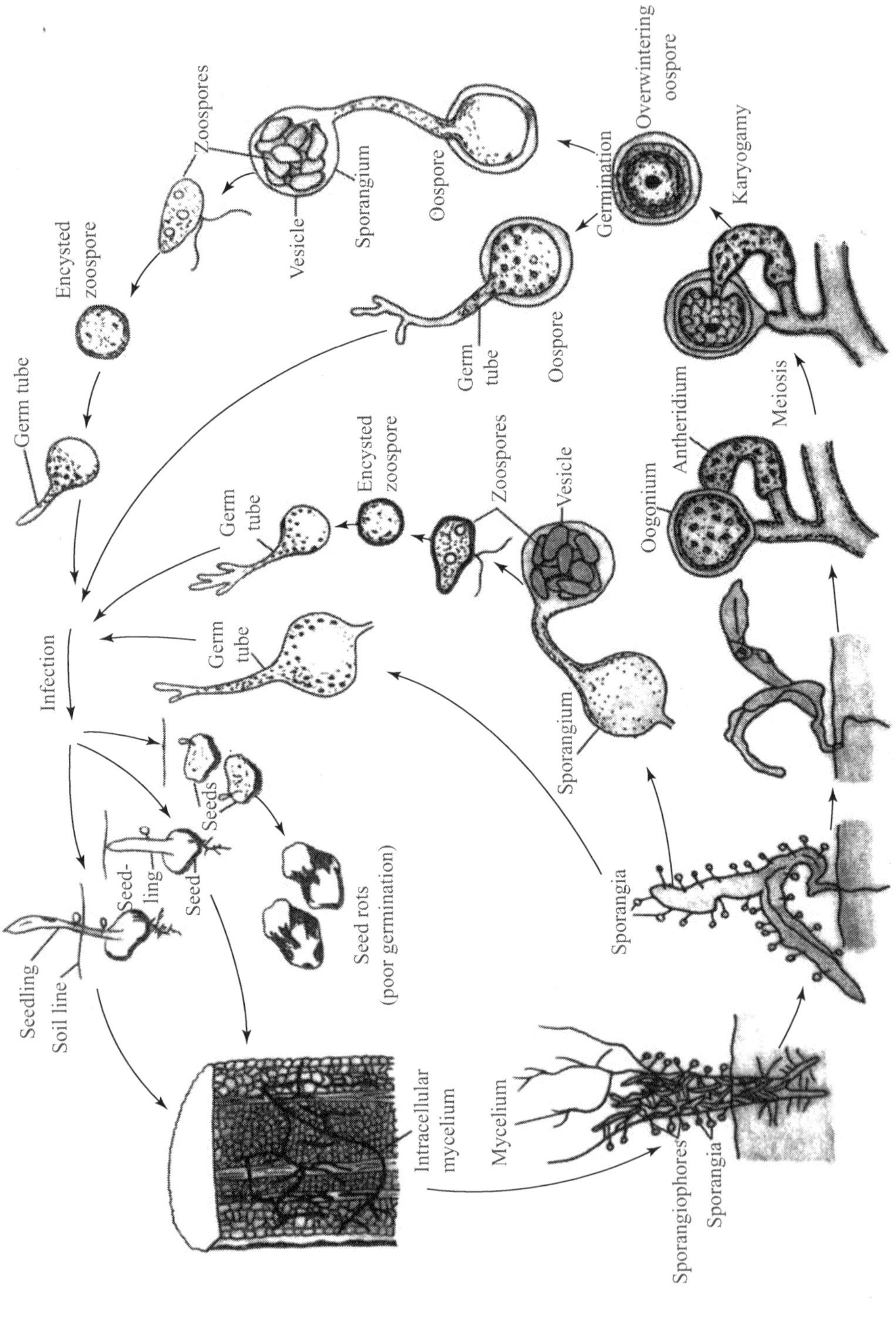

Fig. 18.1 Disease cycle of damping-off and seed decay caused by *Pythium* sp. (Redrawn From G N Agrios, *Plant Pathology*, Academic Press, New York, and London, 1969)

with suitable fungicides, such as Cheshunt compound, 1% Bordeaux mixture, 0.1% Thiram, 0.3% Blitox–50, or 0.2% Esso Fungicide 406 has been found useful in eliminating soil-borne infection.

Dexon, and other soil fungicides, applied at low rates, have been successfully used to retard the rate of reinfestation of fumigated soils by *Pythium.* The use of healthy seeds in crops (ginger and turmeric) which are vegetatively propagated by rhizomes, etc. forms an important control measure in such diseases.

Crop rotation is another method of reducing the populations of soil pathogens, but because the more common and destructive *Pythium* spp. have a wide host range, this practice is not of much help.

Seed treatment with spores of *Trichoderma harzianum, Penicillium oxalicum,* and cells of *Pseudomonas cepacia, P. fluorescens* have been found to be very effective biological control agents. So combination of seed, and soil treatment is best for management of damping-off. Some cultural practices recommended are thin sowing of seeds, light sandy soil for nurseries, use of well-decomposed manure, light but frequent irrigation, and sterilization of soil by burning a 30 cm thick stack of farm trash on the nursery beds. Some new methods such as osmopriming of seeds (seed priming with inorganic salts) before planting have been found promising. Some of the systemic fungicides usually or in combination with broad spectrum fungicides, give excellent control of damping-off, seedling blights, and root rot caused by *Pythium* spp.

Application of talc-based formulations of *T. viride* and *P. fluorescens* reduces damping-off of tomato caused by *Pythium aphanidermatum. Trichoderma* spp. colonize the wounds and broken roots. As a result, these produce glucanase and cellulase enzymes which in turn inhibit germination of encysted zoospores of *Pythium.* Strains of *Pseudomonas fluorescens* produce an antibiotic 2, 4-diacetylphloroglucinol which inhibits the mycelial growth of *P. ultimum* and protects tomato seedlings against damping-off. Other microbial biocontrol agents against damping-off are binucleate *Rhizoctonia* strains, *Pseudomonas putida,* and *Bacillus amyloliquefaciens.* Seed treatment with the microbial strains of *Bacillus cereus, Pantoea agglomerans,* and *Erwinia rhapontici* also control damping-off diseases of safflower, pea, and canola respectively.

Amendment of the growth medium for the greenhouse-grown watermelon with *Brassica* seed pomace (cake in which oil has been extracted) along with *Pseudomonas boreopolis* gives good control of damping-off caused by *Pythium* sp. The mechanism of control is attributed to the hydrolysis products of glucosinolates into allyl and butenyl isothiocyanates which are toxic to *Pythium* species, *Sclerotium rolfsii, Sclerotinia sclerotiorum,* and *Phytophthora capsici.* Cucumber seed treatment with *Pseudomonas putida* bacteria or with the mycoparasite *Verticillium lecanii* resulted in the systemic production of phytoalexins, and other host-defense reactions that protect cucumber seedlings from attack of *Pythium.* Similarly, treatment of spruce (*Picea*) seedling with methyl jasmonate induced accumulation of free salicylic acid, chitinase that protected the seedlings from infection by *Pythium.* Soil treatment with low dose of formalin (one part in 50 parts of water) at a depth of 10 cm is quite effective. The low dose of formalin boosted up the population of *Trichoderma* in soil, and which in turn showed antagonistic behaviour against *Pythium* sp. Use of formalin dust is also effective. The dust consists of 85 parts of charcoal ash, and 15 parts of formalin. Soil solarization by polyethylene sheet mulching for control of damping-off of some vegetables has been reported. As a result of solarization, the temperature of mulched soil is 49.7°C which is quite effective in killing *Pythium* and *Fusarium.*

5. Integrated Disease Control

Integration of cultural practices, chemical, and biological control methods can be useful for management of damping-off. Cultural practices include avoidance of nursery sowing in the same bed repeatedly year after year, and frequent heavy irrigation. Soil solarization with transparent polyethylene mulch has been found effective for control of the disease. Application of Captan as seed dresser, and for soil drenching 15 days after planting has been recommended for healthy plant stand in chilli, cabbage, and onion crops.

Seed coating with conidial suspension of several biocontrol agents like *Trichoderma koningii, T. harzianum,* and *T. viride* have been found more effective than soil application in protecting the tomato seedlings from damping-off. An integrated management module comprising of change of nursery site every year along with pre-sowing, soil sterilization with Formalin (5%), seed treatment with Captan or Thiram, and drenching of beds after emergence with Captan (0.3%) or mixture of Mancozeb (0.25%), and Carbendazim (0.1%) at 8-10 day intervals coupled with avoidance of heavy irrigation have been suggested.

18.3 FRUIT ROT OF CUCURBITS

The disease occurs on bottlegourd (*Lagenaria vulgaris*), spongegourd (*Luffa cylindrica* and *L. acutangula*), snakegourd (*Trichosanthes anguina*), cucumber (*Cucumis sativus*), bitter gourd (*Momordica charantia*), and other cucurbitaceous fruits. Cottony leak disease is very widespread in cucurbit fruits. Sometimes 50% or more of muskmelon or watermelon undergo rotting after rains due to this disease.

1. Symptoms

A watery soft rot develops, especially in the fruits touching the soil surface. The disease is common in the field during and after the rains. It also spreads among fruits during storage and transit. As a result of infection, the diseased fruits show formation of a luxuriant wooly mycelial growth. The decaying fruits also emit a bad smell.

2. Causal Organism

The causal organisms are mainly *Pythium aphanidermatum* and *P. butleri* but other fungal pathogens such as *Fusarium, Rhizoctonia,* and *Phytophthora* spp. may also be involved. The pathogen *P. aphanidermatum* has been described earlier. *P. butleri* and *P. aphanidermatum* are responsible for cottony leak disease (or fruit rot) of cucurbits in north Indian conditions. The susceptibility of several cucurbits, chilli, tobacco, and papaya for *P. butleri* had been reported. *P. bulteri,* a papaya isolate, causes fruit rot of many cucurbits on inoculation.

3. Disease Cycle and Predisposing Factors

The pathogen *P. aphanidermatum* and *P. butleri* are soil-inhabiting ones, and survive as a saprophyte on dead organic matter, causing the disease when a suitable host is available.

4. Disease Management

Efforts should be made to keep the fruits away from the soil surface. Soil disinfection, as recommended for damping-off, is not practicable due to the prohibitive costs involved.

Some strains of *Pseudomonas*, i.e. *P. corrugata* and *P. aureofaciens*, when applied to roots induce systemic resistance against the pathogen. The mode of action is perhaps through accumulation of endogenous salicylic acid, and stimulation of peroxidase and polyphenol oxidase activity. Soil solarization during summer has been found quite effective against *P. aphanidermatum* causing root rot as the pathogen is soil-borne.

18.4 RHIZOME ROT OF GINGER (ZINGIBER OFFICINALE)

The disease is quite severe in south India. *Pellicularia* spp. and *Fusarium* spp. are also known to cause rhizome rot of ginger. The disease is not confined only to the crop in the field but also causes 80–90% loss of corms in storage.

1. Symptoms

The first visible symptoms are seen at the basal portion of the plant which becomes watery and soft. The leaves become slightly pale in colour, their tips turn yellow, and the yellowing spreads down the leaves. The rhizomes start rotting, and are converted into a pulpy mass giving out a putrefying smell.

If the plants are raised from infected rhizome, the young shoots show damping off, and if infection occurs after emergence of shoots, the pseudostem withers and dies, thus resulting in poor crop.

2. Causal Organisms

Pythium myriotylum, P. aphanidermatum, Pythium monospermum, and *Pythium gracile* have been isolated from the diseased plants. The pathogenicity of *P. myriotylum* and *P. aphanidermatum* has been proved so far.

At least 8 species of *Pythium* are known to attack ginger in different parts of the world. According to Rajan and Agnihotri (1989) in India, *P. aphanidermatum, P. butleri, P. myriotylum, P. graminicola, P. deliense, P. pleroticum*, and *P. vexans* have been reported on ginger.

3. Disease Cycle and Predisposing Factors

It is a soil and seed-borne disease, and it is transported over long distances on ginger rhizomes.

4. Disease Management

The first prerequisite for checking this disease is to choose healthy seed pieces. Soft rot of ginger has proved amenable to control in India by the preplanting treatment of rhizomes and the soil with copper fungicides.

Dipping of seed rhizomes in 0.25% Antracol or Propineb or 0.3 percent Fycop or 0.3 percent Blitox-50 in controlling rhizome rot has been recommeded. Since the pathogen is present deep into the tissue of the host (corm or rhizome), surface disinfection of seed is not effective. Spraying of the soil with Bordeaux mixture at the rate of 7 liters per square foot once before sowing and second after germination gives partial control of the disease. Control of the disease by seed and soil treatment with Metalaxyl (Ridomil 5g or Apron 35 WS) has also been reported.

Regarding the biological control agent, *Trichoderma*—a known biocontrol agent—provides significant control of rhizome rot of ginger. Dual inoculation of *T. harzianum* along with fluorescent pseudomonads is also reported to improve plant growth and yield. Cultivars Moran and Riodegeneria are found to be resistant to rhizome rot.

18.5 LATE BLIGHT OF POTATO (*SOLANUM TUBEROSUM*)

In India, the disease was first introduced into the Nilgiri Hills between 1870 and 1880. Very soon it spread to the Darjeeling district in the Himalayas after the introduction of the English potatoes. Losses due to the ravages of this disease have been recorded in different parts of our country by several workers. These have been estimated at 20–25% in Punjab, 40–50% in Haryana, 15–50% in UP, and 5–10% in Bihar and West Bengal. The overall losses range from 20–25% amounting to 4 million tonnes of potatoes worth ₹4 crores. The disease is endemic and more severe in eastern than northern plains.

1. Symptoms

The first symptoms of late blight in the foilage of potato and tomato are hydrotic areas with indefinite margins at the tips or on the margins of the leaflets (Fig. 18.2a). These lesions then become necrotic, and turn brown to almost black. Very often a chloranemic border develops around the necrotic areas.

If moist weather prevails, the entire leaf may be killed within a couple of days. If dry weather intervenes, following the appearance of lesions, the infection advances slowly, and the affected areas curl, and shrivel. However, under moist conditions, they remain limp, and soon decay, producing a characteristic offensive odour. If the weather remains dry after the development of spots on the leaflets, the diseased areas seem hard and easily break away from the leaf blade. At this stage, symptoms are very similar to those due to *Alternaria solani,* which is responsible for early blight. The wet rot in warm humid weather advances rapidly, and affects all the leaves and the stem but if does not go as far down as to the tubers. Tubers are infected separately.

Slightly sunken hydrotic areas develop in the superficial tissues of the potato tubers. These areas become necrotic, and if conditions remain favourable, the infection spreads inwards so that the entire tuber turns brown and decays before harvest. In drier soils, the progress of the disease is slower, and a rot may result.

During wet and humid conditions, the sporangiophores and sporangia can be seen as a downy growth, particularly on the abaxial surface of the leaflets (Fig. 18.2b).

(a) (b)

Fig. 18.2 (a) Late blight symptoms due to *Phytophthora infestans* (b) Sporangiophores and sporangia on leaf

2. Causal Organism and Disease Cycle

The causal organism of late blight of potato is *Phytophthora infestans* (Mont) deBary. The causal organism of late blight of potato is carried through tubers used as seeds (Fig. 18.3).

P. infestans is a heterothallic fungus, and requires two mating types (A^1 and A^2) for sexual reproduction. It has now been reported that both A^1 and A^2 mating types occur in India. However, what role these two mating types play in the annual recurrence of diseases is still to be established.

In India, the main potato-growing season is followed by hot summer months in the plains, and therefore, the fungus is eliminated from the soil. Potato tubers which are stored at ordinary temperatures also become free from disease.

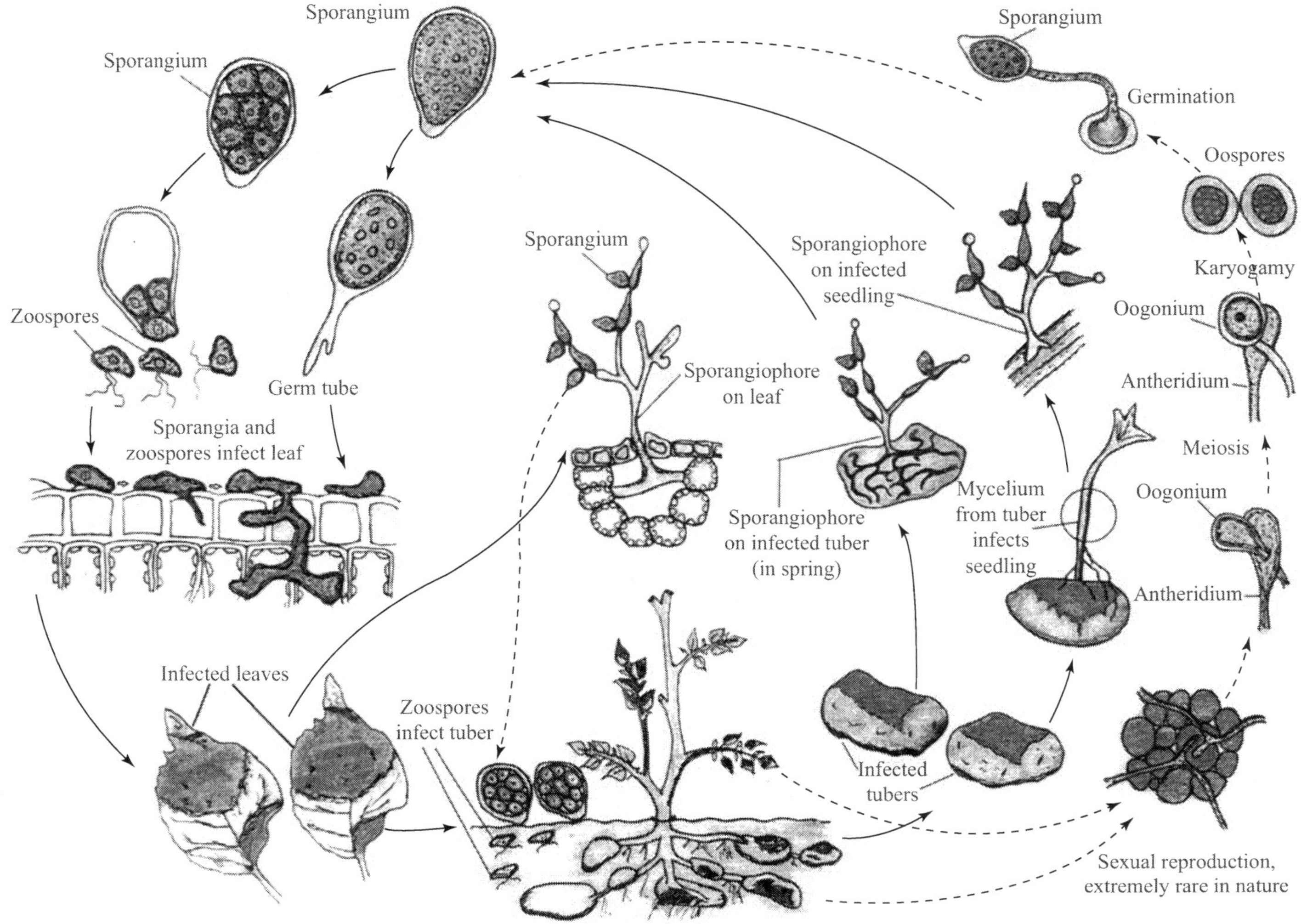

Fig. 18.3 Disease cycle of late blight of potato caused by *Phytophthora infestans* (Redrawn from G N Agrios, *Plant Pathology*, Academic Press. New York, and London 1969)

3. Epidemiology and Forecasting

In the plains, the initial source of the disease is in the tubers placed in cold stores, or the seed tubers have been found to carry the late-blight infection.

4. Disease Management

Certified and disease-free tubers are used for cultivation. Seed tubers should be obtained from uninfected areas.

Harvesting of a diseased crop should be delayed until the plants are fully mature. The danger of infection of tubers by sporangia falling on to them from foliage at lifting time can be minimized by ensuring that all foliage is destroyed beforehand. This is achieved by spraying the foliage two to three weeks before harvesting with sprays, such as sulfuric acid, copper sulfate, tar acid compounds or sodium chlorate. High ridging at the time of earthing helps reduce chances of infection.

In the Nilgiri Hills, Bordeaux mixture provided a good protection to the crop, and helped to increase the yields. In the eastern Himalayas, experiments were conducted in Darjeeling with a few fungicides and Dithane Z-78 was found to give good protection to the crop. In the western Himalayas, at Kufri in the Simla hills, Burgundy mixture was found to be effective.

In the northern hills, a protective spray of Dithane M-45 at 2 kg/ha is essential in the last week of June before the disease appears. This should be followed by a spray of Bordeaux mixture 5:5:50 at intervals of ten days. In the eastern hills, the crops should be sprayed with Bordeaux mixture. In the Nilgiri hills, spraying may be done with Brestan or Difolatan. For spraying in the hilly regions some sticker like Triton HE or Sandont (0.1%) should be added to the fungicides. In the plains, any of the fungicides, for example, Dithane M-45 (2 kg/ha), Brestan (1 kg/ha), and Difolatan 80 WP (2.5 kg/ha) can be sprayed.

The discovery of Metalaxyl, a systemic fungicide, gave a new dimension to late-blight control strategies.

Oxadixyl (a phenylamide fungicide), a new systemic fungicide has also been found effective against the disease. At Kufri, Shillong, and Darjeeling, the formulation of Oxadixyl (Oxadixyl + Mancozeb; Oxadixyl + Copper Oxychloride) were very effective. This treatment did not reduce tuber infection. The pathogens in the tubers can be readily killed if these are dipped in water at a temperature of 45°C for 4–5 hours or at 40°C at 24 hours (Butler, 1903), and according to Singh and Bhattacharya, (1988) dipping the tubers suspected to be carrying infection up to 5% in a solution of Oxadixyl or Metalaxyl for 30 minutes in 1500–2000 ppm of concentration of these systemics is very effective in removing infection from the tubers, but this holds true only when the tubers have overcome the dormancy.

The Metalaxyl-resistant population of *P. infestans* is resistant to Ofurace (acylalanine fungicide) and Patafol. These two fungicides have similar modes of action as that of Metalaxyl. According to Thind (2002), there is no resistance to Strobilurins and Cymoxanil. Two applications of Propiconazole and Chlorothalonil are sufficient to restrict lesion expansion and sporulation of late blight of potato. Combination of Mancozeb with beta-aminobutyric acid (BABA) has synergistic effect against *P. infestans* in potato and tomato.

However, the best control measure lies in cultivating resistant varieties. Hybridization work has been done with *Solanum tuberosum* (South American variety), *S. neoantipoviezi,* etc. showing a high and promising degree of resistance. *Solanum demissum,* a Mexican species, shows complete resistance with commercial susceptible varieties. Kufri Kundan and Kufri Kumar were the first set of varieties, possessing resistance of late blight, released for commercial cultivation. Later on, light hybrids were released for commercial cultivation as Kufri Jeevan, Kufri Neela, Kufri Muthu, Kufri Jyoti, Kufri Khasigaro, Kufri Naveen, Kufri Badshah, and Kufri Sherpa.

5. Cultural Methods of Controls

The following are some of the cultural methods which have been found useful in controlling the disease.

(a) Tubers from infected fields should be discarded for seed purposes or very carefully inspected at planting time to remove all those tubers which might be suspected to be carrying or harbouring (mature tubers) infection.
(b) The tubers should be harvested after proper skin curing.
(c) In an infected field, harvesting should be delayed until the plants are mature. The plants should be allowed to dry completely before harvesting. Alternatively, the infected haulms be cut, and buried or burnt so that the tubers do not get in touch with infected portions of the potato plants.
(d) Infected tubers and vines should be collected and buried deep in the soil.
(e) All sanitary precautions must be enforced to keep the crop free from weeds. Under hill conditions, ground keepers or volunteer potato plants harbor the infection, and therefore it is essential to remove these as soon as these make their appearance.
(f) Tubers should be planted on high ridges. High ridges will cover the tubers thoroughly with soil, and thus prevent growing out of the fungus from mother tuber, and also infection in daughter tubers.
(g) The tubers should be stored either in cold storage at low temperature or in a cool, dry, and well-aerated store. The best storage temperature is 38–40°F.
(h) In the plains, the disease normally appears with the onset of winter rains. Therefore, irrigation should be restricted during this period.
(i) Planting of alternate beds of susceptible and resistant varieties reduces spread of the disease.
(j) In order to reduce humidity in the field spacing and reduced nitrogen are recommended in endemic areas.

6. Biocontrol

In small field trials, some biocontrol agents such as *Trichoderma viride, Penicillium viridicatum,* and *Chaetomium brasiliense, Myrothecium verrucaria, Penicillium aurantiogriseum, Epicoccum purpurascens, Stachybotrys atra, Trichoderma koningii, Bacillus*, *Pseudomonas*, *Rahnella*, and *Serratia* spp. have also found success in controlling the diseases.

18.6 BLIGHT OF COLOCASIA (*COLOCASIA ESCULENTA*)

In India, two *Colocasia* types, viz. *C. esculenta* var. *esculenta,* and *C. esulenta* var. *antiquorum,* are generally cultivated, and both are equally susceptible to *Phytophthora* blight. Infection in both taro leads to heavy yield losses.

1. Symptoms

The disease first shows itself ordinarily on the leaf, the earliest attack occuring in August or September, in the form of small, dark, roundish specks, which increase in area rather quickly by centrifugal growth (Fig. 18.4). The spots may remain circular or assume an oval or irregular appearance. Several such spots may appear on the leaf, and the whole leaf lamina may get infected. Clear, yellow liquid drops ooze from the surface of the spots in the initial stages; later the central portions become dry and drop off, and thus holes in the leaf. The periphery of the spots has a clear zonation pattern of different shades of brown, green, and yellow.

(a)

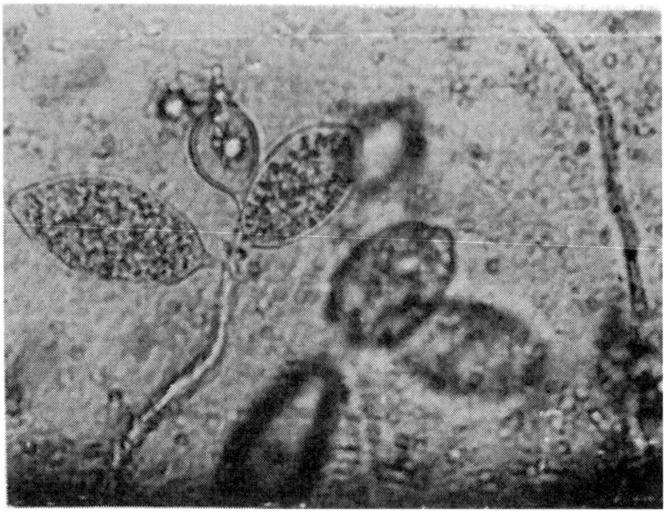

(b)

Fig. 18.4 (a) Symptoms on *Colocasia* due to *Phytophthora colocasiae* (b) Sporangiophores and sporangia

Towards the periphery, these zones reveal on careful observation, a delicate white haze which is caused by the sporangial stage of the pathogen. As the spots increase in size or diameter, sporangia continue to be produced, in a zonate fashion at progressively increasing distance from the center. In severe cases of infection when humidity is high, and temperatures are congenial for the development of blight symptoms, the petioles of the leaves get infected. The tissues at the infection region often become so softened as to be unable to bear the weight of the leaves, which fall over and even break off completely from the plants. Extension of infection may continue down to the sheathing base of the petiole, and probably into the short stem. Inflorescence is also attacked. In severe attacks, all the leaves are completely destroyed, and the underground corms are infected and rot completely.

2. Causal Organism

P. colocasiae Raciborski is the pathogen responsible for the leaf blight and corm rot of *Colocasia esculenta* (Linn.) Schott.

3. Disease Cycle

It is presumed that the disease mainly perpetuates through infected corms and also through oospores persisting in the soil wherever these are produced.

4. Disease Management

Since the disease is mainly carried over from season to season through corms, selection of disease-free seed should be given due priority. Sanitation of the field, including destruction of badly affected plants, should be done. Copper fungicides and Metalaxyl have proved to be quite effective. Metalaxyl (Apron 35F), Captafol (Difolatan), and Chloroneb (Demosan 65w) used as spray at 15-day intervals gave good control of the disease (Aggarwal *et al.*, 1990).

Deshmukh, and Chhiber (1968) have reported field resistance to this disease in certain varieties of *C. esculenta* var. *antiquorum* (*C. antiquorum*). Variety Ahina is fairly resistant.

18.7 "KOLEROGA" OR "MAHALI" OF ARECA PALMS (*ARECA CATECHU* L.)

Koleroga (*Kole* = rotting, *Roga* = disease) as is called in Karnataka is otherwise called Mahali (heavy devastation) in Kerala, or fruit rot in a number of places. The disease was first recorded in India in 1906. It occurs in severe form in heavy rainfall areas of the country wherever the arecanut crop is being grown.

1. Symptoms

'Koleroga' usually makes its appearance two to three weeks after the beginning of the rains. Water-soaked areas develop on the nuts, starting from the base. This is the chlorosis stage (Neergole); in advanced stages, the fungus develops on the outside of the fruits (Bhusargole). Attacked nuts soon begin to shed, and this is a sure sign of the disease in the plantation. Fallen nuts show a felty, whitish mycelial mass on their surface, and very soon the entire nut is enveloped by it. While the disease is predominantly on the nuts, the tops of trees are occasionally attacked. The infection travels up the stalks and sheaths, and reaches the growing tops. As it advances on the crown of the plant, the trees may dry up, resulting in a withering of the leaves and bunches. The main loss is due to the nuts, and there is hardly a 1% annual loss due to the death of the plants.

2. Causal Organism

The fungus that is responsible for Koleroga is *Phytophthora areacae* (Coleman) Pethybridge.

3. Disease Cycle and Predisposing Factors

The fungus is endophytic. The areca palms harbour latent infection in their crowns during the non-monsoon period. Sexual spores are not produced in nature. Oversummering is carried out by mycelium persisting on tree tops and chlamydospores in fallen nuts. Regional sex strains are present, and produce in combination, oospores in culture. Heavy rainfall and constant moist conditions are undoubtedly the chief factors for 'Koleroga'. Intermittent rainfall and sunshine are highly favourable for infection.

4. Disease Management

The earlier practice of controlling Koleroga was to cover the arecanut branches, made of arecanut leaf sheath called Kotte in Malnad region or a kind of grass in other parts named 'Karada', but this did not help in controlling the disease. However, covering the arecanut branches with polythene bags as a prophylactic measure give cent percent control of fruit rot though it is costlier than fungicide spray.

Various workers tested the efficacy of Bordeaux mixture with several adhesives, and spreaders such as potash alumn with casein, vegetable oil such as groundnut, gingelly, coconut, or sunflower oil. However, it was shown that plain Bordeaux mixture (1%) without any adhesive was equally effective in controlling the disease. It has been recommended that prophylactic sprays of Bordeaux mixture two–three times a year were quite effective in keeping the disease under control. Copper oxychloride checked koleroga. However, there were reports that copper oxychloride or nickel chloride were not effective. Systemic fungicides such as Aliette and Ridomil at 0.2% are also good in checking fruit rot of arecanut. Adopting phyto-sanitary conditions in the plantations go a long way in reducing the incidence of the disease. The measures include removal and destruction of fallen nuts, diseased bunches of tree tops, and other plant parts in the field.

18.8 LEAF ROT AND FOOT ROT OF 'PAN' (*PIPER BETLE*)

Root rot collar rot and leaf rot of *P. betle* (*pan*) is a serious disease throughout the betelvine-growing areas in the country, and is a great threat to the cultivation of this crop in our country.

1. Symptoms

The pathogen infects almost all portions of the plant. The most common symptom in an orchard is the appearance of wilted vines. This is not the typical vascular wilt caused by *Fusarium* species. It is due to the fact that the root and basal foot rot, or collar rot, occurs which results in wilting symptoms. The first symptom of a diseased plant is the loss of lustre, a little later the upper leaves, and tender shoots of the plant begin to show signs of drooping. The whole plant shows loss of green colour, which is replaced by yellow. The plant later rapidly dries up. The underground parts are more or less completely rotted (Fig. 18.5b). If a newly diseased plant is uprooted, it is found that the smaller roots are infected first, and gradually the rotting spreads to the bigger roots, and then the collar region of the plant. As the rotting develops, the soft tissues are readily eaten up by the fungus and the fibrous remains give the parts a shredded appearance.

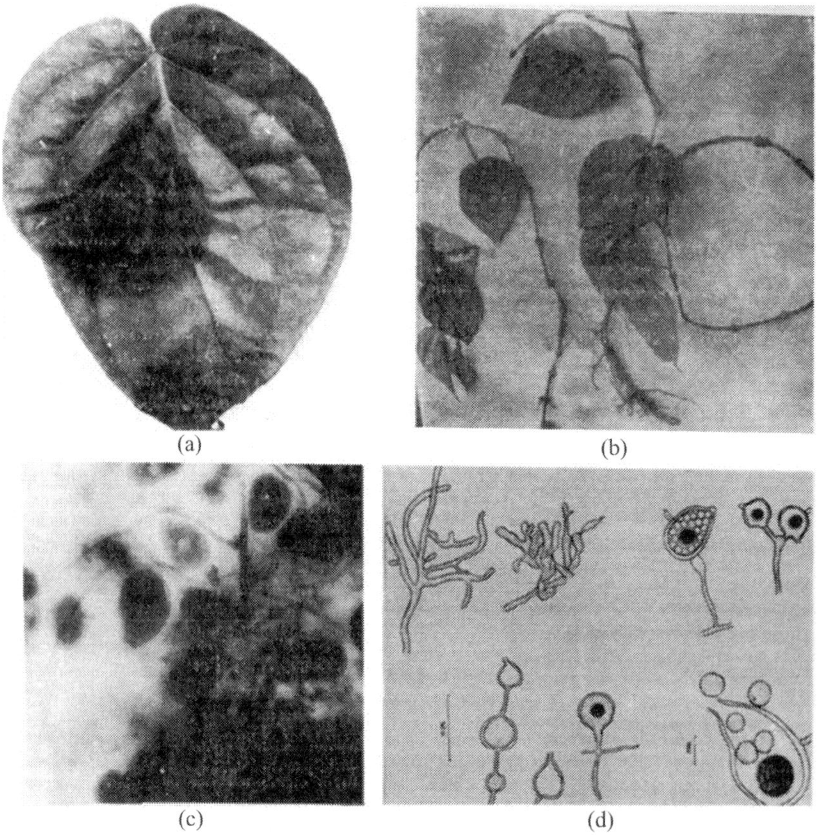

Fig. 18.5 Leaf rot and foot rot of *Piper betle* caused by *Phytophthora parasitica* var. *piperina*: (a) Leaf rot (b) Root rot and foot rot (c) Sporangia on leaves (d) Mycelial knots, sporangia, and chlamydospores

In the leaf-rot symptoms (Fig. 18.5a), the first symptom of the disease is the presence of a circular black or brownish spot which are more or less wet in appearance. Under continuous wet and humid conditions, the black spot rapidly increases in diameter, and develops a wet rot of a major portion to the leaf. The rot extends to the petiole and from the petiole to the stem. It has been observed that when the infection reaches a midrib or one of the main veins, it spreads very rapidly along their length than between the veins. If the atmospheric conditions do not remain humid and wet continuously, and if the rain lasts only for a day or so with warm dry periods in between, the diseased areas show distinct concentric puckerings and wrinkles on account of the unequal growth of the tissues. If the condition becomes dry, the progress of the disease is checked and the black spots usually surrounded by a brownish halo or ring remain dry and localised. More than one spot are often found on a leaf and the spots may coalesce with each other. The leaf at any position in the plant is attacked. It may be noted that in an orchard, the leaves within a couple of feet from the ground surface are particularly liable to infection in the beginning of the season, and start of the disease. The topmost leaves are generally seldom infected.

2. Causal Organism

There is no unanimity with regard to the pathogen/pathogens responsible for the diseases (Fig. 18.5b and c). Dastur (1935) named the pathogen as *P. parasitica* var. *piperina* which has not been accepted by Waterhouse (1963) as it was invalidly published without Latin diagnosis, and therefore, it became *P. nicotianae* var. *parasitica.* Maiti and Sen (1977) considered it as *P. palmivora* while Marimuthu (1991) called the pathogen *P. palmivora* MF_4. In recent literature, *P. palmivora* MF_4 has been called *P. capsici.* It appears that more than one species of *Phytophthora* are infecting this crop in different parts of the country.

3. Disease Cycle

Mahmud (1952) claimed the isolation of the pathogen from the soil and concluded that the fungus lives saprophytically in the soil and can survive for years.

The survival of *Phytophthora* as mycelium in the host tissue was studied by Mehrotra and Tiwari (1967). The pathogen could be isolated up to 17 weeks from the host tissues.

Chowdhuri (1944) and Asthana and Mahmud (1945) found that the disease is primarily carried through the plantation of infected cuttings. The work of Mehrotra (1961), Tiwari (1968), and Saksena and Mehrotra (1970) clearly shows that the infected part rots rapidly and that the mycelium cannot be detected in the healthy parts to any appreciable distance on either side.

Mehrotra and Tiwari (1967) demonstrated that after five weeks, chlamydospores or resting sporangia were formed, and the pathogen could be isolated 17 weeks after burial.

Besides chlamydospores, oospores of the pathogen have also been detected in nature for the first time in India by Chaurasia (1976).

4. Disease Management

(a) Effect of Manure on the Disease The mortality of vines was more when high dozes of nitrogen was applied but was considerably lowered with phosphate alone or in combination with low dosage of nitrogen.

(b) Host Resistance Betelvine is propagated vegetatively. In most places, the vines do not flower. A living female vine has been located in Bongaon (West Bengal) and therefore, there is a possibility of evolving resistant lines using various breeding methods. Generally, the vines produced only the female flowers, and therefore there is the difficulty in the breeding programme. Some varieties are tolerant to the disease in different parts of the country.

(c) Chemical Control Earlier workers recommended the use of Bordeaux mixture of 2:2:50 strength for controlling the disease by treating the ridges before planting, and later subsequent application at monthly intervals or bimonthly intervals. Chaurasia (1976) obtained best results by dipping the cuttings in streptomycin sulfate solution followed by Bordeaux-mixture application twice a month. Johri and Chaursia (1995) have recommended Akomine which is an inorganic source of plant nutrient with phosphorus as an active ingrediant. Being inorganic in nature, there are no residual or side effects. The product has been tested for its efficiency in management of betelvine *Phytophthora* along with Aliette and Captafol. About 80% mortality of the vines was checked. Ayyavoo and Samiyappan (1981) reported the efficacy of Deconil-2787, 0.1% during winters (October to February) and 30 litres fungicidal solution for every 65 sqm at 15-day intervals.

Botanicals for the management of the betelvine diseases were tried with varying successes. Tiwari and Mehrotra (1973) obtained promising results by using carbon disulfide (CS_2) and formalin as soil fumigants in controlling the disease in a limited scale in pots. The fumigant formalin was found to be phyto-toxic at the dosages used to control the disease. Fumigation of soils brings about the dominance of the antagonists *T. viride* in soil, which indirectly acts as a controlling agent.

(d) Cultural Practices The importance of phyto-sanitary approach to reduce the disease incidence in the orchards is emphasized. Collection, removal, and destruction of dead and dying plants or burning or burying with lime, deep in the soil is recommended. Since it is believed that there are chances of the disease being carried through contaminated cuttings, carrying the infection both internally as well as externally, planting of healthy cuttings had been recommended.

Since the pathogen overwinters or oversummers in the soil, judicious rotation of the crop in betelvine orchards may check the spread of the disease. Paddy and banana rotation in Salem and Tanjore districts in Tamil Nadu has been found to considerably safeguard the crop against the incidence of disease. According to an anonymous report in Maharashtra, the cultivars to grow onion, gram, chillies, tobacco, wheat or jowar as rotational crops before planting betelvine. Combined effects of flooding, fallowing hot weather, and deep ploughing has been found to reduce the incidence of the disease (Tiwari and Mehrotra, 1974).

(e) Biological Control Tiwari and Mehrotra (1968) recommended dipping of cuttings in a suspension of *Trichoderma viride* spore suspension prior to plantation. Mehrotra and Tiwari (1976) experimented with organic amendments, and antagonists in soil to control foot rot of *P. betle.* Incorporation of corn straw into the soil gave best control. The effect of different antagonists growing on corn straw, and til (*Sesamum indicum*) oilcake amendments showed that *T. viride* gave best control. Corn straw supplemented with nitrogen gave better control. Chaurasia (1976) demonstrated the effectiveness of *Aspergillus flavus, A. oryzae,* and *Penicillium* sp. in the control of the disease. Tiwari and Mehrotra (1973) obtained promising results by using CS_2 (carbon disulfide) and formalin as soil fumigants in controlling the disease in pots. Formalin, however, is phytotoxic under the dosages used for controlling the disease.

18.9 ABNORMAL LEAF FALL OF RUBBER

There is growing evidence from India, Southeast Asian countries of Indonesia, Malaysia, Philippines, Thail, and, and Vietnam on the onslaught of plant diseases in rubber. However, the severity of the incidence of crop loss in rubber due to abnormal leaf fall has been found more pronounced in the rubber growing regions in Kerala in India than elsewhere.

1. Symptom

Abnormal leaf fall occurs in adult rubber-tree plantations in rainy winter season, after leaf renewal, when plants have mature leaves. It brings in consequence the fall of these leaves. The symptoms of the disease also include rotting of tender shoots and die back of twigs. The impact of Phytophthora disease on rubber production is a reduction in latex yield, caused by the tapping panel and collar diseases and a reduction in growth due to leaf fall. As the leaves are the source of photosynthesis for growth in rubber, the low-leaf retention leads to poor girth, and thereby low latex and timber output.

2. Causal Organism

In the case of rubber (*Hevea brasiliensis*), several diseases including abnormal leaf-fall disease are attributed to a number of species of *Phytophthora* including *P. botryosa*, *P. heveae*, *P. meadii*, *P. palmivora*, and *P. nicotianae*. However, *P. palmivora* and *P. meadii* are isolated most frequently as the causal agent of abnormal leaf fall.

3. Disease Cycle

Besides rubber tree, *Phytophthora* spp. have many other hosts, whose parts when fallen on the ground in conditions of high humidity, enable the sporulation of pathogen. Resistant structures, chlamydospores, and oospores may also germinate and form sporangia. The entry of the pathogen is through stomata.

4. Management

Bordeaux mixture using high-volume sprayers is used to control the disease. The current available technologies for spraying are mostly oil-based and involve high dose of chemical inputs (copper sulfate plus oil). Systemic fungicides as Aliette and Ridomil were found effective, and tree injections were not feasible for adoption on a large scale. Large scale screening of germplasm material to locate sources of resistance of *Phytophthora* should be intensified. It has been found that resistant clones budded on high yielding susceptible clones has experimentally provided a measure of control in southern India.

18.10 DOWNY MILDEW OF GRAPES (*VITIS VINIFERA*)

The disease is prevalent in all the grapevine areas of the world. It also affects other species of *Vitis*. In India, the disease is quite common in Maharashtra state and South India (Karnataka and Andhra Pradesh).

1. Symptoms

All the succulent tender parts of the vines are attacked, leading to a partial or total destruction of the foliage, dwarfing, and killing of the shoots, and rotting, and checking of the berries. The fungus spreads rapidly, with profuse downy growth on the undersurface, and corresponding greenish patches on the upper surface, which turn chlorotic with age. The mildew growth may cover the entire leaf blade which turns brown and withers. The disease may also involve floral parts and the fruit. Infected flowers die and drop off. The fruit becomes grayish, and the skin hardens and shrivels. Very often they are mummified while still attached to the bunch (Fig. 18.6).

2. Causal Organism

The pathogen responsible for grape downy mildew is *Plasmopara viticola* (Berkely, and Curtis) Berlese and de Toni.

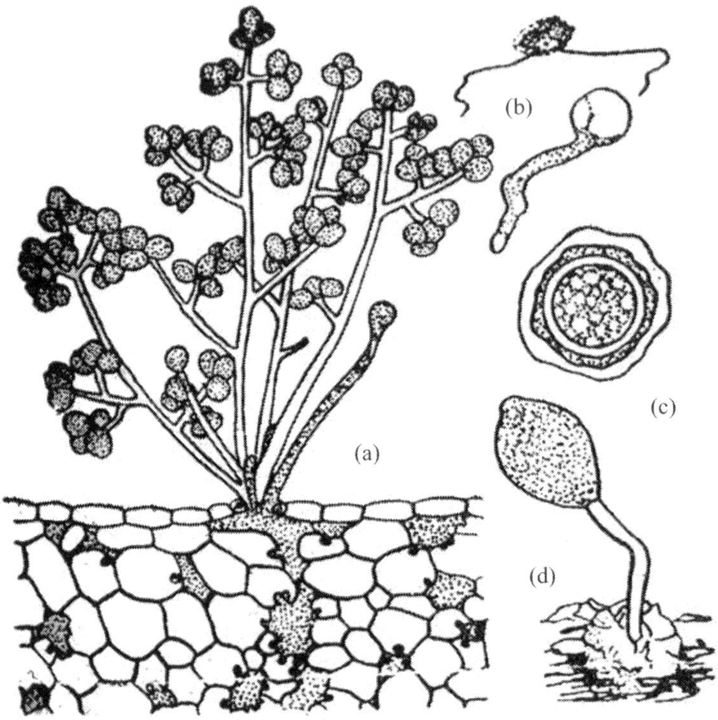

Fig. 18.6 Downy mildew of grapes caused by *Plasmopara viticola*: (a) Sporangiophores and sporangia emerging out of the grape leaves, intercellular mycelium with haustoria (b) Zoospores (c) Oospore (d) Germinating oospore

3. Disease Cycle

In the areas where grapevine is evergreen, the fungus produces sporangia (conidia-like) throughout the year, otherwise the disease is carried from year to year in the form of oospores formed on leaves and leaf shoots.

4. Disease Management

Sanitary measures, such as cutting the primary source of infection by deep cultivation, and ploughing, thus preventing the germination of oospores, are desirable practices. Removing and burning the diseased leaves, shoots, and berries that may contain hibernating oospores, helps in preventing epidemics.

The disease can be effectively controlled by giving three to five prophylactic sprays with 1% Bordeaux mixture.

The applications of copper and dithiocarbamate fungicides have been successfully used against the disease. In addition to Bordeaux mixture, the disease has been controlled with well-timed application of Blitox 50(0.3%), Maneb, Zineb, and Mancozeb (0.2%). According to Singh (1998), fruit sprays have been recommended at the following stages:

(a) Immediately after pruning the grapevine leaves
(b) 3-4 weeks after pruning the leaves
(c) Before the vine buds open
(d) When berries have formed
(e) Finally during growth of roots

A combination of systemic and nonsystemic fungicides is one such strategy for disease control. Hence, application of Metalaxyl (Ridomil) with either copper or Mancozeb fungicides has been found to give good control of the disease. Aureofungin (50 ppm) has been reported to be quite effective. Spraying of Ridomil (0.1%) and Aliette (0.2%) among the systemic group, and Daconil (0.2%) and Delan (0.2%) among the nonsystemic group have been found to be effective.

The use of Metalaxyl fungicide along with either copper fungicides or dithiocarbamates has been found quite effective in controlling the disease. This practice is followed to avoid resistance development in the pathogen against Metalaxyl fungicide. Metalaxyl fungicides (Ridomil) inhibit the mycelial growth and sporangial formation through action of its volatiles. Application of Dimethomorph has been found better than Metalaxyl–Dithiocarbamates combination because of its protective and curative activity. Another systemic fungicide phosphorus (phosphonic) acids or its derivative Phosethyl-Al is considered better than the Metalaxyl group of fungicides because this fungicide is translocated from leaves via xylem or phloem to shoots, and fruits where this fungicide breaks down to phosphorus acid and carbon dioxide. The phosphorus (phosphonic) acid acts directly on the fungus or indirectly effects the fungus, thereby inhibiting the fungal growth.

According to Wong and Wilcox (2001), Azoxystrobin—a strobilurin wide spectrum systemic fungicide—when applied at 0.5 mg/ml gives 61% control of the disease. Even the mixture of Azoxystrobin with Cymoxanil was more effective. Trifloxystrobin, another Strobilurin fungicide, is also effective against downy mildew of grapes. According to Reuveni (2003), Benthiavalicarb is a more recent fungicide effective against Oomycetes, and has been used in the control of downy mildew of grapevine. This fungicide inhibits zoospores encystment, mycelial growth, and a sporangial production. It has a strong prophylactic and local activity.

Induction of resistance to downy mildew of grapes by treatment of DL-β-amino-*n*-butyric acid (BABA) has been reported by Reuveni *et al.* (2001). Elicitation of defence mechanism in grapevine against downy mildew by the application of Laminarin (beta-1-3 glucan laminarin), a phytochemical derived from *Laminaria digitata* (brown algae), is reported by Aziz *et al.* (2003).

It has been observed that the application of water extract of compost horse manure and straw soil mixture develops resistance in plants against *P. viticola.* The mode of action is the result of biological control through enhanced microbial activity. Even *Fusarium proliferatum* has been identified as a biocontrol against *P. viticola*. The hyphae of the mycoparasite (*Fusarium* sp.) coil around and penetrate the sporangiophores of *P. viticola*. Cultivars Amber Queen, Champion, and Digraset have been found highly resistant while cultivars Athens, Buckland Sweet Water, Goethe, James, Malaga are moderately resistant to downy mildew of grapes.

18.11 DOWNY MILDEW OF BAJRA (*PENNISETUM TYPHOIDES*)

Downy mildew is now considered a major problem in the pearl millet crop, not only in India, but in the African countries too. Losses may be as high as 30% in high-yielding varieties in India.

1. Symptoms

The infected plants are dwarfed because of shortening of the internodes, and tillering is excessive. The foliage becomes pale and chlorotic, and on the underside the leaves are covered by the whitish cover of the sporangia (Fig. 18.7a) which are formed in abundance during dewy nights. The leaves may become distorted and wrinkled, and tend to split into threads. Shredding of the leaves is more common in the jowar downy mildew caused by *Sclerospora sorghi*. The tearing and shedding of sorghum leaves has been shown to be due to the activity of pectolytic enzymes of *S.sorghi.* The oosporic stage of the fungus is more predominant, and the ears are transformed into leafy bearded structures. Three types of symptoms (Fig. 18.7b) are seen: (1) normal cob length, and the whole cob converted into green leafy bearded structure, (2) normal cob length but the lower half converted into leafy bearded structures, and the upper half bearing normal grains and (3) the whole inflorescence reduced and converted into leafy bearded structures without bearing any grains. The infection converts the various floral parts, including glumes, palea, stamens and pistils into green, linear, leafy structures of variable lengths.

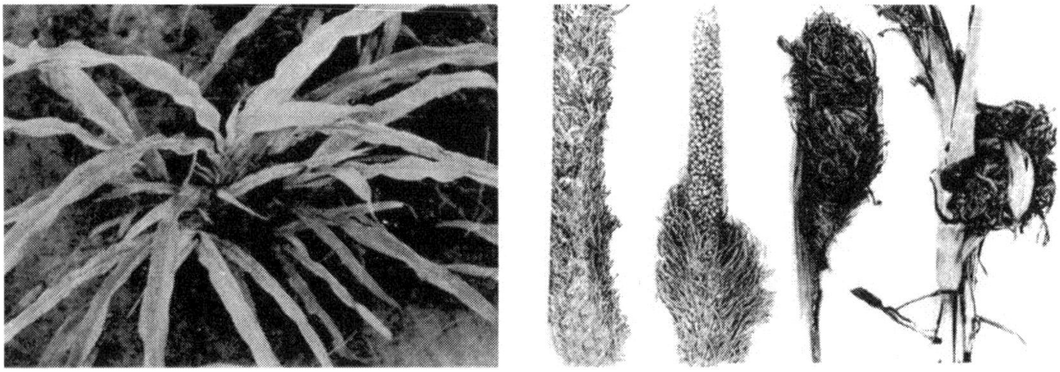

(a) (b)

Fig. 18.7 Downy mildew symptoms on bajra due to *Sclerospora graminicola* (a) On young seedlings, and a (b) Ear showing symptoms

2. Causal Organism

Sclerospora graminicola (Sacc.) Schroet is the causal organism of the downy mildew and green ear disease of bajra.

3. Disease Cycle

The disease is primarily soil-borne. The oospores which are abundantly present in the diseased leaves fall on the ground, and perennate during summer. It is reported that oospores can survive from eight months to ten years. Oospores give increased infection (therefore germinate better after weathering), and one-year-old oospores give more infection (98%) than fresh or more than one-year-old oospores.

4. Disease Management

Alterations in the date of planting (early planting), modification in planting methods (transplanting instead of direct sowing), avoidence of ratooning, roguing, and gap filling, deep ploughing, and sun-baking of soil, removal of infected crop residues, avoidance of monoculture, crop rotation, avoidance of low-lying fields, and waterlogging will reduce the disease incidence. Application of balanced fertilizers, particularly nitrogen and cultivation of trap crops helps considerably in disease management. Since the disease may be carried partly with the seeds, seed treatment may be of help in destroying the spores sticking to the surface of the grains. Seed treatment with Ceresan and Captafol is recommended. The All India Coordinated Millet Improvement Programme had recommended seed treatment with Agroson, GN (0.1%) + Thiram (0.4%) which gave about 50% control of the disease. However, organo-mercurials are not recommended. Recently metalaxyl compounds have been found very effective. Seed treatment with Ridomil (8 g/kg of seed) followed by one spray of 0.1% Ridomil 20 days after planting have been recommended. Appaji *et al.* (1989) recommended that seed dressing with Apron SD-35 (8 g/kg of seed) followed by a foliar spray 25 days after sowing, with Ridomil-MZ-72 WP (Metalaxyl

+ Mancozeb) at 500 ppm concentration or with Ridomil-MZ-280 FW (Metalaxyl + Zineb) at 1000 ppm concentration was the best combination. There are some reports that seed treatment with crude extracts of *Vinca rosea, Ocimum sanctum, Allium sativum, Datura stramonium, Azadirachta indica,* and *Thuja sinensis* reduces incidence of disease.

Some resistant varieties are HB-5, NHB-10, and NHB-14. Cultivars WC-C75 and ICTP-8203 have shown durable resistance.

18.12 WHITE RUST OF CRUCIFERS

White rust of crucifers is common on several cruciferous crops, such as cabbage, turnip, mustard taramira (*Eruca sativa*), and radish, and occurs throughout India. The yield loss due to floral infection may be up to 54.5 percent.

1. Symptoms

All aerial parts of the plant are attacked. White or cream-yellow pustules of various sizes and shapes appear on the surface of the leaves (Fig. 18.8a). In some crucifers, the leaves may become thick, fleshy, and inrolled, and when infection is severe, the size of the leaves may be reduced and even the entire plant may be stunted. The spots are generally produced on the lower surface of the leaves. When young stems and inflorescence are infected, the fungus becomes systemic in the tissues and stimulates various types of deformities of which hypertrophy is most pronounced. Due to the hypertrophy and hyperplasia of the tissues, the floral parts show swelling and distortions. The peduncle, pedicel, sepals, petals, and carpels are hypertrophied. The petals may become like sepals, and the stamens may become leaflike or occasionally like carpels (Fig. 18.8b). The latter may be open, while the ovules are usually atrophied. The parasite also stimulates cell activity, leading to an increase in cell size, cell division, and formation of chlorophyll and starch at places where none is usually formed.

(a) (b)

Fig. 18.8 White rust of crucifer: (a) Symptoms on the leaf (b) Hypertrophied inflorescence

2. Causal Organism

Albugo candida (Lev.) Kunze, also known as *Cystopus candidus* Lev., is an obligate parasite. The mycelium is intercellular, forming knob-shaped haustoria in the host cells. Hyphae from the endophytic mycelium collect beneath the epidermis. These dense masses of hyphae form the sporangial beds (sori). The sporangiophores are formed by the vertical growth of broad, short, clavate stalks under the epidermis which is raised up due to the pressure of these structures. The raised epidermis is separated from the underlying cells. The sporangiophores are free from each other laterally and are very thick, especially towards the base. Sporangia are formed in basipetal succession in chains (Fig. 18.9A). The oogonia and antheridia are formed from the intercellular mycelium in the intercellular spaces [Fig. 18.9 B, C, D (a)]. Oospores are formed only in some hosts, and their formation is conditioned by certain environmental factors. They are globular, provided with a tuberculate brown epispore. Their germination takes place after a long period of rest, and is by the formation of zoospores [Fig. 18.9C (a)]. The rest period can be shortened if they are chilled at 4°C for a week.

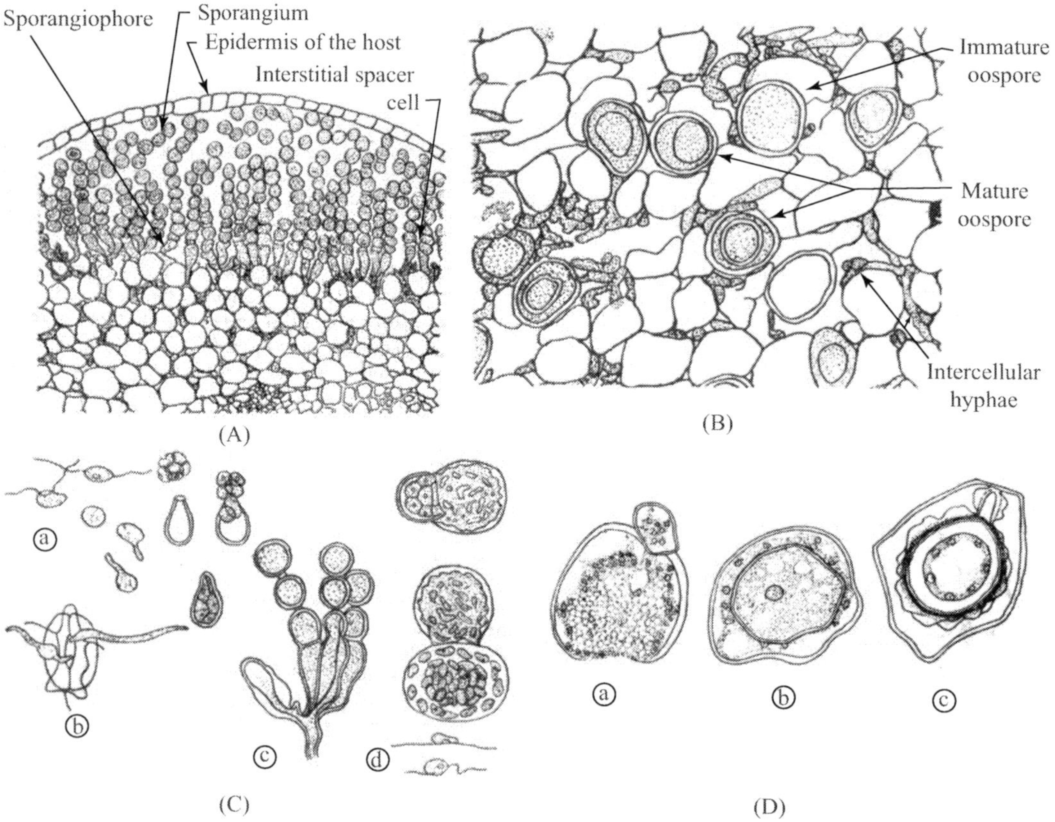

Fig. 18.9 (A) *Albugo candida* section through a conidial (sporangial) pustules showing conidiophores, conidia, and intercellular mycelial with haustoria (B) Intercellular mycelial and various stages of oospore development (C) Stages in the life cycle of *A. candida* (a) Zoospore formation (b) Stomatal penetration by germ tube (c) Conidiophores (d) Oospore germination (D) Formation of oospores (a) Oogonium and antheridium (b) Formation of oospore wall (c) Mature oospore in section

3. Disease Cycle

The pathogen perpetuates through the oospores lying in the soil in diseased plant debris. Perennial weed hosts serve as a source of primary inoculum. Secondary spread of the pathogen is carried out by sporangia and zoospores.

4. Disease Management

(a) Cultural Methods

- Clean cultivation should be encouraged.
- Destruction of weeds in and around the field should be given top priority.
- The disease can be controlled by crop rotation.

(b) Chemical Control Apart from Ridomil and Aliette for white rust of crucifers, other fungicides which are also effective include Bordeaux mixture (0.8%), Difolatan (0.3%), Daconil (0.1%), Dithane M-45 (0.2%).

Sources of resistance to the disease in *Brassica napus* and *B. juncea* are reported by Saharan and his associates (1988).

Good sources of resistance have been listed in some species, and genotypes in rapeseed-mustard having multiple disease resistance, and these include *Brassica alba*, *B. carinata* (HC-1, PCC-2), *B. juncea* (DIR-1507), and *B. napus* (Midas, Tower). According to Singh US *et al.* (2002), pre- or co-inoculation with *Peronospora parasitica* or avirulent strain of *Albugo candida* provides protection against the development of *Albugo candida*.

18.13 SOME OTHER DISEASES CAUSED BY OOMYCOTA

Name of the disease	Causal organism	Symptoms	Disease cycle	Management
Stem or foot rot of papaya	*Pythium aphanidermatum*	Spongy, water-soaked patches on the bark girdle the stem causing the tissue to rot.	Soil-borne (Oospores in soil)	Roguing of infected plants, soil treatment with Bordeaux mixture, Calixin, Topsin M, Daconil, and seed treatment with Thiram or Difolatan are effective.
Bud rot of Palms	*Phytophthora palmivora*	Discoloured spots on leaves, parts of crown rots, emitting foul smell.	Soilborne if oospores are formed or survival of the pathogen on dead fallen leaves	Cutting and burning of diseased leaves, Bordeaux mixture or copper fungicide, Ridomil, and Mancozeb are quite effective.
Gummosis or gum disease of citrus	Several species of *Phytophthora* are responsible—*Phytophthora palmivora*, *P. nicotianae* var. *parasitica*, and *P. citrophthora*	Water-soaked area, and later brownish gummy substance exudes, bark completely rotten, tree dies.	Soil-borne	Use of resistant root stocks, appropriate cultural practices, Bordeaux mixture, Difolatan (3%), Ridomil (0.2%), and Aliette (0.02%) are effective.

(Contd.)

Name of the disease	Causal organism	Symptoms	Disease cycle	Management
Downy mildew of peas	*Peronospora pisi*	Downy growth of the pathogen on the lower surface of the leaf, pale green elliptical blotches appear on pods and finally infected seeds abort.	Soil-borne	Rotation of crops is effective in controlling the disease.
Downy mildew of crucifers	*Peronospora parasitica*	Purplish brown spots on the underside of leaves.	Soil-borne, seed contaminated with trash may also carry the pathogen	Crop rotations, deep ploughing in summer, Dithane Z-78 (0.3%), Dithane M-45 (0.3%), Ridomil (0.1%) are effective. Some resistant varieties are *Brassica alba*, *B. carinola* (HC-1), B. juncea (DIR-1507, DIR-1522).
Downy mildew of cucurbits	*Pseudoperonospora cubensis*	Pale yellow, angular patches on lower surface of leaf.	Pathogen survives on wild host in the off season	Fungicides like Dithane M-45, Zineb, Difolatan, Chlorothalonil, Ridomil, and Aliette are effective.
Downy mildew of maize	*Peronosclerospora sorghi*	Development of long, rather broad, chlorotic stripes on the leaf which are initially yellow in colour, and finally turn brown.	Pathogen can perennate through oospores in soil, and on alternate host	Long crop rotation, burning of crop debris, Dithane M-45, Ridomil-25WP, Apron-35 (as seed treatment) are effective. Phil-DMR-1 is a resistant variety.
Brown stripe downy mildew of maize	*Sclerospora rayssiae* var. *zeae*	Narrow chlorotic spots with well-defined margins which extend in parallel fashion between the veins appear on the leaf blades.	Soil-borne	Fungicides like Dithane M-45, Demosan in combination with roguing of infected leaves, neem oil, and systemic fungicides like Ridomil 25WP are effective.
Sugarcane downy mildew of maize	*Sclerospora sacchari* (now called *Peronosclerospora sacchari*)	Main symptom is the development of long rather broad chlorotic stripe along the entire length of leaf, young leaves show downy growth more clearly than older leaves.	Pathogen can perennate through oospores in soil and on alternate host	Diseased crop residue should be destroyed. Dithane M-45 is effective in checking the disease.

(*Contd.*)

Name of the disease	Causal organism	Symptoms	Disease cycle	Management
Philippine downy mildew of maize	*Sclerospora philippinensis* or *Peronosclerospora philippinensis*	Leaves become pale yellow with wooly white growth on undersurface, plants possess shorter internodes, infected plants may not produce any cob.	Pathogen survives on alternate host (*Saccharum spontaneum*, *S. bicolor*, *S. halepense*)	Long crop rotation is recommended.
Downy mildew or leaf shredding disease of sorghum	*Peronosclerospora sorghi*, earlier known as *Sclerospora sorghi*	White soft downy growth on the lower surface of leaf, and discolouration of leaves, and finally plant dies before the ears are formed.	Soil- and seed-borne	Deep ploughing, long crop rotation for soil-borne pathogens but seed treatment of Ridomil (1 g/kg) is recommended. Resistant varieties are IS-1547, IS-4696, IS-5665.
Downy mildew of onion	*Peronospora destructor*	Downy growth of the fungus all over the leaf surface, growth of the plant stunted, distorted, and pale green, small bulbs are formed.	Seed (bulb)-borne	Dry heat treatment of bulbs, proper drainage, Zineb, Mancozeb, Difolatan, Metalaxyl are effective. Resistant varieties are IC 48045, IC 32149, and IC 33617.
Downy mildew of lettuce	*Bremia lactucae* f. sp. *lactucae*	Leaves are infected, pale yellow areas appear on leaves, and growth of the plant is dwarfed.	Rarely perpetuates through seeds, mostly oospores in crop debris, and on wild host	Well-drained fields, eradication of wild host, use of healthy seeds, Mancozeb, and Metalaxyl are effective.
Stem blight of pigeon pea	*Phytophthora drechsleri* f. sp. *cajani*	Brown to dark brown spots are formed on stem, enlarged in size, and girdle the stem. Portion of plant above the lesions dries, and the stem drops at the point of infection.	Soil-borne	Crop rotation, Use of potassium fertilizers, mixed cropping are recommended for cultural practices. Metalaxyl + Mancozeb provides maximum disease control. *Trichoderma hamatum*, *Pseudomonas fluorescens*, and *Bacillus subtilis* are biocontrol agents.

REVIEW QUESTIONS

1. What are downy mildews and how do they differ from powdery mildews?
2. What do you understand by physiological specialization? Discuss the phenomenon, giving suitable examples from downy mildews.

3. Suggest the various methods of controlling downy mildews.
4. Name some important plant diseases caused by *Pythium* and *Phytophthora* spp. Describe the disease cycle of late blight of potato.
5. Enumerate the major differences between the symptoms of downy mildew and rots.
6. Describe the mechanism of disease development or the host-parasite relationship in each case.
7. What is obligate parasitism? Describe the life cycle of *Albugo candida* along with its methods of disease control.
8. Write notes on the following:
   (i) Late blight of potato (ii) White rust of crucifers
   (iii) Downy mildew of grapes (iv) Downy mildew of cucurbits
   (v) Leaf blight of Colocasia (vi) Damping-off of vegetable seedlings

MULTIPLE-CHOICE QUESTIONS

1. The cell wall of Oomycota mainly consists of
   (i) chitin (ii) chitin-glucan
   (iii) glucan-cellulose (iv) chitin-cellulose
2. White rust of crucifers is caused by
   (i) *Albugo candida* (ii) *Albugo bliti*
   (iii) *Albugo tragopogonis* (iv) All of these
3. Which of the following enzymes play an important role in damping-off of vegetables?
   (i) Pectolytic (ii) Lignolytic
   (iii) Cellulolytic (iv) Both (i) and (iii)
4. Which of the following pathogen is responsible for damping-off of seedlings?
   (i) *Pythium aphanidermatum* (ii) *Phytophthora palmivora*
   (iii) *Peronospora parasitica* (iv) *Sclerospora sacchari*
5. Which of the following practices are used in Integrated Disease Control?
   (i) Cultural methods (ii) Chemical methods
   (iii) Biological methods (iv) All of these
6. Which of the following does not act as a biocontrol agent against damping-off?
   (i) *Trichoderma viride* (ii) *Trichoderma harzianum*
   (iii) *Trichoderma koningii* (iv) *Phytophthora parasitica*
7. Which of the following pathogens are involved in fruit rot of crucifers?
   (i) *Pythium aphanidermatum* (ii) *Fusarium* sp.
   (iii) *Rhizoctonia* sp. (iv) All of these
8. The mode of action of *Pseudomonas* sp. against damping-off is
   (i) accumulation of endogenous salicylic acid
   (ii) stimulation of peroxidase and polyphenol oxidase
   (iii) production of sugars
   (iv) both (i) and (ii)
9. Which of the following are responsible for rhizome rot of ginger?
   (i) *Pythium myriotylum* (ii) *Pythium aphanidermatum*
   (iii) *Pythium gracile* (iv) All of these

10. The greatest famine in Ireland in 1840's is associated with
    (i) late blight of potato (ii) early blight of potato
    (iii) damping-off of seedlings (iv) rhizome rot of ginger
11. The causal organism of late blight of potato in the northern plains of our country is carried through
    (i) tubers used as seeds (ii) true seeds
    (iii) diseased leaves in soil (iv) sporangia in soil
12. The fungus responsible for the Koleroga or Mahali of areca palms is :
    (i) *Phytophthora colocasiae* (ii) *P. drechsleri*
    (iii) *P. infestans* (iv) *P. areacae*
13. Which of the following fungicides are effective in controlling *Phytophthora* diseases?
    (i) Aliette (ii) Ridomil
    (iii) Bordeaux mixture (iv) All of these
14. Which of the following fumigants are effective in control of foot rot of *Piper betle*?
    (i) Formalin (ii) CS_2
    (iii) DD Mixture (iv) Vapam
15. The discovery of holy solution, i.e. Bordeaux mixture, is associated with
    (i) downy mildew of grapes (ii) powdery mildew of grapes
    (iii) late blight of potato (iv) early blight of potato
16. Which of the following is a systemic fungicide used to control downy mildew of grapes?
    (i) Bordeaux mixture (ii) Azoxystrobin
    (iii) Vitavax (iv) Zineb
17. Which of the following is an obligate parasite?
    (i) *Alternaria solani* (ii) *Cercospora arachidicola*
    (iii) *Sclerospora graminicola* (iv) *Phytophthora colocasiae*
18. The use of plant extracts to control plant diseases is called:
    (i) control by Botanicals (ii) cultural practices
    (iii) chemical methods (iv) integrated disease control
19. *Albugo candida* responsible for white rust of crucifers was earlier called
    (i) *Alternaria solani* (ii) *Plasmodiophora brassicae*
    (iii) *Phytophthora nicotianae* (iv) *Cystopus candidus*
20. The pathogen responsible for white rust of crucifers belongs to
    (i) Oomycota (ii) Ascomycota
    (iii) Basidiomycota (iv) Zygomycota
21. The pathogen responsible for downy mildew of cucurbits is
    (i) *Peronospora parasitica* (ii) *Pseudoperonospora cubensis*
    (iii) *Sclerospora sacchari* (iv) *Plasmopara viticola*
22. The stem blight of pigeonpea is caused by
    (i) *Phytophthora drechsleri* (ii) *Phytophthora nicotianae*
    (iii) *Phytophthora parasitica* (iv) *Phytophthora palmivora*

Answers

1. (iii) 2. (i) 3. (iv) 4. (i) 5. (iv) 6. (iv) 7. (iv) 8. (iv) 9. (iv) 10. (i)
11. (i) 12. (iv) 13. (iv) 14. (i) 15. (i) 16. (ii) 17. (iii) 18. (i) 19. (iv) 20. (i)
21. (ii) 22. (i)

19

DISEASES CAUSED BY PLASMODIOPHOROMYCOTA

The following are the three important diseases caused by this group.

1. Club root of crucifers caused by *Plasmodiophora brassicae*
2. *Polymyxa,* causing root disease of cereals and grasses
3. Powdery scab of potato caused by *Spongospora subterranea*

The pathogens are obligate parasites but can survive in soil as resting spores for many years but can grow and multiply in only a few host. The Plasmodium lives off the host cell it invades but does not kill these cells for a long time. In some diseases the pathogen invades the host, and adjacent cells are stimulated by the pathogen to enlarge and divide. The pathogen spread is from one plant to another by zoospores. *Polymyxa* and *Spongspora* are vectors of several viruses.

CLUB ROOT OR FINGER-AND-TOE DISEASE OF CABBAGE (*BRASSICA OLERACEA* VAR. *CAPITATA*)

The club root is the most dangerous disease of the mustard family. The disease is prevalent in India but common in Sri Lanka. Once the disease appears in a particular field, there is complete loss of the crop. It has been reported that this disease is spreading at an alarming rate on rape seed and mustard (yellow sarson variety) in Bengal.

1. Symptoms

The disease may progress to a considerable extent before above-ground signs are noticeable. Infected roots enlarge relatively rapidly to form "clubs" which assume a variety of shapes shown in figure. The clubbing of roots is of many types such as (a) clubbing is complete on main and lateral roots, (b) clubs are present on only main roots, (c) clubs are present on lateral roots, (d) rootlets are free from clubs, (e) clubs appear as tumors on the roots, and (f) sometimes dark spots are present on the root. These various types of symptoms are due to specific roots infected by the pathogen.

Seedlings, when infected, show flagging and symptoms of chronic water deficiency. The leaves assume a pale green or yellowish colour and after infection, the diameter of the roots swells to 10-12 times the diameter of normal plants resulting in their death. In the case of later infections, the disease

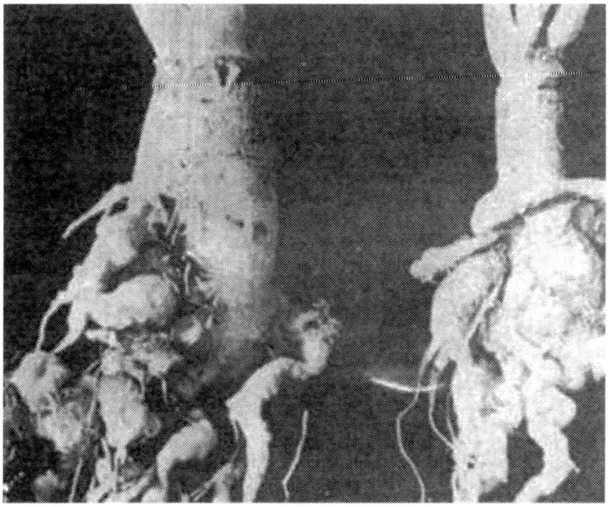

Club-root diseases of crucifers (*Plasmodiophora brassicae*): Symptoms (Courtesy of BASF India Ltd.)

does not assume such a severe form. An affected plant if pulled up shows a grossly distorted mass of large and small, single or many, irregular galls. Translocation of food materials and water is interrupted which results in the starvation of the plants. Heads are normally not formed but if they are, they remain very small.

2. Causal Organism

Plasmodiophora brassicae Wor. is the pathogen. The pathogen produces resting spores inside the host cells.

The resting spore germinates to produce a single zoospore with two flagella of unequal lengths, both of the whiplash type. Germination is stimulated by substances diffusing from cabbage roots.

The primary zoospores infect the roothairs or young epidermal cells of the host. Before infecting the host, the zoospores become amoeboid. The amoeba becomes attached to a roothair and penetration occurs. Aist and Williams (1971) described the mechanism of zoospore infection and entry of the pathogen with the help of electron micrographs. The entire body of the amoeba enters the host cell: within the roothair or epidermal cell, a small uninucleate stage is first seen. Later the nucleus divides and a small plasmodium is formed. The plasmodium divides to form zoosporangia. Each zoosporangium finally produces 4–8 uninucleate zoospores. The mature zoosporangium becomes attached to the host cell wall and a pore develops at this point through which the zoospores escape. Sometimes the zoospores are released into the lumen of the host cells. The behaviour of the released zoospores is not properly known but it is possible that these may function as gametes.

Hypertrophy of the host cells is apparently brought about by the blocking of the mechanism for cell division and is accompanied by enhanced DNA synthesis. Starch accumulates in the infected cells.

3. Disease Cycle

The fungus persists in the soil as resting spores in diseased host tissues. When a susceptible cruciferous host is planted in the soil, it infects the roots. It multiplies and spreads in the plasmodia stage causing the hypertrophy and hyperplasia of infected cells. Later, resting spores are formed. Thus, the disease cycle is continued.

4. Disease Management

Control of club root of crucifers is difficult because of about 15 years longevity of resting spores in soil and ineffectiveness of most chemical treatments. Eradication by partial soil sterilization is expensive and difficult, and liming the soil and crop rotation are less effective and often impractical. Fungicidal treatment can be effective for transplanted crops. However, the following cultural methods have been recommended.

(a) Cultural Methods The disease can be controlled by the following methods.

- Long rotation for crop
- Use of pathogen-free plots
- Eradication of crucifer weeds from the vicinity of the plots
- Seedlings should be raised in pathogen-free soil

(b) Chemical Control As the disease is severe in soil with acidic pH, addition of hydrated lime to raise the pH slightly above neutral (usually about pH 7.2) is quite effective in containing the disease. At that pH spores of the club-root pathogen germinate poorly or not at all.

Although a large number of chemicals have been recommended such as Chloropicrin, Methyl bromide or Metam sodium or PCNB, to eradicate the soil-borne inoculum, all these chemicals are hazardous, pollute the environment, and hence are on the banned list now.

Bhattacharya and Pramanik (1998) isolated a number of rhizobacteria, i.e. *Pseudomonas fluorescens* H 237, *Bacillus subtilis* T 99, and *Streptomyces graminofaciens* G1H, from the rhizosphere of diseased crucifers and tested against *Plasmodiophora brassicae*, both *in vitro* and *in vivo* conditions. They found that *P. fluorescens* and *S. graminofaciens* reduced the growth of club formation as well as of roothair infection while *B. subtilis* stimulated club formation and roothair formation. It was further found that neem (Margosa) oil suspension when applied in soil was phytotoxic but neem cake and Nemin, a gelatinous neem product produced by Godrej India Ltd., reduced disease severity under laboratory and greenhouse conditions.

5. Resistant Varieties All exotic cultures belonging to *Brassica napus*, *B. nigra*, and *B. carinata* are either free from the disease or are resistant. Resistance to rapeseed-mustard isolate of the pathogen in West Bengal has been found in cabbage cultivars KK Cross, Drumhead; Cauliflower cultivars Dania, Snow ball and Khol-Khol cultivar White Vienna (Chattopadhyay *et al.*, 2001)

REVIEW QUESTIONS

1. Name the important diseases and causal organisms of Plasmodiophoromycota.
2. Write the symptoms and disease cycle of club root or finger-and-toe disease of cabbage.
3. Briefly discuss the management of club root or finger-and-toe disease of cabbage.

MULTIPLE-CHOICE QUESTIONS

1. The main symptom of club root or finger-and-toe disease of cabbage is
   (i) hypertrophy (ii) hyperplasia
   (iii) clubbing of roots (iv) all of these
2. Which of the following members of Plasmodiophoromycota act as vector of several viruses?
   (i) *Polymyxa* (ii) *Plasmodiophora*
   (iii) *Spongospora* (iv) Both (i) and (iii)
3. Which of the following cultural practices are effective in controlling club root or finger-and-toe disease of cabbage?
   (i) Crop rotation (ii) Eradication of weeds
   (iii) Sanitation (iv) All of these
4. Which of the following organism act as biocontrol agent against *Plasmodiophora brassicae*?
   (i) *Pseudomonas fluorescens* (ii) *Phytophthora infestans*
   (iii) *Spongospora subterranea* (iv) All of these
5. Which of the following varieties are resistant to club root or finger-and-toe disease of cabbage?
   (i) *Brassica napus* (ii) *Brassica nigra*
   (iii) *Brassica carinata* (iv) All of these

Answers

1. (iv) 2. (iv) 3. (iv) 4. (i) 5. (iv)

20 PLANT DISEASES CAUSED BY MOLLICUTES: PHYTOPLASMAS AND SPIROPLASMAS

Mycoplasmas, the free-living or parasitic organisms, are not new to microbiologists and veterinary pathologists. *Mycoplasma mycoides* was cultured *in vitro* some hundred years ago in Pasteur's laboratory (Nocard *et al.*, 1898) from animals suffering from bovine pleuropneumonia. Later on, similar types of organisms were isolated from a number of other animals and birds, and these were named **Pleuro-Pneumonia-Like Organisms (PPLO)**. Borrel *et al.* (1910) gave the first binomial to these organisms, and called them *Asterococcus mycoides,* and later they were put under the genera *Mycoplasma* and *Borrelomyces.* Since *Asterococcus* was an algal genus, the Editorial Board of the International Bulletin of Bacterial Nomenclature and Taxonomy (1955) discarded the genus *Asterococcus,* and recommended the use of either of the other two genera. Edward and Freundt (1956) proposed to retain only the valid genus *Mycoplasma* on grounds of priority. It is now an established and accepted genus of the order Mycoplasmatales under the class *Schizomycetes.* Although it resembles bacteria to a great extent, it cannot be classified with the so-called true bacteria.

Edward *et al.* (1967) proposed a new class which they called *Mollicutes.*

The morphological features of mycoplasmas which distinguish them from most other organisms are the following:

1. They are pleomorphic. They range from small, spherical bodies of about 10 nm diameter (elementary bodies) to large, irregular bodies with bigger diameters. There are also elongate forms, and their forms vary with the cultural conditions.
2. They are both parasitic and saprophytic.
3. Cells are nonmotile, and usually form typical "poached-egg" shaped minute colonies with a central nipple on agar media.
4. Cells are delimited by a unit lipoprotein membrane and lack rigid mucopeptide. The unit membrane is about 10 nm thick.
5. Within the cytoplasm are ribosomes which are of similar size (14 nm diameter) as those of bacteria (cytoplasmic ribosomes of eukaryotes are about 18 nm in diameter). The mycoplasmal ribosomes resemble those of bacteria in the sedimentation characteristics of both the nucleoprotein and nucleic acid. The ribosomes are mostly 70S.
6. Also, within the cytoplasm, a thin, stranded material (3 nm diameter) can be recognized which has been shown to be DNA. Extracted DNA is circular (a prokaryotic character) but can be distinguished from bacterial DNA by its low guanine and cytosine content.

7. Animal mycoplasmas also contain densely staining bodies of about 20–60 nm diameter. These are thought to be connected with lipid metabolism.
8. Cells of mycoplasmas are usually resistant to the antibiotics which act on cell walls, for example, Penicillin, Cephaloridine, Vancomycin, etc. Mycoplasmas are sensitive to Tetracyclines.

Although mycoplasmas were earlier associated with animal diseases, in 1967 it became apparent that proliferation diseases, yellows type, big bud, witches' broom, and certain stunting diseases are not of viral origin (Doi *et al.*, 1967; Nasu *et al.*, 1967). This inference was drawn on the basis of electron-microscope observations and the recovery of the disease symptoms by Tetracycline antibiotics.

Since 1967, mycoplasma-like organisms have been associated with as many as 200 plant diseases based on (1) the constant presence of Mycoplasma-Like Organism (MLO) in the phloem tissue of diseased plants and their complete absence from healthy plants, (2) the presence of similar microorganisms in insect vectors which carry the infectious organisms and the absence of MLO from non-infective individuals, (3) the recovery of plants after the application of tetracyclines and the reappearance of MLO when the disease recurs, and (4) the disappearance of MLO from the plants and vectors which have been cured by heat therapy. Attempts were made in many laboratories to obtain cell-free cultures of the various MLOs.

Properties of True Mycoplasmas

Mycoplasmas are prokaryotic organisms that have no cell walls. They are members of the class Mollicutes which has one order, Mycoplasmatales, which has further been divided into three families each with one genus: Mycoplasmataceae, genus *Mycoplasma*; Acholeplasmataceae; genus *Acholeplasma;* and Spiroplasmataceae, genus *Spiroplasma*. Mycoplasmas lack true cell wall. They measure from 175 to 250 nm in diameter during reproduction but grow into various sizes and shapes. Mycoplasmas reproduce by budding and by binary transverse fission of cells. They have no flagella, produce no spores, and are Gram negative. Nearly all Mycoplasmas are parasitic to humans and animals, and all saprophytic ones can be grown on more or less synthetic media. Mycoplasmas have been isolated from healthy and diseased animals, and humans suffering from diseases of respiratory and urogenital tracts. According to Agrios (2005), they have been associated with some arthritic and nervous disorders of animals, and some have been found to exist as saprophytes.

Phytoplasmas

The organisms observed in plants and insect vectors, i.e. the Phytoplasmas, which do not include the Spiroplasmas, resemble mycoplasmas of the genus *Mycoplasma* or *Acholeplasma.* According to Agrios (2005), Phytoplasmas are more related genetically to *Acholeplasma* than to *Mycoplasma.* Phytoplasma lack cell wall, are bounded by a unit membrane, and have cytoplasm, ribosomes, and strands of nuclear material. The size of their chromosome varies from 350 Kilobases of DNA to 1130 Kilobases. The shape of phytoplasmas is usually spheroidal to ovoid or irregular tubular to filamentous. Phytoplasmas and Spiroplasmas are generally present in the sap of a small number of phloem sieve tubes of the infected plants. Most plant mollicutes are transmitted from one plant to another by leaf hoppers but some are transmitted by psyllids and plant hoppers. Plant mollicutes also grow in the alimentary canal, hemolymph, salivary glands, and intracellular in various body organs of their insect vectors.

There is an incubation period for the multiplication and distribution of the mollicutes within the insect. Despite several attempts by several workers, no Phytoplasmas could be cultured on artificial media so far. Phytoplasmas have been extracted and their antisera, including monoclonal antibodies, have been prepared. Some of the modern techniques like specific antibodies, DNA probes, RFLP profiles, and analysis of 16s rRNA genes using PCR amplification have been utilized in the detection and identification of the Phytoplasmas.

Spiroplasmas

Spiroplasmas are helical mollicutes. Spiroplasmas are known to cause the stubborn disease in citrus plants, the brittle-root disease in horseradish (*Spiroplasma citri*), stunt disease in corn plants, and a disease in periwinkle. They are helical in shape, and hence called Spiroplasmas. *Spiroplasma citri* has also been found in several other dicots. Several other kind of Spiroplasmas have been found to infect honeybees and several other insects, and several other live saprophytically on flowers and other plant surfaces.

As stated earlier, since 1967, wall-less microorganisms were seen with the electron microscope in the phloem of plants infected with one of several yellows-type diseases, and in insect vectors of these diseases. These organisms were called **Mycoplasma-Like Organisms** (MLO) because of their superficial resemblance to mycoplasmas. It was later shown that these are not mycoplasmas. Although all belong to the class Mollicutes, order Mycoplasmatales of prokaryotes. The true nature of mycoplasma like organisms (MLO), now called phytoplasmas, is still not certain. Their genomes are only distantly related to true mycoplasmas (Agrios, 1996). Phytoplasmas have not been grown on an artificial nutrient medium so far, and in no case Koch's proof has been obtained by directly inoculating with phytoplasmas on healthy plants.

However, Spiroplasmas, as already indicated, have been cultured in the laboratory, and their pathogenicity had been proven. At the moment, Phytoplasmas are considered to belong to the class Mollicutes which include true mycoplasmas, but no family or genus name has been assigned to them. The citrus stubborn, and the corn stunt organisms are placed in the genus *Spiroplasma* in the class Mollicutes.

Spiroplasmas are helical mollicutes. Spiroplasmas are cells that vary in shape from spherical or slightly ovoid, 100–240 nm or larger in diameter, to helical, and branched nonhelical filaments. Unlike mycoplasmas, Phytoplasmas have been cultured on nutrient media. They produce mostly helical form in liquid media, and multiply by fission. They lack a true cell wall, and are surrounded by a unit membrane. The helical filaments are motile, moving by a rapid rotatory or screwlike motion. There are no flagella. Spiroplasma colonies on agar media have a typical fried-egg appearance, and are around 0.2 mm in diameter.

Two distinct groups of spiroplasmas have been reported to cause diseases in plant. One group, represented by corn stunt Spiroplasma, has a narrow host range limited only to species of *Zea*. The other group, represented by *Spiroplasma citri*, has a much wider host range, and has been found to cause stubborn diseases in citrus as well as in crucifers, and other crop plants. Tully and Whitcomb (1983) have reviewed the work in Spiroplasmas.

More than 200 plant diseases, grafts or insect-transmissible with characteristic symptoms such as abnormalities in floral and vegetative parts (virescence and phyllody), yellowing, proliferation of axillary buds, reduction in leaf lamina, general stunting etc. have been reported to be due to mycoplasma-like bodies from various parts of the world (Agrios, 2005).

20.1 SANDAL SPIKE

This disease has become very serious, and is threatening the sandal industry. It has spread to the entire southern part of Karnataka as well as to the border areas of the neighbouring states. It was earlier thought to be caused by graft transmissions and an insect-borne virus. The sandal (*Santalum album*) tree is of great economic importance for its heart wood and oil which are very fragrant. The tree is a semiroot parasite, grows in the forests of southern Karnataka and northern Tamil Nadu. A recurring annual loss of ₹ 194 million has been reported.

1. Symptoms

Trees of all ages get infected. Sandal spike is of two types: rosette and pendular type. **Rosette spike** is more common, and commonly referred to as spike or spike disease. The disease is characterized by an extensive reduction in the size and the internodes. The whole shoot looks like a spike bearing four rows of spiked bristles. Spiked plants do not bear flowers or fruits. Only occasionally phylloid or abortive flowers are produced. Diseased trees display virescens. The disease also affects fine roots and causes their disconnection from the roots of the host plants causing death of the trees.

Pendulous spike is less common. Due to elongation of infected shoots disproportionate to their thickness, they droop. Rosette symptoms are absent in these branches. In this case, the root ends and haustoria do not die.

2. Causal Organism

The MLOs are found in the phloem tissues of the leaves, petioles, stems, and roots of diseased plants. These bodies are mostly 40 to 750 nm in diameter; however some of the bodies were ovoid to ellipsoidal in shape. Typically, they are devoid of cell walls, and the cytoplasm is bound by unit membranes 10–12 nm thick. The organism contains a fibrillar network of DNA and ribosomal bodies.

3. Disease Cycle

The disease is transmitted through root contact and insects. More than one species of leaf hoppers are involved in its transmissions such as *Nephotettix virescens* and *Moonia albimaculata*. The MLO infects a large number of collateral hosts such as *Eucalyptus grandis, Vinca rosea, Zizyphus oenoplia,* and *Dodonaea viscosa*, etc. The diseases can be transferred from *Vinca rosea* to Sandal, and vice versa through dodder. Epidemiology of the disease is poorly understood. *Lantana camara* renders the trees susceptible to the disease and may be acting as a pathogen reservoir.

4. Disease Management

So far no effective method of controlling the disease is known. In the past attempts have been made to cure the trees by raising the temperatures above 38°C by building fires in trenches around the affected trees, but it was neither successful nor practicable. On the assumption that the incidence of the disease and its spread may be lessened by eliminating the source of infection, attempts were made in the past to remove or kill the spiked sandal trees by arsenical poisoning, and also by removing the diseased branches. However, this resulted in the removal of large trees without significantly reducing the disease incidence.

Different host plants of sandal impart resistance or susceptibility to the disease. Among the host plants through which sandal acquires resistance are *Azadirachta indica, Bambusa arundinacea, Cassia siamea, Casuarina equisetifolia, Dalbergia sissoo, Dodonaea viscosa, Erythroxylum monogynum, Ficus benghalensis, Murraya koeningii, Ruta graveolens, Sarcostemma acidum, Semecarpus anacardium, Strychnos nux vomica,* and *Zizyphus mauritiana*. Among the hosts which render sandal susceptible to spike are species of *Acacia, Caesalpinia coriaria, Cajanus cajan, Lantana camara, Ocimum sanctum,* and *Pongamia pinnata*. It may, therefore, be possible to raise sandal in association with those trees which impart relative resistance to the sandal trees.

In nature, sandal generally remains free from infection under a top canopy of host species such as *Casuarina, Eucalyptus, Cassia siamea,* and *Azadirachta indica.*

5. Chemical Control

Sandal spike shows remission of symptoms after treatment with Tetracycline. The systemic fungicide Benlate has been reported to bring about temporary remission of symptoms. Application of the antibiotic by the girdling method has shown the recovery of these 25 to 30 days after treatment with antibiotic Terramycin either alone or in combination with Benlate, and 30-35 days after treatment with another antibiotic Ledermycin alone or in combination with Benlate (Raychaudhuri *et al.*, 1972). However, these treatments provide only temporary relief for 50-150 days, and then the disease reappears (Singh, R S 2007). The production of spike-free plants via somatic embryogenesis, and breeding for resistance will be useful for long-term management of the disease (Sita and Ram, 1981).

20.2 SESAMUM PHYLLODY

This disease is quite common in sesamum-growing areas. It is the most destructive disease of sesamum (*Sesamum indicum*), and causes huge losses in yield wherever it appears earlier in the crop season.

1. Symptoms

Vein clearing of the leaves, phylloid flowers, stimulation of axillary buds with result in profuse branching, reduction in leaf size, and shortening of internodes are some characteristic symptoms of the disease. In affected plants, floral parts are transformed into green, leaflike structures resulting in complete sterility.

2. Causal Organism

Cousin *et al.* (1970) reported pleomorphic **Mycoplasma-like Bodies** (MLBs) in the phloem sieve tubes of phyllody-affected plants. MLBs have also been found associated by others in this disease.

3. Transmission

Diseases are transmitted in nature by a leaf hopper *Orosius albicinctus* in a persistent manner. Sahambi (1970) also studied the 'pathogen vector' relationship, and the host range. The causal MLO can infect 90 plant species belonging to 38 genera in 12 families.

4. Disease Management

Late sowing of the crop results in a reduction in the incidence of the disease. Leaf extract of *Vitex negundo* var. *purpurascens* and *Sesbania rostrata* reduced phyllody incidence by 68 percent. An appropriate sowing time so as to circumvent insect transmission may be helpful for minimizing severity of the disease.

A high degree of control can be achieved by spraying Metasystox (0.1%) along with soil treatment with Phorate or Thimet. Antibiotics like Tetracycline sprays at 500 ppm at the flower initiation stage have proved effective against phyllody but the recovery is temporary.

A possibility of biochemical control by spraying manganese chloride has been indicated. It appears that $MgCl_2$ oxidizes phenols, and protects or inhibits the enzymes, bringing the auxin level to normal. Once hyperauxin is oxidized, the plant can gain its normal conditions. Soil treatment with Thimet 10G @ 10 kg per hectare or Phorate 10G @ 11 kg per hectare or Temik 10G @ 25 kg at the time of sowing are effective. A number of lines, i.e. S1289, S951, S2194, S2201, S2373, S2635 have been found to possess resistance to phyllody.

20.3 GRASSY SHOOT DISEASE (GSD) OF SUGAR CANE

This is one of the most serious diseases of sugar cane, and is especially widespread in Maharashtra, from where it was first detected in 1949. The disease is also prevalent in Andhra Pradesh, Bihar, Uttar Pradesh, Madhya Pradesh, and Tamil Nadu.

The disease is now known to occur in India, Burma, Sri Lanka, and Sudan, and appears to be synonymous with the white-leaf disease of Taiwan and Thailand. In India the disease has been reported from most of the sugar cane growing states namely, Punjab, Uttar Pradesh, Haryana, Bihar, West Bengal, Madhya Pradesh, Maharashtra, Andhra Pradesh, Karnataka, and Tamil Nadu.

1. Symptoms

GSD is characterised by the production of numerous thin, lanky tillers from the base of the affected plants. The tillers bear pale yellow, thin, narrow leaves resembling grasses. The internodes of these tillers are much reduced. If the crop is infected in the early stages of growth, the loss may be up to 100 percent.

Premature and excessive tillering gives a crowded, gramlike appearance of the clump. Affected clumps are usually stunted, and in many instances exhibit premature proliferation of axillary buds. Diseased plants exhibit varying degree of loss of chlorophyll, ranging from total green to white.

In a GSD infected clump, hardly one or two thin, weak, and small milleable canes are produced in a clump having primary infection, and these do not produce inflorescence or arrow.

The disease displays different types of symptoms at various stages of plant growth, and because of this, the disease has been referred to as 'new chlorotic disease', 'albino disease', 'yellowing disease', and 'bunchy disease' or 'leaf tuft' by various workers.

2. Causal Organism

Workers in the Sugarcane Research Institute, Lucknow, in collaboration with scientists from the USA, found the association of mycoplasma-like bodies (MLB) in the sieve cells of infected plants. Rishi *et al.* (1973) found two types of MLB spherical bodies, 300–400 nm diameter, and filamentous bodies, 30–53 nm in diameter associated with this diseases.

Concentration of MLO is high in canes growing at high temperatures (30°C or so). Such canes show severe symptoms. At lower temperatures the symptoms are less severe. Staining of hand sections of diseased canes with Diene's stain give dark blue colour to infected phloem while the healthy phloem remains unstained.

3. Transmission

The primary transmission of the disease is through diseased seed material, while the secondary transmission is through insect vectors. GSD was reported to be sap and aphid transmitted. This is probably the only known example of sap- and aphid-transmitted yellows disease known so far. In addition to sugar cane, the disease also occurs on *Sorghum* which serves as a collateral host. The grassy shoot MLO perennates through diseased canes used for seed and spreads through diseased cane setts and cane cutting knives. Ratooning of diseased crop is an important means of transmission of the disease. The pathogen is introduced into new fields through diseased cane setts. The vectors are *Rhopalosiphum maydis (Aphis maydis), Aphis sacchari*, and *Aphis idiosacchari.* Transmission by dodder (*Cuscuta campestris*) has also been reported.

According to Agnihotri (1990), there is enough circumstantial evidence for the presence of three different types of grassy shoot disease in nature, perhaps caused by the following different organisms:

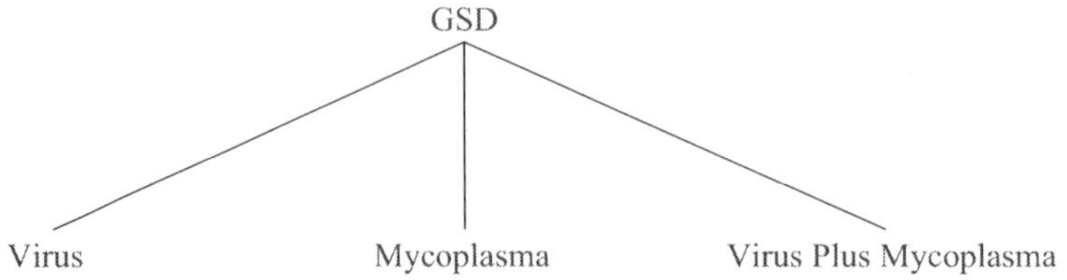

The successful transmission of the disease through sap as also by aphids, reported by Chona *et al.* (1960) and Varma *et al.* (1966), is a pointer that they were dealing with plants affected by virus alone. Edison (1973) failed to transmit the disease symptoms of tillering and albinism using aphids but succeeded with hoppers indicating that his test plants were infected by mycoplasma-like organisms. However, Singh and Singh (1966) were successful in transmitting the albinism indicating that their leaf plants were probably having the infection of both virus and mycoplasma, the aphids transmitting only the virus.

4. Disease Management

Healthy setts should be used for plantation purposes, and should be selected from areas where the grassy shoot is not present at all. Rouging may help consistent reduction in inoculum in the area. Both heat therapy and chemotherapy have been found effective against the GSD pathogen. In heat therapy, the setts should be kept in hot water at 50°C for 2 hours or in hot air at 54°C for 8 hours. According to Shukla and Singh (1990), moist heat at 54°C (90–95 per cent humidity) from one-and-half to 2 hours is enough to inactivate the MLO in an airtight chamber especially designed for the treatment of canes.

Insect vectors can be controlled by weekly sprays of 0.16 per cent Malathion. These treatments are possible for seed nurseries. Sattar and Ali (1976) successfully controlled complete cure of the disease in CO 419 when its single bud setts were treated with 500 or 1000 ppm of Ledermycin.

20.4 LITTLE LEAF OF BRINJAL (*SOLANUM MELONGENA*)

Little leaf of brinjal is a mycoplasmal (phytoplasmal) disease. This graft-transmissible disease was first reported from Coimbatore. It is very widespread in India and can cause losses up to 90% in certain cases.

1. Symptoms

In the beginning, young leaves show chlorosis followed by axillary-bud proliferation. Diseased plants show extreme reduction in the size of their leaves and nodes, giving a bushy appearance. In heavily infected plants, flower, and fruit setting is negligible; virescent and phylloid flowers are very common, and these remain green, and there is a total loss in yield.

The plant gives a bushy appearance because the internodes of the plant are shortened, and a large number of axillary buds are stimulated to grow into short branches with small leaves.

Accumulation of metabolites occurs in the diseased leaves presumably due to obstruction in their translocation to the roots. The diseased plants have reduced growth.

2. Causal Organism

Prior to 1968, the causal agent of little leaf of brinjal was considered to be a virus. Varma *et al.* (1969) showed MLB, spherical to ovoid (40–300 nm) structures, in the phloem cells of diseased plants. The disease is thus shown to be caused by mycoplasma-like organisms.

3. Transmission

The disease is transmitted in nature by insects (leaf hopper) such as *Hishimonus phycitis* which is common on the brinjal crop. MLOs have also been reported in the haemolymph and fat bodies of the vector. The same MLOs occur on *Datura fastuosa* and *Vinca rosea*. Bitter gourd, ground gourd, watermelon, methi, sun hemp, carrot, and *Solanum xanthocarpum* are good hosts of the vectors while beet, castor, soybean, and *Chenopodium amaranticolor* are suitable hosts for oviposition.

4. Disease Management

No effective method of controlling the disease is known. Temporary maskings of symptoms by using Terramycin, Achromycin, Aureomycin, and Ledermycin have been reported. The severity of the disease can be reduced by destruction of affected plants and spraying of insecticides. New crops should be planted only when diseased plants in the field, and its neighbourhood have been removed. Methyldemeton 25 EC (2 ml/L), Dimethoate 30 EC (2 ml/L), Malathion 50 EC (2 ml/L) have been recommended for vector control. Although mycoplasmas are reported to be suppressed by Tetracyclines, yet field application of this method has not been recommended.

Cultural practices like eradication of weed hosts, destruction of diseased plants, and control of insects which are vectors of these diseases by suitable insecticides will go a long way in controlling this disease. Cultivars BB-7, BWR-12, Pant Rituraj, Pusa Purple cluster, Akra sheel, Aushy, Manjari gota, Banaras giant, and H-8 are reported to have shown some resistance to little leaf.

20.5 COCONUT (ROOT) WILT

"Root (wilt) disease of coconut" is of national importance as the disease causes an annual loss of 968 million coconuts in Kerala. Loss of husk per unit nut was 25.8%, and loss of oil copra and oil was 9.08 and 11.3% respectively. According to Koshy (1999), 60% of the leaves were found damaged due to the disease.

1. Symptoms

Wilting and drooping of the leaves, flaccidity, paling and yellowing, and necrosis of the leaves were considered to be the typical symptoms of the disease. Characteristic bending and ribbing of the leaflets of the leaves in the central and outer whorls are the earliest symptoms. Foliar yellowing and marginal necrosis, the other two symptoms set in with the advancement of the age of the palm trees. Flaccidity is considered the most consistent symptom in palms of all ages.

Inflorescence necrosis, lack of ability to produce female flowers, and pollen sterility render the plants unproductive. Shedding of immature nuts even before the appearance of other visual symptoms or after is another important character. In seedlings below 5 years of age, rotting of spear leaf (spindle) is the first symptom to appear. Rotting of roots was considered one of the important symptoms by many workers.

2. Causal Organism

The presence of phytoplasma was identified in the sieve tubes of roots, tender stem, petiole, and developing leaf bases of root (wilt) diseased plants. Tissues of healthy palms were free of such organisms.

Pleomorphic forms ranging circular to oval, and occasionally beaded or filamentous, were observed. The coccoid forms were in the size range of 250–400 nm. Intracellular presence of the organisms, their absence from healthy palms, and the cellular changes noticed further strengthened the etiological role of phytoplasma with the disease. Phytoplasmal infection could be seen under optical microscope employing Dien's staining or fluorochromes such as DAPI of free hand sections of fixed tissues.

3. Disease Cycle

It is now believed that the lace-bug *Stephanitis typica* has been shown to transmit this phytoplasma from diseased to healthy palms. Phytoplasma has also been observed in the salivary glands of the plant hopper *Proutistia moesta* given an acquisition feeding period of more than 30 days. Transmission of phytoplasma to periwinkle (*Catharanthus roseus*) was also achieved through *Cassytha filiformis*. Phytoplasma from coconut could not be cultured though various methods were tried.

4. Disease Management

Field trial was initiated towards controlling the disease by means of Oxytetracyclin, Neomycin, and Penicillin. Spraying the leaves with Monocrotophos (0.05 ml/L) or Endosulfan (1 ml/L) is effective in killing the insects. Application of 50 kg of farmyard manure/palm/year is also effective in controlling the disease. The disease can be controlled by spraying 1% Bordeaux mixture or 0.3% copper oxychloride or Mancozeb. Proper drainage of water during rainy season is also quite effective.

One redeeming feature of this disease is that it is not lethal but a declining malady which responds to ideal management. Based on the experimental evidences, twin strategies one for the highly diseased contiguous areas, and another for mildly affected areas has been formulated. In the former, the yield of palms can be sustained or even improved through adoption of the integrated disease-management practices such as removal of diseased advanced, and juvenile palms, application of balanced fertilizer, addition of organic matter, raising green manure crops in the basin, and their incorporation into soil, irrigation during summer months, adopting inter, and mixed cropping, and practising plant-protection methods. The strategy in the mildly affected areas is to contain the disease from spreading, and this involves the eradication of all diseased palms.

Since there are no curative or prophylactic measures available to control the disease, the long-term option is to breed for disease resistance. However, not much progress has been made so far in this direction.

REVIEW QUESTIONS

1. What are mycoplasmas?
2. Who discovered mycoplasmas as plant pathogens?
3. How would you control diseases caused by mycoplasmas?
4. List some important mycoplasmal diseases of your area.
5. Name any mycoplasmal disease in which Koch's postulates have been proved.
6. Why are diseases caused by phytoplasmas or spiroplasmas not controlled by antibiotics like Penicillin which inhibit cell-wall synthesis?
7. Do spiroplasmas move by means of flagella?
8. Differentiate between mycoplasmas, L forms of bacteria, true bacteria, and viruses.
9. How would you control the following diseases?
   (i) Sandal spike
   (ii) Grassy Shoot Diseases (GSD) of sugar cane
   (iii) Little leaf of brinjal
   (iv) Coconut (root) wilt

MULTIPLE-CHOICE QUESTIONS

1. Pleuropneumonia of cattle is caused by
   (i) virus (ii) fungus
   (iii) bacteria (iv) mycoplasma
2. Mycoplasma differ from viruses in that these
   (i) possess both DNA and RNA (ii) have a definite cell wall
   (iii) infect plants only (iv) lack metabolic activity
3. The outermost limiting layer of typical cell of mycoplasma is
   (i) capsule (ii) slime layer
   (iii) cell wall (iv) cell membrane
4. Mycoplasma are resistant to
   (i) Streptomycin (ii) Penicillin
   (iii) Vancomycin (iv) Nystatin
5. Which of the following diseases are caused by mycoplasma?
   (i) Alfalfa witches' broom (ii) Sandal spike
   (iii) Sugar cane grassy shoot (iv) All of these
6. Which of the following staining method is used to see mycoplasmal in infected plant part?
   (i) Lactophenol cotton blue staining (ii) Gram staining
   (iii) Dien's staining (iv) Safranin staining

Answers

1. (iv) 2. (i) 3. (iv) 4. (ii) 5. (iv) 6. (iii)

21

BACTERIA, AND BACTERIAL DISEASES

21.1 STRUCTURE, REPRODUCTION, AND CLASSIFICATION OF BACTERIA

In 1675, Anton von Leeuwenhoek, for the first time in history, made notable observations on bacteria using crude microscopes that could magnify 150 times. The importance of these observations was, however, realized only after Pasteur (1876) had demonstrated their role in fermentation and decay, and Koch (1876) had proved that bacteria can cause such diseases as anthrax, tuberculosis, and Asiatic cholera.

It is now well recognized that of all the organisms, bacteria are the most closely related to human life. T J Burrill of the University of Illinois (1878) was the first to prove the association of a bacterium with a plant disease. There are about 1600 species of bacteria of which 50% are true bacteria. About 100 species of bacteria cause diseases in plants (Agrios, 2005). Excepting in pits of volcanoes and deep strata of rocks, bacteria are present almost everywhere. They include the photosynthetic and chemosynthetic autotrophs, saprophytes, capable of decomposing all imaginable organic materials, parasites, pathogens, symbionts of plants and animals, and other protista and even predators. Very few bacteria are autotrophic. The average size of bacteria is about 1.25 μm in diameter. The smallest is rod-shaped, measuring 0.15 μm, almost touching the last limit of visibility. The largest is *Spirillum volutans*, measuring up to 15 μm in length and 1.5 μm in diameter. Some interesting calculations have been made to drive home the smallness of bacteria. A single drop of water may contain 50 thousand million bacteria.

1. Forms of Bacteria

Bacteria are of various shapes. They may be spherical (*Coccus*), rod-shaped (*Bacillus*), curved (*Vibrio*), or spiral (*Spirillum*).

2. Flagella and Locomotion

Hair like projections called flagella help bacteria to move. On the basis of the number and arrangement of flagella, bacteria may be classified as follows:

(a) Monotrichous A single flagellum at one end of the cell.

(b) Lophotrichous Two or more flagella at one or both ends of the cells.

(c) Amphitrichous One flagellum at each end.

(d) Peritrichous Large number of flagella surrounding the cell.

Bacterial flagella differ structurally from those of the Eucaryotes. They range in thickness from 15 nm to more than 24 nm, and consist of single or entwined structures that resemble microtubules but are composed of a single protein called **flagellin**. It is interesting to note that similar proteins occur in animal muscle. The flagella of some bacteria are anchored in a specialized peripheral region of the cytoplasm.

3. Pili

A structure somewhat analogous to the flagellum but which is not involved in motility is the pilus (plural, pili). Pili are considerably shorter than flagella, and are more numerous on the cell. They may be chemically similar to flagella. Not all organisms have pili. The ability to produce pili is an inherited character. Certain pili are involved in the transfer of genetic material from one bacterial cell to another; they are the sex pili. The functions of other types of pili are not known. In some cases, they may enable the organism to stick to inert surfaces or to form pellicles on the surface of liquids.

4. Capsules, Slime Layers

Some bacteria secrete slimy or gummy materials on their surface which can sometimes by seen by negative staining. When the material is around the cell surface in a compact manner, it is called a **capsule** but when the material is diffused, it is called a **slime layer**. Capsules and slime layers are usually composed of polysaccharides, polypeptides or polysaccharide protein complexes.

Bacteria are prokaryotic in their organization, that is they do not have membrane-bound organelles or a well-organized nucleus, and their cell-wall composition is entirely different.

5. Outer Covering of the Cell

The outer covering consists of three layers. The inner layer is the cytoplasmic membrane, and consists of lipids and lipoproteins. The middle layer of cell wall is the rigid part of the outer covering of the cell. The purified cell wall contains lipids, carbohydrates, proteins, phosphorus, and other inorganic substances. Here the rigidity is due to a unique polymer, the **mucopeptide**, which is based upon a backbone of alternating molecules of N-acetyl glucosamine and N-acetly-muramic acid. Attached to the muramic acid are short-chain molecules of amino acids which are cross-linked to other muramic acid molecules in adjacent chains. The cross-linking enables the complex to form a rigid network. The third layer is known as the slime layer. If the slime layer surrounding the bacterial cell is thick, gummy, and mucilaginous, it is called a capsule which protects the cell against adverse conditions. Capsules and slime layers of certain bacteria are made up of polysaccharides, a material composed of dextrans, levans, and so on, and appear like plant gums. Capsules or slime of certain other bacteria are composed of polypeptides and peptides.

6. The Cytoplasm

The cytoplasm is granular, viscous, and lies between the nuclear material and plasma membrane. Brock (1970) has called the cytoplasm the "cell pool". The protoplasm of bacterial cells is similar to that of other living organisms. The moisture content is 70–85%, and is colloidal in nature. The cytoplasm is densely packed with ribosomes of 70S.

7. Mesosomes

These are invaginations of the plasma membrane. Their function is a matter of conjecture. They have a role in wall synthesis, and secretion of extracellular products. It has also been shown that transforming DNA is taken up by all the cells apparently via the mesosomes. The nuclear body gets attached to the mesosome at the time of replication. However, recent evidence indicates that mesosomes are cell artifacts, and have no role of any nature whatsoever.

8. Nuclear Region

The difficulty of demonstrating the nuclear material in bacterial cell is due to the extreme basophily of the cells because of the presence of RNA in the cytoplasm. It is now believed that the nucleus in bacteria is present in the cell as a discrete body. Feulgen tests applied in the case of bacteria have revealed the presence of purple-stained bodies in a large number of cases. Prokaryotic organisms do not possess a true nucleus as do eukaryotes. In electron micrographs of thin sections, the nuclear region is seen as a weakly contrasting area that contains thin fibrillar material but which lacks a definite membrane. The fibrillar material is DNA, the genetic material of the cell. When a bacterial protoplast is lysed gently, this DNA pours out of the cell in a long, single fibre; probably a single molecule, whose length when stretched out would be many times the length of the bacterial cell itself. For instance, the length of an *E. coli* cell is about 2 μm, whereas the length of *E. coli* DNA is 110–1400 μm, so the DNA must be highly folded. Using radio autographic techniques Cairns (1963) has demonstrated the circular DNA of *E. coli*.

9. The Gram Reaction

One of the most important cytological features of bacteria is their reaction to a simple procedure called, after its discoverer, the **Gram stain**. The procedure involves staining the cells with the stain crystal violet, and all bacteria are stained blue. The bacteria are then treated with an iodine solution and decolorized with alcohol. Gram-positive bacteria retain the stain crystal violet, whereas the gram-negative bacteria are decolourized. Thus, there is a fundamental difference between gram-positive and gram-negative bacteria. The structure of cell wall of gram-positive and gram-negative bacteria is quite different. The differences between the two types of bacteria are given below:

Gram-positive bacteria	Gram-negative bacteria
1. Cell wall thicker, contains only traces of lipids	Cell wall thinner, may contain up to 20 percent lipids
2. Striking simplicity of amino acids, glutamic acid occurs in large amounts	All amino acids present
3. Diaminopimelic acid in some	Diaminopimelic acid in some
4. More muramic acid	Less muramic acid
5. Teichoic acid present	Teichoic acid absent

10. Acid Fast Stain

Some bacteria are surrounded by a covering of fatty or waxy substances. These organisms are not readily stained, but when stained are able to retain the colour even after treatment with drastic decolourizing agents. They are called acid fast because the stained bacteria are resistant to decolourization with acid alcohol. Two well-known members of the group are *Mycobacterium tuberclosis* and *M. leprae*. In Ziehl Neilson's method, the primary stain is carbol fuchsin.

11. Nutrition

Bacteria can be divided into two groups on the basis of the way in which they derive their nutrition. Those which can utilize carbon dioxide as the sole source of carbon are called **autotrophs**. Very few bacteria fall into this category, the limited number being separated into phototrophs and chemotrophs. **Phototrophs** (e.g. *Chlorobium limicola*) can use the energy of sunlight, and **chemotrophs** (e.g.

Thiobacillus thioxidans) can use the chemical energy of inorganic compounds during the synthesis of their own protoplasm. The photosynthetic process in bacteria differs from the photosynthesis of higher green plants. The photosynthetic bacteria possess chlorophylls (bacteriochlorophylls) different from those of plants in structure as well as in light-absorbing properties. Bacteriochlorophylls absorb light in the near-infra red region (660–870 nm). They are not contained in chloroplasts but are found in extensive membrane systems. Oxygen is not liberated during the photosynthetic process, and the hydrogen source is not water but usually a sulfur compound (e.g. hydrogen sulfide). Most of the bacteria derive their energy from the organic compounds contained in or produced by other organisms. These are the **heterotrophs**. Heterotrophic bacteria may be further subdivided into **saprophytes**, organisms which live on organic matter, and **parasites**, organisms which grow upon living materials. Most heterotrophs are saprophytes, and the laboratory media upon which they grow reflect to some extent the environments in which they are found in nature. A commonly used medium is a nutrient broth (when agar is added, it is called "nutrient agar") or sometimes "beef extract-peptone" broth (or agar). There are, of course, various other media for growing various types of bacteria.

21.2 GROWTH AND REPRODUCTION IN BACTERIA

21.2.1 Growth of Cells

If conditions are favourable, cell division is normally followed by a period of increased growth. Through absorption of water and food manufactured by the protoplasm, the cell enlarges to its original size. Under favourable conditions of moisture, nutrition, pH, and temperature, some kinds of bacteria for example, *E. coli*, may double in mass about every 20 minutes their generation time. If this rate of increase were to continue for even a few days, enormous numbers of bacteria would result. However, there are several factors which prevent this from happening. In some cases, there may be an exhaustion of the food supply or an accumulation of waste products.

Different bacterial species show various shapes of growth curves depending upon the generation time, and the maximum population attainable under the prevailing growth conditions. During a period of one to several hours, there may be a lag phase in which there is little or no increase in cell number. During the first part of the lag phase (which may be called the initial stationary phase), the cells adapt themselves to the new environment.

When the growing cells begin to divide, they usually continue to do so over a regular interval until the maximum growth that can be supported by their environment is approached. This period of rapid cell division is known as the logarithmic phase of growth. The stationary phase occurs when rapid growth is halted by the depletion of nutrients, accumulation of waste products, or other factors. Unless the cells are transferred to a new environment capable of supporting continued growth, they will eventually die.

21.2.2 Cell Division

Cell division requires the doubling of all cell constituents, and their orderly partitioning into two daughter cells. The first step in cell division is the doubling of the DNA complement of the nuclear body which occurs by the separation of two strands, and the replication along each strand of a new complementary strand. Due to the enormous length of the DNA, the cell faces a problem in partitioning the doubled DNA between the two daughter cells. It is thought that the DNA molecule is attached at some point to the cell membrane, and after replication, each one of the double strands, while still attached, swings towards one of the cell halves. Only after this slip has been accomplished, can the new cross-wall form. At the time of the formation of the new cross-wall, a mesosome of the membrane system can be seen attached to the growing cross-wall, and it is likely that this mesosome plays some role in the synthesis of the cross-wall. However, recent evidence indicates that mesosomes are artifacts, and are not true cell structures, and therefore play no part in the synthesis of cross-walls or energy function by the cell. The DNA is probably attached to the other end of the same mesosome.

The highly coordinated process—DNA doubling/DNA partitioning/cross-wall formation—is the equivalent of the mitotic process in eukaryotic organisms.

21.2.3 Bacterial Endospores

Though no plant pathogenic bacteria produce endospores, they are produced in some other bacteria. Endospores are highly resistant structures, and are an important means of perennation for tiding over unfavorable conditions. Such spores are characteristically produced by two genera of bacteria—*Bacillus* and *Clostridium*—both of which are widely distributed in soil.

21.2.4 Sexual Reproduction

In all fundamental aspects, the genetics of bacteria is the same as that of other organisms. Conjugation is most closely related to sexual reproduction Conjugation occurs in several different bacteria but is best known in *Escherichia coli*.

21.2.5 Bacterial Conjugation

This method entails directly transferring genetic material from a donor to a recipient bacterium. It was first discovered in 1946 by Lederberg and Tatum in *Escherichia coli*. It has also been observed to occur in strains of other bacteria, such as *Salmonella*, *Pseudomonas*, *Shigella*, and *Vibrio*. After cell contact, actual chromosome transfer occurs; the process is unique to bacterial conjugation, and is vastly different from chromosome transfer in higher organisms. Sexual conjugation in bacteria involves the union of cells designated as male and female. Electron micrographs of paired cells show them connected by a short tube, 50 nm wide, and it has been assumed that this pairing represents conjugation. It has also been suggested that the fimbriae attached to the surface of male cells are responsible for the transfer of DNA from the male to the female cells. It has been suggested that a sex factor is attached to the plasmalemma at the point where a fimbria forms.

Recombination frequencies from $F^+ \times F^-$ crosses were never particularly high. In the early 1950s, some *Escherichia coli* strains were isolated in which the frequency of recombination was up to 10,000 times greater than observed in F^+ strains. These were called **Hfr strains**. The discovery of Hfr strains made the fact of the actual physical act of conjugation feasible, and a number of fine electron micrographs showing a conjugation bridge between bacterial cells have been obtained.

The discovery of Hfr strains also enabled Jacob and Wollman (1962) to determine the mechanism of transmission of the bacterial chromosome. When a female cell receives a sex factor, either by pairing or by fimbrial conjugation, it changes to a male cell, and subsequently can conjugate with another uninfected female cell. During the conjugation process, the male cell slowly injects a genophore into the female cell so that the genes enter the female cell in a linear way. The episome becomes attached to the female genophore, and is the last part to enter the receptor cell. It takes about 100 minutes for a complete transfer of the genophore to occur. The transfer is usually not complete so that the resulting zygote is usually a partial diploid, of course, crossing over between bacterial chromosomes, and exogenous DNA, which then occurs, thus resulting in recombination.

21.2.6 Bacterial Transformation

In 1928, Griffith (in England) injected mice with noncapsulated and heat-killed capsulated pneumococci; neither of these proved fatal when infected separately. When both these bacteria were injected together, some of the mice died, and living capsulated bacteria were recovered from their blood. The transformation could be brought about by the cell-free extract of the capsulated bacteria. Griffith at first thought that the transforming principle was the bacterial polysaccharide. But Avery, MacLeod, and McCarty (1944) demonstrated that the transforming principle was DNA.

21.2.7 Bacterial Transduction

Zinder and Lederberg's (1952) experiment in a U-tube on *Salmonella typhimurium* indicated that bacterial viruses or phages are responsible for the transfer of genetic material from one bacterium to the other. Zinder, a student of Lederberg, was studying various strains and mutants to see if recombination would occur. An experiment was performed in such a manner that direct cell contact was not possible. This was done by physically separating the bacteria by a sintered glass filter impermeable to bacteria. High-frequency recombinants were still obtained. It became clear that a filterable agent (phage) passed from one cell to another, transferring genes between them. This phenomenon is called transduction. Transduction has been demonstrated in many bacteria.

21.3 CLASSIFICATION OF BACTERIA (PROKARYOTES)

The prokaryotes, Eubacteria and Archaebacteria, evolved along different lines. Modern classification systems for the prokaryotes reflect a phylogenic relationship, placing some prokaryotes in the Kingdom Eubacteria, and the other in the Kingdom Archaea. Some of the important criteria for classifying bacteria are cellular characteristics (cell shape, cell size, arrangement of cells, arrangement of flagella, capsule, endospore), staining reaction (Gram stain, acid fast stain), growth, nutritional characteristics (colonial morphology, pigmentation, energy sources, carbon sources, nitrogen sources, fermentation products, modes of metabolism), biochemical characteristics (cell-wall constituents, pigment biochemicals, storage inclusions, antigens, RNA molecules), physiological and ecological characteristics (temperature range, and optimum, oxygen relationships, pH tolerance range, osmotic tolerance, salt requirements, tolerance, antibiotic sensitivity), and genetic characteristics (DNA molecular, % G, +C, DNA hybridization).

Bergey's Manual of Determinative Bacteriology was published in 1923. From 1923 to the turn of the century, a lot of information has come on various groups of bacteria. Bergey's manual was revised a number of times. After the publication of the eighth edition of the manual in 1974, a new series under the title *Bergey's Manual of Systematic Bacteriology* has been started, and has been published in four volumes.

The eighth edition of *Bergey's Manual of Determinative Bacteriology* had all prokaryotic organisms in the Kingdom Procaryotae. Prokaryotae is distinguishable into two divisions, the Cyanobacteria (includes all the blue-green algae), and the Bacteria. The division Bacteria was further subdivided into 19 parts. The systematics of prokaryotes is always in a state of flux, with descriptions of new bacterial and archaebacterial genera and species, and revisions of older taxonomic classifications frequently appearing in the *International Journal of Systematic Bacteriology*. Periodically, the status of bacterial and archaebacterial taxonomy is summarized in a comprehensive volume, *Bergey's Manual of Determinative Bacteriology*. The ninth edition of *Bergey's Manual*, published in 1994, is the most comprehensive volume. In addition to *Bergey's Manual of Determinative Bacteriology, Bergey's Manual of Systematic Bacteriology* gives the description of the Eubactria and Archaebacteria. The latter manual is a multi-volume treatise: Volume 1 covers the gram-negative bacteria of general, medical, plant pathogenic (*Pseudomonas*, *Xanthomonas*, *Erwinia*, *Agrobacterium*, and the mollicutes, i.e. MLOs, and *Spiroplasma*) or industrial importance. Volume 2 covers the gram-positive bacteria other than actinomycetes, Volume 3 covers the Archaebacteria, Cyanobacteria, and remaining gram-negative bacteria, and Volume 4 covers the Actinomycetes.

At present the bacteria (Eubacteria) are classified into groups or parts. Twenty-two groups or parts are recognized of Eubacteria while Archaebacteria are classified into five groups.

A. Eubacteria

Group or Part- I	Phototrophic bacteria
Group or Part-II	The gliding bacteria
Group or Part-III	The sheathed bacteria
Group or Part-IV	Budding and appendaged bacteria
Group or Part-V	The Spirochaetes
Group or Part-VI	Spiral and curved bacteria
Group or Part-VII	Gram-negative aerobic rods and cocci
Group or Part-VIII	Gram-negative facultatively anaerobic rods
Group or Part-IX	Gram-negative anaerobic bacteria
Group or Part-X	Dissimilatory sulfate or sulfur-reducing bacteria
Group or Part-XI	Anaerobic Gram-negative cocci
Group or Part-XII	Gram-negative chemolithotrophic bacteria
Group or Part-XIII	Gram-positive cocci
Group or Part-XIV	Gram-positive endospore-forming rods and cocci
Group or Part-XV	Gram-positive asporogenous rods of regular shape
Group or Part-XVI	Gram-positive rods of irregular shape
Group or Part-XVII	Mycobacteria
Group or Part-XVIII	Rickettsias and Chlamydias
Group or Part-XIX	Actinomycetes
Group or Part-XX	Mycoplasmas MLOs and Spiroplasmas
Group or Part-XXI	Anaerobic nonfilamentous or filamentous rods
Group or Part-XXII	Endosymbionts

B. Archaebacteria

Group or Part–1	Methanogens
Group or Part–II	Archae sulfate reducers
Group or Part–III	Extremely halophilic aerobic archaebacteria halobacteria
Group or Part–IV	Cell-wall less archaebacteria
Group or Part–V	Extremely thermophilic and hyperthermophilic sulfur metabolizers

21.4 PLANT PATHOGENIC BACTERIA

The following is the list of important plant pathogenic bacterial genera—*Pseudomonas, Xanthomonas, Xylella, Xylophilus, Acidovorax, Agrobacteriurn, Erwinia, Clavibacter, Curtobacterium*, and *Streptomyces.*

Plant Pathogenic Bacterial Genera

Gram negative	Gram positive
1. *Pseudomonas*	1. *Clavibacter*
2. *Xanthomonas*	2. *Curtobacterium*
3. *Xylella*	3. *Streptomyces* (other relatively unimportant plant pathogens of genera *Arthrobacter*, *Bacillus*, and *Rhodococcus*)
4. *Xylophilus*	
5. *Acidovorax*	
6. *Agrobacterium*	
7. *Erwinia*	
8. *Ralstonia*	

21.5 SYMPTOMS OF BACTERIAL PLANT DISEASES

In contrast to fungal pathogens, bacteria are incapable of mechanically penetrating the cutinized plant tissues, cuticle, periderm, and so on. The only ways they can gain entry into the plants are through (1) noncutinized areas, such as root hairs, stigmas, (2) natural openings, such as stomata, and (3) incidental wound or those initiated by insects, nematodes, and other microbes. On entry into a suitable host, the bacteria may establish themselves intercellularly, intracellularly, intravascularly, or in more than one combination. The host tissues react differently to the invasion, and exhibit different types of disease symptoms. The various types of disease symptoms caused by phytopathogenic bacteria are leaf spot, canker, fasciation, dwarfing of plants, distortion of plant parts, discolouration of tissues, rotting, gall formation, wilting, and others. Chlorosis and halo induced by phytopathogenic bacteria retard the photosynthetic activity of the host plant. The following are the major types of plant-disease symptoms commonly observed due to bacterial pathogens.

21.5.1 Local Lesions

Bacterial lesions or localized spots occur on the leaf blade, and sometimes in the petiole, stem, and other parts. Such spots may also occur on fruits, resulting in the discolouration of fruits. The leaf spots start as minute water-soaked specks, spreading rapidly to form circular, irregular, or angular spots, bound by veins and veinlets. The pathogen which enters the host tissues through stomatal openings establishes in the substomatal space and remains so in the parenchymatous tissues causing necrotic lesions. This necrosis may spread rapidly in certain cases, such as in fire blight of apples and pears.

21.5.2 Soft Rots

The soft-rot symptoms are found mainly on fleshy parts. The major effect is the softening of tissues due to disintegration of cells, dissolution of middle lamella as a result of the action of enzymes, and very often a dirty liquid oozes out of the affected parts. The enzymes diffuse in advance to dissolve the middle lamella of the cells. Plasmolysis and death of cells follow, and the bacteria grow upon the dead plant tissues rather than upon the living cells.

21.5.3 Vascular Disease

In some of the bacterial leaf spots, the organism moves into the vascular tissues, and becomes systemic. In others, the invasion is concentrated in the vascular tissues, causing typical wilt of the affected plants. Due to vascular infection, there may be hypoplasia, leading to the stunting of the entire plant. Common examples are cucurbit bacterial wilt, brown rot of potato, ring rot of potato, bacterial wilt of tomato, etc. Certain polysaccharides produced by phytopathogenic bacteria *in vitro* possess the wilt-inducing properties. In other studies, it has been found that certain pectinolytic and celluloytic enzymes possess the wilt-inducing properties. Pathogenic infection causes the movement of water in the vessels to decline due to an increase in the viscosity of vessel fluid. This is because of the formation of a slimy or gummy substance surrounding the bacterial mass. Tylose formation and vessel collapse may also contribute to wilting.

21.5.4 Tumours and Galls

In many bacterial diseases, the effect of invasion by the pathogen is the hyperplasia and hypertrophy of invaded tissues. As a result, tumours develop on the affected organs. The formation of root nodules in leguminous roots in response to bacterial action has also a similar effect. Crown gall (*Agrobacterium tumefaciens*) is the best-known example of this. The pathogen seems to trigger a chain of reactions to produce some chemicals in the vicinity which cause hyperplasia and hypertrophy of the host cells.

21.5.5 Scabs, and Cankers

Scabs and cankers are corky outgrowths which are formed on leaves, twigs, and all other plant parts above the ground. These outgrowths are the result of the reactions of the host tissues to the pathogen. Such reactions are mostly localized, and often confined to the parenchymatous tissues of the host plant. The incubation period normally required for such infections is about ten days after infection. Scab is formed by epidermal infection, and is not deep-seated. The powdery scab of potatoes is a familiar example. Cankers, on the other hand, are deep-seated, and involve the cambium layer. They entail the destruction of woody tissues, and are common infections on stems of citrus, mango, and many fruit and forest trees.

21.6 BACTERIAL PLANT DISEASES

21.6.1 Bacterial Blight of Paddy

Bacterial leaf blight is widespread in South-East Asia. Since the introduction and cultivation of new, high-yielding but susceptible rice varieties over a large acreage, the disease has become one of the most serious problems of rice cultivation in India (Srivastava *et al.*, 1967). Srivastava (1972) had given a detailed account of the disease in his Presidential Address to the members of the Indian Phytopathological Society. Devadath (1992) has reviewed the work done on this disease up to that period. According to Mew *et al.* (1993), Bacterial leaf blight is one of the most destructive diseases of rice in Asia. In India the losses had been estimated to vary from 6–60 percent (Srivastava and Rao, 1966).

1. Symptoms

In the temperate regions, the disease usually becomes noticeable in the field at the heading stage. In severe cases, it may appear earlier but it is rare in the seed bed. The disease is typical vascular wilt, leaf blight is only the mild phase of the disease which results from secondary infection (Singh, 1998). There are three phases of diseases: (1) the leaf blight phase (2) the most destructive form of the disease is "Kresek", or wilt phase, and (3) the pale yellow leaf phase in the tropics.

In the seed bed it first appears as tiny, water-soaked spots at the margin of fully developed lower leaves. As the spots enlarge, the leaves turn yellow, dry rapidly, and wither. On leaf blades, lesions usually begin at the margin; a few are formed at the tip as water-soaked stripes. The lesions enlarge both in length and width, have a wavy margin, and turn yellow within a few days (Fig. 21.1a). The region adjoining the healthy parts shows water-soaking. Lesions may start at one or both edges of the leaves. As the disease advances, the lesions cover the entire blade, turn white, and later become grayish due to the growth of saprophytic fungi. In susceptible varieties, the lesions extend to the leaf sheath where they may reach the lower end. The infected blades in susceptible varieties wilt and roll as the diseased portions enlarge while the leaves are still green. The entire blade may soon become involved, and gets dried up. On more resistant varieties, a yellow stripe appears just inside the margin of the leaf blade with no formation of necrotic lesions. On the surface of young lesions, milky or opaque dew drops may be observed in the early morning. These dry up in the form of yellowish spherical beads (Fig 21.1b) which fall down in the field. In severely diseased fields, the infection may reach the grains. The glumes get discolored, and water-soaked spots develop on them.

Fig. 21.1 Bacterial blight of rice caused by *Xanthomonas oryzae*: (a) Symptoms (b) Portion of blighted leaf magnified to show bacterial exudates

In Indonesia, the disease has been known for a long time by the name 'Kresek'. It affects young plants, the outer leaves of which wither, become grayish green or light brown, and often float on the surface of water. Some of the younger leaves roll up and wither. The bacterium spreads through the xylem vessels to the growing point of the young plant, and infects the base of other leaves. As a result, the entire plant dries. Bacterial exudate can be seen from the leaf blades or sheaths.

In India, symptoms become visible one week after infection. The disease is characterized by the drying of the leaf tip, and inward rolling and twisting of the leaf blade. Linear, yellowish stripes may develop along the midrib or on the blade; marginal necrosis may also occur. The disease extends to the leaf sheaths and culms, killing single tillers or the whole plant.

2. Causal Organism

The disease is caused by *Xanthomonas oryzae* pv. *oryzae* (Ishiyama) Swings *et al.* (1990). It is a rod-shaped bacterium with rounded ends, 1–2 × 0.8–1 μm, and monotrichous flagellum of 6–8 μm. It is Gram negative and a nonspore forming bacterium.

3. Disease Cycle and Predisposing Factors

The bacterial leaf blight is principally a vascular disease. It enters the leaf through wounds, for example, by the chafing, and striking of leaves against one another in the wind, through the cut ends of seedling leaves, through broken roots at transplanting or through the hydathodes. Once inside the leaf, the bacterium multiplies and moves in both directions. Lesions may be seen about ten days after artificial inoculation. Bacterial exudates are produced on the surface of leaves, and during rainstorms these bacteria disperse, and come in contact with other leaves or fall into the water. Bacteria are also carried by irrigation water from one field to another; this has been demonstrated by using the bacteriophage technique of estimating their numbers.

The bacterium appears to be soil-borne in Japan. The organisms may live in the soil for one to three months depending upon soil moisture and acidity. However, the soil is not considered an important source of infection as the short period of survival is not sufficient to carry over the disease to the next year's crop.

In Japan, the bacterium can also overwinter in wild grasses (such as species of *Leersia* and *Zoysia*) growing along irrigation ditches, and in swampy areas. Several workers have demonstrated seed infection, and the survival of the bacterium in the seeds for varying lengths of time up to the next growing season, and the disease from such seeds. Detailed studies conducted by Kauffman and Reddy (1975) on the seed transmission of pathogen in rice indicate that the bacterium remains viable in infected seeds for only two months. In India, the pathogen is propagated from season to season through infected seeds (Devadath, 1992).

The question of weeds being sources of primary inoculum has evoked considerable interest. Several workers have been successful in infecting several weed hosts but the ones considered to function as active natural hosts are *Leersia oryzoides* and *Leersia oryzoides* var. *japonica*.

4. Disease Management

As *X. oryzae* pv. *oryzae* is suspected to be seed-borne in India, seed treatment is advocated. Soaking the seeds for 12 hours in Ceresan (0.05%) plus Agrimycin 100 (0.025%) followed by hot-water treatment for 30 minutes at 52-54°C has been recommended. Later it was found that the preliminary soaking of seeds can be done in plain water. Jain *et al.* (1966) suggested an alternative method which did away with hot-water treatment. This involves an eight-hour soaking in 0.1% Ceresan wet and Streptocycline at 0.3 g/10 litres.

Spraying in the field is also advocated, particularly in Japan. A combination of 60–120 ppm of antibiotic Chloramphenicol with an organomercurial, such as phenyl mercury acetate at 10–20 ppm metallic mercury, has been found effective. Another mixture used is 100 ppm each of Dithianone and Cellucidin. Plants should be sprayed as soon as the first signs of the disease appear. A disease-forecasting service is necessary. A method has been devised in which the bacteriophages of the leaf-blight organism give an indication of the intensity of the disease.

In Japan, the seed beds and fields are kept from being too deeply flooded, but the value of this method in the tropics is doubtful. The burning of crop remains to destroy the bacteria present in infected straw and stubble is recommended in India (Sulaiman *et al.*, 1965). Eliminating volunteer plants, and ratoon rice is recommended in Phillipines. Srivastava (1969) and Kauffman (1971) have suggested that the sowing of nursery beds in disease-free, isolated areas to prevent the inflow of primary inoculum into the fields might reduce the disease and arrest the build-up of an epidemic. Injury of the host plays an important role in infection as injury is a prerequisite for the entry of the bacterium. Pruning of top portions of all varieties, which is a common practice in many parts of India, should be avoided in order to reduce disease occurrence, especially in places where the disease has established itself.

It is well established that the application of nitrogenous fertilizers is an important predisposing factor for the occurrence of bacterial blight. The method of application of nitrogen also influences disease. Nitrogen application at the tillering stage increases the incidence of the disease but a similar application at the time of transplanting, or a later application at the time of flower initiation, results in lower incidence of the disease.

In Japan, no varieties of rice have been identified to be highly resistant to the most virulent strains of *X. oryzae* pv. *oryzae*. More than 6,500 varieties were screened for resistance at the International Rice Research Institute, Manila, Phillipines, between 1964 and 1966. About a dozen or more varieties have a broad range of resistance to the organism, and some of them are being used in the institute's breeding programme.

In India, two varieties (W529 and W348) out of the 16 tested in a trial of early maturing varieties were highly resistant in the field (Padmanabhan, 1965). Srivastava *et al.* (1967), using five virulent isolates of the pathogen, showed that 13 out of the 128 varieties tested showed resistance. The resistant varieties included Tainan 3 and Chainung-292. TKM 6, TR 42, and Chinsurah Boro II are tolerant to both leaf blight and Kresek phase of the disease (Chaube and Pundhir, 2005).

Singh, K (2004), using a molecular marker, studied diversities present in *Xanthomonas oryzae* pv. *oryzae*, and based on these, pyramided three bacterial-blight-resistant genes (Xa5, Xa13, and Xa21) in the background of rice varieties PR106 and Pusa44. Some of the resistant cultivars available in India are IR64, Sona, Mahsuri, Swarna, Janaki, Sujata, Asha, Udya, etc. (Reddy, 2004).

21.6.2 Citrus Canker

In India, citrus canker was first reported from Punjab (Luthra and Sattar, 1942). Its occurrence was further recorded in Tamil Nadu, Andhra Pradesh, Karnataka, Rajsthan, Assam, and Uttar Pradesh. Further, the disease appears as a serious problem wherever acid lime (*C. aurantifolia*) is grown on a large and commercial scale, and has become a permanent major problem to citrus growing of this country (Das, 2003).

1. Symptoms

Citrus canker may affect any part of the tree above the ground, although the leaves, twigs, young branches, and fruits are the most susceptible parts. On the leaves, canker first appears as small, watery, translucent spots, usually of a darker green colour than the surrounding tissue, and with a raised convex surface. The spots first become evident on the lower surface, and as the disease advances, the surfaces of the spots become white or grayish, and finally rupture, exposing a light brown, spongy, central mass developed in a crater like formation. These spots usually become surrounded by a halo which persists in very old lesions. Old lesions become corky and brown, and may appear pinkish. The development of the pink colour may also be due to scab disease, in susceptible varieties, lesions on the leaves but acquire a cankerous appearance, and the surface membrane completely disappears. The canker lesions on the fruits appear similar to those on the leaves, except that the yellow halo is usually absent, and the craterlike appearance is more noticeable. The spongy rough eruptions may be scattered over the surface, or several may coalesce, forming an irregular, scurfy, and scabby mass (Fig. 21.2a and b).

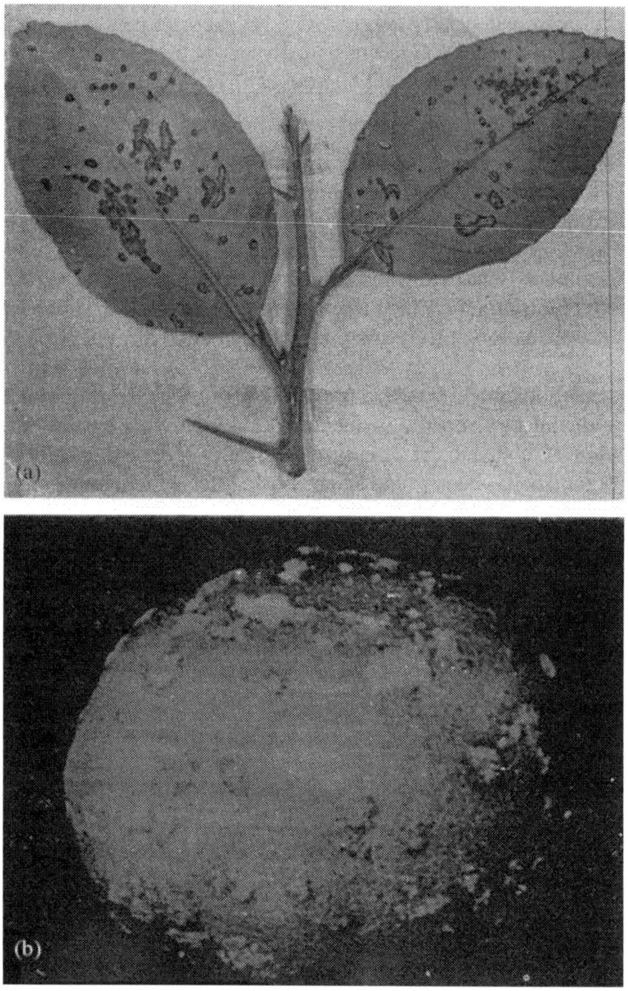

Fig. 21.2 Citrus canker (*Xanthomonas axonopodis citri*): (a) Symptoms on leaves (b) Cankers on fruit

2. Causal Organism

The pathogen responsible for citrus canker is *Xanthomonas axonopodis* pv. *citri* (Hasse) Vauterin [Syn. *X. citri* (Hasse) Dowson and *X. campestris* pv. *citri* (Hasse) Dye]. The bacterium *Xanthomonas* is rod-shaped measuring 1.5–2.0 × 0.5–0.75 μm, gram negative, and has a single polar flagellum.

3. Disease Cycle

The inoculum is provided by the diseased twigs and leaves. The pathogen enters the host through various types of wounds, natural openings, and so on. On penetration into the host, it multiplies in the intercellular spaces, dissolves the middle lamella, and establishes itself in the cortex. Cankerous outgrowths now develop within which bacteria multiply and are released with exudations. The disease is chiefly disseminated by wind, rain, and insects.

Citrus-leaf miners are very important for the dissemination of the disease. However, humans who seem to be the chief agent of dissemination of the pathogen transport the disease through infected nursery trees.

4. Disease Management

Citrus-canker incidence can be reduced considerably by taking an integrated management approach. Control measures include exclusion by quarantine, eradication by the destruction of infected trees, use of canker-free nursery stock, sanitary precautions, and the growth of resistant varieties. Because wind greatly facilitates wound infection, the establishment of windbreaks around citrus orchards is effective in preventing infection. Spraying the trees before planting in new orchards with a copper fungicide, and in old orchards pruning of the affected twigs, and spraying with copper fungicides at periodic intervals, especially during the rainy reason, is recommended. Spraying of neem cake at the rate of about 80 kg per acre has been reported to be highly effective in checking the citrus canker as well as leaf miners.

It has been shown that the disease can be controlled under field conditions with a spray of antibiotics, such as streptomycin sulfate at 500–1000 ppm or crude agricultural preparation containing an equivalent amount of the antibiotic at intervals of 15 days. Other effective antibiotics were Agrimycin and Streptocycline in combination with Bordeaux mixture. Application of neem-cake solution on the foliage reduced the canker in nurseries. Integrated application i.e., pruning of infected twigs, copper oxychloride (0.3%), Streptocycline (100 ppm), and neem-cake suspension was found very effective in controlling the disease. Grapefruit, sweet oranges, limes, and some lemon varieties are very susceptile. Varieties of *Citrus reticulata*, including tangerines, king, mandarin, and satsuma oranges are resistant to canker.

Studies on the biological control of citrus canker are still in a preliminary stage. Some strains of bacteria, viz. *Pseudomonas syringae*, *Erwinia herbicola*, *Bacillus subtilis*, and *P. fluorescens*, were reported to be antagonistic *in vitro* to the canker pathogen.

21.6.3 Angular Leaf Spots and Black Arm of Cotton

The disease occurs in all cotton-growing regions of the world. In India, the disease was first observed in Tamil Nadu in 1918. Since then it has been found to occur in various degrees in all the major cotton-growing areas of the country where the exotic American cotton is grown.

1. Symptoms

The bacterium is capable of infecting all the portions of the plant which are above the ground. There are four major types of symptoms: (1) seedling infection, (2) angular leaf spot of cotton, (3) black arm of cotton, and (4) boll rot of cotton.

2. Seedling Infection

The earliest symptom of the disease is noticed on the cotyledons of the germinating seed. Water-soaked, circular lesions appear first on the lower surface of the cotyledons. The infection soon spreads and forms angular patches. The diseased tissues turn brown, distort the shape of the cotyledons, and finally cause the drying and withering of the seeding. The whole seedling may collapse and wilt in severe infections.

3. Leaf Spots

On the leaves, the spots are usually angular. These first appear on the under surface, and then on the upper surface. The disease advances along the main vienlets, and viens to cause fingerlike lesions on the leaf blade, and then the disease is identified as 'vein blight'. The leaves affected by such vein blights become crinkled, and twist inwards and dry. The infection may also spread through the leaf blade to the petiole, causing severe defoliation. On the surface of the affected tissues, there may be profuse exudation of the bacterial slime which may dry and form a thin film.

4. Black Arm

The pathogen causes dark brown black lesions on the stem and fruiting branches. The lesions are linear, and sunken from the surface, after covering the whole circumference of the stem or branch. Under favourable climatic conditions, most of the branches and the main stem of the plant are affected by the disease. As the disease advances, the branches dry, and soon the plant withers.

5. Boll Lesions

The bacterium affects the boll or fruit also. The first symptom appears as water-soaked lesion on the surface. Such lesions turn dark brown, and finally black. They are invariably sunken on the surface. When young balls are attacked with several such lesions, they may drop off from the plant. In several cases the bacterial infection spreads inside the ball or the fruit, and causes severe reduction in size, quality, and quantity of the lint formed inside. As the disease advances, the seeds invaded by the bacterium suffer a reduction in their size and loss of viability.

6. Causal Organism

The disease is caused by *Xanthomonas axonopodis* pv. *malvacearum* (Smith) Vauterin *et al.* (Formerly *X. malvacearum*). It is an aerobic, Gram-negative rod, 0.3–0.6 μm × 1.3–2.7 μm. The bacterium is motile by a single polar flagellum.

7. Epidemiology

Under north Indian conditions, trash, self-sown seeds and soil were not as important as the seeds in providing the primary source of inoculum (i.e. carry over the infection from one season to another), specially because cotton is a not a continuous crop, and generally a cotton-wheat cropping pattern is followed.

8. Disease Cycle

The pathogen is both externally as well as internally seed-borne. The primary infection starts mainly in the seed where the bacterium may be present as a slimy mass on the fuzz or inside the seed. Infected cotton bolls, leaves, and twigs which fall on the soil surface also form an important source for the carryover of the disease. The secondary spread of the pathogen occurs mainly through the agency of wind-driven rain.

Two insect pests, *Earias* spp., and *Dysdercus koeningii*, transmit the disease mechanically. A large number of collateral hosts are also known but their role in the disease cycle is not well defined.

9. Disease Management

Seed treatment for the control of bacterial blight is an effective means of reducing crop loss. Methods vary widely between different countries, for example, delinting with sulfuric acid followed by dusting with mercury or copper compounds, the former being more effective but more harmful to seed germination than the latter. Delinting by flaming is claimed to be both effective, and nondangerous to seed.

It has been observed that when seeds are treated with organo-mercurials, external infection is checked but internal infection persists, and when such seeds are sown, the bacterium multiplies to cause infection. Treating the seeds with antibiotic substances, such as Streptomycin sulfate, results in complete eradication of both the internal as well as external infection of the seed.

Certain antibiotic mixtures (Streptomycin + Tetracyclin) could control bacterial blight disease effectively, particularly when combined with copper fungicides. However, the use of antibiotics in human use is hazardous, and their use should be minimized to seed treatment. However, certain fungicides (e.g. Carboxin, Oxycarboxin, Busan-72) used are as good as antibiotics or even better under certain conditions. These have an additional advantage because the danger of selection of antibiotic-fungicide resistant mutants is eliminated.

For control of secondary spread spray of the Copper oxychloride/Vitavax/Plantavax alone or in combination with Streptomycin sulfate is very effective. Spraying is usually started when the crop is 6–8 weeks old, and subsequent sprays are made at 20-day intervals. Management of insect pests by spraying of insecticide also reduces the disease incidence up to some extent (Chopra, 1998).

Verma *et al.* (1977) claimed that hot-water treatment of seeds at 56°C for ten minutes eliminates both external and internal infection of *X. campestris* pv. *malvacearum*, without affecting seed germination.

Breeding varieties resistant to the disease has attracted the attention of several scientists all over the world. Knight (1946) had done considerable work on the genetics of black-arm resistance in Sudan. *Gossypium herbaceum* and *G. arboreum* are mostly immune to the disease, whereas *G. barbadense*, *G. herbaceum* var. *typicum*, and *G. hirsutum* are susceptible. Five main genes governing the resistance of *Gossypium* spp. have been isolated and identified: B1, B21 B3, B4, and B5.

Several resistant sources to bacterial blight are available. These are 101-102 B, BJA-592, P- 14-T128, Raba B-50, HG-9, Tamcot-CAMDE, Tx Bonham, etc. which have been used in breeding programmes for development of resistant varieties.

In India, a variety BJR (bacterial blight jassid resistant) has been developed. It is serving as an excellent source for combined resistance to bacterial blight pathogen, and jassid attack in the breeding programme. At Punjab Agriculture University, Ludhiana, a *hirsutum* variety LH 900 has good resistance against bacterial blight, and is also a good yielder. Neem-based ecofriendly botanical pesticides were found to be quite effective. These are Agricare and Plantotype. A few phylloplane bacteria of cotton (namely species of *Aeromonas*, *Flavobacterium*, *Pseudomonas*) on preinoculation protected the leaves from infection by a challenge dose of the pathogen.

21.6.4 Yellow Ear Rot or Tundu Disease of Wheat

The disease was first described by Hutchinson (1917) from Punjab in India. Milne (1919) observed that the disease incidence was dependent on mechanical injury to the plant part by eelworms, and that the bacterium could grow only on the excretions of the eelworms. The disease has subsequently been reported from several countries. Experiments have shown that the nematode, *Anguina tritici*, responsible for transforming the developing grains in wheat ears to galls (popularly called 'gegle', and 'mamni'), or ear-cockle disease is associated with the bacterium in causing the rot.

1. Symptoms

The 'tundu' disease is characterized by the twisting of the stem, distortion of the earhead, and rotting of the spikelets with a profuse oozing of yellow liquid from the affected tissues. Hence, the name of the disease is yellow ear rot (Fig. 21.3). The ooze contains masses of bacterial cells. The nematode alone causes wrinkling, twisting, and various other distortions of the leaves, and stem, and also forms small round galls on the leaves. The infected plants are shorter and thicker than the healthy plants. In the distorted earheads, dark galls are found in place of kernels. When the bacterium is associated with the nematode, the disease symptoms are intensified at the flowering stage, and the yellow ear rot sets in. Due to the combined action of the nematode and bacterium, the earhead becomes chaffy, and the kernels are replaced by dark nematode galls which are also contaminated with the bacterium. The disease is severe during the rainy season.

The infected plants produce more tillers than do the healthy ones. Another interesting feature is the early emergence of ears in the nematode-infected plants which is about 30 to 40 days earlier than in the healthy ones. Suryanarayana and Mukhopadhyaya (1971) observed that the kernels are modified into galls, and bear a superficial resemblance to normal kernels. The number of nematode larvae per gall may be from 80–32,000.

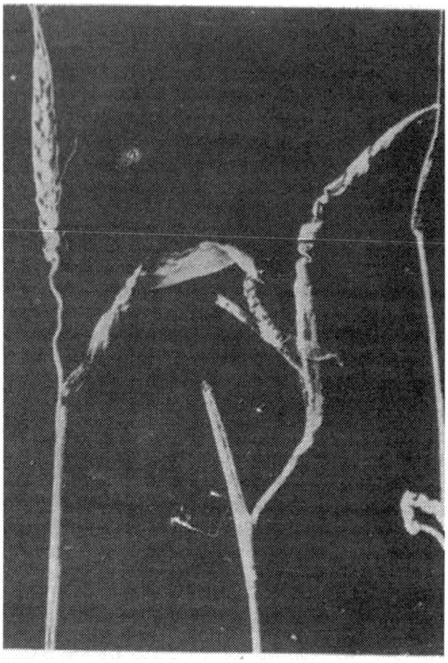

Fig. 21.3 Yellow ear rot of wheat (*Rathayibacter tritici formerly Clavibacter tritici*): symptoms

2. Causal Organism

Rathayibacter tritici (ex. Hutchinson) Zgurskaya *et al.* 1993 is the causal organism, and it was earlier known as *Clavibacter tritici* (Carlon, and Vidaver) Davis *et al.*, and prior to this it was named *Corynebacteriuin tritici*. The bacterium is rod-shaped, and 2.3 × 1 μm in size. It is motile by a single polar flagellum. It is gram positive. It may be mentioned that the disease occurs due to the combined action of the wheat nematode *Angunia tritici* and the bacterium *Clavibacter tritici*.

3. Disease Cycle

The disease starts from seeds contaminated with the nematode galls. When such contaminated seeds are sown in the field, they absorb moisture from the soil. The larvae escape from the galls, and shortly climb up the young wheat plants. The nematodes have a tendency to seek the tender growing points of the plants where they remain as ectoparasites. After the plants flower, the nematodes enter the floral parts, and form galls in the ovaries. While within the galls, sexual differentiation occurs, copulation takes place, and the female lays eggs, up to 2000 in number. The eggs hatch into larvae under favourable conditions, and become active parasites. Once the nematode is inside the tissues of the ovary, the bacterium becomes active and causes rotting. The yellow ooze coming out of the rotting earhead provides the inoculum for the secondary spread of the disease which is favoured by wind and rain. The nematode probably functions as a vector, transporting the bacterium to the otherwise inaccessible meristematic regions of the host. The nematode secretes some substance in the presence of which the bacterium can cause the disease.

4. Disease Management

The sowing of gall-free seeds in non-infested soil will help reduce the incidence of disease. The seeds can be freed from the galls by being floated in brine, at 160 g of Sodium chloride per litre of water. Wheat, barley, oat or other susceptible crops should not be sown in infested soil. Solar heat during May and June in the northern wheat belt of our country can be utilized for destroying the galls and nematodes. Infected plants should be carefully taken out and burnt. Seeds for sowing purposes should be taken from disease-free areas only. Soil treatment with Aldicarb or Carbofuran destroys the nematode in the soil and prevents infection.

21.6.5 Bacterial Soft Rot and Black Leg of Potato

Bacterial soft rot and black leg of potato is caused by *Pectobacterium carotovorum* var. *carotovora* (earlier known as *Erwinia carotovora* subsp. *carotovora*), and *Pectobacterium carotovorum* var. *atrosepticum* (earlier known as *Erwinia carotovora* subsp. *atroseptica*). It occurs in most parts of Europe, the USA, and Canada. According to Arora (2012), the pathogen affects potato tubers under poor storage conditions, especially in warm and wet climate. Losses under bad handling of the produce, poorly ventilated storage or transit may go up to 100 percent.

1. Symptoms

The disease appears sporadically in the field. The plants turn pale green or yellow, and soon wilt and die. The affected haulms are jet black in colour at the soil level; later the discolouration extends into the old seed tuber also. The cortical tissue of the basal stem becomes black, and shrivels, and gives the appearance of a black leg. Sometimes the young seedlings which emerge from the diseased tuber may be destroyed soon after emergence or they may not emerge at all. During the wet season, the disease progresses rapidly, and the entire stool may rot. When the infection takes place late in the crop season, the organism is harboured on the tubers, and the damage continues in storage godowns. In soft rot of tubers, which may occur in the field if the soil is moist and temperature high, or during transit and storage, the tubers are transformed partly or wholly in a soft decayed pulpy mass covered by the corky epidermis which is not attacked.

In storage, bacteria cause severe rot of the tubers, sometimes resulting in almost cent percent damage. When tubers are kept in storage under damp conditions, decay sets in, and the tubers become soft with a watery mass of bacterial ooze accompanied by a sulfurous odor.

2. Causal Organisms

Bacterial soft rot and black leg of potato is caused by *Pectobacterium carotovorum* var. *carotovora*, and *Pectobacterium carotovorum* var. *atrosepticum*.

3. Disease Cycle

The organisms live saprophytically in decaying plant debris, and enter the tubers through bruises, sun scald, insect wounds, etc. The seed corn maggot, *Hylema platura,* is a common pest of potato, and it helps in the infection of seed potato by the bacterium. Infection also takes place through enlarged lenticles in tubers.

4. Disease Management

Only healthy tubers should be sown when cut tubers are sown. The cut tubers used for seed purposes should be stored at a temperatures of 12–15°C in a dark room for 4 days. This will allow cork formation at cut surfaces, and will prevent entry of bacteria through cut surfaces. The cut tubers used for seed purposes should be dipped in a fungicidal solution.

The wounding and bruising of tubers should be minimized at the time of harvest and storage. Chlorinated water should be used for washing the tubers.

Crop rotation following green manure-potato-wheat reduces soft rot. Calibration of equipment to minimize bruises during harvest, avoiding direct exposure of tubers to sunlight, proper aeration in transport, and storage are some measures which help in the reduction of soft rot. Planting of whole seed potato, well-suberized seed pieces, in well-drained soil with temperature around 10-13° C at low depth reduce incidence of black leg (Arora, 2012).

Control of soft rot through botanical products and biopesticides such as essential oils, and extracts of hemp flower or microbial antagonists such as strains of *Pseudomonas putida*, *P. fluorescens*, *Bacillus subtilis* strain BS107, and *Erwinia. herbicola* Eh 252 has been found effective. Certain endophytic bacteria as *Pseudomonas* species promote plant growth and induce resistance of potato plants towards infection by *Pectobacterium atrosepticum* (Pavlo, *et al.* 2011).

21.6.6 Crown Gall Of Stone Fruits

1. Distribution

The disease has been known in the USA since 1870. Cavara from Italy isolated the bacterium from stem galls on grapes, and proved its root pathogenicity. The disease has been reported from USA, South Africa, New Zealand Australia, Asia, Canada, and Europe. In India, the occurrence of crown gall was reported from Shimla. In India, the disease has been reported from cherry (*Prunus avium*), plum (*P. domestica*), almond (*P. dulcis*), apricot (*P. armenica*), peach (*P. persica*), apple (*Malus domestica*), and pear (*Pyrus communis*). Crown-gall bacterium is a natural genetic engineer, and has been utilized in many biotechnological transformations (genetic engineering). Plant biologist are using it as the main tool (vehicle) for transferring all kinds of genes between related and unrelated plants, and even other organisms, e.g. insects or viruses, plants, and animals.

2. Symptoms

The galls are usually 7 to 100 mm in diameter, spherical to irregular with a hard rough surface, generally present on the stem more often at the graft union of fruit tree, and also on roots or just below the soil line. Tumours can appear on any portion of the plants (roots, stem, branches, petiole, and leaf veins). Secondary tumours can appear far away from the primary tumor. The secondary tumours do not contain bacteria, and may not be connected with the primary one. In early stages, the tumors are more or less spherical in shape, white in colour, and soft in texture. Later the outer tissues become dark brown or black. The exterior of the hypertrophied growth is corky. The attacked plants may remain stunted, and in advanced stages may be killed. Unlike other galls, the crown gall (Fig. 21.4) is an example of a non-self-limiting tumor.

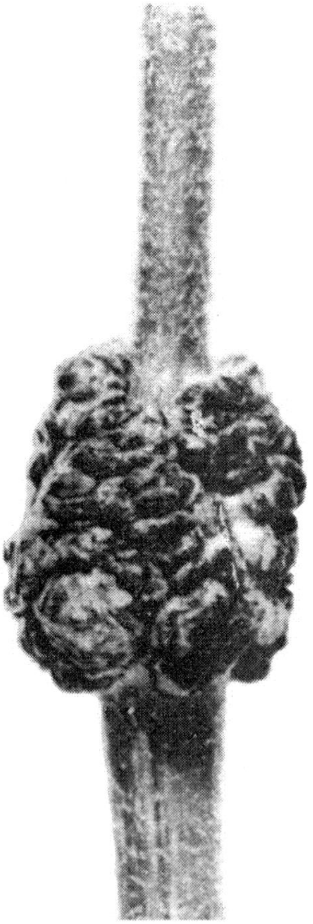

Fig. 21.4 Crown gall (*Agrobacterium tumefaciens*): symptoms

3. Causal Organism

Crown gall is caused by the bacterium *Agrobacterium tumefaciens*. Another species, *A. radiobacier*, is saprophytic, and does not produce galls. *A. tumefaciens* and *A. radiobacter* are identical except for the plant pathogenic nature of the former. The bacteria contain one to several plasmids (small chromosome-like bodies composed of circular double-stranded DNA). One of these plasmids carries the genes for tumour induction, and is called Ti (tumour-inducing) plasmid. The Ti plasmid also carries the genes that determine the host range of the pathogenic bacterium and the kinds of symptoms that will be produced. The most characteristic property of this bacterium is its ability to introduce part of the Ti plasmid (T-DNA) into plant cells, and transform normal plant cells to tumor-producing cells. The Ti plasmid contains *vir*-genes, which code for proteins that are required for the transfer of T-DNA (transforming DNA). The *vir*-genes are expressed after induction by plant-specified phenolic compounds such as *p*-hydroxybenzoic acid and vanillin. These phenolic-inducer molecules are produced by damaged plant tissues. Once inside the plant cells, the T-region-containing DNA is transported into the nucleus, and integrates into plant chromosomes at a number of sites. The T-DNA contains oncogenes that code for either octopine or nopaline (two opines that serve as carbon sources for the infecting bacteria). The bacterium can grow on relatively simple medium but its isolation from the soil requires selective media (Patel, 1926).

After the inducing bacteria have been in direct contact with the plant cells at the wound site for several days, viable bacteria are no longer necessary for tumour formation. The plant cells are transformed, and formed cells become highly irregular. The gross morphology of the crown gall is dependent upon the inducing bacterium strain of *A. tumefaciens*. Some are highly irregular and leaflike which are only partially transformed, while unorganized tumors are completely transformed.

4. Disease Cycle

The pathogen overwinters or oversummers in the soil. There are several hosts. All the root stocks used for the propogation of the temperate fruit plants except *Cotoneaster bacillaris*, *C. microphylla*, and *Malus baccata* were found to be highly susceptible to crown gall (Durgapal, 1977).

5. Disease Management

For controlling the disease and destruction of the infected plant material, adoption of rigorous sanitation in the nursery, rotation of nursery site, use of well-pitted and properly wrapped (with antiseptic tape) grafts, avoidance of highly susceptible root stocks, and preference of budding over grafting are recommended. However, the most effective, and cheap method of controlling the disease is by dipping the seeds, and cutting of roots of young plants before transplanting in a suspension of *A. radiobacter* strain K84. This strain is antagonistic to most virulent strains of *A. tumefaciens*. It is postulated that the antagonistic strain (*A. radiobacter* K84) produces the bacteriocin agrosin 84 which is inhibitory to most virulent strains of *A. tumefaciens*. Unfortunately, some strains of *A. tumefaciens* inherited from strain K84 its resistance to agrosin 84. Therefore, a new strain K-1026 is now recommended for control of the disease. The strain K-1026 lacks the ability to transfer its resistance to pathogenic *Agrobacterium* strains. There is an immediate need to test these strains under Indian conditions (Jindal, 1990).

Other antagonists reported for biological control of crown gall in temperate fruits are *Bacillus subtilis* and *Streptomyces* spp. Gupta and Khosla (2007) observed antagonistic activity of *Bacillus subtilis* against *A. tumefaciens*. According to Pathak *et. al* (1996), painting with the chemical Elgetol (Sodium dinitro-o-cresylate) helps considerably in protecting the plants from infection. Avoid wounding the crown and control of root-chewing insects is also recommended.

21.6.7 Common Scab of Potato

1. Distribution

This disease occurs throughout the world. It is most prevalent, and important in neutral or slightly alkaline soil, and light sandy soils. In India, since 1976 the disease has become widespread in the plains of Uttar Pradesh, Haryana, and Punjab (Nagaich, 1983). In India, the disease has been reported from UP, Assam, Meghalaya, West Bengal, Lahaul Valley of HP, and the plains of Bihar,. Punjab, and Maharashtra. The losses due to this disease can be up to 40 percent. The greatest loss from the disease is the reduction in market value of tubers due to rough, blemished skin, and deep scabs increase the waste in peeling. The disease was earlier considered to be of minor importance.

2. Symptoms

The symptoms of common scab of potato are seen mostly on tubers. There are two types of lesions, the shallow, and the deep scab formed on the tubers. In shallow scab, the affected tubers show superficial roughened areas, sometimes raised above, and often slightly below the plane of the healthy skin. The lesions consist of corky tissue which arises from the tuber periderm. The lesions vary widely in size and shape, sometimes darker than the healthy skin. In deep scab, the lesions measure 1 to 3 mm or more in depth, and are darker than the shallow scab.

3. Causal Organism

Common scab of potato is caused by *Streptomyces scabies* (Thaxter) Waksman and Henrici.

4. Disease Cycle

The pathogen is spread through soil water, wind-blown soil, and on infected potato-seed tubers. However, infected tubers play the main role of its distribution over larger areas. The bacterium can survive through the digestive tract of animals and can be disseminated by manure.

5. Disease Management

(a) Control of the common scab of potato is through the use of certified scab-free seed potatoes.
(b) Tubers with deep lesions should not be used for seed.
(c) The tubers can be disinfected by dipping them for five minutes in a 0.25 per cent suspension.
(d) A fair degree of disease control may be obtained by using certain crop rotation, and bringing the soil pH to about 5.3 with sulfur. Rotational crops such as maize, cotton, sorghum, wheat, cabbage, and onion have been observed to reduce incidence of common scab.
(e) Lapwood and Hering (1970) controlled scab in field plots by maintaining soil moisture near field capacity during tuber formation, until all the internodes of the tubers are formed. Thus maintaining the soil moisture at field capacity from the fifth week after planting up to the ninth week by frequent light irrigation is recommended. In our country, Singh and Singh (1982) have recommended to maintain the available soil moisture at 75.0 per cent, and above from tuber initiation to maturity to reduce tuber infection.
(f) Green manuring and cover crops of soybean promote activity of antagonistic bacteria. Therefore, in infested fields, green manuring before planting or cultivation of green manuring before potato may be useful. It may be mentioned that Millard and Taylor (1927) reported control of scab of potatoes grown in sterilized soil, and inoculated with *S. scabies* through simultaneous inoculation with *S. praecox*, a saprophytic species which is antagonistic to pathogenic *S. scabies*. The disease was reduced by incorporation of rapeseed meal into the soil, and by adding a mixture of *T. viride* isolates in furrow at planting (Hilton *et al*., 2006).
(g) Soil solarization with transparent polyethylene mulching during hot summer season was found effective for control of russet scab. The disease severity by soil solarization reduced to almost one third as compared to the unsolarized plots (Arora, 2005).
(h) Nonpathogenic, antibiotic-producing *Streptomyces* spp. have been reported to reduce severity of potato scab when added to disease-conducive soil (Hiltunen *et al*., 2009). *Bacillus subtilis*, nonpathogenic *Streptomyces* spp., and botanicals/plant products such as *Geranium pretense* have been reported to control common scab. The *Pseudomonas* strain R1 when applied with vermicompost has been reported to show reduction of common scab (Singhai *et al*., 2011).

(i) Resistance has been discovered in *S. andigena* × *S. tuberosum* hybrids, and in red tuber varieties like Kufri Sindhuri (Nagaich, 1983). Thus, the disease can be controlled by using resistant or tolerant potato varieties.

21.6.8 Ratoon Stunting Disease (RSD) of Sugar Cane

It is found in almost all the sugar cane areas of the world. For many years this disease was thought to be caused by a virus. But now it is established that it is caused by a bacterium. In India this disease was first observed in COS 510 at Golakokarannath (Uttar Pradesh) by Professor S J P Chilton, in 1956 (Chona, 1956). Later on it was suspected in the variety Co738 by Rafay (1957).

1. Symptoms

Ratoon stunting does not have well-described external symptoms. Stunting, though present in many cultivars, is absent in others. In mature diseased canes of many cultivars, there is a reddish discolouration of the vascular bundles at the level of nodes. Typical symptoms are seen only when the canes are split open longitudinally. The disease is present in both plant and ratoon crops, although the effect has been more pronounced in ratoon crops. Affected crops look stunted with less vegetative growth in spite of all the inputs and proper agronomical practices. The disease, although present in the stalks, is not expressed under good growing conditions in many instances. Such suspected canes if split open longitudinally show pinhead-like pink coloured spots at the nodes, and most towards the upper nodes.

2. Causal Organism

For many years, the ratoon stunting disease was thought to be caused by a virus as the disease had many characteristics of viral diseases but now it is established that it is a bacterial disease. Ratoon stunting disease caused by the xylem inhabiting slow growing, recently named *Leifsonia xyli* subsp. *xyli* is one of the most important sugar cane diseases in the world. Work by Gillaspie *et al.* (1973), Maramorosch *et al.* (1973), and Teakie *et al.* (1973) indicated that a bacterium of the coryneform group might be the causal agent. Subsequently, the causal organism was cultivated on artificial media by Davis *et al.* (1980), and Lio and Chen (1981). Gillaspie *et al.* (1989) have given a comprehensive account of the disease. In culture the bacterium is rod, 0.25-0.35 × 2-5 μm, gram positive, nonmotile, lacking endospores, aerobic, and of clublike or V-form morphology. Davis *et al.* (1980) have named the pathogen *Clavibacater xyli* subsp. *xyli*. The bacterium is filamentous; the length of the filaments may be 10 microns or even more. The cells may be straight, bent, curved, and/or swollen at the tip, or in the middle.

The bacterium can be observed in nodes, internodes, leaves, young tillers, and also in tissues just below the growing point. This gives an indication that infection is systemic. Currently, serological detection of the pathogen is a common and reliable method by which the disease can be diagnosed easily.

3. Disease Cycle

The pathogen is found in all parts of the sugar cane plant, and is easily transmitted by mechanical means. Juice from diseased plants at dilutions of 1/25,000 is infectious. Cutting knives used for preparing cuttings or setts (seeds), and cutter planting machines transmit the bacterium readily. Humans are the primary agents in spreading the bacterium. Rats have been shown to transmit the bacterium by chewing. The pathogen does not survive in the soil; there is no evidence of transmission of the pathogen through true seeds. No insect vector is known to transmit the bacterium. A number of grass hosts have been found infected by experimental inoculation. Elephant grass (*Pennisetum purpureum*) is a very useful indicator plant. The role of alternate hosts in the epidemiology of the disease is not clear.

4. Disease Management

The control measures involve healthy planting material, preventing disease spread in the field, and the use of resistant cultivars. Canes meant for planting should be treated either by hot water (50°C for 2 hours), aerated steam (53°C for 4 hours), or hot air (54°C for 8 hours). Afterwards, this clean cane is planted, increased, and maintained disease-free. Seed setts should be taken from robust plants only, and ratooning of weak or stunted crop should be completely avoided.

Relatively little work has been done on breeding for resistance to the ratoon stunting disease.

21.6.9 Citrus Greening Disease (CGD)

The disease has been reported from Jammu and Kashmir, Punjab, Haryana, Rajasthan, Uttar Pradesh, Madhya Pradesh, and Karnataka. The disease has completely wiped out the citrus trees from the Coorg region. Greening is more serious in Punjab followed by Assam where the incidence of the disease is 64.8 and 41.1 per cent respectively.

1. Symptoms

Citrus greening is characterized by yellows type of symptoms which are highly variable. There is yellowing of the midrib and lateral veins of mature leaves. Young leaves of infected plants are mottled and chlorotic as though they have been affected by zinc deficiency, and the yellow tissues of the leaf lamina show scattered green islands. Since chlorosis can be caused by mineral deficiency such as zinc, mironutrient spray temporarily halts or masks the symptoms of greening. A characteristic symptom is that yellow areas are delimited by the midrib and expand towards the leaf margins. Soon after a tree becomes infected with citrus greening, tree growth slows down, trees become stunted, and twigs and branches die back. But the trees do not die, and in some cases trees may appear to recover. Leaf size is reduced; leaves become thicker, and usually remain more erect. The internodes and branches are shortened, and the branch usually gives a bushy appearance. Fruit of the affected trees are smaller. The diseased trees flower earlier. The root number is considerably reduced.

2. Causal Organism

Citrus greening disease was first reported by Reinking in 1991 by the name *Huanglongbing*, meaning yellow shoot. The pathogen is a phloem-limited uncultivable Gram-negative bacterium. Initial reports suggested it was a virus but later on it was considered a mycoplasma organism. This pathogen was defined as a gram-negative bacterium in 1984 by Garnier *et al.* (1984), and was further characterized by Jagoueix *et al.* (1996). The genus *Liberobacter* from Asia and Africa were recognised as two widely distributed geographical distinct species, *L. asiaticum*, and *L. africanum*. They are generally rigid, rodlike, and are pleomorphic bodies during growth.

The disease was earlier considered to be caused by a virus. Later a mycoplasma-like organism was suspected to be involved. It was in 1970 that Lefleche and Bove reported that the disease in caused by a phloem-inhabiting fastidious bacterium with a double-membraned cell wall distinct from MLOs cells. Since the causal organism has not been cultured in the laboratory so far, its characterization has not been possible. The cells of the phloem-restricted fasidious bacteria are Gram-negative, nonmotile, nonpleomorphic, rigid rod measuring 1.10–2.0 μm. The bacterial cells occur in mature sieve elements, and irregularly distributed among vascular bundels.

3. Disease cycle

The citrus-greening organism is transmitted through vegetative propagation, and by two species of citrus psylla, *Diaphorina citri* and *Trioza erytreae*. In South Africa, electron microscopic studies have confirmed the role of *T. erytreae* in transmission of the organism (Moll and Martin, 1973). In our country these two vectors are more common in North and Central India than in other regions of the country. The vector once infected remains viruliferous throughout its life. Gamier and Bove (1983) have demonstrated the transmission of the bacterium from sweet orange to *Vinca rosea* (periwinkle), *Cuscuta*, or dodder plant.

4. Disease Management

The disease can be controlled by a combination of the following measures:

(a) Removal of diseased plants from the orchards.
(b) New plant should be raised from indexed stock or nuceller seedlings.
(c) Spray Diazinon (0.02%), Eldrin (0.02%), or Parathion (0.02%) to control the insect vectors.
(d) Spray mixture of Bavistin and Ledermycin (500 ppm each) six times at ten-day intervals. Tetracycline hydrochloride showed good results in reducing leaf symptoms.
(e) Thermotherapy is another approach found to be very effective in the control of citrus greening. Graftwood is heated to 48–50°C for several minutes for complete elimination of the pathogen.
(f) Attempts have been made to control the vectors through different approaches, e.g. by means of chemical, cultural, or biological measures. Systemic insecticides like Dimethoate and Monocrotophs eliminate *Diaphorina citri.* Methomyl or Melathion sprays on citrus trees at 10–12 day intervals were not effective against citrus greening.
(g) Biological control has been tried in India. *Tetrastichus radiata* is reported to be drastically reducing the psylla population (Singh, R S, 2007)

21.7 SOME OTHER IMPORTANT BACTERIAL DISEASES

Black rot of crucifers	*Xanthomonas campestris* pv. *campestris.* It is an aerobic gram-negative rod.	The disease appears near the leaf margin, is characterized by chlorosis, and in severe cases the leaves wither. Vascular blackening may appear on the main stem. From the affected plant parts, bacterial ooze may also be seen.	Soil-borne or seed-borne	Three-year crop rotation, strict sanitary precaution followed by seed treatment with 0.1% Mercuric chloride is effective. Agrimycin and Streptocycline were found effective in checking the disease.
Wilt and brown rot of potato	*Ralstonia solanacearum* (earlier called *Pseudomonas solanacearum*). It is an aerobic gram-negative rod.	Both above and underground parts of the crop are affected. Sudden wilting, yellowing of lower leaves, and stunting of plants are observed.	Soil-borne	Use of disease-free seeds, soil disinfection (bleaching powder), and soil solarization. Crop rotation and proper date of sowing seeds are effective control measures.
Fire blight of apples and pears	*Erwinia amylovora.* It is a gram-negative rod and facultative anaerobes.	Symptoms appear on flowers, leaves, twigs, and fruits. On leaves, brown-black blotches followed by curling are seen. The bark of twigs turns brownish black. The fruits become water-soaked, shrivel, mummify, and finally turn black.	The pathogen overwinters at the margins of cankers formed during the previous year.	Destruction of overwintering cankers and twigs. Application of 1% Bordeaux mixture or Streptomycin 100 ppm is quite effective.

(*Contd.*)

(*Contd.*)

Stalk rot of maize	*Erwinia chrysanthemi* pv. *zeae* (earlier called *E. carotovora*)	Basal portion of leaves rots, internodes become soft and discolored.	Soil-borne	Application of Streptocycline and Agrimycin 100 as well as bleaching powder has been found effective in controlling the disease.
Angular leaf spot and wildfire of tobacco	Angular leaf spot is caused by *Pseudomonas angulata*, and wildfire is caused by *P. syringae* pv. *tabaci*	Spots are small, angular, and black or dark brown in color. In wet weather the disease spreads so wildly and rapidly that this phase is called wildfire of tobacco.	Plant debris harbours primary infection	The site of nursery should be changed, and refuse of tobacco should be removed. Seed treatment with silver nitrate gives some protection.
Bacterial blight and canker of mango	*Xanthomonas campestris* pv. *mangiferae-indicae*, rod-shaped, gram-negative bacterium	Groups of minute, water-soaked dark brown to black lesions appear on tips of leaves, young fruits, petioles, and tender stems. Badly affected fruits drop prematurely.	The bacterium harbours on the leaves throughout the year.	Disease can be controlled by application of Streptocycline spray.
Bacterial spots of chili and tomato	*Xanthomonas vesicatoria* (syn. *X. campestris* pv. *vesicatoria*)	Small, circular to irregular water-soaked spots appear on leaf. Later these spots become brown to dark and rough by rolling back of the leaf surface, and leaves may be twisted.	Seed-borne	Mixture of Streptomycin and copper sulfate gives good control of disease. Actigard (Acibenzolar-S-methyl) induces SAR in tomato. Neem oil also reduces the disease.
Red stripes of sugar cane	*Acidovorax avenae* subsp. *avenae* Gram-negative, and rod-shaped bacterium	Water-soaked elongated streaks, chlorotic, long, narrow, sometimes stripes fusing to form large bands appear on leaves.	Sugar cane plants grown in adjacent fields provide uninterrupted supply of the bacterium.	Cutting down and burning of affected sugar cane shoots reduces the spread of disease.
Bacterial wilt of tomato, and egg-plant	*Ralstonia solancearum* Gram-negative, and rod-shaped bacterium	Wilting, stunting, yellowing of leaves, and finally collapse of the entire plant are the main symptoms of the disease.	Soil-borne	Crop rotation, soil-solarization are effective cultural practices. *Bacillus* sp., *Pseudomonas fluorescens*, and *P. putida* are potential biocontrol agent.
Bacterial canker of tomato	*Clavibacter michiganensis* subsp. *michiganensis*	Small, light-coloured spots appear on fruits, leaves, and stem.	Seed- and soil-borne	Soaking of seeds in 0.25% solution of cupric acetate acidified with acetic acid for 20 minutes are quite effective. Crop rotation helps in reduction of soil-borne inoculum.

(*Contd.*)

(Contd.)

Bacterial leaf streak of rice	*Xanthomonas oryzae pv. oryzicola*	It appears as fine interveinal long or short lines on the leaf blade which are water-soaked and grayish in color.	Seed-borne	Seed treatment with Thiram or Cerasan or Streptocycline is effective. Vitavax prevents leaf infection and spread. IR20, Krishna, and Jaganath have shown good tolerance to the disease.

REVIEW QUESTIONS

1. What are prokaryotes and eukaryotes?
2. Name some important genera of phytopathogenic bacteria.
3. What is Gram stain? Give examples of gram-positive and gram-negative bacteria causing plant diseases.
4. How will you differentiate plant pathogenic bacteria?
5. How will you control the following diseases?
    - (i) Bacterial blight of paddy
    - (ii) Citrus canker
    - (iii) Angular leaf spot of cotton
    - (iv) Black rot of crucifers
    - (v) Brown rot of potato
    - (vi) Yellow ear rot or tundu disease of wheat
    - (vii) Black leg and soft rot of potato
    - (viii) Crown gall of stone fruits
    - (ix) Stalk rot of maize
6. What are the different antibiotics used for controlling plant bacterial diseases?

MULTIPLE-CHOICE QUESTIONS

1. In bacteria, cell wall is made up of
    - (i) mucopeptide (ii) cellulose
    - (iii) hemicellulose (iv) chitin
2. Who was the first scientist to observe bacteria under a microscope?
    - (i) E J Butler (ii) Anton von Leeuwenhoek
    - (iii) Robert Koch (iv) Louis Pasteur
3. Which one of the following bacterial characters is a plant character?
    - (i) Heterotrophic nutrition (ii) Presence of flagella
    - (iii) Rigid cell wall (iv) Prokaryotic cell
4. Gram-negative bacteria lose crystal violet stain because of
    - (i) walls with more lipids (ii) walls with less lipids
    - (iii) walls with more permeability (iv) thinner cell wall

5. In lysogenic cycle, the virus is
   (i) temperate (ii) virulent
   (iii) harmless to the host cell (iv) reproductive only
6. Bacterial transduction in phage-mediated genetic transfer was discovered in *Salmonella* by
   (i) Beadle and Tatum (ii) Zinder and Lederberg
   (iii) Griffith (iv) Avery, Macleod, and McCarty
7. Pili are present in
   (i) mycoplasma (ii) cytoplasm
   (iii) bacterial donor cell (iv) archaebacteria
8. Bacteria used in the biogas fermentation are
   (i) halophytes (ii) cyanobacteria
   (iii) thermoacidophiles (iv) methanogens
9. In which type of bacterium are flagella are uniformly distributed around the cell?
   (i) Monotrichous (ii) Lophotrichous
   (iii) Peritrichous (iv) Amphitrichous
10. Which of the following stains is used to identify *Mycobacterium tuberculosis*?
   (i) Grams stain (ii) Lactophenol stain
   (iii) Safranin stain (iv) Acid fast stain
11. The mycoplasmas are classified as Mollicutes because they
   (i) lack a cell wall (ii) are symbionts in plants
   (iii) are endosymbionts (iv) none of these
12. Which of the following is a plant pathogenic Gram-positive bacterial genus?
   (i) *Clavibacter* (ii) *Curtobacterium*
   (iii) *Streptomyces* (iv) All of these
13. The pathogen for angular leaf spot and black arm of cotton is
   (i) externally seed-borne (ii) internally seed-borne
   (iii) both (i) and (ii) (iv) air-borne
14. The pathogen responsible for yellow ear rot or tundu disease of wheat is
   (i) *Clavibacter tritici* (ii) *Pseudomonas solanacearum*
   (iii) *Xanthomonas campestris* (iv) *Ralstonia solanacearum*
15. Which of the following bacterium is used as a biocontrol agent against crown gall of stone fruits?
   (i) *Agrobacterium tumefaciens* (ii) *A. radiobacter*
   (iii) *Pseudomonas solanacearum* (iv) *Xanthomonas campestris*

Answers

1. (i) 2. (ii) 3. (iii) 4. (i) 5. (i) 6. (ii) 7. (iii) 8. (iv) 9. (iii) 10. (iv)
11. (i) 12. (iv) 13. (iii) 14. (i) 15. (ii)

22 VIRUSES, VIROIDS, AND THEIR DISEASES

22.1 PLANT VIRSUES AND PLANT VIRAL DISEASES

Virus diseases of plants, although of course not recognised as such, were known long before the discovery of bacteria. Breaking of flower colour of tulips was mentioned as early as 1576. Transmission of leaf variegation from the scion to the stock of woody plants was recorded as early as 1700. By 1870, potatoes were known to degenerate in some parts of Europe. Anderson (1778) showed it to be transmitted with the seed tuber, and likened the disease to small pox in nature in its infectiveness. Peach yellows were recognised as early as 1791 in the eastern USA. Tobacco mosaic was identified by Swieten in Holland in 1857. German-born Adolph Mayer (1886), a Dutch investigator in Holland, was the first to point out that tobacco mosaic is readily transmissible, and infectious. He showed that the juice of the infected plants, if applied to healthy plants, could cause the disease. He could not see or isolate any bacteria in the clarified sap. But he still believed that bacteria are the cause of the disease.

The word *virus* was used by the ancient Romans to mean poison, venom or secretion—all unpleasant meanings. When the field of bacteriology began to develop, the medical use of the term *virus* implied any microscopic etiologic agent of disease. The field of virology really began to develop after the invention of bacterial filters by Charles Chamberland in 1884. After a few years, a most remarkable discovery was made by a Russian botanist named Dimitri Iwanowski (1892). He was also involved in the task of investigating mosaic disease of tobacco. He was specifically trying to find out the nature of the agent causing the disease. He carried out most of Mayer's experiments, and confirmed some of the results. For example, he also found that the sap obtained from the leaves of infected plants was capable of causing fresh infection in healthy plants. He also demonstrated that the power to infect was lost if the sap was previously heated. He, however, strongly refuted Mayer's observation that the capacity to infect was lost when the sap was passed through a pad of filter papers. He reported that infectiousness was retained even when the sap was passed through more sophisticated filters (bacteriological filters) small enough to retain bacteria. He called these agents filterable which indirectly meant that they were of a different nature. However, Iwanowski could not think of a different type of organism altogether. He simply hinted at the possible presence of bacteria smaller than they were supposed to be. It was M W Beijerinck, a distinguished Dutch microbiologist, Professor of Microbiology at the Technical University of Delft, who got interested in these filtrable agents causing the mosaic disease of tobacco. He demonstrated in 1896 that the tobacco mosaic infectious agent could diffuse through an agar membrane, and concluded that the tobacco mosaic was caused by a noncorpuscular 'contagium vivum fluidum' which he called virus.

Despite all these developments, however, the controversy regarding these agents persisted, and branched into a new direction. In the year Beijerinck proposed that new concept, two German scientists, Loeffler, and Frosch, suggested that these filterable agents were particles, perhaps of a new type. They were searching for the causal agent of foot-and-mouth disease of cattle during these years. Beijerinck was opposed to their idea. He observed that he "cannot agree with Mr. Loeffler as regards the corpuscular nature of the virus of foot-and-mouth disease". This sort of uncertainty regarding the nature of the filterable agents was to continue for more than thirty years. Gibbs and Harrison (1976) likened the search for viruses in plant extracts even in the early 1930s to trying to find a black cat in a dark cellar without being certain of its presence. This was true inspite of the fact that many more diseases of plants, animals, and human beings could be traced to such agents. Baur (1904) showed that the infectious variegation of *Abutilon* could be transmitted by grafting but not by mechanical inoculation. Beijerinck and Baur used the term 'viruses in describing the causative agents of these diseases to distinguish them from bacteria. As already stated, the term *virus* had been used almost synonymously with bacteria by earlier workers. As more and more diseases of this sort were discovered, the unknown causative agents came to be called 'filterable viruses'.

The discovery of the relationship between viruses and insects was not made in a day, and a period of years elapsed between the time when insects were first suspected of transmitting plant viruses, and the actual demonstration of this method of transmission. In 1894, a Japanese farmer, Hashimoto, was the first to prove the relationship between an insect and a virus. He showed that the leaf hopper *Nephotettix apicalis* var. *cincticeps* is connected with the dwarf disease of rice. However, this was not known to the western world. About 1907, three workers in America—Ball, Adams, and Shaw—suggested that there was some connection between curly top, and the leaf hopper *Eutettix tenella.* In 1926, the first enzyme was isolated, crystallized, and identified as a protein. The isolation of others soon followed. In the early 1930's, workers in various countries began attempting to isolate, and purify plant viruses, using methods similar to those that had been used for enzyme assays. Following detailed chemical studies suggesting the infectious agent of TMV might be a protein; Stanley (1935) announced the isolation of this virus in an apparently crystalline state. At first Stanley (1935, 1936), working in the USA, considered that the virus was a globulin containing no phosphorus. For this discovery, he was later awarded the Nobel Prize. But Stanley overlooked the essential component of TMV. Bawden and Pirie (1936), working in Britain, described the isolation from TMV-infected plants of a liquid crystalline nucleoprotein containing nucleic acid of the pentose type. They showed that the particles were rod-shaped, thus confirming the earlier suggestion of Takahashi and Rawlins (1932) based on the observation that a solution containing TMV showed anisotropy of flow. Bernal and Fankuchen (1937) applied the technique of X-ray analysis to the purified preparations. They obtained accurate estimates of the width of the rods and showed that the needle-shaped bodies produced by precipitating the virus with salt were regularly assayed on only two dimensions, and, therefore, were better described as paracrystals than as true crystals. The isolation of other rod-shaped viruses and spherical viruses which formed these crystals so on followed. All were shown to consist of protein and pentose acid.

Electron microscopy and X-ray crystallography have been the major techniques used to explore the structure of virus particles. Early electron micrographs confirmed that TMV was rod-shaped, and provided approximate dimensions, but the micrographs were not particularly revealing because of lack of contrast between the virus particles, and the supporting membrane. The development of shadow casting with heavy metals greatly increased the usefulness of the method for determining the overall size and shape of the virus particles. High-resolution microscopy, and negative staining in the 1950s became important tools for studying viruses.

Until about 1948, attention was focussed on the protein part of the viruses. There are probably several reasons for this. Quantitatively, protein constitutes the large part of the virus particle.

Markham and Smith (1949) isolated TYMV, and showed that purified preparations contained two classes of particles—one, an infectious nucleoprotein with about 35% of RNA; and the other, an apparently identical protein particle which contained no RNA, and which was not infectious. The results clearly indicated that the RNA of the virus was important for biological activity.

Experiments conducted by Hershey and Chase (1952) showed that when *Escherichia coli* was infected by a bacterial virus, the viral DNA entered the host cell, while most of the protein remained outside, which emphasized the importance of nucleic acids in viral replication.

However, the classic experiments of Gierer and Schramm (1956), Fraenkel-Conrat and Williams (1955), and Fraenkel-Conrat (1956) demonstrated the infectively of naked TMV-RNA, and the protective role of the protein coat. Subsequent work has shown that for viruses, such as TMV, the RNA is present as one strand, and a single cleavage in this strand leads to inactivation of the virus particle.

Until a few decades ago, all the plant viruses that had been examined contained single-stranded RNA as their genetic material. It is now known that wound tumor, and rice dwarf viruses contain double-stranded RNA, while cauliflower mosaic virus contains double-stranded DNA. Viruses containing single-stranded DNA have been found (Gemini viruses and banana bunchy-top virus).

Kassanis (1962) was the first to describe the Satellite Virus (SV) which has been found only in association with Tobacco Necrosis Virus (TNV), the latter being the activator. TNV is often formed alone, and multiplies indefinitely in plants without the production of SV, which has to be introduced as such. SV is a very small virus measuring only 17 nm in diameter. It has an RNA content of 20% by weight, and its maximum molecular weight is 340,000. Both viruses occur in the roots of apparently normal plants, and both are transmitted by the zoospores of *Olpidium brassicae*. SV cannot multiply by itself. TNV alone can multiply indefinitely without causing the production of SV, and the two viruses are antigenitically unrelated, and their antigenicity is not influenced by the activator. The antigenicity depends on the strain of SV used in the inoculation which shows that the protein of SV is coded by its own RNA (Kassanis, 1964, 1966). Since SV codes for its own protein, the help that is needed by SV from the activator is not known at present. Dodds (1998) has reviewed the work done on satellite tobacco mosaic virus.

The term 'viroid' had been introduced to denote a newly recognized class of subviral plant pathogens (Diener, 1971). The viroids currently known consist solely of a short strand of RNA with a molecular weight of about 75,000–125,0000 daltons. The introduction of this RNA of low molecular weight into susceptible hosts leads to replication of RNA, and in some hosts to disease. The first viroids came to light during efforts to purify and characterize the agent of Potato Spindle Tuber Disease (PSTV), a disease which for many years had been assumed to be of viral etiology (Diener and Raymer, 1971). Diener and Raymer (1967) reported that the infective agent of this disease is a free RNA, and that virus particles apparently are not present in the infected tissues. Later sedimentation and gel-electrophoretic analyses conclusively demonstrated that the infectious RNA has a low molecular weight (Diener, 1971), and the agent, therefore, differs basically from conventional viruses. Citrus exocortis (Semancik and Weathers, 1968) and cucumber pale fruit (Van Dorst and Peters, 1974) are known also to be caused by low-molecular-weight RNA that is by viroids. The recognition of viroids as a newly identified class of pathogens raises several questions that have potential important implications for microbiology, molecular biology, and plant pathology as well as for veterinary, and human medicine. Cadang-cadang (dying) disease of coconut and other palms had been found to be of viroidal origin.

To conclude this introductory section, some definitions of a virus are appended. The following characteristics separate viruses from other agents:

1. Only one type of nucleic acid—either DNA or RNA—is present in virions, other agents possess both types of nucleic acids.
2. Virions are reproduced from their sole nucleic acid—either DNA or RNA—whereas other agents reproduce form the integrated sum of their constituents.
3. Virions are incapable of growing or undergoing binary fission.
4. The genetic information for the synthesis of **Lipman System** (the system responsible for the production of energy with high potential) is absent in viruses.
5. Viruses make use of the ribosomes of their host cells. Bawden (1964) defined viruses as submicroscopic, infective entities that multiply only intracellularly, and are potentially pathogenic. On the other hand, Hahon (1964) defined viruses as bits of infectious heredity in search of a chromosome.

22.2 SYMPTOMS DUE TO VIRAL INFECTIONS IN PLANTS

Virus diseases of plants exhibit a variety of symptoms. The most common is the mosaic symptom. Interspersion of various degrees of chlorosis with the normal green color of the leaf results in a mosaic with yellow and green mosaic pattern. This is the most common symptom encountered in viral infections, and the virus diseases with this type of common symptom are popularly called **mosaic diseases**. On the contrary, when there is a uniform reduction of chlorophyll, the symptom is referred to as the **yellows type**. It should, however, be emphasized that these are now believed to be caused by mycoplasmas rather than viruses. Some of the other symptoms due to viral diseases are as follows.

1. Vein Clearing and Vein Banding

These commonly appear in systemically infected leaves before mottle or mosaic clearing or chlorosis of the tissue in or immediately adjacent to the veins. This pattern is referred to as **vein clearing**. **Vein banding** consists of broader bands of green tissue in that portion, set off by chlorosis or necrosis in the interveinal parenchyma, Vein clearing or vein banding may be transient, and precede mottle or it may remain as the major symptom of the disease concerned, for example, potato vein-banding diseases on tobacco.

2. Ring Spots

On inoculated or systemically infected leaves, the symptoms appear in localized spots. These spots consist of various types of chlorosis, and necrosis. They may be circular chlorotic areas, in which case they are called chlorotic ring spots. In other cases, necrosis may appear in rings alternating with normal green. Such spots are called necrotic ring spots. The centres of either type of spot may eventually also become necrotic. Ring spots are formed due to tobacco ring-spot virus on *Nicotiana rustica.*

3. Necrosis

In diseases of both the yellow and mosaic groups, necrosis appears in various forms. The most drastic effect of viruses on their host plants, through not the commonest one, is to kill cells, tissues or the whole plant. Necrosis is not confined to chlorophyll-containing tissues. Chlorophyll-less leaves, hypocotyls, or even roots can develop necrotic lesions similar to those in green leaves.

4. Wilt

In certain virus-infected plants, the outer whorls of leaves droop and become yellow, and the inner whorls also become pale yellow in color. The leaflets soon turn brown in colour, and start drying up from their tips.

The leaves start shedding in quick succession after a plant is infected. The leaves become smaller and stunted is size. There is also considerable reduction in root growth.

5. Stunting

It may affect all parts of a plant, involving a reduction in the size of the leaves, flowers, and fruits, and a shortening of the petioles, and internodes, for example, and bean yellow mosaic virus on bean. The extent of stunting varies according to when infection occurs in the developmental stage of the plant. Sometimes some parts of plants may be more stunted than those same parts in other plants.

6. Changes in Leaf Form

In mosaic diseases there is often uneven growth of leaf lamina. Leaves become curled, brittle (crinkling), and show prominences and depressions (puckering). In some cases, there may be upward and inward rolling of the leaf, such as in potato leaf roll. In certain other cases, the lamina gets extremely reduced, giving a 'fern-leaf' effect (filiform) or shoestring symptoms, such as TMV or CMV on tomato. In cherry leaves, the lamina develops perforations, giving a 'tatter-leaf' appearance.

7. Production of Outgrowths

Foliar outgrowths and vein swellings are generally formed on the underside of the leaves of infected plants, for example, tobacco leaf curl disease, Fiji disease of sugar cane, etc.

Tumors or swellings are formed on stems and roots in the case of sweet clover (*Melilotus* spp.) infected with wound tumour virus, cocoa infected with swollen shoot virus, and woody gall of citrus. Less commonly, the stem is affected, for example, grapevine fan-leaf virus is responsible for fasciations in grapevine stems.

8. Flower Breaking

Variegation in flower color due to viral infections is very common. This is called breaking of flower color, for example, tulip breaking. Similar symptoms are produced by turnip mosaic virus in wallflower (*Cheiranthus cheiri)*.

9. Phyllody, Sterility, and Witches' Broom

The most characteristic symptom is the virescence or greening of floral parts. The auxillary buds are stimulated to grow, and these give a bushy appearance to the plants as in Sesamum phyllody. Flowering and seed production may be decreased or even stopped. In the case of broad bean (*Vicia faba*) infected with bean-leaf roll virus, all the flowers abscises, no more are produced, and the ovules already fertilized produce small seeds.

10. Fruit Abnormalities

A variety of symptoms may develop on the fruits in certain viral infections, for example mottling on cucumber and papaya due to mosaic type of viruses, and star crack in apples and plum due to plum pox virus.

11. Internal Symptoms

The histology deviations, although especially seen in tissues, are actually of purely cytological origin. The effects of hyperplasias, such as tumours or enations, and of necrotic reactions are of course shown in the histology of the plant, but other changes can also be seen. Many viruses affect the vascular tissues. The xylem elements of grapevine infected with grapevine fan-leaf virus contain characteristic lignified strands which are known as endocellular cordons (these are uncommon in virus-free plants).

Tyloses are formed in the xylem elements in the barley yellow dwarf, and other leaf-yellowing viruses. The phloem cells degenerate or die, and callose deposition occurs on the phloem sieve plates.

According to Esau (1967), the phloem tissue degenerates followed by hyperplasia of phloem and xylem, and supernumerary sieve tubes are produced in sugar beet, tobacco, and tomato plants infected with curly top virus. In the cacao swollen-shoot virus disease, the swellings are the result of over-production of xylem elements, whereas in the wound tumours caused by wound tumour virus, the tumour is the result of excessive and abnormal meristematic activity of phloem parenchyma cells.

When the fine structure of the infected cell is studied, many alterations are seen. Some viruses affect particular organelles. Tymovirus induces the formation of marginal vesicles in the chloroplasts. Some isolates of tobacco rattle virus greatly modify mitochondria, and aggregate them to form the so-called inclusion bodies. The most characteristic change is the presence of virus particles. These are usually confined to the cytoplasm but they may also be found in other parts of the cell. The particles of TMV are only found in the cytoplasm whereas those of a strain, U5, can be found in the chloroplasts or nuclei. Particles of some rhabdoviruses are found in the space between the outer and inner nuclear membranes, nucleocapsids of the virus are found in the nucleus, and it appears that they acquire their outer membranes when they bud through the inner nuclear membrane into the perinuclear space.

The virus particles may be scattered in the cytoplasm or they may be aggregated in various ways. Particles of tomato spotted wilt virus are found in membranous bags, those of beet yellows virus occur in bundles, while those of tobacco rattle virus are found packed around mitochondria.

A major cytological effect of virus infection is the development of inclusion bodies, visible by light microscopy in infected cells. One school of thought considered them to be organisms while others thought them to be the side effects of disordered metabolism. There seems to be little doubt that some of these inclusions are aggregates of virus particles large enough to be seen in the light microscope.

There are two main types of inclusions—crystalline and amorphous; the latter are known to occur in the cytoplasm of infected cells, and have bizarre structures in the shape of 'pinwheels', and 'cat-o-nine-tails'. These bodies occur in infections by flexuous rod-shaped viruses of the potato virus Y group. The exact nature of these bodies is not known. They have a very close connection with the endoplasmic reticulum and it is unlikely that they are virus particles.

Generally, the inclusion bodies are most common in the epidermal cells of leaves and stems, but they also occur in roots and flowers, and in most tissues, except the phloem sieve elements.

22.3 TRANSMISSION OF PLANT VIRUSES

Plant viruses are more dependent on the organism for their transfer from infected to healthy plants than any other pathogen. Five important methods are known for the transmission of plant viruses. These are grafting, mechanical transmission, seed transmission, pollen transmission, and vectors.

22.3.1 Graft Transmission

Theoretically, any virus can be transmitted by grafting. The only limiting factor is the tissue compatibility of the stock and scion. Virus diseases which are systemic can be transmitted by grafting. Graft transmission is generally practised in experimental virological work. There are several such procedures, such as cleft or wedge grafting, bud grafting approach or in arch grafting, patch or bark grafting, and core grafting.

22.3.2 Mechanical Transmission

Transmission by this means consists of the removal of the plant sap from a diseased plant by artificial or natural means, and its transfer to a healthy plant by rubbing or injecting. Most of the basic knowledge of plant viruses has resulted from mechanical inoculation. The earlier workers used a variety of methods for the mechanical transmission of viruses. Mayer (1886) sucked the juice of diseased plant, and inserted it into the midribs of healthy plants. Beijerinck (1898) injected juice with a syringe, and inserted pieces of dry leaf into the wounds in the stem. Schultz, and Folsom (1923) inoculated potatoes by brushing leaves with their fingers, and then applying the juice of the virus-affected potato on healthy leaves. The efficiency of the inoculum is greatly improved by the addition of the abrasive, such as carborundum powder (silicon carbide), 300–500 mesh or celite, sand or animal charcoal. The abrasive causes slight injury to the cells, and thus facilitates the penetrations of the juice. It removes the cuticle, and hair, and makes the leaf highly susceptible to infection. Yarwood (1955) considers that a brief washing of inoculated leaves immediately after inoculation increases infection, but reduces it if prolonged for more than 20 seconds. There are very many modifications, and special methods with unstable viruses or those which are difficult to transmit mechanically. Some plants which are difficult to infect by these methods are easily infected when the inoculum containing a fine abrasive is sprayed into the plant by a pressurized air brush. This is a relatively quick way of inoculating a large number of plants. The success of mechanical inoculation depends on a number of host factors such as the host plants, inoculum, inhibitors of infections, and virus inactivators present in the host plants.

The ease with which a virus is transmitted mechanically usually has a bearing upon the degree of natural spread. The tobacco mosaic virus, for instance, is seldom spread by insects but it is readily transmitted by the hands of workmen in the process of removing 'suckers' in tobacco, and in greenhouse-grown tomatoes. Pickers spread cucumber mosaic virus when they remove green fruit from the plants if their hands or implements are contaminated. Spindle tuber virus (veroid) of potato is spread by the knife blades used for cutting seed tubers. Viruses of the mosaic group are in general more easily transmitted mechanically. Potato virus X spreads in potato crops by contact when the leaves of a virus-infected plant brush against those of virus-free plants.

22.3.3 Seed Transmission

It is believed that more than 100 viruses are transmitted through seeds. Seed transmission can occur in two ways: (1) virus particles are attached to the surface of the seeds, and during germination the seedlings are infected through the little wounds caused by soil particles. This is called **surface-seed transmission**. (2) Virus particles enter the reproductive organs in some phase of the generative process. This is known as **internal seed transmission**.

Two things are needed if a virus has to be externally transmitted. These are (1) the high stability of the virus, and (2) peculiar structure of the fruit. Tobacco mosaic virus infecting tomato is carried externally in the seeds of tomato while the same virus is not carried in the seeds of the tobacco. The reason for this is that in the former case, the seeds are embedded in a gelatinous material in the fruits. The virus containing substance in a dried state adheres to the seeds, and infects them during germination. In the case of tobacco plants, the seeds are not enveloped in any gelatinous substance, and hence tobacco mosaic virus is not carried in the seeds of tobacco. With red pepper, a mechanism similar to that existing with tomatoes is responsible for the transmission of tobacco mosaic virus by red pepper seeds.

The transmission of viruses through seeds is dependent on the host virus combination. Tobacco ring-spot virus is transmitted by bean seeds but not by tobacco seeds. Bean plants transmit bean mosaic virus through seeds but the same host plant infected by bean yellow mosaic virus does not transmit the virus through the seeds. Many factors affect seeds transmission; seed-borne viral infections are common in the family Leguminosae. Viruses belonging to nepoviruses are typically seed-borne. If seed transmission has to take place, the plants must be infected before the ovules are fertilized. Certain viruses are transmitted through the pollens. Blakeslee (1921) pointed out that the mosaic of *Datura stramonium* was transmitted through the pollen to the seed of 79% of the offspring. Reddic (1931) offered evidence of the fact that the bean mosaic virus is transmitted by this method. Das and Milbrath (1961) infected squash plants with stone fruit ring-spot virus through pollen.

22.3.4 Vectors

Walker (1969) claimed that no other group of diseases is so generally dependent upon insects for dissemination of the infectious causal entity as are viruses. Most viruses are transmitted by insects which belong to the Hemiptera. From an epidemiological point of view, insects play the most important role in the transmission of plant viruses. Virus-transmitting insects are called vectors. For an insect to be a vector, it is necessary that its mouth parts are formed in such a way that they can pick up and eject viruses. The most suitable mouth parts for this purpose are those of Hemiptera.

No other type of insect was implicated in virus spread until Doolittle (1916) showed that the aphid *Myzus persicae* could transmit potato leaf roll virus. About 400 species of insects are now known to transmit over 200 different viruses. These include the biting and sucking insects (aphids, leaf hoppers, plant hoppers, white flies, mealy bugs, thrips), mites, beetles, nematodes, and fungi belonging to Chytridiales.

22.3.5 Dodder Transmission

Dodder (*Cuscuta* spp.) are parasitic plants with slender straggling stems, which where they touch the stems of the host plant, produce rootlike haustoria that connect with the vascular bundles of the host. Bennett (1940) observed that dodder can transmit viruses from plant to plant, such as sugar-Beet Curly-Top Virus (BCTV) with *Cuscuta subinclusa* and CMV with *C. californica.* Johanson (1941) transmitted aster yellow, CMV, DCTV, Tomato Bushy Stunt Virus (TBSV), and TMV with *Cuscuta campestris.* Dodders are used for the transmission of those viruses which are otherwise difficult to transmit.

22.4 CONTROL OF VIRAL DISEASES

The control of viral diseases, which is a complicated programme, can be classified under the following headings:

1. Eliminating the sources of virus infection
2. Avoiding the vectors
3. Directly attacking the vectors
4. Breeding resistant varieties
5. Curing virus-infected plants
6. Adopting special methods of propagation

22.4.1 Eliminating the Source of Virus Infection

It is obvious that there will be no virus problem if the crop is free of virus when planted, and if there is no source of infection in the field or none near enough to allow its spread into the crop. There can be several sources of infection: (1) perennial weed hosts in which the virus is seed transmitted or annual weed hosts which have several overlapping generations throughout the year, (2) perennial ornamental plants which often harbour infection in a mild form, (3) unrelated crops, (4) volunteer plants of the same species remaining from a previous crop, (5) biennial seed crops of sugar beet, and many vegetables may approach maturity about the time the annual crop emerges.

Elimination of the sources of infection will reduce viral infections but in practice such elimination is not possible.

Two discoveries in 1960s, have emphasized the importance of weeds as sources of virus infection. One is the implication of nematode worms as vectors of several viruses affecting raspberry and strawberry plants. The other is the fact that these viruses are transmitted through the seeds of many common weeds which carry the viruses without symptoms. This is doubly important due to the fact the overwintering nematodes lose their infectivity but regain it in the spring from germinating infective weed seed. The survival of the virus through the winter is thus ensured.

Cooper and Harrison (1973) studied the importance of weed hosts, and the distribution and activity of vector nematodes in the ecology of tobacco rattle virus. *Viola arvensis and Stellaria media* acted as overwintering hosts of the virus. The virus is retained for long periods in the vector *Trichodorus pachydermus*.

Plant remains in the soil may harbour mechanically transmitted viruses, and act as a source of infection for the next season crop. In the case of TMV, general cleanliness is very important for control.

Sometimes it may be necessary to remove infected plants from a crop. The removal of all diseased plants as soon as they are spotted has proved advantageous in crops like potato and sugar cane.

22.4.2 Obtaining Virus-free Planting Material

Virus-free planting material which has been inspected carefully at regular intervals, and from which all infected plants have been removed is recommended. In the case of vegetatively propagated crops, the main source of virus is chronic infection in the plant itself. One of the most successful forms of control with such crops lies in the development of virus-free clones.

Where a vegetatively propagated cultivar is not totally virus-infected, test can be made to find virus-free plants to be used as mother plants, but where the cultivar is totally infected, special techniques have to be applied to obtain virus-free plants. These include heat therapy (35–54° C) meristem tip cultures, cold treatment, and chemotherapy.

A number of fruit trees, such as apple, cherry peach, strawberry, raspberry, etc. have also been cured by heat therapy. Grant (1957, 1958) obtained virus-free bud wood by treating key lime with T_3 strain of tristeza virus at 36.6–40°C for 86–100 days.

Considerable interest has been shown by growers in other countries in the heat therapy of certain crops, such as strawberry, raspberry, chrysanthemum, carnation, sugar cane, citrus, pome, stone fruits, etc. Thirumalachar (1954) showed that potato tubers stored in thatched huts in Bihar for six months at an average temperature of between 29–36°C were freed from leaf roll, whereas others maintained in cold storage were all found to be infected.

22.4.3 Apical Meristem and Tip Culture

The apical meristem of systemically infected plants may contain little or no virus. Morel and Martin (1952, 1955) reported that virus-free dahlia, and potato plants could be grown from apical meristem excised from infected plants. The meristem tips are dissected from plants in sterile conditions, and grown into small plants on nutrient agar or on filter-paper wicks dipped into a nutrient solution.

Apical meristem or tip culture is now widely used to produce virus stocks of vegetatively propagated plants, and is often combined with heat treatment or chemotherapy. Kassanis (1957) freed the King Eward variety of potato from potato paracrinkle virus with which all the stocks of this cultivar were infected.

A very interesting technique is used to obtain virus-free plants of various citrus species. Citrus seedlings are of two types—one is sexually produced, and the other is produced from the nucellus, and thus derived from the female parent. Nucellar seedlings are rarely infected with the viruses that occur in the parent plant. After suitable selections, virus-free mother plants can be obtained.

22.4.4 Avoidance of the Vectors

This can be achieved in various ways—by isolation, by breaking the cycle of the vector, virus, and host plant, and by creating artificial barriers to exclude the vector.

The growing of seed potatoes in particular areas of Scotland and Ireland is a practical illustration, on a large scale, of the control of potato virus diseases by avoiding the vector. Sometimes it is possible to avoid a bud infestation of an aphid vector by early sowing. Rotation of crops may reduce soil-borne virus diseases. Whitefly transmitted virus of tobacco, known as leaf curl, was successfully controlled by breaking the crop cycle though whitefly had bred on the suckers of the previous year's crop. A fallow period of three weeks after the destruction of the entire crop controlled the disease. Screens or cages may be used to ward off flying insects, such as aphids, and leaf hoppers, or a living screen or 'barrier crop' is sometimes quite effective.

22.4.5 Control of the Insect Vectors

Various pesticides may be used to kill or inhibit the activity of plant virus vectors but for various reasons, pesticides have been successfully used to control only a few viruses. Pesticides seem to be useful in controlling viruses that are acquired and transmitted by their vectors only during long feeds.

There have been many reports of the successful use of insecticides for controlling virus diseases. The results are not very convincing. Broadbent *et al.* (1956) tried several insecticides, such as DDT, Endrin Sohradan, Mipufax, Malathion, and Systox, and successfully prevented the spread of potato leaf roll virus from infected plants within the crop. In the case of nonpersistent viruses, for example, potato virus Y, these insecticides could neither protect the entry of the virus into the host nor stop its spread even though the incidence had decreased slightly.

22.4.6 Antiviral Agents

Antiviral agents have also been tried for the control of virus diseases. Antiviral agents may be divided into two categories: (1) inhibitors against virus infection, and (2) inhibitors against virus multiplication. Some of the inhibitors against virus infection are inhibitors from plant extracts, from microorganisms, and from insects, oxidized phenolic compounds, Ribonuclease, photosensitising dyes, and milk. Inhibitors against virus multiplication are the antimetabolites (such as purine and pyrimidine analogues). A number of antibiotics have also been tested for their inhibitory effects on plant-virus multiplication. In Japan, antibiotic Blasticidin S isolated form *Streptomyces griseochromogenes* has been found most inhibitory against TMV multiplication in tobacco leaves. There are several other antibiotics with inhibitory effects in viral multiplication, such as Trichothecin against Potato Virus Y (PVY) infections of tobacco, Gentamycin against TMV multiplication, and Actinomycin D (AMD) against cowpea yellow mosaic virus. Ribavirin, applied as a spray or injected into plants reduce symptoms drastically, and may eliminate the virus from the treated host (Agrios, 1996).

22.5 STRUCTURE OF PLANT VIRUSES

Viruses are the smallest biological structures that have the ability to replicate only inside living cells. Essentially they consist of a shell of protein which encloses a core of nucleic acid—either ribonucleic acid (RNA) or deoxyribonucleic, and (DNA).

Most viruses fall in the size range of 10-200 millimicrons; in other words, between the fortieth of a wavelength, and half a wavelength of violet light. Since objects smaller than the wavelength of light cannot be seen in an ordinary microscope, viruses can be observed only with the aid of an electron microscope. These instruments employ a beam of electrons whose wavelength is much smaller than the dimensions of a virus. Viruses can also be studied indirectly by placing crystals of a pure virus preparation in an X-ray beam, and recording the diffraction patterns produced when the X-rays are reflected from the planes of atoms in the crystal. An analysis of such X-ray diffraction patterns suggested that the protein sub-units forming the virus shell are arranged symmetrically. The tobacco mosaic virus, for example, showed up in early electron micrographs as a slender rod without visible subunits. When the virus was examined by X-ray diffraction, however, one could see patterns suggesting that the sub-units are arranged in a helix.

Morphology and Biochemistry of Plant and Animal Viruses

All virions, even if they possess other constituents, are constructed around a nucleocapsid core (indeed, some viruses consist of only of a nucleocapsid). The nucleocapsid is composed of a nucleic acid, either DNA or RNA, held within a protein coat called capsid, which protects viral nucleic acid, and aids in the transfer of the nuclear material between host cells.

There are four general morphological types of capsids and virions structure:

1. Some capsids are icosahedral in shape. An icosahedron is a regular polyhedron with 20 equilateral triangular faces, and 12 vertices (such as potato leaf roll virus or barley yellow dwarf).
2. Other capsids are helical, and shaped like hollow protein cylinders, which may be either rigid or flexible (such as tobacco mosaic virus).
3. Many viruses have an envelope, an outer membranous layer surrounding the nucleocapsid. Enveloped viruses have a roughly spherical but somewhat variable shape even though their nucleocapsid can be either icosahedral or helical (such as tomato spotted wilt virus, and potato yellow dwarf virus).
4. Complex viruses have a capsid symmetry that is neither purely helical nor icosahedral. They may possess tails and other structures (e.g. many bacteriophages) or complex multilayered walls surrounding the nucleic acid (e.g. pox viruses such as Vaccinia).

Nucleic Acids of Viruses

All cellular organisms contain two types of nucleic acids, the DNA which is found in the chromosomes, and RNA which is present in the cytoplasm. Viruses are not cellular organisms, and neither have nucleus or cytoplasm but possess only one type of nucleic acid, either DNA or RNA but never both in the infective particle. Either DNA or RNA serves as the genetic material in case of viruses.

Before 1968, all plant viruses that were studied contained only RNA but in 1968, cauliflower mosaic virus was found to contain DNA, and not RNA.

Now we know that plant viruses can be arranged into 6 groups. These are (1) single-stranded positive RNA (2) single-stranded negative RNA (–) ssRNA, (3) double-stranded RNA (dsRNA) (4) double-stranded DNA (dsDNA), and (5) single-stranded DNA (ss DNA) and single stranded (RT).

22.6 VIRUS PURIFICATION

Virologists must be able to purify viruses, and accurately determine their concentration in order to study virus structure, and reproduction. There are four methods used for purification of viruses.

1. Differential and density gradient centrifugation
2. Precipitation of viruses
3. Denaturation of contaminants
4. Enzymatic digestion of cell constituents

22.6.1 Procedure

Old host cells that contain mature virions are used as the source of material. Infected cells are first disrupted in a buffer to produce an aqueous suspension or homogenate containing cell components and viruses. Viruses can be then isolated by differential centrifugation, the centrifugation of a suspension at various speeds to separate particles of different sizes. The homogenate is first centrifuged at high speed to sediment viruses, other larger cellular particles, and the supernatant which contain the homogenate's soluble molecule is discarded. The pellet is next resuspended, and centrifuged at a low-speed centrifugation which then sediments the viruses. This process may be repeated to purify the viruses particles further.

22.6.2 Purification by Gradient Centrifugation

Here, viruses can be purified based on their size, and density by use of gradient centrifugation.

Procedure

A sucrose solution is poured in a centrifuge tube so that its concentration smoothly and linearly increases between the top, and the bottom of the tube. The virus preparation, often after purification by different centrifugation, is layered on the top for the gradient, and centrifuged. The particles settle under centrifugal force until they come to rest at the level where the gradient's density equals theirs (isopycnic gradient centrifugation). Viruses can be separated from other particles only slightly differential in density. Gradients can also separate viruses based on differences in their sedimentation rate (rate zonal gradient centrifugation). When this is done, particles are separated on the basis of both size, and density; usually the largest virus will move most rapidly down the gradients. Viruses differ from one another and cell components with respect to either density (gram per millilitre) or sedimentation coefficient (s). Thus, these two types of gradient centrifugation are very effective in virus purification.

22.6.3 Virus Purification through Precipitation

Procedure

Viruses like many proteins can be purified through precipitation with concentrated ammonium sulfate. Initially, sufficient ammonium sulfate is added to raise its concentration to level just below that, which will precipitate the virus. After any precipitated contaminants are removed, more ammonium sulfate is added, and the precipitated viruses are collected by centrifugation. Viruses sensitive to ammonium sulfate are purified by precipitation with PolyEthylene Glycol (PEG).

22.6.4 Denaturation of Viruses

Viruses are less easily denatured than many normal cell constituents. Contaminants may be denatured, and precipitated with heat or a change in pH to purify viruses. Because some viruses also tolerate treatment with organic solvents like butanol and chloroform. Solvent treatment can be used to both denature protein contaminants, and extracts any lipids in the preparation. The solvent is thoroughly mixed with the virus preparation, and then allowed to stand, and separate into organic and aqueous layer. The unaltered virus remains suspended in the aqueous phase while lipids dissolve in the organic phase. Substances denatured by organic solvents collect at the interface between the aqueous and organic phases.

22.6.5 Removal of Cellular Proteins and Nucleic Acids

Cellular proteins and nucleic acid can be removed from any virus preparations through enzymatic degradation because viruses are usually more resistant to attack by nucleases and proteases than are free nucleic acids, and proteins. For example, ribonuclease and trypsin often degrade cellular ribonucleic acids, and proteins while leaving virions unaltered.

22.6.6 Viroids

At least 40 plant diseases have been shown to be caused by viroids (Agrios, 2005). The term 'viroid' has been introduced to denote a comparatively recently recognized class of subviral plant pathogens (Diener, 1971). The first viroid came to light during efforts to purify, and characterise the agent of the Potato Spindle Tuber Disease (PSTV), a disease which, for many years, had been assumed to be of viral etiology (Diener and Raymer, 1971). Diener and Raymer (1967) reported that the infectious agent of this disease is a free RNA, and that virus particles are apparently not present in infected tissue. Later, sedimentation and gel electrophoretic analyses conclusively demonstrated that the infectious RNA has a very low molecular weight, and therefore the agent differs basically from conventional viruses.

Several plant diseases—citrus exocortis, chrysanthemum stunt, chrysanthemum chlorotic mottle, cucumber pale fruit, and cadang-cadang disease of coconut—are caused by low-molecular-weight RNAs that are viroids. As already indicated, viroids are low-molecular-weight ribonucleic acids (RNA) that can infect plant cells, replicate themselves, and cause disease. Viroids differ from viruses in at least two respects: (i) the virus RNA is enclosed in a protein coat while the viroids lack any protein coat, and exist as free (naked) RNA, (ii) the size of the RNA, which consists of 250 to 370 bases in viroids as compared to that in viruses which is 4 to 20 kb. During viroid replication, the circular (+) strand of the viroid is replicated while it acts as a rolling drum producing multimeric linear strands of (–) RNA. The linear (–) strand then serves as a template for replication of multimeric strands of (+) RNA. The (+) RNA is subsequently cleaved by enzymes that release linear, unit-length viroid (+) RNA's which circularize, and produce many copies of the original viroid RNA. How viroids replicate is still not exactly known, and how viroids cause disease is also not known so far properly.

Transmission of viroids from diseased to healthy plants is primarily by mechanical means mainly through sap carried on tools or hands or other implements. Potato spindle tuber, cadang-cadang, tomato bunchy top are believed to be transmitted through pollens and seeds. But the rate of such transmission is very low.

In nature, viroids survive in dead plant matter for varying periods of time. Perennial hosts are a very good means for the perpetuation and overseasoning of the viroids.

Viroids are quite resistant to high temperatures, and cannot be inactivated in infected plants by heat treatment. Use of viroid-free planting material, rogueing of infected plants, clean cultural operations can help in controlling the viroid.

22.7 INFECTIONS, MULTIPLICATION, AND MOVEMENT OF VIRUSES WITHIN PLANTS

In their mode of replication, viruses differ greatly from cellular microbes which multiply by the division of mature cells, often in a binary fashion. Hence, bacterial cells, for example, are intact at all stages of division, and viable, and recognizable as bacteria. By contrast, soon after their entry into host cells, viruses partially or wholly disassemble themselves into their constituent macromolecules. The virus genome or messenger RNA transcribed from it is translated by the host's translation apparatus into new virus proteins, the nucleic acid genome is replicated, and the progeny nucleic acid together with the molecules of one or more of the new virus proteins assemble to form new virus particles. Thus, intact virus particles are not present during part of the replications cycle of a virus. The virus genome is incorporated into the host's protoplasm, and is totally dependent upon it.

The first stage of infection is the entry of viruses into the cells, and this happens in nature through breaks in the cell walls made by vectors or by abrasion. There is no unequivocal proof that viruses enter through wounds made during mechanical inoculation, although viruses have been found in wounds caused by abrasion. Ectodesmata have been implicated as possible routes of infection for plant viruses. When intact virus particles enter a cell they disassemble ("uncoat"), and uncoating may in part control host range.

When infective brome mosaic virus RNA was coated with TMV protein, the particles were not infective for barley leaves which are susceptible to infection by intact particles and by RNA of brome mosaic virus. According to Bancroft *et al.* (1971), the ability to uncoat depends upon the structure of the virus-coat protein, and can be greatly changed by a single amino acid replacement. Evidence that uncoating takes place is found by comparing the sequence and timing of events after inoculating French-bean leaves with either intact tobacco necrosis virus particles or with the infective RNA extracted from them. When inoculated with RNA, the virus appears earlier, and the symptoms also appear earlier than when inoculated with intact virus particles.

It is also possible to stop infection by inactivating the leaves with ultraviolet light immediately after inoculation if inoculation has been done with the nucleic acid. On the other hand, resistance to ultraviolet inactivation starts to increase almost at once in leaves inoculated with nucleic acid, whereas no such increase occurs in leaves inoculated with intact virus particles until three hours later (Kassanis, 1960). Mandels (1967), in his studies of polio virus (a simple RNA containing animal virus), came to the conclusion that uncoating might be initiated at the cell surface but completed intracellularly by means of pinocytotic movement. Cocking (1970) calls the movement of macromolecules into the cellular environment after attachment to the cell membrane (plasmalemma), rhopheocytosis, while Dales (1965) calls it **viropexis**. The situation in the case of plant virus infection is very unsettled, and if uncoating takes place on the cuticle, the viral RNA has to get into the cell either by direct (active) transport through the plasmalemma or through viropexis. Viropexis has been inferred to exist in plant protoplasts. Cocking (1970) and Otuski *et al.* (1972) found this to be true from TMV moving into tomato-fruit protoplasts, and for TMV, PVX, and the Cucumber Mosaic Virus (CMV) moving into tobacco protoplasts. Motoyoshi *et al.* (1973) have reported successful tobacco protoplast infection, probably by the viropexis of Cowpea Chlorotic Mottle Virus (CCMV), and its RNA.

22.8 TOBACCO AND TOMATO MOSAIC

Tobacco mosaic, the first recognized virus disease of plants, is worldwide in distribution. It is most common on tobacco crops in India, appearing in every tobacco tract of the country. The same virus occurs on many other plants. Over 20 different virus diseases occur on tobacco. The mosaic disease is most important.

1. History

There are many facts associated with tobacco mosaic virus as listed by Agrios (2005)

(a) Adolph Mayer began to study the tobacco mosaic disease in 1886 for the first time.
(b) In 1898, Beijerinck proposed that tobacco mosaic was caused by an infectious fluid which he called a virus.
(c) TMV was the first virus to be shown that plants infected with it contained a specific antigen (Beale, 1928).
(d) It was the first virus to be quantified by Holmes (1929) by the number of local lesions produced on healthy leaves by different concentrations of sap of an infected plant.
(e) Then in 1935, it was the first virus to be crystallized by W Stanley.
(f) Bowden and Pirie (1936) reported that TMV actually consisted mostly 95% of protein but it also contained 5% of RNA.
(g) Kausche (1939) took the first electron microscopic photographs of TMV.
(h) In the mid 1950s, Gierer and Schramm (1956), and Fraenkel-Conrat (1955) showed that RNA is responsible for infection.
(i) The first virus where protein coat was fully sequenced into its 158 amino acids (Anderer, 1960; Tsugita, 1960), and the sequence of the amino acids of several natural and artificially induced mutant TMV strains was instrumental in establishing the universality of genetic code, and the chemical basis of mutation.
(j) In 1969, Takebe used TMV for the infection of suspended tobacco-leaf protoplasts, thereby proving the basis for a synchronous infection system for studying virus replication.
(k) TMV was the first plant RNA virus of which the complete genome was sequenced (Goelet, 1982), and to which monoclonal antibodies were produced.
(l) TMV was the first plant virus to be shown that introduction and expression of its coat protein gene in plants protected those plants against TMV (Powell-Abel, 1986; Beachy, 1986).

2. Symptoms

The first symptoms are light discoloration along the veins of the youngest leaves. Soon the leaves develop a characteristic light and dark-green pattern, the dark-green areas usually being associated with the veins. This typical mosaic symptom is sometimes accompanied by the blistered appearance of leaves due to the more rapid growth of the dark-green tissue. At times leaf spots may also appear. Plants that become infected early in the season are usually very much stunted, while those infected late in the season show little reduction in size. In severe infections, the leaves are narrow, puckered, and thin, in general, malformed beyond recognition. While it is easy to recognize a diseased plant in the main field, it is almost impossible to differentiate the diseased seedlings in the nursery. The disease causes extensive damage by way of reduction in yield and quality of the product. Frequently, yellow patches develop on the leaves due to the presence of more than one strain produced by mutation. The disease causes partial sterility of the pollens.

3. Causal Organism

Tobacco mosaic virus is caused by *Nicotiana* virus 1 (Smith). The virus is rod-shaped, measuring about 280 × 15–18 nm with a central hollow tube of about 20 Å in diameter. It is made up of centrally placed ribonucleic acid molecules covered with a protein coat (Fig. 22.1).

This virus particle has a molecular weight of 39 million daltons. It is cylindrical with dimensions of 160 × 3000 Å.

The protein coat (called **capsid**) of this virus contains 2130 identical protein sub-units, each with a molecular weight of 18,000. The sub-unit is composed of a single chain of 158 amino acid residues of known sequence. The folding of the chain produces an ellipsoid with an axial ratio of 3: 1. The RNA of the virus is single-stranded, and the molecular weight is 2.4 millions. It consists of 6400 nucleotides. The RNA strand forms a helix with a radius of 40 Å that can be seen under an electron microscope (Fig. 22.2), with a special staining technique when phospho-tungstic acid is used as a negative stain. The RNA has four Open Reading Frames (ORF), and is translated into four proteins, one of which (17.6 kDa) is the coat protein, two (126 and 183 kDa) are components of the RNA polymerase enzyme, and the fourth (30 kDa) is associated with cell-to-cell movement of the virus (Agrios, 2005).

Each plant virus consists of at least a nucleic acid and a protein. It appears that first new RNA produced is not the viral RNA but a complementary strand that is a mirror image of the virus, and which, as it is formed, is temporarily connected to the viral strand. Thus, the two form a double-stranded RNA which soon separates to produce the original virus RNA and complementary (–) strand, the latter serving as a template for more virus (+) strand RNA synthesis.

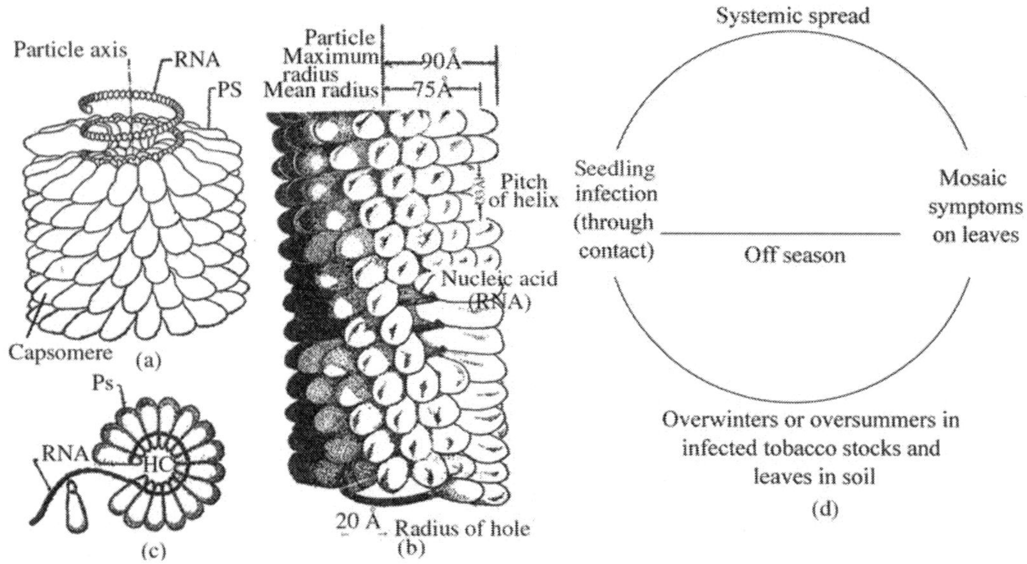

Fig. 22.1 Structure of TMV: (a) Drawing to show the relationship of the RNA and the protein sub-units (capsomeres) (b) Photograph of a model of TMV with major dimensions indicated (c) Cross section of (b) (d) Disease cycle of TMV

4. Virus Protein Synthesis

The viral RNA itself (or the part of the viral coding for the viral protein) plays the role of messenger RNA. The virus apparently utilizes the amino acids, ribosomes, and transfer RNAs of the host, but it becomes its own blueprint (messenger RNA), and the protein formed. The RNA is translated into four proteins one of which is the coat protein, two are components of the RNA polymerase enzyme, and the fourth is associated with the cell-to-cell movement of the virus.

During virus synthesis, it appears that part of its nucleic acid also becomes involved with the synthesis of protein other than the viral coat protein. Each protein sub-unit of TMV consists of 158 amino acids with a notch at 40 Å. Amino acids of each TMV protein sub-unit add up to a total of 17,531.

Since each TMV virus particle weighs approximately 39,000,000 molecular weight units of which 95% or 37,000,000 units are protein, it is easily calculated that the protein of each virus particle consists of approximately 2100 protein sub-units called **capsomeres**.

5. Disease Cycle

The virus is sap transmissible, and enters the host through wounds. From the point of entrance (wounds), the virus moves from cell to cell through plasmodesmata, multiplies in, and infects each cell, and when it reaches the phloem, travels systemically through it, and infects the whole plant. No insect vector is known. The virus is easily transmitted in the field by mechanical means, and by wind and water. Various cultural operations cause wounds on the plant which become the portals of entry of the virus. Such operations as topping or clipping the shoots spread the virus form diseased to healthy plants. Even the workers in the field may transmit the virus to the standing crop. Thus, there is unlimited scope for the virus to set in and cause damage to the crop. The disease cycle of TMV is depicted in Figure 22.1d.

6. Disease Management

The various physical and biological properties which impart highly resistant qualities to the virus, and the varying modes of spread of the virus, make disease control almost impossible. The only method seems to be to grow resistant or at least tolerant tobacco varieties and practice field sanitation. The field should be weeded thoroughly. After weeding, the workers should wash their hands thoroughly with soap and running water. Regular rouging of diseased plants and weed in the field should be carried out. Utmost care should be taken while handling the diseased plants.

Sanitation and the use of resistant varieties are the two main means of controlling TMV. It has been shown that infection by TMV is inhibited by whole milk. Some states in the USA are now recommending the spraying of plants with milk before transplanting or otherwise handling them, or dipping the hands in milk during transplanting and handling. Since this greatly reduces the spread of tobacco mosaic virus.

In the case of tobacco mosaic virus, cross protection had been reported if the plants are inoculated with milder strain first. Genetic engineering has made it possible to produce tobacco plants with the gene coding of the TMV coat protein. The N gene is the most widely used gene conferring resistance to TMV. The N gene was transferred into *Nicotiana tabacum* from *N. glutinosa*. Once a plant containing the N gene is inoculated with TMV, a hypersensitive reaction occurs at the viral points of entry. The interaction of the N gene product with the TMV virus causes the plant to kill its own cells at any point where the virus entered. Subsequently, lesions form on any inoculated leaf, restricting the movement of the virus to other plant parts, and more importantly, to other plants.

22.9 VIRAL DISEASES OF POTATO

22.9.1 Leaf Roll of Potato

Leaf roll is of great economic importance, being one of the main causes of potato degeneration.

1. Causal Virus

It is caused by the Potato Leaf Roll Virus (PLRV) (Syn. = phloem necrosis virus), also known as **potato virus 1** or ***Solanum* virus 14**. The virus is a member of luteo-virus group.

The name 'leaf roll' is partly descriptive of the disease. In many varieties, the margins or the tips of the leaves sometimes become yellow. The leaves of an infected plant roll up along the margins, leaving the midrib at the bottom of the trough. In secondary infection (in plants with diseased seed tubers), this rolling of leaves starts in the lower leaflets and progresses upward throughout the entire plant. In some varieties individual leaflets have a tendency to be more erect, giving the diseased plant an upright appearance. The infected leaves turn leathery brittle, pink or brown pigmented. Infected plants remain stunted chlorotic, and make a rattling noise when shaken. As a result of infection, necrosis occurs in the phloem, and there is excessive accumulation of starch. Often nonviral or pseudo "leaf roll" symptoms are seen in the field induced either by reasons like purple-top roll (phytoplasmic), cutworm canker, black scurf (fungi) or calcium deficiency or waterlogging (Khurana, 2000).

2. Damage

The number of tubers produced per plant, and their size in a diseased crop are greatly reduced.

3. Transmission

The virus is not transmitted through the sap. Transmission of the virus in nature occurs through infected seed tubers, and through the agency of insects. The main insect vector is an aphid, *Myzus persicae*. The virus persists in the body of *Myzus persicae* for many days even if the vector has fed on many immune hosts after feeding on the diseased plants.

4. Disease Management

Potato seed certification and roguing are recommended for disease control. According to Kassanis (1950), if tubers are stored at 37.5°C in a humid atmosphere for 20-25 days, they become free of the leaf-roll virus. In India, tubers can be free from PLRV infections by storing in ordinary stores in summer months at a temperature of 29–35°C. Hot-water treatment (50–55°C for 15–20 minutes) has also been found effective in eradicating the virus. Resistance to PLRV is rather difficult to incorporate, yet it has been identified in *S. brevidens* and *S. acaule* OCH 13823. Build-up of incoming aphid vectors be controlled by application of suitable contact/systemic insecticides and virus spread in storage.

22.9.2 Mild Mosaic of Potato

Several types of mosaic virus affect potatoes, but in India, latent or mild mosaic, rugose mosaic, and crinkle are most important.

1. Mild Mosaic

Mild mosaic is caused by potato latent virus, potato mottle virus, potato virus X or Solanum virus 1. The virus contains a large number of strains.

The disease, although occurring widely in most of the seed stocks, causes very slight or negligible symptoms of interveinal mottling or mosaic. Under favourable growth conditions of the host, the symptoms are masked, thus making recognition of the disease difficult. The mosaic symptoms are almost completely masked at high temperatures (above 21°C). In many cells the virus is carried by the plants without any visible symptoms. There is only slight dwarfing of the plant or deformation of the foliage (Fig. 22.2). The virus produces top necrosis in the varieties Key Edward, Arran, Crest, and Epicure.

The damage is not much as compared to leaf roll or other viruses occurring on potatoes in India.

2. Transmission

The virus is easily transmitted by sap inoculation, and is one of the few viruses which spread in the field by contact between healthy and diseased plants. Virus can be transmitted by core grafting, and by

Fig. 22.2 Mild mosaic of potato (Potato virus X or Solanum virus 1): symptoms (Courtesy Dr. B.B. Nagaich)

dodder, *Cuscuta campestris*. No insect vector is known. However, the source of parenation and spread in the field under natural conditions is mostly the diseased seed stock.

22.9.3 Rugose Mosaic of Potato

The disease causes very severe damage to individual plants. The foliage is not only mottled but is also severely wrinkled, puckered, and markedly reduced in size. The leaflet margins are rolled downwards, and the entire plant is dwarfed. The lower leaves generally have black necrotic veins. While the symptoms of mottle may be masked by high temperature, roughness or rugosity, and abnormal hairiness of leaves, and dwarfing of the entire plant persists. In severe cases, plants may even die before producing any tuber.

The disease is caused by a combination of two viruses—potato virus X (mild or latent mosaic) and potato virus Y. The latter causes blightening of the leaves in severe attacks, and is also known as vein-banding virus or *Solanum* virus 2.

Under field conditions, the transmission of the virus is through small-sized tubers. On diseased plants, *Myzus persicae* is capable of efficiently transmitting virus Y which is one of the components of the complex.

22.9.4 Crinkle of Potato

The disease resembles rugose mosaic very closely but the yellowish patches on the foliage are bigger, and hence, more prominent. As the plants begin to die, this colour becomes more pronounced, and is accompanied by rust spots, beginning near the tip of the leaves. The foliage is brittle and is easily injured.

Potato virus X and potato virus A (*Solanum* virus3) cause super mild mosaic. This virus complex is also tuber-borne. Sap inoculation transmits only potato virus X. Potato virus A particles are 730 × 13 nm flexuous rods transmitted by aphids.

Disease Management of Potato-Virus Diseases

1. Systemic roguing of diseased plants as soon as they are noticed and their destruction by burning is recommended.
2. Indirect roguing of diseased seed tubers is recommended at planting time by using large-sized potatoes (about 3.75 cm in diameter).
3. Raising a mosaic-free seed crop by planning large-sized seed tubers during the coldest season helps check the disease considerably.

The Central Potato Research Institute, Shimla, has suggested the following way in which disease-free crops for seed tubers can be raised:

(a) Obtain certified seeds from a reliable source and plant them early in the season.
(b) Rogue out the sick-looking plants as soon as located.
(c) Detop (removal of foliage) the plants in the third or fourth week of December when aphid infection starts, and leave the tubers in the soil to mature. The yield may be less but the tubers will be free of virus.
(d) Use of True (botanical) Potato Seed (TPS) also holds promise for areas where quality seed is not available. TPS offers various advantages, viz. reduced cost, early storage and transportation of seed material, low incidence of diseases, but it requires more labour, lacks uniformity, etc.
(e) In the late crop, use of systemic insecticides as spray is recommended to control the insect vectors (such as 0.1 percent Metasystox). In recent studies it has been found that if at the time of planting, systemic insecticides applied to the soil in the furrows delays the progress of the build-up of aphid population.

22.10 LEAF CURL OF TOMATO

The disease has been reported from many parts of the world. It is fairly common in the winter season in India.

1. Symptoms

Dwarfing, puckering, twisting, and curling towards the dorsal side of leaves, mottle, vein clearing, excessive branching, shortening of the entire plant, and partial or complete sterility are the characteristic symptoms (Fig. 22.3). The curl of tomato is caused by the tobacco leaf-curl virus which belongs to the gemini virus group. The geminate particles of the virus measure 25–40 × 15–20 nm. In gemini viruses, the single-stranded DNA (ssDNA) molecule is circular.

2. Transmission

It is seed-borne, but the seeds of freshly infected fruits may be contaminated. The principal insect vector is *Bemisia tabaci.*

3. Disease Management

There is no definite control measure. *Lycopersicon peruvianum* shows a high degree of resistance. Spraying with Ekatox 0.20% and Rogor 0.05% at intervals of ten days helps reduce the incidence of the disease. There are indications that growth regulators like gibberellic acid, and 2,4-D reduce disease incidence and increase the yield of fruits.

Fig. 22.3 Tomato leaf curl: symptoms

22.11 VIRAL DISEASES OF PAPAYA

22.11.1 Leaf Curl of Papaya

The disease is found commonly in kitchen gardens and commercial orchards around Delhi. The disease has also been reported from Bihar and other parts of the country. Crinkling and curling of the leaves accompanied by vein clearing and reduction in the size of the leaves are the marked symptoms. Leaves become leathery and brittle, and the interveinal areas are raised on the upper surface due to hypertrophy which produces rugosity. The most prominent symptom is the downward and inward rolling of leaves, and the veins get thickened, and turn dark green in color (Fig. 22.4). Sometimes all the leaves at the top of the plant get affected by these symptoms. The petioles are twisted in a zigzag manner. The severely affected plants fail to flower or bear fruit. In the advanced stages of the disease, defoliation takes place and plant growth is stunted.

1. Causal Virus

Tobacco virus 16 or Nicotiana virus 10 is responsible for leaf curl of papaya. It is a Geminate (twin) virus containing single-stranded DNA (ssDNA).

2. Transmission

The virus of papaya leaf curl is not mechanically transmitted. It can be readily transmitted by grafting or juice inoculation. In nature the important agent of its transmission is the white fly *Bemisia tabaci*.

3. Disease Management

No suitable control method is known. Roguing is helpful in reducing the incidence of the disease. No other control measure of the disease can be recommended except that insect vectors should be kept under control by spraying of suitable insecticides.

Fig. 22.4 Papaya leaf curl (Tobacco virus 16 or Nicotiana virus 10): symptoms

22.11.2 Mosaic of Papaya

The disease is widespread in India in states such as Madhya Pradesh, Uttar Pradesh, Bihar, West Bengal, Andhra Pradesh, and Maharashtra. The mosaic is caused by papaya ring-spot virus (Potyvirus group). It is a flexuous filamentous virus 11–16 nm in diameter.

1. Symptoms

The disease is characterized by mosaic symptoms on the leaves with blister-like patches of green tissue distributed all over the yellowish green lamina.

2. Transmission

The papaya mosaic virus is not seed-borne. It can be transmitted by mechanical sap inoculation, and wedge and bud grafting. Aphids such as *Aphis malvae, Aphis gossypi,* and *Aphis medicaginis* are known as vectors. It is transmitted under natural conditions, only through insect vectors.

3. Disease Management

Direct control by the use of insecticides is not possible. According to Capoor and Varma (1961), *Carica cauliflora,* a wild papaya, has been found to be immune to the virus. It can be utilized in the breeding programme. The resistance is gene controlled. According to Bhargava and Paul Khurana (1969), weekly sprays with 1% groundnut oil help to prevent infection.

22.12 YELLOW VEIN MOSAIC OF BHINDI (*ABELMOSCHUS ESCULENTUS*)

Yellow vein mosaic disease of bhindi is a serious problem wherever this crop is grown in India and Sri Lanka. When the infection is early, the entire crop is destroyed.

1. Symptoms

The main symptom of the disease in bhindi is vein clearing followed by veinal chlorosis of the leaves. The yellow network of veins is very conspicuous, and the veins and veinlets become thick. In severe cases, the chlorosis may extend to interveinal areas, and may result in the complete yellowing of leaves. Fruits are dwarfed, malformed, and are yellowish green in colour (Fig. 22.5).

Fig. 22.5 Yellow vein mosaic of bhindi: symptoms

2. Causal Virus

The virus responsible for the disease is *Hibiscus virus* 1 or yellow vein mosaic virus. Not much is known of the nature and properties of this virus.

3. Transmission

The virus is transmitted in nature by white flies (*Bemisia tabaci*), and the okra leaf hopper *Empoasca devastaus*.

4. Disease Management

At present, no definite control measures are known. Variety Pusa Sawani is reported to be tolerant (Joshi *et al.*, 1960). Ekatox (0.02%) or Rogor (0.05%) sprays have been reported to reduce disease incidence (Sastry and Singh, 1973). Spraying should be started soon after the emergence of the plants. Soil application of Thimet or Disyston granules is also recommended. Use of trap crops and crop-free season can reduce vector population. Destruction of weed hosts should be done.

22.13 VIRUS DISEASES OF SUGAR CANE

Edgerton (1958), in his book, *Sugar Cane and its Diseases*, has discussed the following virus diseases of sugar cane.

1. Fiji diseases
2. Sereh disease
3. Chlorotic streak
4. Mosaic disease
5. Australian dwarf
6. Ratoon stunting diseases of sugar cane, now believed to be caused by xylem-inhabiting fastidious bacteria

In addition, two more diseases have received attention in India. They are grassy shoot and stripe. Agnihotri (1990) has discussed only two important viral diseases, i.e. mosaic and spike of sugar cane.

Mosaic Diseases of Sugar Cane

It is widespread in all cane-growing areas of India. Chona (1944) observed that even cent percent infection of a crop with mosaic does not cause more than 10% loss in yield, and that the juice quality remains unimpaired.

1. Symptoms

Six weeks after planting, pale patches or blotches in the green tissue of the leaf appear. Diseased leaves show a characteristic mottling of chlorotic or light-colored, elongated, more or less irregular stripes or streaks. The cells of affected leaves show, in one part of the cytoplasm, an area of proteolysis. This can be seen under a microscope as being more heavily stained than the rest, and consisting of a vacuolated mass representing the X body.

2. Causal Virus

The virus responsible for sugar-cane mosaic is called Sugar cane Mosaic Virus (SCMV) of *Saccharum* Virus 1 or Sugar Cane Virus 1.

3. Transmission

The virus is transmitted from sugar cane to sugar cane by large number of species of aphids. Transmission is in the nonpersistent manner.

4. Disease Management

The use of selected healthy setts for seed, systemic roguing out of the infected canes if infection is not very severe, elimination of weed hosts, wherever possible, and use of resistant varieties are recommended for controlling the disease. However, due to continuous evolution of strains of the virus, resistant varieties are not a permanent solution of the mosaic problem in sugar cane. Heat therapy is effective against certain strains, and can be used for raising disease-free nurseries.

22.14 BUNCHY TOP DISEASE OF BANANA

In India, the disease is believed to have been introduced through infected suckers brought from Sri Lanka into the state of Kerala. It has now spread to almost all the banana-growing areas in India.

1. Symptoms

In the infected plants, the leaves become bunched together at the apex of the pseudostem to form a rosette or bunchy top. The margins of the leaves become waxy, and in the later stage roll upwards. As a result of the reduction in the size of the petioles, the whole plant gets stunted. Due to the failure of the leaf stalks to elongate, the leaves invariably show dark green streaks.

2. Causal Virus

Bunchy top of banana is caused by Banana Virus 1 or *Musa* Virus 1.

3. Transmission

The virus is not transmitted by mechanical inoculation. The chief insect vector is *Pentalonia nigronervosa*. The virus perpetuates through the infected suckers.

4. Disease Management

Diseased plants should be thoroughly and carefully destroyed. They should be carefully dug out and destroyed by burning. An alternate procedure could be to kill the plants along with the aphids feeding on them by pouring kerosene over them. The plants can also be killed by some hormone weed killer, such as MCPA (2-methyl-4 chloro-phenoxy acetic acid).

Control of banana bunchy top depends primarily on adopting control measures that help avoid or minimize virus infections. Such measures will include quarantine to keep the virus out from an area where the disease does not occur. The using of virus-free propagating material will go a long way in controlling the disease. Only certified virus-free suckers should be used for planting.

22.15 TRISTEZA OF CITRUS

1. Distribution

Tristeza affects all types of citrus plants but mainly orange, lime, and grapefruit. Losses from tristeza vary because it depends upon the virulence of tristeza virus, and the abundance of the vectors. Citrus tristeza virus exists as many strains having different biological activities. These biological activities belong to one or more of the following five major categories: (i) mild, (ii) seedling yellows (SY), (iii) decline on sour orange, (iv) stem pitting on grapefruit, and (v) stem pitting on sweet orange.

2. Symptoms

The typical symptoms appear as a quick or chronic decline of trees budded on susceptible rootstocks, but seedlings or budded trees may also develop stem pitting. In older trees, the symptoms appear as suppression of new growth, and change of color of leaves from bronze to yellow. With the passage of time, the older infected leaves begin to fall, and later twigs begin to die back from the tip. Twig growth becomes weaker every season until the tree dies. The infected plants show root symptom consisting of a marked depletion of starch, death, and decay of the feeder rootlets. Seedling and budded trees of many plants develop stem pitting as a result of infection. Stem pitting consists of longitudinal grooves in the stem. Stem pitting decreases vigor of the trees, poor bushy growth, and small, distorted fruits. Seedlings when attacked with tristeza develop the so-called seedling yellows symptoms. As a result of seedling infection, the first leaves are small and yellow.

3. Causal Organism

The causal organism is Citrus Tristeza Virus (CTV). It is a phloem-limited closterovirus (threadlike virus), and is one of the largest plant viruses, appearing as a threadlike particle (2000 nm long, and 11 nm in diameter). The virus contains a single-stranded RNA (ssRNA). The virus is transmitted by grafting as it is only present in phloem cells. The most important vectors are all aphids particularly citrus aphid, *Toxoptera citricida,* which is a stylet-borne vector. The vector become viruliferous after feeding for a few seconds and transmit the virus after equally short feedings.

4. Disease Cycle

The spread of the pathogen is through the use of tristeza-infected propagated material, and through several aphid species in a semipersistent manner. The virus is also transmitted by dodder.

5. Disease Management

It is more satisfactory and economical to remove susceptible trees after they become infected, and to replant with a resistant plant. In recent years more satisfactory control of tristeza virus has been obtained by inoculating young citrus trees with selected mild strains of the virus in cross protection or immunization. According to Agrios (2005), considerable efforts are being made to express CTV genes such as the coat protein-gene that might make the trees resistant to Tristeza.

22.16 CITRUS EXOCORTIS (VIROID DISEASE)

1. Distribution

Exocortis is worldwide in distribution. The pathogen (viroids) affects oranges, sweet lime, some lemons, grapefruit, and other citrus trees grafted on exocortis sensitive root-stocks. The pathogen causes great reduction in growth and yields are reduced up to 40 percent.

2. Symptoms

The infected plants show narrow vertical splits in the outer bark that give the bark a cracked and scaly appearance. Infected plants may also show yellow blotches of twigs on young infected stems, and some plant show leaf and stem epinasty, and cracking and darkening of leaf veins and petioles. All infected plants show stunted growth, and have lower yields. Under field condition the symptoms of cracking and scaling are evident on bark below the union of various *Citrus* species such as grapefruit, mandarins, and sweet oranges budded above root stocks.

3. Causal Organism

The disease is caused by Citrus Exocortis Viroid (CEV). Viroids are small, low-molecular-weight ribonucleic acid (RNA) without protein coats that can infect plant cell, replicate themselves, and cause disease. CEV is transmitted from diseased to healthy plants by budding knives, pruning, by hand, by animals, and by dodder (*Cuscuta* sp.). On contaminated tools, the pathogen retains its infectivity up to 8 days, and can remain active at room temperature for several months.

4. Disease Cycle

The pathogen (CEV) survives in citrus plants, and many herbaceous hosts, and is spread to healthy plants by budding or wedge grafting, and by contaminated cutting tools. No insect vector is known. The viroid enters the phloem elements and spreads in the plant.

5. Disease Management

The diseases can be controlled by propagating exocortis-free nursery trees. Tools should be disinfected by dipping in a 10–20 per cent solution of sodium hypochlorite. Removal and destruction of viroid-infected plants are effective cultural practices.

22.17 CADANG-CADANG DISEASE OF COCONUT (VIROID DISEASE)

The name *cadang-cadang* means dying. The disease was first reported from the Philippines in 1937, and it has killed an estimated 30 million trees since that time in that limited area. According to Agrios (1996), the disease is of great economic importance to the Philippines because of the great dependence of the local people for food and lumber, and as a major cash crop from the export of coconuts and copra (the dried coconut meat) from which oil is extracted.

1. Symptoms

The symptoms of the disease develop very very slowly over 8 to 15 years, and the tree may not die from the disease for 5–10 years after first symptoms appearance. Coconut palm trees usually become infected after they start flowering. The first symptoms appear on the coconuts, which develop scarifications on their outer surface, and the leaves start to show bright yellow spots. Three to four years later, the inflorescences are killed which stops production of any further fruits on the palms. A few new leaves may develop on which more and larger spots develop leading to the chlorotic appearance of the disease from a distance. Five to seven years from the beginning of the symptoms, the constantly increasing number of leaf spots gives the whole crown of the leaf bronze colour, and the number and size of the leaves progressively reduced. Ultimately, the growing bud dies which falls off, and leaves the trunks of the trees looking like telephone poles.

2. Causal Organism/Entity

The disease is caused by a viroid known as Cadang-Cadang Viroid (CCCVd). The viroid is known to consist of 246 nucleotides, and in fact is the smallest viroid known so far.

3. Disease Cycle

The viroid is believed to survive in most palm tissues including the husk and embryo of coconut. It is also transmitted to a small extent through seeds of coconut. Pollens of affected palms carry the viroid. However, it is not very definite or clear as to how the viroid spreads from tree to tree in nature. The viroid may be carried on mouth parts of several chewing insects, mechanically on the machetes used to cut steps at the base of the palm, to dislodge the nuts, and to make cuts to the inflorescence for tapping the sap, and also possibly through the infected pollens.

4. Disease Management

No effective method of control of the disease is available so far, and no resistant cultivars are available against this disease. Tissue culture can help in the production of viroid-free seedlings. Recent techniques of electrophoresis and nuclear techniques that help in the rapid detection of the viroid in parental material will go a long way in solving the problem of production of healthy seedlings free of any viroid through the use of tissue-culture techniques or through healthy seeds.

22.18 POTATO SPINDLE TUBER DISEASE (VIROID DISEASE)

1. Distribution

It attacks all varieties of potatoes, and is one of the most destructive diseases of potatoes because it spreads rapidly.

2. Symptoms

The infected potato plants appear erect, spindly, and dwarfed. The infected leaves are darker green, and sometimes show rolling and twisting. The infected tubers are elongated with tapering ends. Yields are reduced up to 25 percent.

3. Causal Organism

The causal organism is Potato spindle Tuber Viroid (PSTV).

4. Disease Cycle

After inoculation of a tuber with PSTV by means of a contaminated knife, or of growing plants with sap from an infected plant, the viroid replicates itself, and spreads systemically throughout the plant. PSTV is also transmitted by several insects including aphids, grasshoppers, and beetles. In potato and tomato, the viroid is transmitted by true seed, and through pollen.

5. Disease Management

The disease can be controlled effectively by planting only disease-free (PSTV) potato tubers in fields. Washing of hand or sterilizing of tools after handling viroid-infected plants before moving on to healthy plants also helps.

22.19 SOME OTHER IMPORTANT VIRAL DISEASES

Sugar cane mosaic	It is caused by sugar cane mosaic virus which belongs to polyviruses group.	Appearance of pale patches or blotches on the leaf which are irregular in shape and size. In some cases, mottling of stem occurs.	Vegetative propagation and aphids are responsible for transmission of disease.	Crop rotation, selection of healthy setts, heat therapy, systematic roguing of infected canes, and use of resistant varieties are effective in disease control.
Yellow mosaic of *urad*, *mung*, and soybean	Mungbean Yellow Mosaic Virus (Begomovirus)	Yellow spots on the leaves which are round in shape.	Transmitted by whitefly (*Bemisia tabaci*)	Use of resistant varieties, application of insecticides like 0.1% Metasystox, 0.2% Anthio, Aldicarb, and oil sprays are effective control measures.
Bean mosaic	Phaseolus virus	The leaves of infected plants develop mottle, blisters, and lamina curls downward.	Seeds, aphids are responsible for spread of disease.	Use of certified virus-free seeds, and cultivars check the disease.
Mosaic disease of cucurbits	Cucumber mosaic virus, cucumber green mottle mosaic virus, watermelon mosaic virus	Alternate green and yellow patches on leaves, which are irregular and enlarge rapidly.	Transmitted by aphids, seeds, and by dodder.	Destruction of weeds and infected plants, use of virus-free seeds and insecticides (Thiodan, Metasystox) controls the disease.
Rice tungro viruses	It is caused by two viruses: the rice tungro spherical virus and the rice tungro bacilliform virus	The main symptoms are stunting with mottling and yellowing of leaves.	Transmitted by green leafhopper (*Nephotettix* spp.)	Use of rice cultivars resistant to disease has been a major approach for control of the disease.
Sterility mosaic of pigeonpea	Pigeonpea sterility mosaic virus	Yellowing of leaves, dwarfing of branches, and reduced or no flowing are main symptoms.	Transmitted by mite (*Aceria cajani*)	Use of resistant varieties is the effective control measure.
Mosaic of cassava	African cassava mosaic virus	Mosaic along with chlorotic areas are scattered on the leaf.	Transmitted by whitefly (*Bemisia tabaci*)	Use of virus-free cassava cuttings, and application of Metasystox (0.1%) is effective in disease control.

REVIEW QUESTIONS

1. What are viruses? How do they differ from other plant pathogens?
2. Viruses do not generally disseminate in regions of high altitude. Explain why this is so and how this feature has been utilized in agriculture.

3. Give in a tabular form some important viral diseases and their vectors.
4. How will you find out whether a disease is due to virus or due to nutritional deficiency?
5. What are viroids? Who discovered viroids?
6. Name some important diseases of viroids.
7. Give one example each of ssRNA (positive RNA), ssRNA (negative) RNA, ssDNA, and dsDNA viruses.
8. Name a few diseases of international importance caused by viruses and viroids.
9. How can tissue-culture techniques help in controlling some of the diseases caused by viroids?

MULTIPLE-CHOICE QUESTIONS

1. Who for the first time pointed out that the juice of tobacco mosaic virus if applied to healthy leaves could cause disease?
   (i) M W Beijerinck (ii) Adolf Mayer
   (iii) D Iwanowski (iv) W M Stanley
2. The first discovered virus was a
   (i) mycophage (ii) bacteriophage
   (iii) animal virus (iv) plant virus
3. Plant viruses generally have
   (i) ssRNA (ii) dsRNA
   (iii) ssDNA (iv) ds DNA
4. Binal symmetry is common among
   (i) viroids (ii) prions
   (iii) bacteriophages (iv) plant viruses
5. Bacteriophages mostly have
   (i) dsRNA genome (ii) ssRNA genome
   (iii) dsDNA genome (iv) ssDNA genome
6. Who coined the term *viroids* for a newly recognized class of subviral plant pathogens?
   (i) T O Diener (ii) J A Dodds
   (iii) D L F Brown (iv) M W Beijerinck
7. Which of the following diseases is caused by viroids?
   (i) Citrus exocortis (ii) Cadang-cadang of coconut
   (iii) Cucumber pale fruit (iv) All of these
8. Which of the following vectors is responsible for spread of viruses?
   (i) Whiteflies (ii) Beetles
   (iii) Mites (iv) All of these
9. Which of the following fungi are responsible for transmission of viruses?
   (i) *Olpidium brassicae* (ii) *Polymyxa graminis*
   (iii) *Phytophthora infestans* (iv) Both (i) and (ii)
10. Viruses are made up of
    (i) proteins and lipids (ii) proteins and nucleic acid
    (iii) starch and carbohydrates (iv) starch, protein, carbohydrates, and lipids

11. Which of the following statements is true for viruses?
    - (i) They occur only inside bacteria.
    - (ii) They are made up of only proteins.
    - (iii) They behave like plants.
    - (iv) They can multiply only in host cells.
12. Who for the first time obtained viruses in crystallized form?
    - (i) D Iwanowski
    - (ii) R Koch
    - (iii) Adolf Mayer
    - (iv) W M Stanley
13. The number of capsomeres in TMV is
    - (i) 1130
    - (ii) 2130
    - (iii) 3130
    - (iv) 130
14. A bacteriophage can be recognized by
    - (i) round shape
    - (ii) tadpole shape
    - (iii) rod shape
    - (iv) irregular shape
15. The genetic material of cauliflower mosaic virus is
    - (i) RNA
    - (ii) DNA
    - (iii) both (i) and (ii)
    - (d) none of these

Answers

1. (ii) 2. (iv) 3. (i) 4. (iii) 5. (iii) 6. (i) 7. (iv) 8. (iv) 9. (iv) 10. (ii)
11. (iv) 12. (iv) 13. (ii) 14. (ii) 15. (ii)

23 DISEASES CAUSED BY NEMATODES AND FLAGELLATES (PROTOZOANS)

23.1 DISEASES CAUSED BY PLANT PARASITIC NEMATODES

Nematodes are the most highly developed of Pseudocoelomates. They belong to the Phylum Nemata, previously named Phylum Aschelminthes. They are widely present in the soil, fresh water or marine water. Nematodes and some flagellates are the only plant parasites belonging to the animal kingdom which are studied in plant pathology. Nematodes, sometimes called **eelworms**, are wormlike in appearance but quite distinct taxonomically from the true worms. Numerous species of nematodes attack and parasitize humans and animals and cause various diseases. Several hundred species are known to feed on living plants as parasites and cause a variety of plant diseases.

All plant parasitic nematodes belong to the phylum Nemata (Nematoda). Most of the important genera belong to the subclass Secernentea: order Tylenchida.

The characteristic features of the subclass Secernentea are excretory canals (protoplasmic) present; terminal excretory duct cuticularized; caudal glands absent; plasmids usually present; amphids usually minute, porelike, and cephalic in position; sensory organs papilloid, seldom setose; hypodermal glands absent; male with or without caudal alae.

23.2 SYMPTOMS CAUSED BY NEMATODES

Nematodes infect the roots as well as the parts of the plant which are above the ground. Root symptoms may appear as hypertrophy, necrosis, or abnormal growth and include the following:

1. Root Knots or Root Galls

These are enlargements of the roots caused by the feeding of the nematodes which may not necessarily be enclosed within them. The swellings may vary in size from 1 mm to more than 2 cm. The feeding of nematodes induces the formation of 'giant cells' in the host tissue and cell division is stimulated. This leads to the formation of galls of various sorts. Bird (1974) has discussed the response of plants to root-knot nematodes under two major headings. First, the responses of entire plants, and second, the responses of their cells. Whole plants respond to infection by reducing their photosynthetic rate, growth, and yield. It is suggested that the nematode influences the physiology of the plant by interfering with the synthesis and translocation of growth hormones produced in the roots. Cells of susceptible plants are changed from normal undifferentiated cells to highly specialized syncytia, also known as **giant cells** or **multinucleate transfer cells**. This process involves a chain of events, starting with nuclear and nucleolar enlargement, followed by cell-wall breakdown, synchronous mitosis, and incorporation of adjacent cells. This highly specialized syncytium is induced and maintained by, and is completely dependent on a continuous stimulus from the nematodes.

2. Root Lesions

There are discoloured and often collapsed portions of the root, consisting of cells on which nematodes have fed. They vary in size, from being as minute as to be almost invisible to the naked eye to lesions girdling the whole root. Necrotic lesions are probably caused by toxic salivary secretions injected during the feeding of nematodes, for example, *Radopholus*.

3. Excessive Root Branching

This is caused by the formation of numerous short laterals in the vicinity of nematode injury.

4. Root Rots

Root rots occur when nematode infections are accompanied by plant pathogenic or saprophytic bacteria and fungi.

5. Injured Root Tips

This happens when the nematodes feed on or near root tips and cause them to stop growing, enlarging, or disintegrating. Nematodes feeding ectoparasitically at the root tips suppress cell division in the apical meristem and result in short roots, as in *Trichodorus,* the stubby root nematode.

Nematode root infections are usually accompanied by noncharacteristic symptoms in the above-ground parts of plants, appearing primarily as reduced growth, symptoms of nutrient deficiencies, such as yellowing of foliage, excessive wilting in hot or dry weather, reduced yields, and poor quality of products.

Nematodes which attack the serial parts of the plant may cause discolouration, necrosis, blotches, spots, distortion, and galls on the leaf, stem, and seed. Buds, growing points, or flower primordia are attacked by some nematodes, resulting in the abnormal growth of the affected plant. These symptoms are thought to result from substance secreted by the eelworm or perhaps by the invaded plant tissue.

23.3 MECHANISM OF NEMATODE INJURY TO PLANTS

The mechanical injury directly inflicted upon plants by the nematodes during feeding is slight. Most of the damage seems to be caused by a secretion injected into plants while the nematodes are feeding. This secretion, called saliva, is produced in three glands from which it flows forward into the esophagus and is ejected through the stylet. Some nematodes are such rapid feeders that in a matter of seconds they pierce a cell, inject saliva, withdraw the cell contents, and move on. Others feed more leisurely and may remain at the same juncture for hours or days.

The sedentary forms (*Heterodera* and *Meloidogyne*) remain attached to one point in the tissue throughout their lives. The males of the species may not penetrate the roots but the females invariably get established in or on the roots in a fixed position. In such species, it is the female which is responsible for the destruction of the host.

Nematode saliva has various functions, depending on the habit of the organism. The saliva of the plant nematodes seems to aid the parasite to penetrate the cell walls and possibly to liquefy the cell contents, making them easier to ingest and assimilate. The saliva, being toxic, proves disastrous to the plant tissues and its effects may reach up to the leaves even if the nematode is present only in the roots. As a result of the action of this saliva, any one of the following host responses may appear as symptoms of the disease: 1. cellular hypertrophy and hyperplasia, 2. suppression of mitosis, 3. cell necrosis, and 4. stimulation of growth.

The feeding of some nematodes produces only slight trauma in host cells. When the nematode *Tylenchorhynchus dubius* feeds upon the root hairs of *Lolium perenne*, a spherical mass is formed at the tip of the stylet within the host cells. For about 30 seconds, the nematode remains quiet, after which its bulb begins to pulsate and the spherical mass starts diminishing in size and finally disappears after one minute (Fig. 23.1a). In some associations, the stylet acts simply as the organ of penetration and suction. The nematode *Paraphlenchus acontiodes* penetrates the host cells of *Pyrenochaeta terrestris* and withdraws their contents within 2 to 3 seconds (Fig. 23.1b).

In certain other associations, such as *Ditylenchus myceliophagus* feeding upon *Botrytis cinerea*, there is delayed removal of contents (Fig. 23.1c). Progressive penetration is shown by the nematode *Rotylenchus uniformis* when feeding on the root hairs and cortical parenchyma cells of *Lolium perenne* as well as by the nematode *Pratylenchus crenatus* when piercing the epidermal cells of *Poa annua* rootlets (Fig. 23.1d). Many plant parasitic nematodes affect only those cells upon which they feed or a limited number of cells. In the immediate vicinity of the feeding site, cells of infected tissues separates undergo hypertrophy and lose chloroplasts (Fig. 23.1e). In certain host-parasite combinations, the pathogen stimulates changes in host cells, bringing about alterations in metabolism vital to the growth and development of the pathogen but do not bring about any cell destruction. The nematode *Tylenchulus semipenetrans* brings about an increase in the size of the nuclei and nucleoli, the cytoplasm of the host cells becomes dense, their wall thickens and the vacuoles of the infected cells disappear (Fig. 23.1f).

Anguina and *Nothanguina* species stimulate gall formation in leaves and flower parts of grasses and other plants. The galls develop here due to hypertrophy and hyperplasia of parenchymatous tissue with a central cavity harboring the nematode. No syncytia develop here.

Members of the genera belonging to the family Heteroderidae, show the greatest morphological adaptation to parasitism. At the site of the feeding of the nematode, a group of cells develop into characteristic syncytia around the head of the parasite. The cytoplasm becomes dense and the size of the nuclei and nucleoli enlarges considerably. The cell walls are usually altered. Species of *Meloidogyne* induce hyperplasia in the pericycle which results in the formation of galls. *Heterodera* species do not induce extensive hyperplasia. *Nacobbus batatiformis* forms galls on the roots of sugar beets and some other plants. Here, galls are formed as a result of hypertrophy of cortex and epidermal cells. The stele is not affected. Syncytial mass is formed by the alteration of cells immediately adjacent to the nematode's head and extends longitudinally along the root (Fig. 23.1g).

In *Heterodera* infection, syncytium is formed as a result of the enlargement of nuclei, incomplete dissolution of cell walls, disintegration of nucleoli, thickening of cell walls, and by the cytoplasm becoming dense and granular (Fig. 23.1h). Giant cells of *Meloidogyne* are formed as a result of the enlargement of nuclei, and the cells become polyploid and undergo synchronous mitosis. The cytoplasm of these cells becomes granular and new cells are incorporated by cell-wall dissolution. The walls of the syncytium become thickened and the cells of pericycle divide repeatedly (Fig. 23.1i).

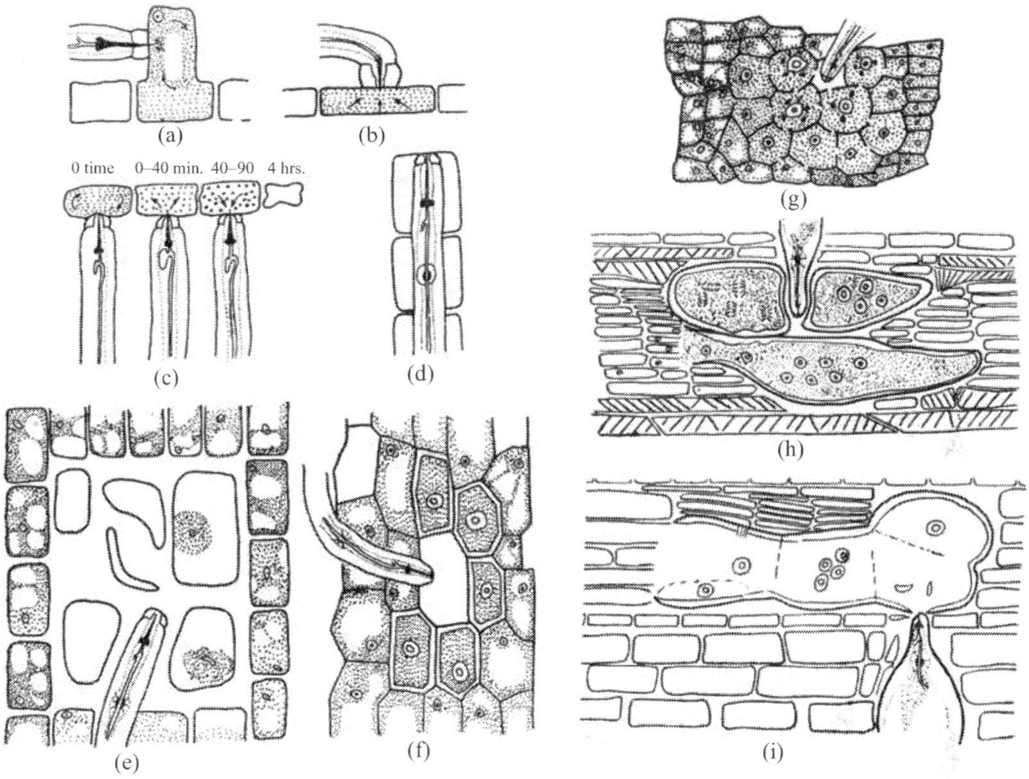

Fig. 23.1 Effect of nematodes on plant cells: (a) No visible destructive effect (b) Immediate removal of cell contents (c) Delayed removal of cell contents (d) Progressive penetration (e) Cell lysis (f) Nurse cells of *Tylenchulus* (g) Syncytium of *Nacobbus* (h) Syncytium of *Heterodera* (i) Giant cells of *Meloidogyne* [Redrawn from V H Dropkin (1969), *Ann. Rev. Phytopath*].

23.4 INTER-RELATIONSHIP BETWEEN NEMATODES AND OTHER PLANT PATHOGENS

The role of certain nematodes as vectors of plant viruses has been studied in recent years and there is an increasing awareness that they may also be involved in the transport and inoculation of other pathogens, notably bacteria and fungi attacking roots or other plant organs in the soil. Several types of nematode-pathogen interaction are suggested by Pitcher (1965), of which the following are examples:

The nematodes transport the pathogen from the soil to the plant or from one part of the plant to another, and the pathogen may fail to develop fully in the absence of the eelworm, as in twist of grasses and cereals caused by the fungus *Dilophospora alopecuri* in association with *Anguina tritici*.

Much more important and more common are the inter-relationships between nematodes and viruses. Several plant viruses, such as grapevine fan-leaf virus, arabis mosaic virus, tomato ring-spot virus, tobacco black-ring virus, raspberry ring-spot virus, and tobacco rattle virus are transmitted by nematodes. Sixteen species of nematodes have so far been thought to be vectors of plant viruses.

Hewitt *et al.* (1958) showed that the fan-leaf disease of grapevines was spread by the eelworm *Xiphinema index*.

Several nematode fungus disease complexes are known. Fusarium wilt of several plants increases in incidence and severity when the plants are also infected by nematodes. Similar effects have also been noted in disease complexes involving nematodes and *Verticillium* wilt, *Pythium* damping off, *Rhizoctonia*, and *Phytophthora* root rots.

Relatively few cases of nematode-bacterial disease complexes are known. In most of these, the role of the nematode seems to be that of providing the bacteria with an infection court and to assist bacterial infection by wounding the host.

In the yellow ear rot of wheat, the role of the nematode is to mechanically carry the bacterium *Clavibacter (Corynebacterium) tritici* inside the ovaries of the wheat flowers. In strawberry infection by the leaf nematode and the bacterium, a more involved interaction between the two pathogens seems to exist. Plants inoculated with both pathogens produce the so-called cauliflower symptom, quite distinct from either of the symptoms produced by each of the pathogens described above.

23.5 ECOLOGY AND SPREAD

Almost all plant pathogenic nematodes lead a part of their lives in the soil. Many of these live freely in the soil, feeding superficially on roots and underground stems, and although they may cause injury to plants, they are not strictly parasitic. Even in the most highly specialized sedentary parasites, the eggs, the pre-parasitic larval stages, and the males are found in the soil for all or part of their lives. Soil temperature, structure (porosity), moisture, and aeration affect the survival and movements of nematodes in the soil. Nematodes occur in greatest abundance at a depth of 0–15 cm in soil. The distribution of nematodes in cultivated soils is irregular and is greatest in or around roots of susceptible plants. The greatest concentration of nematodes is in the region of host roots. This is primarily due to the greater amount of food available and also to the attraction of nematodes to the substances released into the rhizosphere. To these must be added the so called hatching factor—the effect of substances exuded from the root which diffuse into, or are carried to the surrounding soil and markedly stimulate the hatching of eggs of certain species, although most nematode eggs hatch freely in water in the absence of any special stimulus.

Nematodes spread through the soil very slowly and by their own power. The speed of movement in the soil is dependent on pore diameter, particle size, water content of the soil and the diameter and relative activity of nematodes.

In addition to their own movement, however, nematodes can be spread by anything that moves and can carry particles of soil. Farm equipment, irrigation, flood or drainage water, animals, and duststorms spread nematodes.

Temperature has a profound effect on the survival and multiplication of nematodes in the soil and also on their parasitism. The literature on the effect of the soil, temperature, moisture, aeration, soil texture, organic matter, rhizosphere, and various cultural operations on the activity of plant parasitic nematodes is quite substantial.

Meloidogyne spp. have been reported to develop most rapidly at about 27°C. Reproduction in *M. halpa* is extremely reduced at 35°C. The optimum temperature for the hatching of larvae in *M. incognita* is 30°C. The temperature also affects the resistance of resistant varieties and the size of the galls formed by root-knot nematodes on the host.

Soil moisture is important for nematodes in many ways. *Meloidogyne* spp. can enter potato tubers through lenticels which occur on tubers commonly in wet soils. Galling of tubers is less in dry soils. Other factors which influence the activity of plant parasitic nematodes are soil pH, soil texture, aeration, types of plants grown between the susceptible hosts, duration of cultivation of resistant varieties in the field, type of decomposable organic matter added to the soil, inorganic fertilizers, and type of nitrogen and potassium status in the soil.

23.6 CHARACTERISTICS OF PLANT PATHOGENIC NEMATODES

The phytopathogenic nematodes and their relatives comprise a large number of species placed in 15 families and 111 genera. The gross morphology of these nematodes, with few exceptions, is generally very similar. The body is covered by a cuticle. Internally, there are digestive, excretory, and reproductive systems.

The nematode body is more or less transparent. It is covered by a colourless cuticle, which is usually marked by striations, or may show bristles, punctuations, warts, or other markings. The cuticle molts when the nematodes go through their successive larval stages. The cuticle is produced by the hypodermis which consists of living cells and extends into the body cavity as four chords separating four bands of longitudinal muscles. These muscles enable the nematode to move. Additional specialized muscles exist at the mouth, and along the digestive tract and the reproductive structures.

23.6.1 Digestive System

The esophagus (pharynx) which is the anterior end of the digestive system consists in the Tylenchids, of a stylet (stomatostyle), corpus, isthmus, and glandular posterior enlargement. Modifications in the morphology of the esophagus are widely used at all levels of nematode taxonomy. Many of the groupings of higher taxa in the Tylenchida, as well as in those groups of Dorylaimida containing phytoparasitic nematodes, are based largely upon modifications of gross shape and internal morphology of the esophagus. The digestive system is a hollow tube extending from the mouth through the buccal cavity, esophagus, intestine, rectum, and anus. Lips, usually six in number, surround the mouth.

All plant parasitic nematodes have a protrusible hollow stylet or spear. This is the main characteristic that suggests that the nematode is a plant parasite. This structure has been homologized in Tylenchida with the stoma of other nematodes and is used as a piercing organ through which nematodes acquire food. It is actually a three-part structure taxonomically; the stylet is used at all levels of classification.

The body cavity is rudimentary in its development and contains a fluid through which circulation and respiration take place.

The excretory system is not well developed in nematodes. However, the nervous system is well developed and consists of many nerves, ganglia, and sensory structures.

23.6.2 Reproductive Structures

The reproductive systems are well developed. The female nematodes have one or two ovaries followed by an oviduct and uterus terminating in a slit like vulva. The male's reproductive structures are similar to the females but have a testis, seminal vesicle, and ejaculatory duct, and terminate in a common cloaca with the intestine. A pair of protrusible, copulatory spicules is also present in the male. Reproduction in the nematodes is through the eggs and may be sexual, hermaphrodite, or parthenogenetic. Many species lack males.

23.6.3 Life Cycle

The life histories of most plant parasitic nematodes are, in general, quite similar. There are six stages in the life cycle of a nematode—egg, L_1 (Larval stage), L_2, L_3, L_4, and adult. Eggs hatch into larvae, whose appearance and structure are usually similar to those of the adult nematodes except for the development of the reproductive systems. Larvae grow in size and each larval stage is terminated by a molt. After the final molt, the nematodes differentiate into adult males and females. The females can then produce fertile eggs either after mating with a male or in the absence of males, parthenogenetically, or they can produce sperm themselves.

The time for the completion of the life cycle is about 20–40 days under optimum environmental conditions but takes longer in cooler temperatures. In some species of nematodes, the first or second larval stages cannot infect plants and they depend for their metabolic functions on the energy stored in the egg. When the infective stages are produced, however, they must feed on a susceptible host or else starve to death. If suitable hosts are not available, all the individuals of certain nematode species may die but in other species, the eggs may remain dormant in the soil for years. The disease cycle of a typical root-knot nematode of the genus *Meloidogyne* is given in Fig. 23.2 and that of soybean cyst nematode (*Heterodera glycines*) in Fig. 23.3.

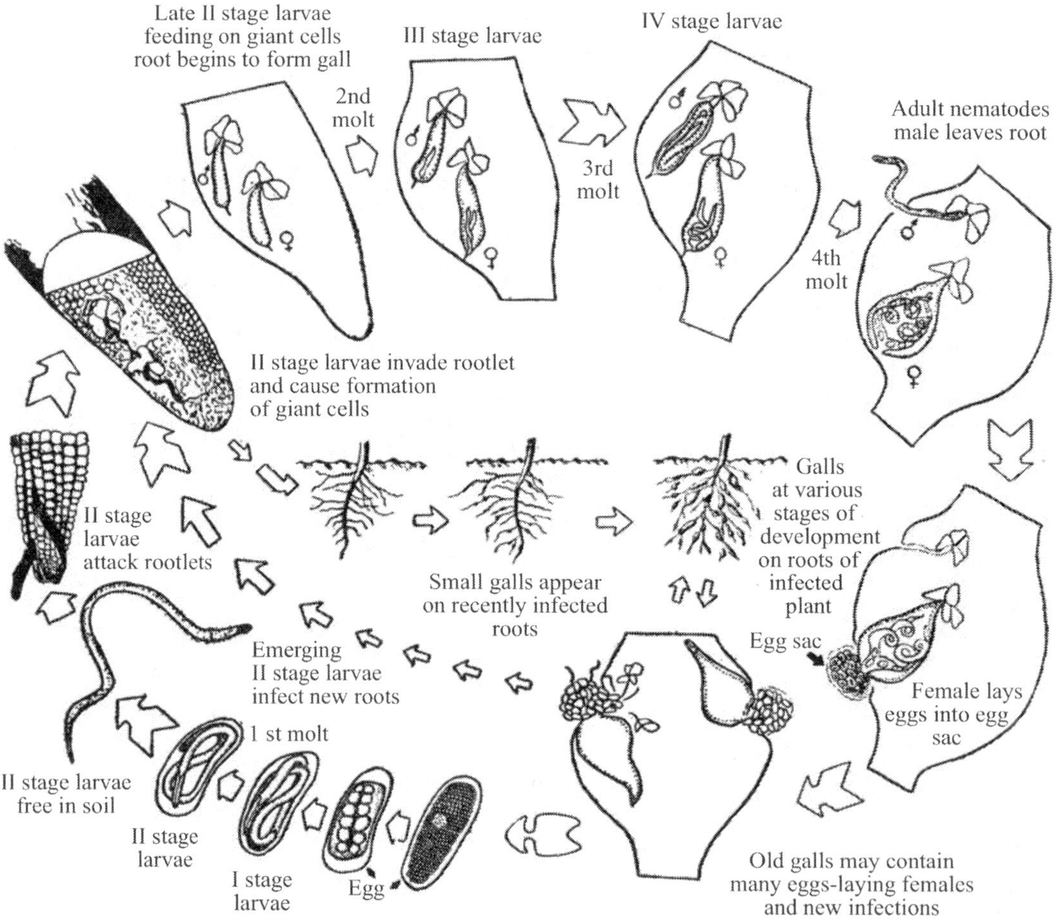

Fig. 23.2 Disease cycle of root knot caused by nematodes of the genus *Meloidogyne* [Redrawn from G N Agrios (1969) *Plant Pathology*, Academic Press, London, p. 519]

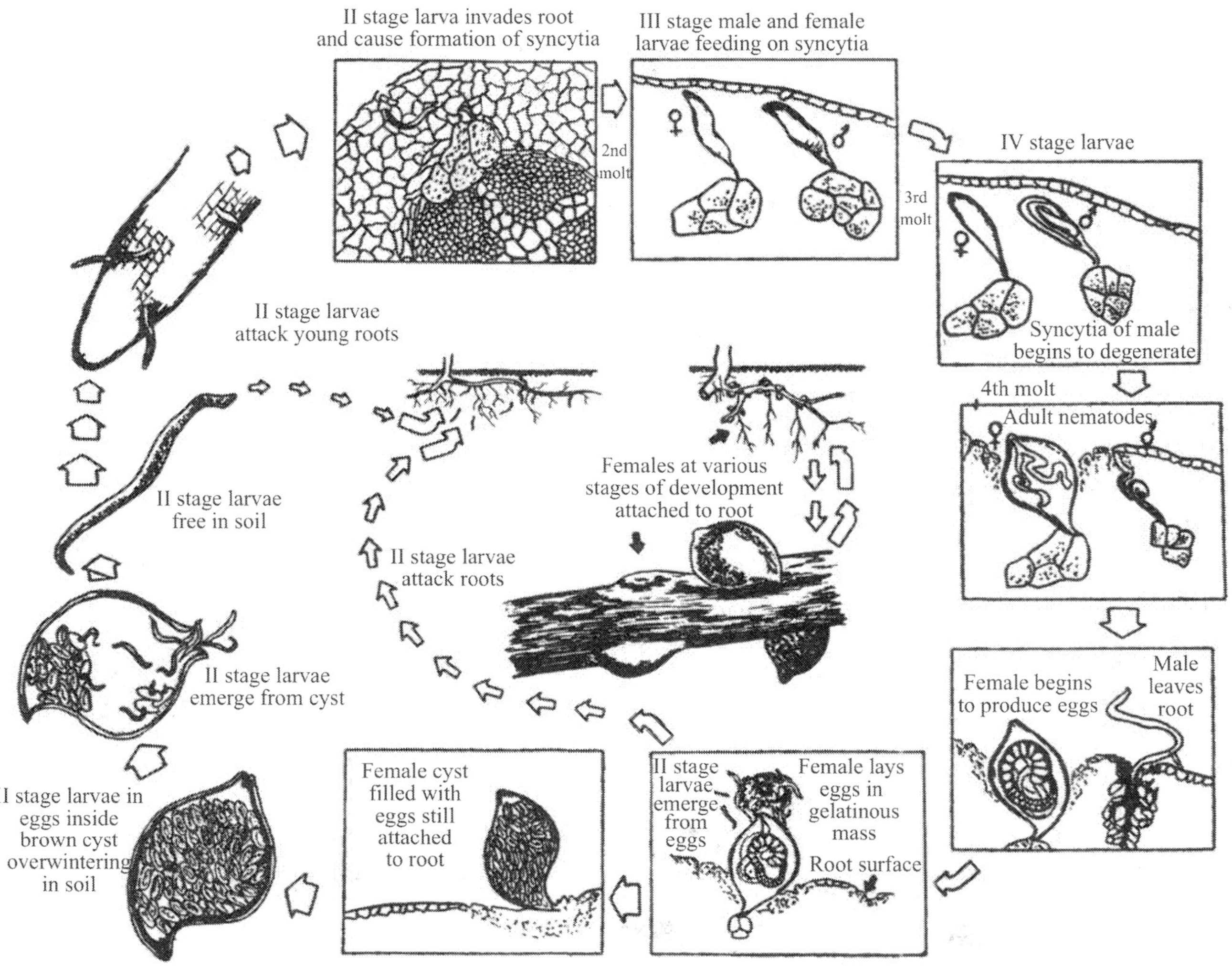

Fig. 23.3 *Disease cycle of soybean cyst nematodes Heterodera glycines* [Redrawn from G N Agrios (1969) *Plant Pathology*, Academic Press, London, p. 525]

23.7 GENERAL METHODS OF CONTROL OF NEMATODE DISEASES

Many accounts have been published of the attempts to control plant nematodes in different countries and climates. These are discussed in brief here.

23.7.1 Chemical Control

Chemicals with nematicidal properties are known as **nematicides**. Many such chemicals are recommended for nematode control. Nematodes are diverse in their chemical and biological activities and in their behaviour in soil. There are two general types of nematicides—fumigants and non-fumigants. Chemically they can be categorized as halogenated aliphatic hydrocarbons, methyl isothiocynate liberators, organophosphates, and organocarbamates. The common pesticides of nematodes are given in Table 23.1.

Table 23.1 Common nematicides

Class	Common name
1. Fumigants	
A. Halogenated aliphatic hydrocarbons	
Bromomethane	Methyl bromide
Trichloronitromethane	Chloropicrin
1,3-Dichloropene	1,3 – D
1,2-Dibromoethane	EDB
1,2-Dibromo-3Chloropropane	DBCP (Nemagon)
B. Methyl isothiocynate liberators	
Methyl isocynate	MIT
2. Nonfumigants	
A. Organophosphates	
O EthylS,S-Dipropylphosphorodithioate	Ethoprop (Mocap)
O,O-Diethyl-O- [(methylsulfinyl)phenyl]-Phosphorothioate	Fensulfothion (Dasanit, Terracur P)
Ethyl 4-(methylthio)-m-tolyl-Isopropylphosphoramidate	Phenamiphos
B. Organocarbamates	
2,3-Dihydro-2,2-dimethyl-7-benzofuranyl methylcarbamate	Carbofuran (Furadone)
2-Methyl-2(methylthio) propionaldehyde O-(methylcarbamoyl) oxime	Aldicarb (Temik)
Methyl N, N-(dimethylcarbamoyl)-N-(methylcarbamoyl) Oxy)-1-thioformimodate	Oxamyl (Vydate)

Use of nematicides is one of the most important and reliable means for management of root-knot nematodes. Nematicides are principally applied to control nematode population in the soil before the crop is grown (Khan, 1997). As the number of nematicides available for safe control of nematode diseases is declining drastically, the other methods of control are becoming increasingly important. Among the fumigants, carbon disulphides, chloropicrin, DD, DBCP (Nemagon), Vapam, methyl bromide, etc. have been banned now due to their polluting effect. More stress is now on systemic, nonfuming nematicides such as Carbofuran, Aldicarb, Oxamyl, Phenamiphos, Phorate, etc. In addition, there are a number of highly effective contact nematicides/insecticides which are being used increasingly for chemical control of nematodes in soil.

23.7.2 Cultural Practices

They result in partial or complete control of nematodes and include the following:

1. Crop Rotation of Nonhost Plants

Several nematode species can infect only a few crops. Since all plant pathogenic nematodes are obligate, the absence of susceptible hosts from the soil for two to three years will result in the elimination of nematodes from that area through starvation and inability to reproduce. The value of crop rotation has long been recognized in the control of *Heterodera* spp. Thus, continuous cropping with potatoes increases the cyst population of *H. rostochiensis* more that does three-year rotations with corn, green beans, red clover, or perennial rye grass. Brown (1961) gives some idea of the effect of crop rotation on the yields of potatoes growing in soil infected with *H. rostochiensis.* Potato yield after three years of continuous cropping was 1,631 kg/hectare, and with a nonhost in the second year, it was 4,515 kg/hectare.

In recent years, a new term, **cropping system**, which is defined as the sequence of growing crop may be spatial or temporal. Arrangement of different crops such as mixed cropping, row intercropping, and strip cropping is known as **spatial sequence**. Distribution of crop cover over a period of time such as crop rotation, relay cropping, discontinuous cropping, and fallowing is called a **temporal sequence**. Both spatial and temporal sequencing have been found effective in controlling root-knot nematodes on vegetable crops (Khan, 1997). A number of nonhost or poor host of root-knot nematodes like wheat, rice, onion, and maize are known to increase yield of vegetable crops in appropriate temporal sequence. Economic consideration is, however, a serious demerit of crop rotation. It can be successfully used when all the grown crops in the sequence are of economic value. Rotation of crops or cultivars with different genes for resistance against the species or races of root-knot nematodes has been suggested.

Some plants appear to be antagonistic to nematodes because they reduce populations to a greater extent than do nonhosts. *Crotalaria* is a good example of this. It has been suggested that the *Crotalaria* root system has a toxic effect on nematodes, and planted in advance of a crop, it may reduce subsequent damage.

There is evidence from Dutch nematologists that *Tagetes* spp. may reduce the populations of some plant parasitic nematodes. Oostenbrink and his colleagues (1957) tested 16 varieties of *Tagetes patula* and *Tagetes erecta* placed between the rows or around other plants. They found that all varieties suppressed the populations of *Pratylenchus* in the roots of other plants as well as in the soil, resulting in a better growth of perennials in the second year of the succeeding crop. The *Tagetes* had to be grown for three to four months and it was suggested that the roots had a nematicidal action. In spite of many criticisms, the use of cover, trap, or enemy crops provides a potentially useful control measure in crop rotations. Such an opinion is supported by the evidence that *Asparagus* exudes a nematicidal chemical from its roots.

In addition to *Crotalaria* or *Tagetes*, there are some other plants such as some grasses, varieties of mustard, Margosa (*Azadirachta indica*) root exudates, and *Asparagus* that have been tested as enemy plants to nematode population. In some vegetable plots and in greenhouse-grown crops, mulching of soil with straw, weeds, grasses, old newspaper sheets, dry pine needles, the bank, etc. have been found effective in reducing nematode attack.

2. Sanitary Practices

They include the cleaning of all machinery thoroughly before moving into an uncontaminated area, taking care not to bring nematodes into a field by means of contaminated nursery stock, seed, containers, etc., keeping soil fallow, free from host plants which deprive nematodes of roots on which to feed. Thus, regulatory methods (quarantine) can be the least expensive methods to control nematode diseases.

3. Flooding

Some species of nematodes have become adapted to living in the normal soil with ordinary moisture content and depend on a certain amount of aeration present in it. Flooding the land for a period of several months results in the death of these nematodes and thus frees the land of these pathogens. However, it is possible or practical to flood the land for long periods in very few fields and, therefore, the applicability of this control method is quite limited.

4. Fallowing

This is another method of controlling the soil-borne nematodes. The fields or plants are ploughed to keep them from vegetation and exposed to dry summer heat of the sun. Starvation and desiccation kill the nematodes.

5. Predators

Although simple observations have shown that predaceous species of nematodes occur in the soil and can be utilized to eat away parasitic forms, so far very little work along these lines has been done. Nevertheless, it has been found that many nematophagous species of fungi are effective means of destroying the nematodes and these can be encouraged in the soil through organic soil amendments, such as the addition of decomposable vegetable matter.

6. Physical Methods

Plant parasitic nematodes can be easily killed in the laboratory by the application of heat irradiation, osmotic pressure, etc. It is more difficult, however, to employ such methods or large quantities of soil, especially if they are under cultivation.

7. Irradiation

There is strong evidence that nematodes are killed or seriously affected by irradiation. Fassuliotis and Sparrow (1955) demonstrated that irradiation of potato tubers with X-rays inhibited sprouting of *Heterodera rostochiensis.* Cysts of this developing nematode exposed to 20,000 roentgens contained only brown and dead eggs; at 40,000 r,. The eggs lost their contents completely. It is, however, difficult to irradiate nematodes. In fact it is impracticable because of the length of time needed to irradiate even small areas and because damage to plant roots occurs at levels of irradiation below those required to disrupt the nematodes' life cycle.

8. Heat

Heat treatment is probably the most successful physical control measure developed so far. It is widely used for the killing of nematodes within plant tissues before planting and has proved useful on nematode-infested bulbs, corms and tubers, and roots of plants such as chrysanthemums, strawberries, bananas, and citrus. This process involves determining the correct time and temperature for killing the nematode but not the host. This entails immersing the nematodes (contained in as small a volume of water as possible) into thermostatically controlled water baths at different temperatures for different lengths of time. After such treatment, the nematodes are examined for activity. These are stained to see if they are dead or introduced into soil containing host plants to assess their ability to invade, develop, and reproduce. Blake (1961) suggested that banana setts less than 13 cm in diameter should be immersed in water at 55°C for 20 minutes in order to kill *Radopholus similis.*

A new approach to heat treatment is the application of dry heat. Sweet-potato seed roots were treated at 45°C for 30 hours and these completely controlled root-knot nematodes, with 60% of the roots surviving. The white tip nematode of rice *Aphelenchoides besseyi* is of widespread occurrence in India in most of the rice-growing areas. Hot-water treatment of the seed at 52–55°C for ten minutes is the best way to ensure complete destruction of the nematodes.

More than 100 fungus species have been designated as nematophagous, a term that they have applied to fungi that "feed" on nematodes either as predators or parasites. These nematophagous fungi, members of either the phycomycetes or the class Moniliales of the Fungi Imperfecti, are abundant throughout the world and are frequently encountered in soil samples.

23.7.3 Biological Control

According to Garrett (1965), "biological control is any condition under which, or practice whereby, survival or activity of pathogen is reduced through the agency of any other living organism (except man himself), with the result that there is a reduction in incidence of the disease caused by the pathogen". Trap crops, resistant varieties, and antagonistic plants that release nematicidal root exudates may be considered useful or potential biological controls for nematodes.

Biological control aims at increasing the parasites and predators of nematodes in the soil, to increase the mortality of plant nematodes. This can be done by changing the environment, adding organic amendments, or introducing other organisms directly.

To increase the concentration of nematode-destroying organisms in the soil, it is necessary to alter the environment so that some of the forces restraining the multiplication of the organisms are weakened. The only serious attempt to biologically control plant nematodes has been concerned with the use of nematode-trapping fungi.

Oil cakes of margosa, castor and peanut at 0.2% (W/W) three weeks before planting significantly reduce infestation of *M. javanica* on okra and tomato. In almost all these experiments, a soil amendment was used, thereby altering the environment in favour of nematode-trapping fungi.

In the late 1930's, Linford and his group successfully reduced the incidence of root-knot of pineapple by incorporating green plant material into the soil. It is generally assumed that the addition of this material stimulates the population of organisms antagonistic to nematodes, which in turn, reduces the parasitic nematode population. There are several nematode-trapping fungi as *Artabotrys robusta, A. oligospora*, and *A. superba.*

Duddington and Duthoit (1960) obtained statistically significant reductions in the mean number of larvae of *Heterodera avenae* on 60 cm square plots to which 675 g of chopped cabbage had been added as compared with untreated plots.

Resistance to nematodes may be due to the production of toxic root exudates as found in asparagus, lack of an attractant or a hatching factor in the exudates, and barrier to penetration or failure of nematodes to develop within plant tissues.

Paecilomyces lilacinus has been found quite effective in controlling root-knot nematode under field conditions. *Pasteuria penetrans*, a bacterial biocontrol agent for the control of *Meloidogyne* spp., has also shown promising results.

The biology, ecology, and potential of biocontrol agents for nematodes have been reviewed in recent years (Trivedi and Sharma, 2012). It must be stressed that several organisms that are effective natural enemies of nematodes in the field may have limited potential as biological control agents for application by growers. The stage in the nematode life cycle attacked by a control agent has a profound effect on the damage to the control and the level of population control. Therefore, trapping fungi and rhizosphere bacteria that attack the second-stage juveniles of cyst and root-rot nematodes may significantly improve crop growth but is unlikely to prevent nematode population increase, especially those species that have more than one generation in a growing season. Root-colonizing fungi such as mycorrhizae and endophytic species such as *Fusarium* may reduce both nematode invasion and development.

Soil amendments may also reduce nematode infestations and by providing an energy source, help increase numbers of facultative parasites. Pre-colonized substrates are most effective in establishing nematophagous fungi in soil. Soils naturally suppressive to some nematodes may be used to shorten rotations of susceptible crops and improve the performance of nematicides and resistant cultivars.

It is established that many fungi are known to produce nematicidal or nematistatic compounds. Dababat and Sikora (2007) investigated the activity of *Fusarium oxysporum* against *Meloidogyne incognita*. The fungus inhibited juvenile penetration, gall formation, and egg-mass production within tomato plants. Bernabeu and Lopez-Llorca (2002) studied the biological factors related with the development and performance of *Chlamydosporium, Verticillium lecanii*, and *Paecilomyces lilacinus* as biocontrol agents for *M. javanica*. Khan *et al.* (2004) could prove that hatching of *M. javanica* eggs were reduced as a result of *Paecilomyces lilacinus* protease and chitinase enzymes either individually or in combination. The chitinase activity was also increased in culture filtrates of *Trichoderma harzianum*, this increment expresses the ability of this fungus to infect *M. javanica* eggs (Sahebani and Hadavi, 2008).

Vesicular-Arbuscular Mycorrhizal (VAM) fungi are beneficial soil fungi which form symbiotic association with roots of many plants. During colonization, VAM fungi can prevent root infections by reducing the access sites and stimulating host defense. VAM fungi have been found to reduce the incidence of root-rot nematodes. Various mechanisms also allow VAM fungi to increase a plant's stress tolerance. This includes the intricate network of fungal hyphae around the roots which block pathogen infections. Higher amounts of amino acids, especially arginine, found in the root exudates of mycorrhizal plants, reduced chlamydospores production in *Thielaviopsis basicola*. Tomato roots inoculated with *G. fasciculatum* had increased concentration of phenylalanine and serine; these two amino acids being inhibitory to root-knot nematode development. Increased peroxidase, localized mainly in the vacuoles, and cell wall and phytoalexins (glyceollin, coumestrol) have been extensively documented in VAM positive plants. Thus, plants infected with VAM fungi are likely to be more resistant to subsequent attack by pathogenic fungi and nematodes.

Nematodes in the soil are subject to infection by bacteria and fungi. This creates the possibility of using soil microorganisms to control plant parasitic nematodes. Bacteria are the most abundant organisms in the soil and some of the genera like *Pasteuria*, *Pseudomonas*, and *Bacillus* have shown great potential for the biocontrol of nematodes. They affect nematodes by a variety of modes: for example parasitizing, producing toxins, antibiotics or enzymes; interfering with nematode-plant host recognition; competing for nutrients; inducing systemic resistance of plants and promoting plant health. Spores of *Pasteuria* can attach to the cuticle of the second-stage juveniles and germinate after the juvenile has entered roots and begun feeding. The germ tubes can penetrate the cuticle and vegetative microcolonies then form and proliferate through the body of the developing female. Finally, the reproductive system of the female nematode degenerates and the mature endospores are released into the soil. Numerous *Bacillus* strains can suppress pest and pathogens of plants and promote plant growth. Rhizosphere *Pseudomonas* strains also exhibit diverse pathogenic mechanism upon interactions with nematodes. The mechanism employed by some *Pseudomonas* strains is the production of antibiotics and the induction of systemic resistance. Other Rhizobacteria reported to show antagonistic effects against nematodes include members of the genera *Agrobacterium, Arthrobacter, Clavibacter, Clostridium, Corynebacterium, Enterobacter, Gluconobacter, Klebsiella.* Rhizobacteria reduce nematode populations mainly by regulating nematode behaviour, interfering with plant-nematode recognition, competing for essential nutrients, promoting plant growth, and inducing systemic resistance. *Bacillus thuringiensis* (Bt) produces one or more parasporal crystal inclusions (Cry or d-endotoxins) which are known to be toxic to a wide range of insect species. Cry protein exerts its effect by forming lytic pores in the cell membrane of gut-epithelial cells of an insect. After ingestion of toxin by target nematode larvae, the crystals dissolve within the gut of the nematode and this is followed by proteolytic activation. Cry toxicity is directed against the intestinal epithelial cells of the midgut and leads to vacuole and pore formation and eventually degradation of the intestine.

23.7.4 Breeding for Resistance

This is the best method of controlling any disease. However, breeding for resistance against soil nematodes is a difficult task. This is further made difficult by the presence of large host range of the pathogens and the occurrence of several races or biotypes of the pathogens. Several varieties of crop resistant to nematodes have been developed in our country. Some important ones are Karnataka Hybrid and VNF-8 in tomato against root-knot nematode, Raj Kiran and RD-2052 in the barley crop against cereal cyst nematode, Anbeern-56 in cotton, Evergreen variety in onion, V-161 in cowpea against reniform nematode, TVAV 6464 in rice against rice lesion nematode Troyer, Carrizo in citrus against citrus nematode, gros Michel in banana against burrowing nematode, Branco muniero in garlic against stems, and bulb nematode and SSM 591 in rice against *Aphelencoides besseyi.*

23.8 ROOT KNOT OF VEGETABLES

The root-knot nematodes, *Meloidogyne* spp., stand out as the most dominant group of plant parasitic nematodes in almost every vegetable field and cause enormous losses every year in the nursery as well as planted fields. The common species are *Meloidogyne arenaria, M. incognita,* and *M. javanica.* They attack such crops as various cucurbits, potato, tomato, brinjal, chillies, ladyfinger (okra), groundnut, carrot, radish, and *Colocasia. M. arenaria* is commonly associated with yellow decline disease of chillies in the district of Coimbatore. As a result of infestation, plant growth is reduced and the leaves turn yellow.

1. Symptoms

In potatoes, the disease is not easy to recognize in the field as often no above-ground symptoms are produced. Sometimes plants are stunted, sickly, and even show signs of premature and sudden drying. The nematode infestation on the tuber appears as tiny tubercles but heavy and localized infestations stimulate excessive cell division leading to gall formation. The disease may be complex with brown rot or bacterial wilt caused by *Ralstonia* (*Pseudomonas*) *solanacearum* and in such cases, plants usually die suddenly. In tomatoes, the symptoms are somewhat more pronounced. The plants develop slowly and appear stunted if the infestation has been early and severe. The leaves are yellowish-green to yellow, tend to droop and wilt suddenly if wilt organisms are present. Sometimes the leaves get scorched from the margin inwards. The main root and the laterals in all cases bear spherical to elongated galls. The presence of these galls is the most characteristic symptom.

Symptoms similar to those on tomato are also noticed on brinjal (eggplant), chillies, ladyfinger (okra), cowpea, and French beans. In the case of chillies, the plant growth is reduced and the leaves turn yellow. The disease cycle of a typical root-knot nematode of the genus *Meloidogyne* is given in Fig. 23.2.

2. Disease Management

Because of the very widespread and serious nature of this problem on vegetable crops, much attention is being given to control these nematodes by the use of chemicals and resistant varieties. The use of chemical fumigants has been attempted in certain cases with marked success. Nematicides, such as DD, DBCP, Phorate and Carbofuran, have been found to be very effective in experiments, but their use is still restricted due to the high cost of the chemicals.

The addition of neem cake and similar non-edible oil cakes at the rate of 25 quintals per hectare has been found to reduce root-knot incidence in repeated field trials. Application of sawdust at the rate of 25 quintals per hectare along with standard NPK fertilizers is cheaper and equally effective in reducing the disease.

Soil solarization has been found effective in control of root knot of tobacco. Krishna Rao and Krishnappa (1995) reported that solarization by covering the soil with clear transparent polythene sheet for 6 weeks during hot summer months showed an increase in soil temperature by 8°C and concentration of moisture (5% more) when compared to unsolarized soils. This resulted in decrease in population densities of *H. incognita* and *Fusarium oxysporum* in chickpea fields.

It is essential that all the sources of nematodes, such as weed hosts, should be eliminated from the field. The soil should be deep ploughed during summer and allowed to dry.

Flooding has been found to effective in eliminating root-knot nematodes from the soil. Lowland rice culture preceding the vegetable crop reduced the incidence of root knot in the latter. Resistant varieties to various species of root-knot nematodes do occur in many vegetable crops and these should be cultivated.

23.9 MOLYA DISEASE OF BARLEY AND WHEAT

It is known to be widespread in the states of Rajasthan, Haryana, and Punjab. The damage may be as 50% in certain infected areas. Barley is more susceptible than wheat. This disease is known as cereal root eelworm in England and occurs on wheat, oats, and barley.

1. Symptoms

The disease occurs in patches in the field. If the same crop is cultivated year after year, the diseased patch increases until the whole field is infected. The infected plants become dwarfed and pale, with a matted root system. Mild swelling occurs near the root tips. Glistening white bodies are seen adhered to the roots. These bodies (cysts) become brown and may remain attached to the roots or fall off in the soil after the roots decay.

2. Disease Cycle

The host range of *Heterodera avenae (H. major)* is confined to the family, Gramineae, including some grasses, and these may serve as alternate hosts (collateral hosts). The main method of survival of the nematode in the soil is by means of cysts.

3. Disease Management

A long rotation of at least four years is recommended. The rotation is more effective if the soil is kept fallow and dry during summer.

Green manuring of infested soil with chopped cabbage leaves heavily reduces the larvae in the soil. The activity of some nematode-trapping fungi, such as *Arthrobotrys oligospora* and *Dactylaria thaumasia*, is increased as a result of green manuring.

Soil fumigation with D-D mixture has been found to be quite efficacious in controlling the nematode population (90% kill) but the remaining larvae flare up in the soil in the presence of susceptible plants. Bhatti *et al.* (1980 a, b) had reported that early sowing of the crop (i.e. early November) and soil application of Aldicarb at the rate of 2 Kg a.i/ha resulted in significant control of the disease. None of the wheat varieties had been found to be resistant to the cereal cyst nematode.

23.10 CITRUS NEMATODE DISEASE

The citrus nematode *Tylenchulus semipenetrans* is invariably present in citrus plantations all over India especially with trees in various stages of decline and dieback.

1. Symptoms

The affected trees are generally stunted with a general reduction in citrus tree growth, leaves, and undersized fruits with little bearing, and exhibit dieback of twigs. The roots of such trees are dark in colour with shortened rootlets, swollen and irregular in appearance.

2. Disease Cycle

The widespread distribution of *T. semipenetrans* in India has been mainly through the infested planting material. Infested soil cause more local and short-distance spread of the free living stages of the nematode. Agricultural implements, animals, humans, wind, and water are common agents in spreading the nematode. It is believed that the nematodes may remain viable in the absence of the host for as long as 10 years in soil.

3. Disease Management

The citrus nematode can be controlled by bare-root dips in hot water at 45°C for 25 minutes or alternatively at 46.7°C for 10 minutes.

Singh and Sitaramaiah (1973) reported that castor cake was highly effective whereas neem, mahua, groundnut, and mustard cakes also reduced the nematode population in the citrus rhizosphere. It has also been reported that irrigation of citrus orchards with sewage water reduces the nematode population in the soil.

Since citrus nematode is easily transmitted to newer sites through the infested seedlings and budded plants, nurseries should never be established near the old citrus orchards. Soils in the nurseries should be properly sterilized before planting.

Use of resistant root stocks or highly tolerant root stocks to nematodes is recommended. Trifoliate orange (*Poncirus trifoliata*) and its hybrids (*Citrumelo*) were highly resistant to citrus nematode.

At IIHR, Bangalore, by crossing Rangpur lime (*Citrus limonia*) with trifoliate orange and its hybrids, viz. CRH3, CRH5 and CRH-41, were quite resistant.

Pasteuria penetrans (bacterium) can be multiplied readily on the citrus nematode. Similarly, inoculation of *Paecilomyces lilacinus* significantly reduced nematode population in soil. Therefore, biological control with the help of these two biocontrol agents is a distinct possibility.

There are several other important nematode diseases, such as the reniform nematode on castor, the root-knot nematode in groundnut and *Heterodera cajani* on sesame amongst the oilseed crops.

Root-knot nematodes constitute another problem in plantation crops, such as cardamom, pepper vine, ginger, and turmeric in the state of Kerala.

Amongst the nematodes affecting tobacco are species of *Meloidogyne, Pratylenchus, Tylenchorhynchus*, *Trichodorus* and *Rotylenchulus* reniforms.

The lesion and burrowing nematodes pose the most serious problems to the coffee areas in South India. *Radopholus similis is* another nematode in banana in Tamil Nadu, Kerala, and some parts of Karnataka.

Tylenchorynchus brassicae has been reported by workers in Aligarh which causes damage to cauliflower in Uttar Pradesh.

23.11 EAR-COCKLE OF WHEAT

The ear-cockle of wheat is a well-known disease caused by *Anguina tritici.* This is known in India as "sehun" disease and is common in different parts of northern India (Uttar Pradesh, Punjab, and western part of Bihar). It is often present in association with the yellow ear rot (tundu or tannan disease) caused by the bacterium *Clavibacter tritici.* In Uttar Pradesh, the disease causes an annual loss of 30%. Apart from loss in yield, the disease has been known to produce toxic effects on humans and animals when quantities of galls are consumed along with the flour.

1. Symptoms

The disease does not produce galls but its effects are visible on the stems, leaves, and floral organs. Affected plants may be dwarfed and their leaves twisted and crinkled which prevents the normal emergence of the younger leaves from within, causing them to be buckled. The infected heads are partially or completely replaced by cockles that are hard, dark brown, or black. These stony structures vary in size from region to region. They are filled with nematode larvae which can be seen by soaking the gall in water and then macerating it.

Severely affected plants may even die. Even if seedling symptoms are absent, plants may bear galls in their ears. The plants show a spreading nature and tend towards more tillering. Early ear formation has also been noticed. Basal enlargement of the stem and formation of galls on awns, glumes, and staminate tissues have also been reported.

2. Causal Organism

The nematode causing this disease is known as *Anguina tritici* (Steinback) Filipjev.

The larvae in the wheat galls have extreme longevity. They can survive up to 28 years under dry conditions and up to 8–9 or even 14 years under moist conditions. In soil, the galls become moist and the larvae break free of the softened walls of the galls. The liberated larvae make their way to the growing point of the wheat plant while it is still near the soil base. These larvae around the growing point feed ectoparasitically and are carried upwards with the lengthening of the culms, until the embryonic flower tissues are formed, when they enter the endoparasitic mode of life by invading these tissues. Plants can become infected with nematodes in the initial stages of growth only, that is, before the seedlings have emerged from the soil. Larvae free from bacteria have been found to produce only the cockle disease. Bacteria are unable to produce the rot independently. However, in combined infection, the nematode larvae are completely killed by the yellow rot bacterium (Gupta, 1966).

As already mentioned, there is a biological relationship of the nematodes with the bacterial ear rot of wheat. Infection of the plant with nematode can occur in the initial stages of plant growth (before the emergence of seedlings out of the soil). The nematode is essential for development of symptoms like basal enlargement of the stem, hoisteng and working of the leaves, formation of gall and development of bacterial yellow rot. Larvae which are devoid of bacterial contamination cause only ear-cockle disease.

3. Disease Management

Since the parasite is introduced into new areas mostly through galls mixed with seeds, proper seed selection is the best method of controlling this disease. The diseased ears should be picked out and burnt.

Healthy seeds can be picked out from a mixed lot by using a sieve that can retain normal plump grains and allow the small round galls to pass through. Infected and healthy seeds can also be separated by being immersed in water or brine solution (20% salt solution). The galls do not sink but float on the surface and can be removed.

In fields where the disease has occurred once, cultivation of wheat should be stopped for two to three years. The disease does not occur on barley and oats. Either of these crops can be grown in such fields.

Nematode infection of wheat plants is influenced by temperature. The incidence of ear cockle is higher with lower temperatures. Hence, crops sown early usually escape infection. In areas where the disease has established itself, it is better to resort to early sowing.

Nemaphos (granular) proved to be most effective when used at the rate of 4.5 kg per acre. Aldicarb or Thionazin can be used for soil treatment.

23.12 SOYBEAN CYST NEMATODE

The most severely affected host is soybean but many other legumes, such as common bean, pigeonpea, black gram, chickpea, mung bean, hyacinth bean, and cowpea are also affected by cyst nematode. Depending on the degree of infestation, it can cause losses varying from slight to complete destruction of the crop. In heavily infested areas, the yield is reduced up to 75 per cent.

1. Symptoms

The main feature of cyst nematode infections is the presence of cysts on the roots and usually the proliferation of roots and production of shallow, bushy, root systems. The adult *Heterodera* female is swollen and flask shaped and is normally found attached to the root. When the female dies, the body wall becomes brown and hardens to a tough cyst, which contains embryonated eggs. The infected second-stage juveniles hatch from the eggs and after leaving the cyst, penetrate suitable host roots and find a feeding site in the stele before developing further. On severely infested plants, the mature, swollen white females are clearly visible covering the surface of the roots. The infected plants show stunted growth and their foliage turns yellow and falls off prematurely. The infected plants bear only a few flowers and a few small seeds. The root system of infected plants appears smaller than that of healthy plants, and infected roots have considerably fewer bacterial nodules than those of healthy plants.

2. Causal Organism

The soybean cyst nematode is *Heterodera glycines*. Other common cyst nematodes and their hosts are *H. avenae* on cereals, *H. schachtii* on sugar beets, crucifers and spinach, *H. tabacum* on tobacco and *H. trifolii* on clover. A cyst nematode severe on potato, tomato, and eggplant is *Globodera rostochiensis*.

3. Disease Cycle

Brown cysts containing viable eggs can survive in the soil for many years.

4. Disease Management

(a) Cultural methods Many cyst nematodes have a limited host range and crop rotation can be successfully employed for their control. Planting the more resistant soybean varieties, which do not allow a quick and excessive build-up of nematode populations, increases the effectiveness of crop rotation. A two or three-year rotation with cotton, maize, or other nonhost plants has been recommended to control *H. glycines* on soybeans.

(b) Chemical Control The infective juveniles of *Heterodera* are well protected in cysts, and chemical control is difficult and expensive. There are a few specific examples of chemical control of cyst nematode on tropical legumes, but the carbamate group of nematicides, namely Aldicarb, Oxamyl, and Carbofuran have given good control of cyst nematode at a rate of 2–11 kg a.i./ha. Soil fumigation of field with Ethylene DiBromide (EDB) temporarily increases plant growth and soybean yield. Even when used, effective and economic control can only be achieved when chemicals are applied in conjunction with crop rotation.

Cysts and eggs of soybean nematode are often seen infected with many fungi such as *Verticillium, Fusarium*, and *Nematophthora*. In the near future, it may be possible to use these fungi as biological control agents against cyst nematodes. Quarantine regulations can be enforced to prevent the parasite from spreading into nematode-free areas by means of contaminated soil, products, and machinery.

23.13 PLANT DISEASES CAUSED BY FLAGELLATE PROTOZOA

Certain trypnosomatid flagellates, belonging to the kingdom Protista, phylum Euglenozoa, order Kinetoplastida, family Trypnosomatidae, have now been known to be parasitic to plants since the beginning of the 19th century. Flagellate protozoa were first found to be associated with plants in 1909 in Mauritius, when Lafont reported that they parasitize the latex-bearing cells (the laticifers) of the laticiferous plants belonging to the family Euphorbiaceae, especially *Euphorbia*.

Protozoa (singular, protozoan), from the Greek *protos* and *zoon*, meaning "first animal" are eukaryotic protists. They occur generally singly and may be distinguished from other eukaryotic protists by their ability to move at some stage of their life cycle and by their lack of cell walls. Organisms in the order Kinetoplastida are grouped together because of the presence of a kinetoplast (an extranuclear region of DNA associated with mitochondrion). It is the members of the family Trypnosomatidae that are known to be parasitic on plants. It may be mentioned that African trypnosomiasis or sleeping sickness is caused by *T. gamibiense*, which is a human disease. To distinguish them from protozoa parasitizing humans, plant parasitizing protozoa have been placed in a separate genus *Phytomonas*. Some of the diseases caused by flagellates are the work done on plant diseases caused by flagellate protozoa. Some of the diseases caused by flagellates are given in Table 23.2.

Table 23.2 Diseases caused by flagellates

Disease	Causal organism
Phloem necrosis of coffee (*Coffea liberica* and *C. arabica*)	*Phytomonas leptovasorum*
Hartrot of coconut palm	*Phytomonas* species
Sudan wilt of (Marchitez) oil palm	*Phytomonas* species
Empty root rot of cassava (*Manihot esculenta*)	*Phytomonas francai*

In addition to these diseases, flagellates have been reported from milkweed (*Asclepias* sp.). *Phytomonas elmassiani* is the pathogen on milkweed. *Phytomonas bancrofti* has been reported from the *Ficus* species.

Although flagellates have been associated with the diseases mentioned above, Koch's rules of pathogenicity have not been provided so far against any of these diseases.

23.14 PHLOEM NECROSIS OF COFFEE

The disease has been reported from Surinam, Guyana, San Salvador, Columbia, and probably Brazil. It has not been reported from India so far.

1. Symptoms

As a result of infection, trees show sparse, yellowing and dropping of the leaves and in advanced cases of infection, only the topmost leaves remain on the branches. The roots begin to die and the trees so infected begin to die. It has also been seen that in the beginning of the dry season, trees wilt and die within 20 to 45 days.

2. Causal Organism

The pathogen is a flagellate, *Phytomonas leptovasorum*. As a result of the pathogen the roots and trunks of the trees internally show multiple divisions of cambial cells and production of a zone of smaller and shorter phloem elements of disorderly structure right next to the wood cylinder.

3. Disease Cycle

The disease is transmitted through root grafts but not through grafts of the aerial parts. The disease spreads in the field from one tree to another and healthy trees often become infected if transplanted into areas where diseased trees have been removed. In nature, the insect vector is the genus *Lincus* (pentatomid insect).

4. Disease Management

In order to control this disease, it is absolutely essential that nurseries should be free from pathogenic flagellate protozoans. Healthy plants should be planted far away from the infected plants.

23.15 HARTROT OF COCONUT PALMS

Hartrot disease has been reported from Surinam since the beginning of the nineteenth century under the names **lethal yellowing** or **bronze-leaf wilt**. It has also been reported from Columbia, Ecuador, and Trinidad.

1. Symptoms

The first symptoms are the yellowing and browning of the tips of the older leaves followed by the younger leaves. Unripe fruits (nuts) of the diseased trees fall off. The recently developing and opening inflorescences are blackened. Petioles of the older leaves break and the spears become necrotic. The root tips of the infected plants start dying. In later stages of infection, the apical region of the crown rots and produces a foul odour.

2. Causal Organism

The pathogen associated with this disease is a species of *Phytomonas*. The flagellates occur in the sieve elements of young leaves and inflorescence of hartrot-infected plants. In later stages of the disease, the flagellates virtually plug the sieve elements.

3. Disease Cycle

The pathogen is transmitted by the insects of the genera *Lincus* and *Ochlerus*.

4. Disease Management

Planting healthy seedlings from nurseries which are free of the pathogen is the first step in protecting plants from infection. Controlling the insect vectors by insecticides will go a long way in managing the disease.

REVIEW QUESTIONS

1. Name some important plant pathogenic nematodes.
2. Name some important nematicides used for controlling plant-nematode diseases.
3. What is the rationale in growing trap crops? How could they be utilized for controlling plant nematodes?
4. Name some important trap crops used to control plant nematodes.
5. What is a syncytium?
6. Give some suitable examples of controlling nematode diseases by means of biological agents.
7. What is soil solarization? How can it be used to control nematode population in soil?
8. What is Integrated Pest Management strategy?
9. Why have some of the earlier recommended nematicides been abandoned now?
10. Name some important diseases caused by flagellates. How will you control them?

MULTIPLE-CHOICE QUESTIONS

1. Which of the following method is used for isolation of nematodes from soil?
   (i) Soil-dilution plate method (ii) Warcup's method
   (iii) Baermann funnel method (iv) All of these
2. Which of the following are the symptoms caused by nematodes?
   (i) Root knots or root galls (ii) Root rots
   (iii) Injured root tip (iv) All of these
3. Which of the following noncharacteristic symptoms are seen in the above ground parts of plants as a result of nematode root infection?
   (i) Discolouration (ii) Necrosis
   (iii) Galls on the leaf and stem (iv) All of these
4. Which one of the following host responses may appear as symptoms as a result of the action of saliva of nematodes?
   (i) Cellular hypertrophy (ii) Cellular hyperplasia
   (iii) Cell necrosis (iv) All of these
5. Which one of the following fungal disease/s increase/s in incidence and severity when the plants are also infected by nematodes?
   (i) Fusarium wilt (ii) Damping off
   (iii) Smuts and rusts (iv) Both (i) and (ii)
6. Which of the following nematicides is used to control nematode population?
   (i) Captan (ii) Kerenite
   (iii) Aldicarb (iv) Chloroneb

7. Which of the following fungus is used as a biocontrol agent in controlling root-knot nematodes?
   (i) *Paecilomyces lilacinus* (ii) *Alternaria solani*
   (iii) *Aspergillus flavus* (iv) *Phytophthora drechsleri*
8. Which one of the following is a nematode-trapping fungus?
   (i) *Entyloma* sp. (ii) *Arthrobotrys* sp.
   (iii) *Cercospora* sp. (iv) *Mucor* sp.
9. Which of the following bacterium is used as a biocontrol agent for the control of *Meloidogyne* spp.?
   (a) *Pasteuria penetrans* (b) *Erwinia amylovora*
   (c) *Agrobacterium radiobacter* (d) *Rhizobium trifolii*
10. Which of the following innovative methods has been found effective in control of root knot of vegetables?
    (i) Use of fumigants (ii) Use of nematicides
    (iii) Soil solarization (iv) Application of NPK fertilizers
11. Which of the following nematodes is responsible for causing Molya disease of wheat and barley?
    (i) *Heterodera avenae* (ii) *Tylenchulus semipenetrans*
    (iii) *Meloidogyne incognita* (iv) *Anguina tritici*
12. Phloem necrosis of coffee is caused by
    (i) bacteria (ii) protozoa
    (iii) fungi (iv) viruses
13. Hartrot of coconut palms is caused by
    (i) *Phytomonas* sp. (ii) *Phytophthora* sp.
    (iii) *Pythium* sp. (iv) *Peronospora* sp.

Answers

1. (iii) 2. (iv) 3. (iv) 4. (iv) 5. (iv) 6. (iii) 7. (i) 8. (ii) 9. (i) 10. (iii) 11. (i) 12. (ii) 13. (i).

24

NON-INFECTIOUS OR ABIOTIC DISEASES

24.1 NON-INFECTIOUS OR ABIOTIC DISEASES OF PLANTS

Non-infectious or abiotic diseases have been recognized as a major source of crop failure ever since humans resorted to cultivation. For a long time such failures were believed to be largely due to unfavourable climatic and edaphic conditions. During recent years, environmental pollution and debris from nuclear weapons and devices have added greatly to the problem of disorders of this type. Major disorders of nonpathogenic origin are given in Table 24.1. Agrios (1997, 2005) has discussed these diseases and has provided useful information.

Table 24.1 Major non-infectious or abiotic disorders of plants

I. Disorders resulting from unfavourable soil environment	
(a) *Nutritional imbalance:*	Deficiency of essential elements, toxicity of essential and non-essential elements, multiple and induced deficiencies
(b) *Adverse water relation:*	Drought, waterlogging, impeded aeration, etc.
(c) *Adverse physio-chemical:*	Salinity, alkalinity, acidity, calcareousness, etc.
II. Disorders resulting from climatic stress*	
(a) *Temperature:*	High and low temperatures
(b) *Light:*	Duration and quality of light
III. Disorders caused by atmospheric impurities (environmental pollution)	
(a) *Harmful gases:*	Sulfur dioxide, hydrogen fluoride, ozone, ethylene, chlorine, nitrogen dioxide, other harmful gases and dusts, smog (peroxyacetyl nitrate), fly ash, brick kiln fumes, etc.
(b) *Pesticides:*	Herbicides, insecticides, nematicides, rodenticides, molluscacides, etc.
(c) *Ionization radiation:*	Naturally occurring radio-nucleides and radioactive fall-out.

*Excluding injury due to climatic factors—lightning, high wind, heavy rain, hail, sleet, ice, mechanical injury, etc.

24.2 TIP BURN OF PADDY OR PANSUKH OF RICE

The names *tip burn*, *dakhina*, *chatra*, *ukra*, *kadamara*, or *pansukh*, have been given to leaf-drying disease of rice known in India for quite some time.

1. Symptoms

The first symptom of the disease is the drying of the leaves. They become less lustrous to bronzing and later become scorched at the top. The drying progresses down along the margin resulting in the drying of the complete leaf. In advanced stages, the entire plant dries and wilts. The leaves and stalks turn brown. This spreads rapidly in the field.

2. Causal Factor

The cause of the disease is not properly understood. The disease is common in wet soils or in poorly drained fields where the water is allowed to remain stagnant for more than a fortnight. This is because water reduces the supply of oxygen to the roots.

3. Disease Management

The use of ammonium sulfate had been recommended to control this disease as the disease is due to the lack of available nitrogen. According to Sinha and Singh (1956), green manuring reduces the incidence of tip burn and increases the yield.

24.3 BLOSSOM END ROT OF TOMATO

This disease of tomato fruit is a nutritional disorder believed to be caused by a derangement of the water balance between the leaves and fruits during the development of the plant. A lesion at the blossom end of the fruit, which occurs, characterizes it while it is ripening on the vine. A water-soaked spot appears at the point of attachment of the senescent petals. It enlarges quite rapidly until it is about 2 cm or more in diameter, and at the same time the diseased tissue shrinks, depressing the surface of the lesion. The colour of the fruit becomes darker and secondary organisms generally invade the tissues. The diseased portion of the fruit becomes delimited as a sunken, leathery, dark-coloured area. If tomatoes are grown at relatively constant low-moisture content, the disease does not develop. If plants are grown at relatively high-moisture content, succulent growth is very rapid, the rate of transpiration is very high, and disease incidence is greater. Sandy soils are more favourable to blossom end rot.

Disease Management

The most obvious method of controlling blossom end rot is to maintain an equable environment and to keep the soil uniformly moist. This should be taken care of, especially when the fruit begins to develop.

24.4 HEAT CANKER OF LINSEED

It is a nonparasitic disease caused by high soil temperatures near the soil surface.

1. Symptoms

Heat injury of the cortical tissues of the hypocotyl and stem near the soil line is common in flax and other succulent herbaceous plants. If the soils are dark coloured and the surface temperature becomes high enough to kill the cells of young cortical tissues before the plants are larger enough to shade the surface, the cortical tissues collapse, which results in the death of the young seedlings. As a result of the collapse of tissues, sunken brown lesions develop on the stems. There is an enlargement of the cortical tissues above the cankers.

2. Disease Management

High temperatures can be avoided by early planting. The rows should be drilled from the north to the south. This results in maximum shedding. Damage from this disease can be reduced by higher rates of seeding using a nurse crop, mulching the soil surface, and irrigation.

24.5 BLACK HEART OF POTATO

Black heart is an important storage, transit, and market disease of potatoes as a result of poor oxygen relation. The conditions under which this disease develops vary greatly. It is known that a number of rather distinct sets of environments produce practically the same result or cause this disease, such as high temperature in transit, poor ventilation in storage and high soil temperature during the growing season. Bartholomew (1913) revealed the fundamental nature of black heart and made a pathological study of the black heart of potato tubers (Barholomew, 1915). He found that he could produce black heart in tubers by exposing them to high soil temperatures for 24–48 hours in an oven heated to 40–42°C. Mann and Joshi (1920) made a chemical study of heart rot or black heart in potato tubers. They could produce this disease by first coating the tubers with collodion or paraffin or replacing the air in a container with carbon dioxide or nitrogen. Placing tubers in sealed containers can also produce black heart. This shows that oxygen is very necessary to prevent the occurrence of black heart on potato tubers.

The centers of potato tubers attacked by black heart show black portions when sliced open, or after being sliced open, the potato tubers undergo a fairly rapid change in colour from white or pink to brown to dark brown or black.

Disease Management

The disease can be significantly controlled by avoiding very high or low temperatures and by properly storing potatoes in well-ventilated storage godowns, and the bags of tubers should not be piled very high upon each. In the fields, if the temperature goes up very high, the fields should be irrigated.

24.6 MANGO NECROSIS OR BLACK TIP

The disease was first reported from Bihar and is now well known in Bihar, Uttar Pradesh, West Bengal, and Punjab. It is generally found near the brick kilns in Uttar Pradesh, Bihar, West Bengal, Punjab, and Haryana in northern India.

1. Symptoms

The necrosis of one third or half of the fruit characterizes black tip of mango. The fruit becomes abnormally long at the tip. Premature ripening of the affected fruit occurs. The most susceptible varieties are 'Daseri', 'Safeda', and 'Langra'. The disease starts as a small, circular, pale patch at the tip of the mango fruit and soon enlarges, becomes black, and covers the tip of the mango completely. The outer skin of the affected portion becomes flat and hard. Internally, in the vessels of the mesocarp, are found certain brown deposits, which have been provisionally identified as belonging to the group of tannins, flavotannins, flavophenols, or their derivatives.

2. Causal Factor

Previously it was thought that brick-kiln fumes are a direct cause of the disease. It has been suggested that the smoke pollutes the air with toxic gases such as sulfur dioxide and that these gases causes the necrosis of the fruit tissues. Later on, Das Gupta and Sen (1960) reported that there is a deficiency of boron in the mango fruit exposed to brick-kiln fumes. The disease was, however, not attributed to a deficiency of boron in the soil but to disturbances in boron metabolism due to the interaction of brick-kiln fumes with certain cell metabolites of mango.

3. Disease Management

The disease can be controlled by spraying borax at 2.7 kg/100 gallons three times in the season, that is pre-flowering, flowering, and immediately after fruit setting

24.7 KHAIRA DISEASE OF RICE (ZINC DEFICIENCY DISEASE OF RICE)

Dr Y L Nene and his associates first discovered Khaira disease of rice in the Tarai region of the then Uttar Pradesh (now Uttarakhand). The disease has also been reported from Mysore, Haryana, and Punjab. Khaira disease of rice is an excellent example of a disease due to non-availability of zinc to the plants in an otherwise normal soil.

1. Symptoms

The disease symptoms usually appear 10–15 days after the transplantation of the paddy crop. This period also coincides with the decomposition of the previous year's stubbles in the flooded fields. The first symptoms seen are the chlorosis of the leaves starting from the base of the leaves. After this, a large number of small brown or bronze-coloured spots appear on the leaf surface which later on coalesce to become bigger in size and cover the whole leaf surface of the older leaves. The leaves become brown or bronze coloured and ultimately dry and die. The infected plants show stunted growth. The underground portion of the plants also shows symptoms. The roots are poorly developed and usually the main roots turn brown. In severe cases, plants fail to grow further and produce ears with few grains. Sometimes natural recovery may take place six weeks after transplantation.

2. Disease Management

The disease can be effectively managed by two sprays or a mixture of zinc sulfate and slaked lime (2 kg zinc sulfate +1 kg of slaked lime) in 400 liters of water per acre of the plot. Care should be taken that the first spray is given as soon as the disease is first seen in the field. Another spray should be given after 10 days of the first spray.

24.8 ZINC-DEFICIENCY DISEASE OF CITRUS

This disease is very common in many citrus gardens in India. It is also known as mottling or foliocellosis.

1. Symptoms

The main symptom is the production of alternative chlorotic regions and dark green patches in the leaf. The portions near the leaf veins are usually brown and the rest of the leaf is chlorotic. As a result of continual deficiency of zinc in the soil, there is severe stunting of the plant, reduction in leaf size, narrowing of the leaf, and severe chlorosis. The fruit size is also reduced. The plants become week and are exposed to root infections and this slowly results in their death.

2. Disease Management

The application of zinc sulfate to the soil gives a very slow response. Spraying zinc sulfate and lime mixture (2.268 kg zinc sulfate and 2.268 kg quicklime in 100 gallons of water) is recommended for a quick response.

REVIEW QUESTIONS

1. What are deficiency diseases?
2. How is proximity to brick kilns responsible for mango necrosis or black tip?
3. What is khaira disease of paddy? What is it due to? Suggest control measures.

MULTIPLE-CHOICE QUESTIONS

1. The main cause of non-infectious disease of plants is due to
   (i) deficiency of essential elements (ii) nonpathogenic microbes
   (iii) high and low temperature (iv) both (i) and (ii)
2. The tip burn of paddy is due to the deficiency of
   (i) lack of available nitrogen (ii) naturally occurring radio-nucleides
   (iii) harmful gases (iv) lack of available phosphorus
3. Heat canker of linseed is caused due to
   (i) duration and quality of light (ii) high soil temperature
   (iii) waterlogging (iv) harmful gases
4. Black heart of potato is caused due to
   (i) poor oxygen supply (ii) radiation
   (iii) herbicides (iv) all of these
5. Khaira disease of rice is caused due to the deficiency of
   (i) nitrogen (ii) phosphorus
   (iii) zinc (iv) boron

Answers

1. (iv) 2. (i) 3. (ii) 4. (i) 5. (iii)

25 APPLICATION OF BIOTECHNOLOGY IN PLANT PATHOLOGY

According to Agrios (1988, 2005), biotechnology is defined as the manipulation, genetic modification, and multiplication of any living organism through novel techniques such as tissue culture and genetic engineering, resulting in the production of improved or new organisms and products that can be used in a variety of ways. The term *genetic engineering* can be broadly defined to include all techniques designed to manipulate the genetic constitution of organisms. Plant biotechnology impinges or helps plant pathology in many ways.

1. It helps us obtain pathogen-free mother plants through rapid clonal propagation.
2. New plants to which genes have been incorporated through genetic engineering are likely to show instability towards environmental conditions and towards the pathogenic microflora of their habitats. Here, plant pathology plays an important part.
3. The main vehicle for transferring genes from donor to recipient plants are plant pathogens, particularly the bacterium *Agrobacterium tumefaciens* and the cauliflower mosaic virus.
4. It helps control plant diseases by inserting resistance genes into plants by genetic engineering techniques.
5. The study of plant genes for resistance to disease and of pathogen genes for virulence to pathogens has already added considerably by genetic-engineering techniques.

25.1 THE USE OF TISSUE-CULTURE TECHNIQUES OF IMPORTANCE TO PLANT PATHOLOGY

Almost all tissue-culture techniques are used in plant pathology. Some of the important tissue-culture techniques of importance to plant pathology are briefly discussed.

25.1.1 Callus Culture

Generalized protocol for the screening of callus tissue for the isolation of toxin-resistant plants is given on next page (Jones, 1990).

Callus tissue
(derived from explants)

↓ Culture medium and sub-lethal concentration of toxin preparation

Selection for resistance

↓ Subculture medium + toxin

Selection for resistance

↓ Subculture medium + toxin (concentration lethal to parent tissue)

Regeneration of surviving callus

↓

Screen plants for resistance to
toxin and pathogen

Fig. 25.1 *Generalized protocol for the screening of callus tissue for the isolation of toxin-resistant plant*

This technique is commonly used to study the following:

1. The behaviour of pathogens, particularly biotrophs such as rusts, powdery mildews, viruses, nematodes, and so on. This potential of callus was first recognized by Morel (1944) who infected callus tissue from vine (*Vitis vinifera*) with zoospores of the downy mildew fungus, *Plasmopara viticola*. Since then methods have been devised for growing a number of other biotrophs or symbionts (Mycorrhiza) in callus/organ culture.
2. The effect of toxins secreted by pathogens and phytoalexins produced as defense mechanism and compared to the reaction of susceptible varieties. The plantlets that survive the toxin are apparently resistant to the particular pathogen and are, therefore, selected for incorporation into the breeding programmes. Phytoalexins have been detected in the tissue cultures derived from a number of plant species. Examples are pisatin from *Pisum sativum* (Bailey, 1970), glyceollin from *Glycine max* (Ebel et al., 1976), phaseollin, phaseollidin, and phaseolinisoflavan from *Phaseolus vulgaris* (Dixon and Fuller, 1978).
3. These infected calli are used for *in vitro* testing of chemicals such as fungicides and antibiotics used for control of pathogenic microorganisms. Table 25.1 lists plants of some crops that have been regenerated from fungal toxin-resistant tissue (callus) culture.

Table 25.1 *Plants of some crops regenerated from fungal toxin-resistant cultures*

	Crop	Toxin obtained from
1.	Alfalfa	*Fusarium oxysporum* f.sp. *medicaginis*
2.	Maize	*Helminthosporium maydis*
3.	Oats	*H. victoriae*
4.	Onion	*Pyrenochaeta terrestris*
5.	Potato	*Phytophthora infestans, F. oxysporum* f. sp. *solani*
6.	Rape	*Phoma lingam*
7.	Tomato	*Alternaria solani, Phytophthora infestans*

25.1.2 Apical or Meristem Tissue Culture

Morel and Martin (1952) published an important paper in which they excised shoot apical meristems of infected Dahlia variety and grew them *in vitro* tissue-culture medium. This differentiation of the meristem led to the production of plantlets which upon transferring to soil gave rise to adult plants which were then multiplied. Most of such plants are found free of the virus, because majority of viruses are unable to attack or reach the apical and axillary meristem tissues of systemically infected plants. If meristem tissue-culture technique is combined with thermotherapy, the combinations are very useful. The technique has been used by a number of workers for curing infected valuable cultivars of a wide variety of plants. Most vegetatively propagated plants and many perennial plants are infected by viruses which remain latent in the plants without expressing major visible symptoms. Hence, meristem tip-culture technique is a very useful and valuable asset in cultivating vegetatively propagated plants.

Explants (from shoot, root tip, apical meristem, seedlings)

↓

Surface sterilized (sodium hypochlorite 2%, 10–15 min)

↓

Wash repeatedly in sterile distilled water

↓

Place in culture media containing growth regulators (auxins and cytokinin)

↓

Incubation under light, 23°C

↓

Numerous shoots or roots or both produced (callus culture)

↓

Subculture by dividing single mother culture into several daughter cultures

↓

Complete plants

↓

Grow in greenhouse or in the field

Fig. 25.2 Protocol for the apical or meristem tissue-culture technique

Table 25.2 Species which have been freed of virus by tissue culture

S. No.	Host	Viruses/viroid eradicated (heat treatment if applied)
1.	Garlic	Mosaic
2.	Cauliflower	CaMV, Turnip Mosaic Virus (TuMV)
3.	*Chrysanthemum morifolium*	Virus-B, vein mottle (35–38°C, 4–37 weeks)
4.	Taro	Dasheen mosaic
5.	*Dianthus barbatus*	Ringspot, mottle, vein mottle
6.	Strawberry	Vein chlorosis, vein banding (3–40°C, 4–7 weeks)
7.	Sweet potato	Internal cork, mosaic, mottle (27-28°C)
8.	Banana	CMV (35–3°C, 14–15 weeks).
9.	Tobacco	Tobacco ring spot, alfalfa mosaic (32–40°C)
10.	Potato	Potato virus A (37°C, 2–4 weeks) spindle tuber (33–38°C, 4–8 weeks)

Some Advantages of this Technique

1. Some completely lost, valuable ornamental and vegetable varieties have been saved.
2. Virus-free clones of a particular variety can be built up.
3. It helps in production of disease-free clones and storage of valuable germ-plasm.

25.1.3 Protoplast Fusion

Disease resistance in a breeding programme may come either from closely related species or from more distantly related species. Problems are generally encountered if an effort is made in crossing distantly related species. Protoplast fusion is one of the methods that can be used to circumvent problems in introducing genes for resistance. By this method, factors that contribute to crossing barriers between species can be avoided, and viable hybrids (Cybrids) have been recovered, even between distantly related species (Harms, 1985). Some of the disease-resistant plants produced from protoplast fusion are given in Table 25.3.

Table 25.3 Disease-resistant plants produced from protoplast fusion (After Eck and Smith, 1996).

Species used for fusion	Disease	Reference
Lactuca sativa and wild *Lactuca* sp.	Downy mildew (*Bremia lactucae*)	Maisonneuve *et al.*, 1995
Brassica oleracea and *Raphanus sativus*	Club root (*Plasmodiophora brassicae*)	Hagimori *et al.*, 1992
Brassica napus and *Brassica nigra*	Black leg *(Phoma lingam)* club root *(Plasmodiophora brassicae)*	Gerdemann-Knorck *et al.*, 1994
Brassica oleracea and *Brassica napus*	Black rot (*Xanthomonas campestris* pv. *campestris*)	Hansen and Earle, 1995
Solanum brevidens and *Solanum tuberosum*	Bacterial soft rot (*Erwinia* spp.)	Austin *et al.*, 1988

An outline of the protocol for the isolation of plant protoplast from intact leaf tissue or from cell-suspension culture is given in Fig. 25.3.

Tissue for protoplast liberation must be surface sterilized and chopped to enable enzyme penetration or the lower epidermis may be removed to improve enzyme access. Enzyme mixtures normally make use of cellulase and pectinase activities to macerate wall polymers. Protoplasts are released into an osmotically stabilized solution to prevent lysis. It is harvested efficiently by gentle centrifugation. Washing in fresh stabilizer solution removes lytic enzymes. The protoplast may be separated from contaminating wall fragments by filtering or by layering in sucrose medium. The protoplast from cell suspension can also be liberated using the same methods.

For subsequent regeneration, protoplasts are placed on agar medium containing stabilizer solution, a carbon source, and essential nutrients.

25.1.4 Chemically Induced Fusion

Isolated protoplasts are sticky, tend to aggregate in suspension and show fusion spontaneously during incubation. Chemicals tend to increase the fusion frequency. Fusion can occur in the presence of high Ca^{+2} and high pH (9–10) but the commonly used chemical (Fusogen) is polyethylene glycol (PEG).

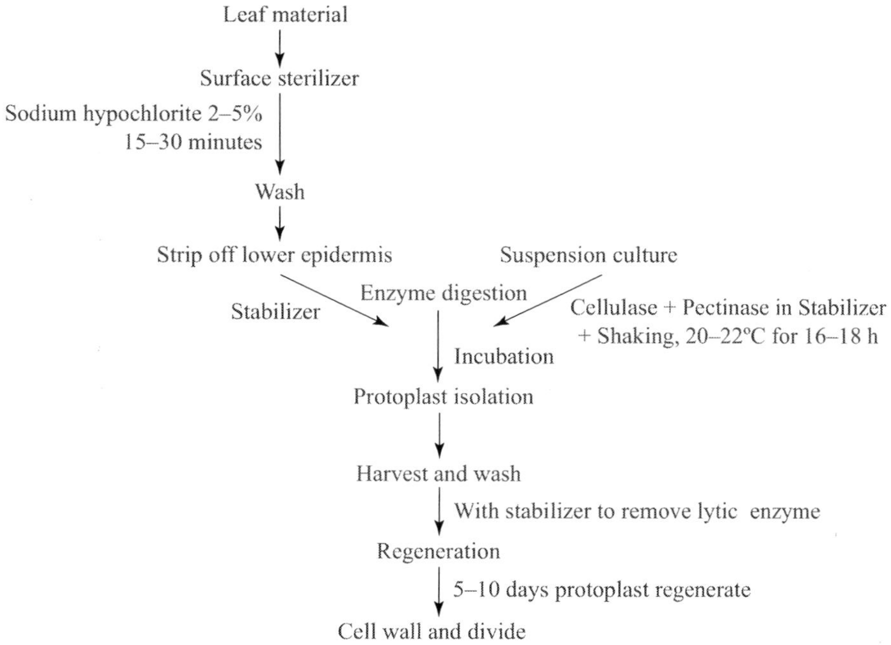

Fig. 25.3 Protocol for the isolation of plant protoplast from intact leaf tissue or from cell-suspension culture

Procedure for the chemically induced fusion of plant protoplast

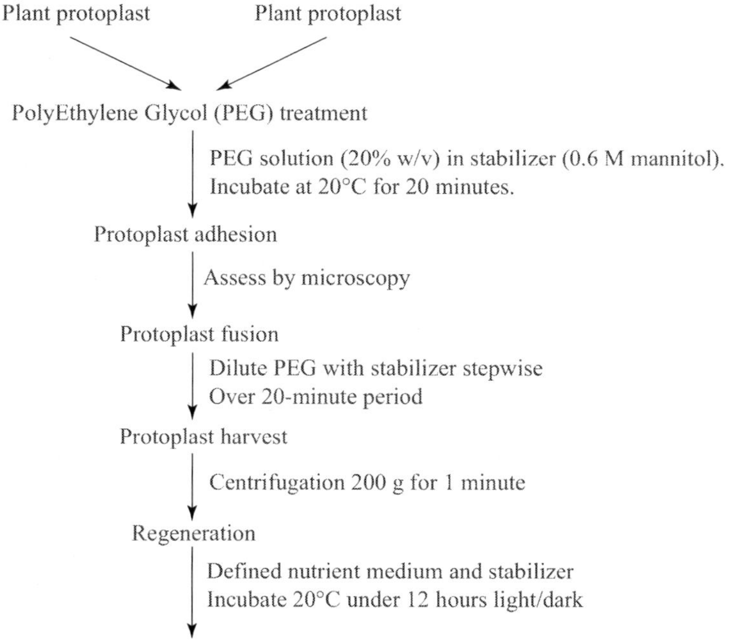

Fig. 25.4 Protocol for the chemically induced fusion of plant protoplast

Due to addition of PEG, there is adhesion of protoplast to their neighbours which can be assessed by microscope. Subsequent dilution of the stabilized PEG, either stepwise or at once, results in fusion and mixing of the cytoplasm. PEG causes slight dehydration of the protoplasts and crinkling of the membrane. The level of fusion is usually 1–10% as chemical fusion agents are toxic and, therefore, damaging to the cell (Fig. 25.4). Higher fusion has been reported for electrically stimulated fusion events (about 80%), than for chemically induced fusion (Zimmerman and Scheurich, 1981).

The most important use of protoplast fusion is the transfer of genes between sexually incompatible plants. In plant pathology, protoplast fusion has been used for transmitting disease-resistant genes into sexually incompatible hosts. For example, wild potato species *Solanum brevidens* is resistant to potato leaf roll virus but is sexually incompatible with cultivated potato *S. tuberosum*. Here, protoplasts of *S. brevidens* were fused with protoplasts of a commercial and agronomically selected variety of potato (*S. tuberosum*) leading to regeneration of somatic hybrid potato plants which were resistant to potato leaf-roll virus (Austin *et al.* 1988, Helgeson *et al.*, 1986).

It has been possible to inoculate the protoplast of plant with viruses and study their replication and physiology. The bacterium *Agrobacterium tumefaciens* or its modified Ti plasmid and the double-stranded DNA virus, cauliflower mosaic virus have been used to introduce foreign genetic material into plant cells. At present, this is the most likely method for introducing genes controlling disease resistance into important agronomic plants.

25.2 SELECTION FOR DISEASE RESISTANCE

In vitro selection has a distinct advantage over other selection systems since it allows significant saving of space, time, and money. For plant diseases that cause damage through toxins, cell selection for toxin resistance in cultures and regeneration of plants from descendants of the selected cell lines can give disease-resistant genotype. For example, disease-resistant crop plants have been produced through *in vitro* selection in potato against *Phytophthora infestans* (late blight of potato, *Nicotiana tabacum* in tobacco) against *Pseudomonas tabaci*. At the biotechnology centre, IARI, New Delhi, plants resistant to toxin produced by *Alternaria brassicae*, a blight-causing organism, have been isolated. Nagarathna *et al.* (1993) have selected resistance from the regenerates of pearl millet for downy mildew.

25.3 RECOMBINANT DNA TECHNOLOGY

Advances in molecular biology have opened up possibilities of identifying and isolating any gene of an organism, and mobilizing and expressing it in a different organism of one's choice.

1. Engineering Plants for Resistance to Disease

A notable success has been made with regard to viral diseases following use of r-DNA technology. For example, a major achievement has been the transfer and expression of coat protein genes of Tobacco Mosaic Virus (TMV) and Alfalfa Mosaic Virus (AMV) in tobacco, resulting in protection against or delay of disease development in the transgenic plants (Beachy *et al.*, 1990). The purpose of introducing coat protein genes to give resistance against the virus is that the multiplication of infecting viral RNA is somehow checked by the coat protein synthesized in the plant cells. With the help of r-DNA technology, the chitinase gene of *Serratia marcescens* has been introduced into tobacco, potato, lettuce, and sugar beet. Engineered plants synthesize chitinase which breaks down the fungal cell wall and this kills the soil-borne pathogen, *Rhizoctonia solani* (Schell and Vasil, 1989).

2. *Engineering Plants for Resistance to Pests*

The best way of insect control has been the use of insecticides. These insecticides were effective but proved to be environmental hazards and forced development of resistant strains of insects.

There are genes in bacteria (*Bacillus thuringiensis*) that encode insecticidal proteins. *B. thuringiensis* strains toxic to dipteran, lepidopteran, and coleopteran insects have been identified and the insecticidal protein gene cloned. Using Ti plasmid vectors of *Agrobacterium tumefaciens*, the gene encoding the insecticidal protein has been transferred to tobacco, potato, tomato, rice, and corn. Such transgenic plants incorporate resistance to specific insects that feed on these crops.

Another way of producing insect-resistant transgenic plants is to introduce into the plant, genes that inhibit enzymes responsible for the breakdown of protein (protease inhibitors). Transgenic plants with cowpea (*Vigna sinensis*) protease-inhibitor genes have shown resistance to tobacco budworm, corn earworm, army worm, and hornworm of tomato and tobacco.

25.4 THE USE OF MONOCLONAL ANTIBODIES IN PLANT PATHOLOGY

The hybridoma technique was developed by Georges Kohler and Cesar Milstein in 1975 at the Medical Residential Council Laboratory in Cambridge, England. This technique was never patented but its commercial applications were recognized immediately. This discovery led to the production of monoclonal antibodies. Kohler and Milstein were awarded the Nobel Prize in 1984 in Medicine. In this technique, there is the fusion of myeloma cells (cancer cells) with antibody-producing white blood cells (β lymphocytes). The resulting hybrid cell is called a hybridoma.

In the last few years, techniques have been developed to produce large quantities of identical antibodies. These new antibody-forming hybrid cells, hybridomas, can now be grown in culture indefinitely. Each hybridoma clone produces only one type of antibody. But via selection techniques the clone that produces the desired antibody can be chosen. Monoclonal antibodies can be obtained from the liquid of hybridoma cultures and can be used to detect, identify, and measure the antigens that induced their production. Monoclonal antibodies are, however, very specific and may not detect strains of the same virus. It is for this reason that mixtures of several monoclonal antibodies are often used in the detection of viruses and in screening tests.

A useful serological technique called **ELISA** developed in the 1970s has been very widely used by all types of pathologists. It has increased the ability of plant pathologists to detect and study plant viruses and other plant pathogens.

25.4.1 Technique for Production of Monoclonal Antibodies

Rats are immunized by injecting the selected antigen either subcutaneously or into the peritoneal cavity. A few days later, the spleen is removed from the rat. The antibody-synthesizing cells from the spleen are then mixed with fast-growing rat cancer cells, called myelomas, in the presence of polyethylene glycol (PEG).The result is a hybridoma—fusion of an antibody-making cell and a cancer cell. All the hybridoma cells that occurred after a fusion were grown in selective culture medium, and then a mixture of antibodies would be released. Individual hybridoma cells are grown as a clone. In this way, only one type of antibody will be secreted by all the cells in a clone. After the desired hybridoma cell lines have been identified and cloned, culture may be expanded and clones may be tested for the desired antibody. Once the cells have been cloned, either selected clones are cultured and frozen or the hybridoma cell may be injected into the peritoneal cavity of the rat, to produce ascitic fluid. From the ascitic fluid, monoclonal antibodies are purified. The stored and frozen cells can be used at different times of the year and in different laboratories. This eliminates variation between laboratories since the same cells can be used to produce antibodies at different times in the same laboratories or in different laboratories. Enzyme-Linked ImmunoSorbent Assay (ELISA) makes it possible to screen a number of organisms.

Hybridoma Technology

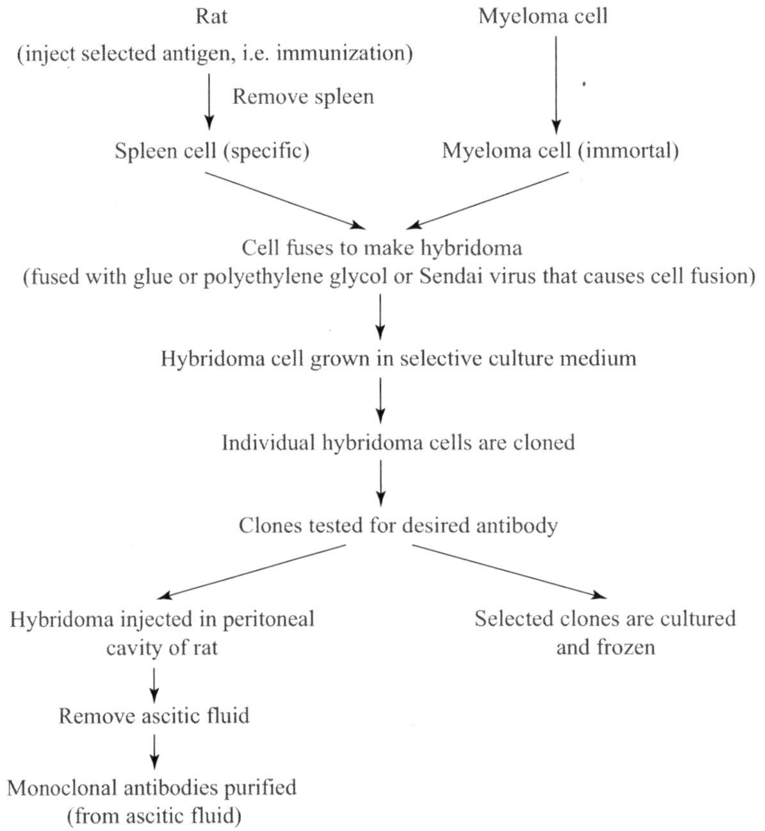

Fig. 25.5 *How monoclonal antibodies are made*

25.4.2 Uses and Applications of Monoclonal Antibodies

1. The most effective application of monoclonal antibodies has been with plant pathogenic viruses. This technique has been used for accurate identification of viruses, for tracing the viruses during epidemiological studies, to distinguish between virus strains, and in isolating and purification of viruses.
2. These procedures are now being used to study bacteria and fungi and will probably be useful in investigations of various populations of rhizosphere organisms. This technology is also very helpful in attempts to identify bacteria in mixed populations. Applications of this technology to fungi are in its infancy, but the potential is great.

25.5 SOMACLONAL VARIANTS AS A SOURCE OF DISEASE RESISTANCE

Plants regenerated from the callus, single cell, or protoplast-derived plants, often show one or more different characteristics than other or parent plants. Plants derived from any form of cell culture are called **somaclones** and the variations shown by such plants, **somaclonal variation**. These variations are due to the genetic instabilities displayed by nonmeristematic tissue explants, cells, or protoplasts once they are removed from the stabilizing environment of the intact plant and cultured *in vitro*. Somaclonal variations are widespread being found not only in asexually propagated species but also in seed-propagated, self-fertilizing species (Larkin and Scowcroft, 1981). It is a useful tool for incorporating additional traits into existing plant varieties.

Initial studies with sugarcane suggested that the clones with disease resistance could be regenerated from the callus obtained from susceptible plants. Many plants resistant to different diseases have been regenerated from susceptible parents in many crops (Table 25.4).

Table 25.4 *Some examples of disease-resistant somaclonal variants in plant species*

	Plant	Disease/Pathogen	Reference
1.	Potato	Late blight, early blight	Matern *et al.* 1978, Behnke, 1980; Shepard *et al.* 1980
2.	Sugar cane	Fizi virus, downy mildew, smut	Heinz *et al.,* 1977; Larkin and Scowcroft, 1981
3.	Maize	Southern corn-leaf blight	Brettel and Ingram, 1979
4.	Tobacco	TMV, potato virus Y, wildfire due to *Pseudomonas tabaci, Alternaria alternata* leaf spots and chlorosis.	Burk & Chaplin, 1978; Thautony *et al.*, 1983
5.	Rapeseed	Black leg	Sachristan, 1982
6.	Pearl millet	Green ear	Nagrathna *et al.*, 1993

In vitro Screening for Disease Resistance

It is an obligatory step in crop-improvement programmes using tissue-culture methods that offer several advantages over conventional glasshouse and field-screening methods with the intact plant. It is quick, and consumes less space and energy. It has many advantages—growing a large number of single cells or cell clumps in a small space, exposing them to different mutagens, diploidization of chromosomes and regenerating plant populations from resistant material. Here, a callus supporting poor or no growth towards a pathogen's toxin(s) is characterized as resistant *in vitro*.

One of the primary advantages of tissue-culture techniques is that environmental conditions can be controlled precisely. Plant protoplasts can be used to study penetration and replication of plant viruses, screening of antiviral chemicals, cell organelles, symptomatology, and maintenance of virus culture. Explants can be used for screening antiviral compounds. Tissue-culture techniques can be used to study host-pathogen interaction for the production of pathogen-free clones, for *in vitro* clonal propagation and for generation of disease-resistant plants. Phytoalexins may be used as selection agents and various procedures can be used to select disease resistance, using specialized tissue-culture procedures. The use of molecular genetic engineering involving isolation and transfer of genetic material offers exciting possibilities. Tissue-culture techniques provide a means to accelerate crop improvement and the production of disease-resistant crop plants.

25.6 INNOVATIVE TOOLS FOR DETECTION AND DIAGNOSIS OF PLANT PATHOGENIC MICROBES

Plant pathogenic microbes are responsible for increasing economic losses worldwide. They cause a large number of symptoms in most cultivated plants which can be affected in their different parts with various agronomic impacts. Viruses and bacteria cause plant diseases that are difficult to control because of the lack of efficient products for chemical treatment under field conditions. As many pathogenic viruses and bacteria remain latent in the plant material and in very low numbers, methods of high sensitivity and reliability are required. The molecular techniques available have evolved significantly in the last few years to achieve rapid and reliable detection of pathogens. Currently, automation and electronic data management are vital to increase the productivity and efficiency of routine analysis for detection of plant pathogenic bacteria and viruses. Several genomes from causal agents of plant diseases, both viral and bacterial, have been completely sequenced and more are underway. In the following sections, a number of important developments and improvements in serological and molecular techniques are briefly described.

25.6.1 Monoclonal and Recombinant Antibodies

The specifity of detection of viruses and bacteria by well-known serological techniques such as IF (ImmunoFluorescence) and ELISA (Enzyme-Linked ImmunoSorbent Assay) has improved greatly with the use of specific monoclonal and recombinant antibodies. Both allow the selection of specific target epitopes to avoid false positives. Specificity problems are frequent when analyzing bacteria in plant material, water, and soil due to large quantities of other microorganisms present in the sample and to the difficulties in obtaining polyclonal antibodies specific for the target pathogen without cross reaction with other microorganisms. Commercial monoclonal antibodies for the detection of plant pathogenic viruses are now available from several companies. Several companies have developed a wide range of kits for rapid and/or *in situ* bacteria detection using polyclonal or monoclonal antibodies based on slide agglutination, IF, and ELISA. The kits based on agglutination or ELISA are appropriate only for diagnosing plants already showing symptoms. In plant pathology, serological detection has been widely used and some recombinant constructs have been produced but only a few of them have been applied to routine ELISA tests.

25.6.2 Enrichment ELISA Protocols

Combining ELISA which uses specific monoclonal antibodies with a prior enrichment step greatly improves the sensitivity of detection of bacterial pathogen. This methodology has been developed because of the low sensitivity of ELISA for bacterial detection using specific monoclonal antibodies and the need to improve this sensitivity, especially to detect latent infection of quarantine bacteria.

25.6.3 On-site Testing: Tissue-Print ELISA and Lateral-Flow Devices

Simple commercial methods of rapid detection are required for testing large number of samples by non-experienced technicians. For this purpose, tissue-print ELISA and lateral-flow devices have been designed for several plant bacteria and viruses. Although detection specificity is very high when using the appropriate monoclonal antibodies, the sensitivity of these methods, although good for virus detection, is relatively low for bacteria, and they are more appropriate for analyzing plants with symptoms. In some cases, tissue-print ELISA has demonstrated potential in, for example, CTV (Citrus Tristeza Virus) detection in plant tissues imprinted on nitrocellulose membranes. Lateral-flow devices use specific antibodies and the tests give results in a couple of minutes, but such kits are available only for a few viruses and bacteria.

25.6.4 Flow Cytometry

It is a technique for rapid identification of cells or other particles as they pass individually through a sensor in a liquid stream. Bacterial cells are identified by fluorescent dyes conjugated to specific antibodies and detected electronically using a fluorescens-activated cell sorter, which measures several cellular parameters based on light scatter and fluorescence. Flow cytometry has excellent potential as a research tool and possibly for routine use in seed-health testing and other fields. Several parameters can be analyzed simultaneously including total particle count, distinction between living and dead cells, and differentiation of target and nontarget bacterial populations associated with the seeds or other plant material. This technique has also been adapted to the analysis of viability, metabolic state, and antigenic markers of bacteria.

25.6.5 Fluorescence *In Situ* Hybridization (FISH)

It is a technique applied for bacterial detection that combines the simplicity of microscopic observations and the specificity of hybridization. Its use in detection of plant pathogenic bacteria is recent and is dependent on the hybridization of DNA probes to species-specific regions of bacterial ribosomes. They are particularly suitable as diagnostic targets because ribosomal RNA contains functional sequences that are common to all species but also sequences that are very specific to individual species and FISH only needs to recognize this specific information.

25.6.6 Polymerase Chain Reaction (PCR)

PCR is a very rapid and sensitive method for the molecular diagnostics or plant pathogens which are otherwise difficult to identify morphologically. Pathogen detection by using PCR is mainly dependent on the design of primers. Specific, semispecific, and arbitrary primers can be used for PCR. However, DNA sequences of plants and other organisms are also amplified with random primers, which lead to ambiguous results. Thus, plant pathogens must be isolated from the host and purified before DNA extraction. Hence, RAPD (Random Amplified Polymorphic DNA) analysis may not be useful for the direct detection of plant pathogens in the infected tissues. However longer and species-specific PCR primers allow detection of target sequences in crude specimens of infected tissues. Primers have been used to develop PCR-based diagnostic assays for *Xanthomonas campestris* pv. *phaseoli*, *Erwinia amylovora*, and *X. campestris* pv. *citri*. Genes controlling the specific properties (e.g. toxin production and pathogenicity) of pathogens have also been used as target sequences for specific detection of plant pathogens. PCR amplification of *Tox* gene region (phaseolotoxin-producing gene) for the detection of *Pseudomonas syringae* pv. *phaseolicola* have been successfully used.

Serological techniques can also be combined with PCR, e.g. immuno-capture PCR and immuno-PCR. In **immuno-capture PCR**, the antigen is concentrated by use of specific antibodies which is then subjected to PCR, while **immuno-PCR** enhances the antigen-antibody reaction. Immuno-capture technique was used for the detection of *Xanthomonas campestris* pv. *citri*.

The co-operational PCR technique has been used successfully for the detection of plant RNA viruses, e.g. cherry-leaf-roll virus, cucumber mosaic virus, and Citrus tristeza virus.

Multiplex PCR allows the simultaneous and sensitive detection of different DNA or RNA targets in a single reaction. Multiplex-PCR is useful in plant pathology because different bacteria and/or RNA viruses frequently infect a single host and consequently sensitive detection is needed for the propagation of pathogen-free plant material. One of the examples is the simultaneous detection of six major characterized viruses described in olive trees which belong to four different genera: Cucumovirus, Nepovirus, Necrovirus, and Oleavirus.

Multiplex-nested RT–PCR has been developed for simultaneous or sensitive detection of cucumovirus, nepovirus, and the bacterium *Pseudomonas savastanoi* pv. *savastanoi* from olive plants using 20 compatible primers in a compartmentalized tube. It enables the simultaneous detection of several viral RNA and bacterial DNA targets in a single analysis performed with woody plants. It also saves time and reagent cost because it can be performed in a single reaction, although accurate design of compatible primers is needed.

25.6.7 DNA Micro-arrays

DNA micro-arrays, or biochips, are made of a surface on which are linked multiple capture probes, each one being specific for a DNA or RNA sequence of the targets. Their purpose is the detection of numerous sequences in a single assay. Various supports are currently in use for the elaboration of micro-arrays including glass, nylon, and different polymers. Up to 30000 DNA probes (gene sequences) can be arrayed on to a single chip. The probes arrayed can be PCR products amplified to high concentrations or relatively short oligonucleotide probes. Once arrayed, the chip can be exposed to fluorescently labeled DNA/RNA from the sample to be tested. The detection system uses one or several fluorophores that can be read with laser technology to reveal the targets present in the samples. Micro-arrays can generate fast results for several pathogens but their cost is still very high.

25.6.8 Characterization of Genetic Variability in Plant Pathogens using DNA Markers

Traditional markers used to study the variability in pathogens are based on the use of differential host, cultural characteristics, morphological markers, and biochemical tests. These markers distinguish pathogens on the basis of their physiological characters, i.e. pathogenicity and growth behaviour and can group them according to their similarity for these particular characters. Moreover differential hosts are not available in most of the host-pathogen systems, thus variability cannot be assessed. In such cases, molecular markers are used for the characterization of genetic variability of the plant pathogens. Using PCR, very close strains of a pathogen can be distinguished without prior knowledge of the nature of polymorphic region by the use of RAPD. In India, 20 isolates of *Alternaria brassicae* collected from geographically distinct regions of the world and different host species with RAPD markers have also been analyzed and identified.

25.6.9 Molecular Mapping of Disease-resistant Genes

The process of locating genes of interest via a linkage to markers is referred as **gene tagging**. The tagging of disease-resistant genes with molecular markers involves the evaluation of classical phenotype of resistance and molecular marker genotype on the same individuals, and the data is analyzed to determine if any of the markers co-segregates with the target phenotype (resistance phenotype). The basic requirement of tagging plant-disease-resistant genes with molecular markers is the development of a mapping population.

25.6.10 Gene Pyramiding

Gene pyramiding is the technique of combining more than one gene for resistance in a common genetic background by the repeated back-crossing and the selection with the virulent races of the pathogen. The technique is highly effective in getting durable resistance to the target pathogen. Gene pyramiding is being used in many host-pathogen systems. Hittalmani *et al.* (2000) pyramided three rice-blast resistance genes by using MAS (Marker assisted selection).

There are mainly three benefits to agriculture and crop improvement programmes from the use of biotechnology. The first of these, the **reduction of the duration of breeding programmes**, makes use of tissue culture. Plant tissue culture also allows creation of new genetic variations from which useful new varieties can be obtained. New methods of hybridization can be combined with cell and tissue culture for selection of plants with desired traits. The last and most promising benefit is from use of recombinant DNA technology. Transgenic plants, engineered for resistance to disease, pests and herbicides, and for better nutritional quality, have already been produced in the laboratory (Chopra and Sharma, 1991).

The panel of techniques for rapid, specific, and sensitive detection of plant pathogenic viruses and bacteria has improved and increased over the last few years. The future will bring more novel tests for detecting plant pathogens and novel demands will be introduced. However, only some of these will be accepted by plant pathologists, with not only the analytical quality of the test results but also their sensitivity and specificity contributing to selection of the best cost-effective diagnostic strategies. These new detection tools will be used not only for rapid, sensitive, and specific diagnosis but also to help understand host-pathogen relationships and structure and function of pathogens and their communities.

REVIEW QUESTIONS

1. What is biotechnology? Give at least two examples where this technology has helped in the production of resistant genotypes.
2. Define the followings terms:
   (i) r-DNA (ii) Hybridoma
   (iii) Transgenic plants (iv) Monoclonal antibodies
   (v) Somaclonal variation
3. Why does apical or meristem tip culture generally free the plantlets from viruses or other pathogens?
4. Give the main contributions of the following:
   (i) P R White (ii) G Morel
5. Is biotechnology the alternative to traditional resistant plant breeding?
6. Give a brief account of innovative tools for detection and diagnosis of bacteria and viruses?
7. Define the following terms:
   (i) ELISA (ii) FISH
   (iii) PCR (iv) DNA microarray
   (v) Gene pyramiding

MULTIPLE-CHOICE QUESTIONS

1. The main vehicle for transferring genes from donor to recipient plants is
   (i) *Agrobacterium tumefaciens* (ii) *Agrobacterium radiobacter*
   (iii) Cauliflower mosaic virus (iv) Both (i) and (ii)
2. Which of the following disease-resistant plants have been produced from protoplast fusion?
   (i) Alfalfa (ii) Oats
   (iii) Potato (iv) Onion
3. The most effective application of monoclonal antibodies has been
   (i) accurate identification of virus
   (ii) to distinguish between virus strains
   (iii) for tracing the viruses during epidemiological studies
   (iv) all of these
4. Which one of the following techniques is used in rapid identification of cells, biomarker detection as well as protein engineering?
   (i) PCR (Polymerase Chain Reaction)
   (ii) Flow Cytometry
   (iii) ELISA (Enzyme-Linked ImmunoSorbent Assay)
   (iv) FISH (Fluorescence In Situ Hybridization)
5. Which one of the following is not a type of PCR?
   (i) Anchored PCR (ii) Nested PCR
   (iii) Multiplex PCR (iv) Taq PCR
6. Which of the following techniques is used for assembling of multiple desirable genes into a single genotype and used particularly for breeding of self-pollinated crops?
   (i) Nested PCR
   (ii) Gene pyramiding
   (iii) ELISA (Enzyme-Linked ImmunoSorbent Assay)
   (iv) DNA micro-arrays
7. How many DNA probes can be arrayed on to a single chip in DNA micro-arrays?
   (i) Up to 30,000 DNA probes (ii) Up to 200 DNA probes
   (iii) Up to 10 DNA probes (iv) Up to 46 DNA probes
8. The most important serological techniques used particularly for detection of viruses and bacteria are
   (i) IF (ImmunoFluorescence) and ELISA (Enzyme-Linked ImmunoSorbent assay)
   (ii) protoplast fusion
   (iii) tissue culture
   (iv) meristem culture

Answers

1. (iv) 2. (iii) 3. (iv) 4. (ii) 5. (iv) 6. (ii) 7. (i) 8. (i)

26 PHYTOPATHOLOGICAL TECHNIQUES IN PLANT PATHOLOGY

26.1 ISOLATION OF FUNGI

Most of the plant diseases can be diagnosed by observation with the naked eye or with the microscope, and isolation of the pathogen is not necessary. If the pathogen is mixed with one or more contaminants and has not produced spores or fruiting bodies, or the disease is caused by a new, unknown pathogen then the isolation of the pathogen is necessary.

26.2 REQUIREMENTS FOR ISOLATION OF FUNGI

To isolate the microorganism from the diseased plant tissue, the following things are required:

26.2.1 Common Culture Media used in Laboratories

There are a large number of natural, semisynthetic, and synthetic media that can be used to grow plant pathogenic microorganisms.

Natural Media

A natural medium is one which is composed entirely of complex natural materials of unknown composition. Among natural substances are plant parts, malts, yeast, peptone, fruits, and vegetables. The following are few natural media used in isolation of fungi.

(a) Cooked Vegetable Agar Broth or decoction of the desired vegetative plant parts can be prepared by 10 to 20% of the tissue in water, steamed for 30 minutes and the contents mashed and squeezed through muslin cloth. To this broth, the required quantity of agar-agar (2%) may be added. It may be required to adjust the pH of the medium to the desired level before autoclaving.

(b) Lima Bean Agar Use to cultivate *Phytophthora* spp. Blend 50 g lima beans, previously soaked in water for 10 hours with 500 ml of water containing 15 g melted agar. Adjust the final volume to 1 litre and then autoclave.

(c) Oatmeal Agar Used for general cultivation of fungi. Blend, in a commercial blender, 50 g of rolled oats in 600 ml of water and then heat to 45 to 55°C, add 12 g agar dissolved in 400 ml of water. Autoclave for 90 minutes.

26.2.2 Semisynthetic Media

A semisynthetic medium is made up of natural substances of unknown composition, and by adding some chemical compounds of known composition.

1. Potato–Dextrose Agar (PDA)

Numerous variations occur in the quantity of potato and dextrose; the standard preparation is to boil 200 g of peeled and sliced potatoes in 1 litre of water until the potatoes are soft. Strain through cheese-cloth and adjust the filtrate to 1 litre. Add 20 g dextrose and 17 g agar. Autoclave it.

2. Malt Extract Peptone–Dextrose Agar

Used to cultivate fungi and bacteria. Malt extract = 20 g; Dextrose = 20 g; Peptone = 1 g; Agar–agar = 20 g and Distilled water = 1000 ml.

3. Martin Rose Bengal Streptomycin Agar

Dextrose = 10 g; Peptone = 5 g; KH_2PO_4 = 1 g; $MgSO_4$ = 0.5 g; Rose bengal – trace; Streptomycin = 0.03 g; Agar-agar = 20 g; Distilled water = 1000 ml. Dissolve 1 g of streptomycin sulfate in 100 ml sterile distilled water, after opening the vial aseptically. Add 0.3 ml of streptomycin in solution to each 100 ml of the basal medium after it is cooled.

4. Soil–extract Agar

Used to isolate and enumerate bacteria from soil. To 1 litre of soil extract add 0.2 g K_2HPO_4 and 15 g agar. Adjust to pH 6.8 and autoclave. To prepare soil extract, autoclave 500 g fertile soils in an amount of water that will yield 1 litre of extract. After cooling, filter through filter paper.

5. Yeast Extract

Glucose = 10 g; Peptone = 1 g; Yeast extract = 3 g; Distilled water = 1000 ml.

26.2.3 Synthetic Media

A synthetic media is one of known composition and concentration. Compounds that are used should be 'chemically pure'.

1. Brown's Medium

Glucose = 2 g; K_2HPO_4 = 1.25 g; Asparagine = 2 g; $MgSO_4 . 7H_2O$ = 0.75 g; Agar-agar = 20 g; Distilled water = 1000 ml;

2. Czapek–Dox-Agar

Used to cultivate many fungi. $NaNO_3$ = 2 g; K_2HPO_4 = 1 g; $MgSO_4 . 7H_2O$ = 0.5 g; KCl = 0.5 g; $FeSO_4$ = 0.01 g; Sucrose = 30 g; Agar-agar = 20 g; Distilled water = 1000 ml.

If glass-distilled water is used, add 1 ml of 1% $ZnSO_4$ and 0.5% $CuSO_4$. Heat the full chemical solution with sucrose in a water bath for 15 minutes. After cooling, add sucrose and agar-agar, melt the agar and autoclave. Do not filter.

3. Galactose Nitrate Agar

Used to isolate *Fusarium oxysporum* from soil. Galactose = 10 g, $NaNO_3$ = 2 g; KH_2PO_4 = 1 g; $MgSO_4.7H_2O$ = 0.5 g; $K_2S_2O_5$ = 0.3 g; Agar-agar = 15 g; Distilled water = 1 litre. Autoclave for 20 minutes.

26.3 SOME SPECIAL MEDIA FOR CULTIVATION OF PATHOGENS

26.3.1 Media for Actinomycetes

1. Basal Mineral Salts Agar

Glucose = 10 g; $(NH_4)_2SO_4$ = 2.64 g; KH_2PO_4 anhydrous = 2.38 g; $K_2HPO_3.3H_2O$ = 5.65 g; $MgSO_4.7H_2O$ = 1 g; Distilled water = 1000 ml; Agar-agar = 15 g.

Adjust pH to 6.8–7.0 with 1N NaOH or 1N HCl. Used for culture of *Streptomyces* with addition of sterile carbon source at a concentration of about 1%.

2. Oatmeal Agar

Oatmeal = 20 g; Agar-agar = 18 g; Distilled water = 1000 ml; ($FeSO_4.7H_2O$ = 0.1 g; $MnCl_2.4H_2O$ = 0.1 g; $ZnSO_4.7H_2O$ = 0.1 g. Dissolve these in 100 ml and out of this solution, add one ml to oatmeal agar solution) steam oatmeal in a litre of water for 0 minutes, filter through cheesecloth, add trace salts, and bring to volume. Adjust to pH 7.2 with NaOH. Add agar and steam for 15–20 minutes.

26.3.2 Media for Bacteria

1. Asparagine Lactose Medium

This medium is used to isolate *Rhizobium* sp.

Lactose = 10 g; Asparagine = 0.5 g; Sodium faurocholate = 5 g; K_2HPO_4 = 1 g; $MgSO_4$ = 0.2 g, Neutral red = 3.5 ml (2/40 ml ethyl alcohol); Agar-agar = 20 g; Distilled water = 1000 ml; pH = 7.5.

This medium is suitable for the growth of all the Rhizobial species and especially for isolating *Rhizobium* from the nodules of groundnut. This medium can be used for distinguishing the *Agrobacterium* and *Rhizobium*. *A. tumefaciens* shows quick growth within 24 hours with pink-coloured colony on Congo red (10 ml of 1/4000 aqueous solution) incorporated in the medium, while *Rhizobium* strains grown late after 72 hours show no absorption of Congo red.

2. Ashby's Mannitol Agar Medium

Used to cultivate *Azotobacter* sp. Potassium acid phosphate = 1 g; $MgSO_4$ = 0.2 g; Ammonium sulfate = 0.2 g; NaCl = 0.2 g; Mannitol = 10 g; Calcium sulfate = 0.1 g; Calcium carbonate = 5 g; Agar–agar = 20 g; Distilled water = 1000 ml.

Dissolve the phosphate in 500 ml and make alkaline to phenolphthalein. Melt agar and other ingredients in the remaining 500 ml, mix both the lots and sterilize for 20 minutes at 15 pounds.

26.3.3 Media for Fungi

1. V-8 Juice Agar

Used to support growth and induce sporulation of many fungi. It can be used instead of PDA. V-8 juice = 200 ml; $CaCO_3$ = 3 g; Agar-agar 15–20 g; Distilled water = 1000 ml; pH = 7 to 7.5 which can be varied by changing the amount of juice or $CaCO_3$.

2. Cassava-Dextrose Agar

Used in place of PDA for routine work of fungi:

Peel cassava (*Manihot utilissima*) tubers and cut into chips. Dry overnight at 55°C and grind to 30 mesh. Soak 135 g of the powder in 500 ml of water for 15 minutes at 60°C. Filter through cheesecloth. Add 20 g glucose and 12 g agar-agar. Complete the volume to 1 litre and autoclave.

Apart from the above-mentioned media for fungi, there are some basic points, which are to be kept in mind while preparing media.

(a) Fungi usually grow best in a rich carbohydrate medium.
(b) Fungi usually prefer a slightly acid reaction, pH 6–6.5.
(c) Agar is slow to dissolve thoroughly. It is advisable to dissolve the agar in half the water and nutrients in the other half and then mix. Agar does not solidify satisfactorily in very acid or alkaline solutions.
(d) Peptone may generally be omitted from fungus culture.
(e) Tap water is often preferable to distilled water as it contains useful trace elements.

26.4 SELECTIVE MEDIA FOR THE ISOLATION OF PLANT PATHOGENS

There are several selective media for the isolation of plant pathogenic fungi. For example, P10VP medium of Ocana and Tsao (1969) is very useful in isolating *Phytophthora* species from plant tissues and soil. The medium contains Difco cornmeal agar-agar (17–20 g/litre) which is autoclaved (cooled to 45°C) and amended with low dosage of Pimaricin (10 ppm), Vancomycin (200 ppm), and Pentachloronitrobenzene (100 ppm). Plates are stored in the dark because of sensitivity of Pimaricin to light. After usual sterilization procedure of the diseased tissues, the tissues are plated on this selective medium. Infested soil or suspected soil containing the *Phytophthora* propagules or mycelium can also be plated on this medium to isolate the pathogen. Hymexazole can be added after autoclaving to suppress species of *Pythium*. There are several other selective media for the isolation of *Phytophthora* species.

26.4.1 Nash and Snyder's PCNB Medium (Nash and Snyder, 1962)

It is a selective medium for the isolation of *Fusarium* species from field soils. In the same way, there are several selective media used for the isolation of plant pathogenic bacteria and other pathogens. The reader is referred to, *Plant Pathological Methods—Fungi and Bacteria* by Tuite (1969), *Plant Pathologist Pocket Book* published by Commonwealth Mycological Institute, England (1968), *Basic Plant Pathology Methods* by Dhingra and Sinclair (1968), and *Laboratory Exercises in Plant Pathology* edited by Boudoin *et al.* (1988), published by the American Phytopathological Society.

26.4.2 Sterilization

Sterilization is achieved by exposing materials to lethal agents, which may be chemical, physical, or ionic in nature or in the case of liquids, physical elimination of cells from the medium. Selection of a method depends upon the desired efficiency, its applicability, toxicity, ease of use, availability and cost, and effect on the properties of the object to be sterilized.

1. Red Heat

Inoculating needles, etc. can be sterilized by heating to redness.

2. Hot Air (Oven)

Glassware, metal instruments resistant to high temperature can be sterilized in a hot-air oven. Sterilization should be complete after the time and temperature shown.

Temp °C	Time (hours)
120	8 h
140	3 h
160	1 h
180	20 min

3. Steam Under Pressure (Autoclaving)

Most media can be sterilized by heating at 10–15 psi for 15 minutes in an autoclave or domestic pressure cooker.

Pressure (psi)	Temperature (°C)	Time (hours/min.)
5	107	
7	110	2.5 h
10	115	51 min.
15	121	15 min
20	126	6.4 min.

4. Gas Sterilization

It is used on objects that cannot be sterilized by heat or liquid filtration. Advantages are that the process can be carried out at low temperature and low relative humidity. The process can be carried out using simple equipment such as rubber or plastic bags and metal or plastic drums. Major disadvantages are that a longer time is required for sterilization over that of heat; materials used are flammable and highly toxic and the cost is higher than heat. Some gases used are ethylene oxide, propylene oxide, formaldehyde, methyl bromide, β-propiolactone, and ozone. The first three are commonly used as alkylating agents.

5. Filtration

Filtration physically separates microorganisms, cells, and debris from liquids but not viruses or metabolic byproducts. Aqueous solutions, oils, and organic solutions can be sterilized by filtration. Filters used are sintered glass, asbestos pads, unglazed porcelain, diatomaceous earth disc or candles, and cellulose ester membrane. Microorganisms and other large particles are retained on the filter in part by the small size of the pores and dry adsorption onto pore walls.

6. Surface Sterilization

Some of the liquids used in lessening of microorganisms from bench tops, instruments, and the surface of plants are given as follows.

Chemical	Concentration	Uses
1. Ethanol	70–95%	Disinfection of instruments and bench tops
2. Mercuric Chloride	0.1%	Disinfection of the plant surface, surfaces for isolation
3. Sodium hypochlorite	1%	Isolation of microorganisms
4. Formaldehyde alcohol	8% formaldehyde	Sterilizing of instruments in 70% isopropyl alcohol
5. Formaldehyde	1 part of formaldehyde to 10 parts of water	Elimination of plant pathogens from greenhouse tools
6. Propylene oxide	1 ml in 1 litre	Filter papers are sterilized
7. Ultraviolet light	15–30 min.	Surface sterilization, 210 nm to 300 nm

26.4.3 Isolation of Fungi

1. From Soil

Here, the method involves the transfer of microbial propagules from soil to the culture medium with the aim of obtaining a pure culture. Because of the diversity of growth requirement of soil microbes, various procedures are developed. Therefore, any isolation technique is relative and can be adopted on the basis of its applicability.

(a) Soil-Dilution Plate Method This is the classical and most widely used method in soil microbiology and has been adopted from the method originally developed for the general isolation of soil bacteria (Waksman, 1927). The main modifications are the use of lower dilution on account of lower number of soil fungi and development of different isolation media adjusted to lower pH (4.5) for suspension of bacteria.

There are many variations of this method. In its simplest form, place 10 g soil (own dry) in a sterile flask with 90 ml sterile water and stir with a magnetic stirrer for 20 to 30 min. To this solution, withdraw 10 ml soil water and add to 90 ml of sterile water in a sterile bottle. Shake for 1 minute and transfer 10 ml of the suspension to another 90 ml sterile water. Repeat the process until the desired dilution is obtained. A proper dilution allows 50 to 150 colonies per culture plate. In general, suitable dilutions are 10^{-3} to 10^{-4} for actinomycetes, 10^{-5} to 10^{-6} for bacteria, and 10^{-4} to 10^{-5} for fungi. Spread 1 ml of a dilution on a suitable agar surface by an inclined rotary motion of the plate.

(b) Soil-Plate Method This is a less complicated method, which tends to yield more kinds of fungi than those obtained from the dilution plate technique. Suitable small amounts of soil (0.005 g to 0.15 g) are taken from the mean soil sample using a sterile needle with a flattened tip. Each small, subsample is placed in a drop of sterile water in a Petri dish and dispersed. Eight to ten ml of molten agar medium are added to each Petri dish, the dish being rotated gently to allow dispersion of the soil particle in the nutrient medium (Warcup, 1950).

(c) Direct Inoculation and Soil Dessication Method Lumps of soil, about 1 cm in diameter, are transferred into the sterile plates of Czapek's solution agar. After incubation for 24 hours, hyphal tips are removed on a slate. The organisms thus isolated are believed to come from the mycelium that is actually found in soil (Waksman, 1961) Warcup (1960), however, observed that while mucorales were isolated from soil lumps, direct microscopic examination of lumps failed to reveal phycomycetous hyphae, thus suggesting that mycelia developing on the agar plate are derived from the spores.

(d) Immersion-tube Method This method is described by Chester (1948). This consists of introduction of glass tube, with 4–6 spirally arranged invaginated capillaries filled with agar into soil. After

7–14 days, the tube is removed and fungi are isolated by removing the soil core and then cutting it into portions, which are plated out, The limitation of the immersion-tube method is that it easily isolates only active spreading mycelium or active localised mycelium, which happens to come in contact with capillaries.

(e) Screened Immersion Plate Thornton (1952) devised the screened immersion-plate method, which permits isolation of microorganisms capable of growing from the soil on water agar placed behind a performed screen.

A glass slide that fits closely into a shallow perspex box carries the medium. A 1 mm thick plastic sheet covers the slide and box without touching the slide. Two rows of five 5 mm diameter holes each are made in the corner. All components are sterilized. After sterilization, the slide in the box is placed in a large culture plate and 8 ml of water agar is poured over it. After solidification, it is covered with the perforated plastic sheet. After a suitable period of burial in the soil, the plates are removed, examined, and fungal growth transferred to PDA.

(f) Isolation by Baiting A variety of microorganisms may be isolated from the soil by the use of some preferential substrate. The desired organism is able to develop on the bait to the exclusion of other organisms, thus allowing isolation by ordinary method. *Agrobacterium tumefaciens* and *Thielaviopsis basicola* are isolated from soil by the use of carrot discs. *Rhizoctonia* with cellophane, and a number of water molds are isolated by the addition of boiled hemp seed to pond water or soil water mixtures. The insertion of diseased materials or soil inside the bait- apple is used to isolate many but not all species of *Phytophthora*. Planting susceptible plants in infested soil is often considered a form of baiting.

2. Isolation of Pathogens from Diseased Plant Tissues

Tissues sampled during the active stage of an infection are likely to have within them only the pathogen responsible for the infection; the surfaces of such tissues, however, are usually contaminated with saprophytic organisms. Thus, surface disinfestation of tissues selected for isolation work is a prerequisite to successful isolation. Many surface contaminants can be removed by washing diseased tissues with soap or detergent and then rinsing them for 1 to 2 hours in running tap water. Tender tissues such as seedling stems containing damping-off fungi are cut into small pieces and planted directly on acidified PDA immediately after they have been rinsed in running tap water.

Additional surface disinfestation of less tender tissues is usually accomplished with 0.5 to 1.0 percent calcium hypochlorite (10 to 20 percent commercial bleach) or with a 0.1 percent aqueous solution of mercuric chloride. Tissues treated with bleach can be planted directly on PDA and those treated with mercuric chloride, however, must be rinsed in sterilized water before they are planted.

To isolate plant pathogenic fungi by the tissue-planting method, the following procedure is recommended:

Pieces of root, stem, leaf, and other diseased tissue showing distinct symptoms are washed and placed in a beaker, which is then covered with cheesecloth and left in running water for 1 or 2 hours. Squares of tissues approximately 0.5 cm across are cut from the advancing margins of lesions and placed in 0.5 to 1.0% calcium hypochlorite after 3 to 5 minutes, the small pieces of tissues are lifted with alcohol-flamed forceps onto hardened agar medium in Petri dishes. Later, the predominating organisms can be transferred to fresh agar slants for growth in pure culture.

When sclerotia are associated with diseased tissue, they can be surface sterilized, cut in half and planted on solidified PDA.

Bacteria and fungi that sporulate in or on diseased tissues can be isolated by a dilution method. Bits of diseased tissues are chopped with a flamed scalpel in a few drops of sterilized water; after a few minutes, a loopfull of the water containing the bacteria or fungal spores are transferred to a drop of sterilized water in a Petri dish. Then serial dilutions are made to several other dishes. Melted agar medium at 37°C is then poured into the dishes. After a day, colonies of bacteria or fungi are visible, and transfers of single colonies can be made to an appropriate medium in tubes or in other Petri dishes.

26.4.4 Inoculation Techniques

1. Soil Infestation

Bits of mycelium are often used as inoculum when the infection material is a root or seedling stem. The mycelium from liquid or solid media can be chopped in water in a commercial blender.

To test the pathogenicity of damping-off, root rotting, or wilt-inducing fungi, a small square of PDA infested with the appropriate mycelium can be buried in soil near the centre of a clay pot and seeds of the suscept can be planted around the centre. Within about a week, damping-off fungi will have induced diseases; more time is required by the other pathogens. A better method is to grow the pathogen on sand maize meal or sand oatmeal medium. The constituents of the medium are 100 g washed sand, 3 g oatmeal or maize meal, 20–25 ml water. The sand is taken in 250 cc conical flask, oatmeal or maize meal is sprinkled on the surface of sand in the flask, and sufficient water is added to wet the sand and oatmeal. The flasks are corked with non-absorbent cotton and sterilized at 15 lb psi for 15–20 minutes. The pathogen is inoculated in these flasks and allowed to grow for 2–3 weeks and then added to soil in pots at the rate of 10% to the soil (w/w). In zoospore-forming fungi or those which produce conidia abundantly, zoosporic suspension can be prepared of the pathogen and similarly conidial suspension can be prepared. The healthy seedlings can be inoculated by dipping the root system in zoosporic or conidial suspension for varying periods of time and planted in healthy soil.

Root-invading pathogens (including bacteria, fungi) can be mixed with potting soil into which transplanted seeds are placed. Suspensions of bacterial cells or fungal spores may be used. Fungal mycelium may be chopped in water in a blending machine and added to the soil; or the fungus may be grown on sterilized cereal grains, which should then be added to soil in which the suspect is to be planted.

2. Inoculation of Leaves and Stems with Fungi or Bacteria

Leaves and stems of susceptible plants can be inoculated with suspensions of fungal spores or bacterial cells in the following manner. Culture dishes or tubes are flooded with sterilized water to which has been added a drop of wetting agent, and the spores or bacteria are gently scraped off the substrate into the water. The suspension is then placed in an atomizer, from which it is sprayed on the susceptible plants. When the fungus is inoculated, the sprayed plants should be kept in a moist atmosphere for 12 to 48 hours. With leaf-attaching bacteria, however, best results are usually experienced when the plants are inoculated in the greenhouse, preferably between 9:00 a.m. to 3:00 p.m., when their stomata are most likely to be open. It is preferable to inoculate the plants in the afternoon.

When dry spores, such as those of cereal rusts, can be collected in mass, dry inoculum can be prepared by mixing one part of the spores with 3–4 parts of talc powder. After plants to be inoculated have been sprayed with water containing a little wetting agent, the spore–talc mixture is dusted onto the plants. Inoculated plants should be kept in a moist chamber for 12 to 24 hours.

Another method of inoculating pathogen is to cut agar discs with the help of sterilized cork borer and transferring them to healthy leaves in injured (needle pricked) or healthy plants under moist conditions covered with glass cages.

Verma and Singh (1980) have described several techniques of inoculation of bacterial pathogens. These are spray inoculation, smear/swab inoculation, leaf infiltration inoculation (the bacterial suspension is injected into the leaf tissue with the help of a hypodermic syringe), pinprick/needle inoculation, root-dip inoculation and some special techniques. The special techniques include (a) clipper with bacterial suspension mounted on it in a plastic bottle allowing continuous release of phytopathogenic bacteria, (b) use of insect vectors in special cages, and (c) immersion of whole seedling in bacterial suspension before transplanting.

26.5 KOCH'S POSTULATES

Diagnosis of plant diseases is a very important step in identifying a pathogen responsible for a particular disease. It is easy to identify many diseases in which the pathogen produces specific symptoms/growth such as powdery mildew, rusts, smuts, etc. but in many diseases, the symptoms are not useful as they are caused by many pathogens. For example, yellowing or chlorosis or leaf spot on leaves is a symptom which can be caused by bacteria, fungi, virus or due to nutritional deficiency. For effective management of any disease, accuracy in diagnosis is important for selecting the proper control measures. If the pathogen is not a biotroph, it can be cultured and tested for Koch's postulates.

Robert Koch (1843–1910), a bacteriologist and a doctor by profession, was the first to show in 1876 that anthrax of sheep was caused by a bacterium called *Bacillus anthracis*. Koch in 1887 set out four steps or rules also called **Koch Postulates** that must be fulfilled before a pathogen isolated from diseased material can be considered as an agent for a particular disease.

1. The suspected causal agent must be present in the diseased material.
2. The suspected causal agent must be isolated and grown in pure culture.
3. When a pure culture of the suspected agent is inoculated into healthy susceptible host, it must produce the typical disease.
4. The same causal agent must be re-isolated from the diseased/infected host and must have the same characteristics as the organism in step (3).

REVIEW QUESTIONS

1. Describe the natural, semisynthetic, and synthetic media with suitable examples.
2. Describe some special media for cultivation of plant pathogens.
3. Why are so many types of microbiological media employed in laboratories for the culture of microorganisms?
4. Describe briefly the various sterilization techniques used in microbiology/plant pathology.
5. What is aseptic transfer technique? Why must you master this technique to work in a plant-pathology laboratory?
6. Discuss the various inoculation techniques used in plant pathology.
7. What is pure culture? Why do we place such importance on obtaining and maintaining pure culture?

MULTIPLE-CHOICE QUESTIONS

1. Which of the following media is used for isolation of fungi?
   (i) Potato dextrose agar medium (ii) Asparagine lactose medium
   (iii) Ashby's mannitol agar medium (iv) All of these
2. Who gave the serial dilution agar-plating method for isolation of microorganisms?
   (i) Warcup (1950) (ii) Waksman (1950)
   (iii) Warcup (1927) (iv) Waksman (1927)
3. Which group of fungi is isolated using the 'baits'?
   (i) Air-borne fungi (ii) Water fungi
   (iii) Both (i) and (ii) (iv) None of these
4. Find the odd one out of the following:
   (i) CDA (ii) YEMA
   (iii) PDA (iv) BOD
5. Which of the following is used as a gas sterilant?
   (i) Ethylene oxide (ii) Propylene oxide
   (iii) Methyl bromide (iv) All of these
6. Autoclave is used for the sterilization of
   (i) common media (ii) plant tissues
   (iii) rubber products (iv) all of these
7. Which of the following chemicals is used as a disinfection of plant surfaces for isolation of microbes?
   (i) Mercuric chloride (ii) Sodium hypochlorite
   (iii) Both (i) and (ii) (iv) Propylene oxide
8. Multipore filters are used for the sterilization of
   (i) vitamins (ii) hormones
   (iii) heat-labile media (iv) all of these
9. The selective medium used for the isolation of *Phytophthora* is supplemented with
   (i) PCNB, Pimaricin, and Vancomycin
   (ii) PCNB, Streptomycin, and Vancomycin
   (iii) PCNB, Pimaricin, and ammonium nitrate
   (iv) PCNB, Pimaricin, Vancomycin, and Streptomycin
10. Borosil or corning glassware are sterilized by
    (i) hot air (oven) (ii) gas sterilization
    (iii) sulphuric acid (iv) ethanol

Answers

1. (i) 2. (iv) 3. (ii) 4. (iv) 5. (iv) 6. (i) 7. (iii) 8. (iv) 9. (i) 10. (i)

SELECTED REFERENCES

1. Abeles, F B (1973); *Ethylene in Plant Biology*, Academic Press, New York
2. Agrios, G N(1997); *Plant Pathology*, Fourth Edition, Academic Press, London, New York
3. Agrios, G N (2005); *Plant Pathology*, Fifth Edition, Elsevier, Academic Press, New York
4. Ainsworth, G C and Sampson, K (1950); *The British Smut Fungi Ustilaginales,* Commonwealth Mycology Institute, Kew, Surrey, England
5. Akai, S (1959); "History of Defence in Plants" in *Plant Pathology—An Advanced Treatise*; Horsfall, J G and Dimond, A E (Eds.); Academic Press, London, pp 435-467
6. Akai, S and Fukutomi, M (1980); "Preformed Internal Physical Defences" in *Plant Disease—An Advanced Treatise,* Vol . IV, Horsfall, J G and Cowling, E B (Eds.); pp 139-160, Academic Press, New York
7. Alexopoulos, C J and Mims, C W (1979); *Introductory Mycology*, Third Edition, John Wiley and Sons, New York
8. Alexopoulos, C J, Mims, C W and M Blackwell (1996); *Introductory Mycology*, John Wiley and Sons, New York
9. Bawden, F C (1964); *Plant Viruses and Virus Diseases,* Fourth Edition, Ronald Press, New York
10. Bedi, P S (1986); "Plant Disease Forecasting" in *Vistas in Plant Pathology*; Varma, A and Varma, J P (Eds.); Malhotra Publishing House, New Delhi, pp 249-261
11. Budoin, A B A M *et al* (1988); *Laboratory Exercise in Plant Pathology*, An Instructional Kit, APS, St Paul, Minnesota
12. Burnett, J H (1976); *Fundamentals of Mycology,* Second Edition, Edward Arnold, London
13. Butler, E J (1918); *Fungi and Diseases in Plants,* Thacker Spink and Co., Kolkata
14. Butler, E J and Jones, S G (1949); *Plant Pathology*, Macmillan, New York, pp 249
15. Chupp, C and Sherf, A F (1960); *Vegetable Diseases and Their Control,* Ronald Press, New York, pp 436-437
16. Fry, W E (1982); *Principles of Plant Disease Management*, Academic Press, New York
17. Gäumann, E (1950); *Principles of Plant Infection*, English Edition; Brierly W B, (Ed.), Crosby Lockwood, London
18. Gibbs, A J and Harrison, B D (1976); *Plant Virology—The Principles*, Edward Arnold, London, pp 292

19. Gnanamanickam, S S (Ed.) (2002); *Biological Control of Crop Diseases*, Dekker, New York
20. Goodman, R N; Kiralay, Z and M Zaitlin (1967); *The Biochemistry and Physiology of Infectious Plant Disease*, D Van Nostran, Princeton
21. Issac, S (1992); *Fungal Plant Interactions*, Chapman and Hall, London
22. Kolte, S J (1994); *Diseases of Annual Edible Oilseed Crops,* Vol. I, *Peanut Diseases,* CRC Press Inc., Boca Raton, Florida, pp 142
23. Kuc', J (1972); Phytoalexins, *Ann Rev Phytopath* ,10 : 207–232
24. Mahadevan, A (1984); *Growth Regulators, Microorganisms and Diseased Plants*, Oxford and 1BH Publishing Co., New Delhi
25. Mehrotra, R S and Aggarwal, A (2001); *Phytophthora Diseases in India*, Bishen Singh and Mahendra Pal Singh, Dehradun: pp 322
26. Nagarajan, S (1983); *Plant Disease Epidemiology*, Oxford and IBH Publishing Co., New Delhi: 267 pp
27. Nene, Y L and Thapaliyal, P N (1993); *Fungicides in Plant Disease Control*, Third Edition, Oxford and IBH Publishing Co., New Delhi, pp 691
28. Rangaswami, G (1972); *Diseases of Crop Plants in India,* Prentice Hall of India Pvt. Ltd., New Delhi, pp 504
29. Rangaswami, G (1962); *Bacterial Plant Diseases in India*, Asia Publishing House, Mumbai, pp 163
30. Siegel, M R and Sisler, H D (Eds.) (1977) *Antifungal Compounds*, Volumes I and II, Marcel Dekker Inc., New York and Basel
31. Singh, R S (1991); *Plant Diseases*, Sixth Edition, Oxford and IBH Publishing Co., New Delhi
32. Singh, R S (1998); *Plant Diseases*, Seventh Edition, Oxford and IBH Publishing Co., New Delhi
33. Singh, R S (2005); *Plant Diseases*, Eighth Edition, Oxford and IBH Publishing Co., New Delhi,
    a. *Plant Diseases,* 76: 545-550
34. Sitaramaiah, K (1984); *Plant Parasitic Nematodes of India*, Today and Tomorrows Printer and Publishers, New Delhi, 197 pp
35. Sitaramaiah, K and Singh, R S (Eds.) (1993); *Handbook of Plant Nematology*, Cosmo Publishers, New Delhi, 336 pp
36. Stakman, E C and Harrar, J G (1957); *Principle of Plant Pathology*, Ronald Press, New York
37. Tuite, J (1969); *Plant Pathological Methods,* Burgess Publishing Company, Minneapolis, pp 239
38. Van der Plank, J E (1963); *Plant Disease Epidemics and Control*, Academic Press, New York
39. Vander Plank, J E (1968); *Disease Resistance in Plants*, Academic Press, New York
40. Waksman, S M (1927); *Principles of Soil Microbiology,* Williams and Wilkins Co., Baltimore pp 897
41. Walker, C J (1969); *Plant Pathology*, Third Edition, McGraw-Hill, New York
42. Wood, R K S (1967); *Physiological Plant Pathology,* Blackwell, Oxford and Edinburgh

Important Journals and Websites

1. *Indian Phytopathology,* published by the Indian Phytopathological Society, Division of Mycology and Plant Pathology, New Delhi; www.Ipsdis.com
2. *Indian Journal of Mycology and Plant Pathology,* published by the Indian Society of Mycology and Plant Pathology
3. *Plant Disease Research,* published by the Indian Society of Plant Pathologists, PAU, Ludhiana
4. *Indian Journal of Plant Pathology,* published by Association of Plant Pathologists of Society of Mycology and Plant Pathology, Rajasthan College of Agriculture, Udaipur, Rajasthan, India,

5. *Phytopathology,* published by the American Phytopathological Society, USA; http://www.Sci-soc.Org
6. *Plant Disease,* published by the American Phytopathological Society, USA; http://www.Aspnet.org
7. *Molecular Plant Microbe Interactions (MPMI,* published by the American Phytopathological Society, USA
8. *Phytoparasitica (Israel Journal of Plant Protection Sciences),* Israel; http://www.phytoparasitica.org
9. Annals of Phytopathological Society of Japan
10. *Netherland Journal of Plant Pathology*
11. *FAD Plant Protection Bulletin*
12. *Journal of Phytopathology,* published by Blackwell Science, UK: http://www.Blacksci.co.uk
13. "*Canadian Journal of Plant Pathology*" published by the Canadian Phytopathological Society, Ottawa, Canada; http://www.cps-sep.ca
14. *Mycological Research* (formerly, *Transactions of the British Mycological society*), published by the British Mycological Society, UK; http://uk.cambridge
15. *Molecular Plant Pathology* online—The British Society for Plant Pathology runs an electronic journal, which can be accessed at http//www.bspp.org.uk/mppol
16. Physiological Plant Pathology, published from UK
17. *Phytopathologia Mediterranea*
18. *Phytopathologische Zeitschrift*
19. *Zeitschrift fur Pflanzenkrankheiten* and *Pflanzenscutz*
20. *Journal of Bacteriology;* www.journals.Asm.org
21. *Journal of Virology*, published by the American society for Microbiology (ASM); http://gvi.Asm.org
22. *Journal of General Microbiology*
23. *Review of Plant Pathology (*formerly, *Review of Applied Mycology*); published from UK; http://uk.cambridge
24. *Tropical Pest Management* (formerly, PANS) published by the Tropical Development and Research Institute, London
25. CABI Publication; http://www.calbi.org
26. CAB international have recently launched a database of plant pathology literature
27. There are several other journals, which contain articles of plant pathological importance such as *American Journal of Botany, Canadian Journal of Botany*, *New Phytologist*, *Journal of Experimental Botany*, *Annals of Applied Biology*, and *Annals of Botany*

INDEX

D